# Mirage in the Sand

# MIRAGE
## IN THE SAND

**REFLECTIONS OF A GRUNT IN THE EARLY DAYS OF GWOT**

## JEFF HARSTAD

Dedication

---

Captain Michael John MacKinnon – Bandit 6
*Thank you for everything, then and now*


Laylah and Alexandria Harstad
*May you never know the brutalities of war*


Infantrymen of the United States Army and United States Marine Corps
*Follow Me*


The Peoples of Iraq
*I long to walk the sandy soils of ancient Babylon once more,*
*To feel the desert sun*
*And say goodbye to the ghosts*
*As-salamu alaykum*

---

TABLE OF CONTENTS

# Introduction

Have you ever wondered what it's like to serve in the military? I'm not talking about working behind a desk or in a JAG courtroom, and I'm not talking about joining some super-secret special operations unit that will never be confirmed to exist, and where no one can tell if anything you say really happened. I'm talking about being a grunt. A ground-pounder. Someone on the front lines, living in filth and exhaustion, fighting tooth and nail to stay alive. Someone who carries the weight, the weapons, and the burden. In other words: an infantryman. If that question stirs something in you, then this book is for you.

What is the training like? Is Infantry School really as challenging and scary as people say? Where will you be sent? What is war actually like? What does it feel like to kill the enemy and to lose friends in combat? These are hard questions. In *Mirage in the Sand*, you'll find honest, unfiltered answers to all of them and more.

However, this book may not be what you expect. It isn't a traditional war story, packed cover to cover with blood and guts, glory, and flag-waving. It's not a celebration of violence or a collection of action scenes for entertainment. Yes, there are moments in these pages that are graphic, unsettling, and painfully real. Many of the stories I wish I could forget. But that is not the heart of this book. To me, it holds a more significant meaning. By the final page, I hope you will feel the same.

So, you may be wondering: *What exactly is this, then?* Honestly, it doesn't fit neatly into one category. In some ways, it's a memoir of my four brief but memorable years in the military, where I recount experiencing September 11th as a high school senior, my struggles at the U.S. Army Infantry School, and life through five training and combat deployments, including two to Iraq. In other ways, it's a historical narrative, because I step back often to lay out the wider story. The politics that set the war in motion, the way the media framed it in real time, and the long shadow Iraq's own history cast over everything we did there. At times, it even leans into a quasi-travel book, highlighting a few foreign places I was fortunate enough to see. Still, it's neither purely a memoir nor a history, and it isn't really a travelogue either. It's much more than that.

I've done my best to weave a collection of memories together into a coherent narrative that flows easily from one topic to the next. Sometimes, they are fleeting in nature, with few details provided, but other times, I linger upon them to thoroughly explore my feelings and memories. Piecing together recollections this way has been quite therapeutic, and I'm pleased with my accomplishment. Gathering a selection of highlights from my life elicited a range of emotions, from terrible sadness to joy to thankful remembrance.

Along the way, I hope to accurately convey what it felt like to live as a grunt in America's Army during the early years of the Global War on Terrorism (GWOT). You will meet some interesting characters, explore fascinating locations, learn of dysfunction and government waste, experience the heartbreak of losing comrades in battle, and hopefully feel as I felt during that rollercoaster ride. This is a tall order, but I hope you'll see those of us who call ourselves combat veterans as fallible yet profoundly resilient human beings. We lived our youthful years in a chaotic world during a crazy time, and this is one small story from that era.

This book also serves as a brief overview of military life in the early 2000s. Many are still curious about training and how we prepared for and fought the enemy. Perhaps you have a relative who served, and you're wondering about their experience. Or maybe you served and wish to relive that camaraderie through someone else's wild adventures, even for a moment. While I can't speak for everyone, as the military is vast and covers hundreds of jobs, I'm confident many of these stories will resonate with shared experiences you or your loved one has had. I hope one of these memories sparks meaningful conversations among your family and friends.

But before we jump into the story, let's go back to the beginning. I believe it is essential to understand the origins of the account you're about to read. I've tried many times, too many to count, to document my experiences in the Army by writing a book. Unfortunately, all those early drafts ended up in the trash. Most were abandoned after a chapter or two, as my focus was constantly interrupted by work, children, or simply the demands of life. I also struggled to convey the events accurately, simply because I lacked the skill to express them well, or because the book's flow never turned out as I had envisioned. There was always some problem to be found.

Every author has a unique style that shapes how readers receive their message. I would write several pages of an interesting memory, only to decide it sounded too immature or sometimes too grim. I struggled to find my voice. To remedy this, I took a page from Benjamin Franklin's approach to self-improvement: I read… *a lot*. I set ambitious yearly reading goals across a variety of topics to increase my vocabulary, learn the art of storytelling, and develop my personalized tone. I did this for years while the idea of this book simmered in my mind. That effort ultimately paid off. It reignited my passion for telling this story. And when I was ready, I had a new set of skills at my disposal.

In early 2024, I briefly moved to Germany to work and continue my education with a PhD in astrophysics at a research center. This all happened as I turned 40, which, in many ways, seems to be a bit late in life for that sort of thing. Unfortunately, the position wasn't a good fit, and I returned to the United States feeling quite dejected. Unemployed and searching for a new job, I suddenly had plenty of time to devote to a demanding project.

As I toyed with a few preliminary passages, I realized I had developed a clear and engaging way to tell my story. It may have come from age, maturity, and experience, or perhaps from the hundreds of books I read in preparation. The college degrees I had worked hard to earn could also have shaped it. Most likely, it was a combination of all these things. Whatever the reason, I had discovered a new way to communicate, and for the first time, I felt ready to tell this story how it deserved to be told. My brothers-in-arms deserve that much.

But then I hit another snag. After almost twenty years, some memories were blurry and others were just… gone. Then, almost miraculously, something happened that helped fix my predicament.

Out of the blue, my dad called to say he'd found a box of mine in his basement while doing some cleaning. When we opened it, we found a treasure trove of letters and correspondence that my family had carefully preserved, mostly by my mom, who has since passed. It felt as though she was reaching across time to help me piece my story together.

Over my years in the Army, I wrote and received hundreds, if not thousands, of letters, emails, messages, and notes, most of which had been wholly forgotten to time. Reading through them was a mixture of delight and discomfort. On the one hand, seeing my youthful perspective on the world and the events that unfolded was fascinating. On the other, I was often embarrassed by my poor use of language, the immature things I wrote, and my general lack of humility.

I once read an interview with Neil Peart, the late drummer and lyricist for my favorite band Rush, where he reflected on their early albums. When asked about their first songs together, he said something that likely resonates with anyone looking back at their past works. "When I look back on that it's an indulgent smile. We would later do better but there wasn't anything wrong with it. I described it once as young, foolish and brave."[1] I don't think I could have said it any better myself.

While I was "young and dumb," as the saying goes, some of my early writings still embarrassed me, and I worried about how they would look to future generations. I could be harsh, dramatic, and at times too judgmental. I confided these concerns to one of my closest friends, Donald Cyr (who you will meet later in the book). After listening to me ramble, Cyr said, "If you hadn't changed in over twenty years, after all that time for self-reflection, *then* you'd have something to be horrified about. Just print it how it was." Well said.

Another striking realization was how recklessly naïve my writing was. Perhaps it still is in some ways. I didn't hold back on my opinions, many of which were later proven entirely wrong. But, as the saying goes, "If only I knew then what I know now." I believe this is something important for the reader to consider. The story you are about to read shouldn't be viewed through today's lens, not entirely, at least. Instead, I hope to transport you back to the early 2000s. Set aside the modern cultural and social perspectives, even if just for a moment, because the world was vastly different in so many ways.

Podcasts, YouTube, streaming services, iPhones, and wireless internet didn't exist or weren't yet mainstream. News stations flooded television and radio, but much of it was steeped in pro- or anti-war propaganda, often cleverly disguised. Research was still done in physical libraries, and the internet was new and expensive for most people. Websites were basic and frequently looked like elementary school newspaper projects. It was a very different time. Though, in hindsight, not long ago. It was just the blink of an eye.

---

[1] Kinos-Goodin, Jesse. "Neil Peart on the 10 Best Rush Songs Ever," *CBC Music*, November 12, 2014. https://www.cbc.ca/music/read/neil-peart-on-the-10-best-rush-songs-ever-1.5000896.

I mention this because today, our world is saturated with conspiracy theories and armchair quarterback opinions about 9/11, the wars in Iraq and Afghanistan, and weapons of mass destruction (WMDs) in the Middle East. Every political faction has its version and interpretation of events. And, with information now available 24/7 at our fingertips, we are constantly inundated. This wasn't the case in 2001.

There was, of course, debate, sometimes vigorous debate, about whether the United States should invade any country in the Middle East after 9/11. The pro-war argument centered on the belief that we had been attacked by terrorists based in that region, that Saddam Hussein was developing WMDs, and that it was better to fight the enemy abroad than at home. The opposing view held that the wars were motivated solely by oil interests and national pride, particularly the desire to remove Saddam Hussein, who had reportedly attempted to assassinate former President George H.W. Bush during a 1993 visit to Kuwait.[2] There were other fringe ideas too, but these were the arguments most Democrats and Republicans focused on, at least as I remember from the age of seventeen. Looking back, it is clear that neither political side fully understood the situation.

At the time, there was also widespread disbelief and chaos as we tried to understand what was happening. The Twin Towers fell, the Pentagon was severely damaged, anthrax was being mailed to a variety of individuals, the Beltway sniper attacks were unfolding on the East Coast, and other panic-inducing events seemed to consume our nation. It was confusion and fear, plain and simple. That is why it would be misguided to read this story, or any historical account, and question why we didn't foresee the subsequent events and outcomes. We simply didn't know. The future is always shrouded in uncertainty, while the past appears so clear.

With that cautionary statement in mind, each chapter is divided into two distinct but related sections. First, I tell the story of my experiences in the Army in chronological order, with the initial chapter covering 9/11 and the opening days of Infantry School. Once the memories are complete, I revisit much of that information in a section titled *Reflections*. There, I often apply interpretive, philosophical, or motivational insights to the story. Basically, it is my perspective now, shaped by age, reflection, and the experience of a life lived since those days. So, in a way, I suppose this section has a "self-help" feel, though I will let you decide what lessons to take away from the story, if any. Again, we are venturing into another genre that blurs the lines beyond memoir, historical narrative, and travel writing.

But there is an important reason I chose the title *Reflections*. There is a well-known oil painting by Lee Teter that depicts a middle-aged man standing at the Vietnam Veterans Memorial. His hand rests against the black granite, his head bowed, his face drawn tight with grief. In the polished surface of the wall, the ghosts of fallen comrades, dressed in their uniforms and battle gear, reach

---

[2] Los Angeles Times. "Did Saddam Plot to Kill Bush? : Retaliation Would Be in Order Only if Proof Is Clear and Convincing." *Los Angeles Times*, May 11, 1993. https://www.latimes.com/archives/la-xpm-1993-05-11-me-33738-story.html.

out toward him, as if to comfort him from the other side. The piece is simply titled *Reflections*[3], and it captures, with aching honesty, the quiet sorrow many veterans carry long after the war is over.

When I was in my late teens, I bought a print of that painting for my uncle, who had been, in many ways, a second father to me. He had not served, but it meant a great deal to him. He was of that generation, had lost friends in the war, and his nephew had just gone off to fight in the Middle East. When he passed away, the painting came back to me. Now it hangs above my front door, where I see it every day. It is a reminder to pause, to remember, to look inward, and to reflect.

So, as a respectful nod to that masterpiece, and to what it represents, I chose to name this most contemplative section *Reflections*. My hope is that I honor the spirit of that painting, my uncle, and the emotions it conveys with the pages that follow.

Sprinkled throughout the book, I also weave in excerpts from the flood of letters I wrote and received, marked by dated passages, to transport you back to those exact moments in time. I believe it's important to capture not just how I see things now, but how I felt *then*, in the thick of it. Some of these letters are humorous (and, admittedly, a bit immature), while others are heartbreaking or deeply informative. Selecting which passages to include was no easy task, as so many genuine messages and unforgettable moments had to be trimmed for the sake of brevity. Inevitably, some memories will fade with time and then be lost entirely, but I hope this book preserves at least a few of them. Perhaps someone will find value in those included.

In addition to including many of my own letters, I sought permission to use excerpts from the correspondence of fellow soldiers, leaders, teachers, family members, and others who were part of my life at the time. They graciously allowed me to share their words, and without their support, this project would not have been possible. Through these documents, you will not only see what was happening in my mind then, but also what they were experiencing alongside me.

I also spoke with others who were deeply involved in these events, and I included paraphrased versions of their recollections and interviews. You will also find these as dated passages throughout the book. I discovered that as I wrote, the story began as mine, but it quickly grew into *ours*. I may be the one telling it, but we all share in it.

With that in mind, I should offer an important disclaimer. Before studying physics in college, I spent eleven years as a police officer, eventually retiring as a detective. That experience taught me a sobering truth: eyewitness accounts are often unreliable. Time and again, I encountered situations described one way by witnesses, only to have video evidence or additional testimony reveal something very different. Research confirms this. Stress can distort memory, affecting the timeline, the details, and even a person's overall sense of what occurred. It is a strange and unsettling reality.

I mention this because this book is my story as I remember it. Even so, I have taken extraordinary care to verify details by speaking with fellow soldiers, researching specific events, conducting dozens of Freedom of Information Act (FOIA) requests, and corroborating accounts

---

[3] Teter, Lee. *Reflections* (1988), oil on canvas. Described in "Reflections – Lee Teter," Kenosha Area Vietnam Veterans, https://sewivets.org/kavv/kenosha-area-vietnam-veterans/reflections/.

whenever possible. Many of the details can be independently checked if the reader chooses to do so. In fact, I encourage that. Unlike many memoirs, I have included citations so you can examine the sources for yourself.

Why go to such lengths for transparency? Because without context, many of the stories that follow may seem unbelievable. So, when something sounds extraordinary, I almost certainly have a source to support it, whether it is an article, a book, a video, an old letter, or the testimony of someone who stood beside me. In this way, you don't have to take my word for it.

However, despite my very best efforts, some memories may be out of order, slightly exaggerated, or, on rare occasions, misinterpreted. If any errors exist, they are unintentional, but I take full responsibility for them. This book is an honest attempt to weave together my recollections, preserved correspondence, research, and the testimonies of others.

Also, much of this happened during a time of great stress, and because of this, I don't wish to claim every detail in this book is flawlessly accurate. It would be unethical to make that suggestion. On top of this, combat is often a confusing and chaotic scene that can permanently alter one's perspective. Yet, I have done my very best to avoid these errors in the name of transparency. Meaning there will be instances of embarrassing and difficult moments, along with tales of heroism and triumph. Take what resonates with you from the book.

I should also acknowledge that there are portions of this book where I'm highly critical, particularly of three specific units, and some of what I say may come across as overly harsh. If you or your loved one served in one of those units with honor, or if someone you cared about was lost while serving in them, please understand: my criticism is not directed at you, nor at those who gave their lives. Those who served with integrity remain my brothers for life. My criticism is solely aimed at the systems, the failures of leadership, and the culture that allowed dysfunction to take root.

In nearly every case, I avoid naming names unless an outside source has already reported them. Why? Because I don't see these people as villains in my story. They're like everyone else: flawed, human, and usually doing their best, even though they made decisions I disagreed with, sometimes vehemently. So, even when I express anger and frustration, nothing I write should be read as ill will toward them.

Still, I feel an obligation to tell the truth, even when that truth is uncomfortable for some people to read. During my interviews, I spoke with a respected leader who served alongside me in Iraq. I explained that part of my hesitation in writing this book was that calling out mistakes, corruption, incompetence, and criminal behavior, even when they are clearly wrong, can still feel like a form of betrayal. He understood. Then he shared a story that has stayed with me. I didn't witness it myself, and I cannot verify every detail, and it may have taken on the shape of a cautionary tale over time, but the lesson it conveyed was unmistakable.

According to him, during one of our deployments, a support element was struck by an explosion while driving down the road. One of the soldiers, a woman, was hit in an artery and began bleeding out during the evacuation. Her team didn't immediately remove her clothing to locate the wound, as they would have done for a male soldier. When later asked why, every single one of them gave the same answer: they were afraid of getting in trouble for exposing a female soldier's body. Unfortunately, she didn't survive.

Afterward, the brigade commander met with her family. He told them the truth, without excuses or cover-ups. He acknowledged the failure and vowed that it would never happen again. It was an awful tragedy, but he chose to confront it so that others might learn from it, rather than burying it to protect reputations. In many ways, that is what I'm trying to do here.

So, some parts of this book may be uncomfortable. Some may make you angry. But the only way we learn is by confronting our failures honestly. My hope is that by telling what happened as clearly and truthfully as I can, we give ourselves the chance to do better and ensure that the same mistakes are never repeated.

But what weighed on me more than anything else, and what gave me countless sleepless nights and long moments pacing my office floor, was the responsibility of writing about the final moments of the men I served beside. Their deaths are not simply events in a story. They are sacred and painful memories. They are heavy truths. And I know that many of the families of these men, including wives, parents, siblings, and children, may one day read these words. That knowledge has never left me. It has been a constant presence throughout this process, and it has never been taken lightly. I have carried that weight every step of the way.

To the families of those who were killed, I want to say this clearly: I wrestled with every sentence. I agonized over what to include and what to leave unsaid. I questioned myself again and again, trying to determine which details were necessary to understand their courage, their humanity, and the reality of what happened. There were times when I removed descriptions that I felt would offer nothing but pain. There were details I chose not to tell, not because they were unimportant, but because they didn't honor who these men were in life or in death. Even still, some passages may be tremendously difficult to read. They were difficult to write.

I wrote about their deaths because I believe I owe them that honesty. Silence would be a disservice to their memory. Their stories are not mine to hide in the quiet corners of my mind. They deserve to be known. They deserve to be remembered as they truly were: husbands, fathers, sons, brothers, friends, soldiers. They were real people with real laughter, real dreams, and real fears. I want the reader to feel, even if only for an instant, the same gut punch we felt when we lost them. I want them to understand what their absence meant, and still means today.

I wrote about their deaths because we loved them, and because we still do. And when those moments appear in these pages, I try to tell them honestly and with the respect they deserve, even when it's difficult to revisit. So, if anything written here about your loved one is exceedingly upsetting or offensive, I offer my sincere apology. That was never my intent.

At the end of specific chapters, when I write about soldiers who died and whom I personally knew or served directly alongside, they are honored in a section titled *In Memoriam*. I encourage you to search for their names, see their faces, and learn their stories. By honoring them here, and by you choosing to seek them out, we help ensure their memory continues to live.

Still, I didn't want this book to become only a record of loss and pain. War is not a single, constant emotion, and neither were the men who lived through it. What you will find in these pages is the rhythm of real soldiering. It is unpredictable, exhausting, and sometimes bizarre. One moment we were laughing over something stupid, and the next we were scrambling under fire. Because that is how it was. That is how it felt.

There is also a uniquely military, gallows humor running throughout, which is often very dark, sometimes ridiculous, and usually the only thing keeping us from falling apart. It grew out of shared hardship and from trying to stay human when everything around us wore us down. But most of the time, the joke is on me. So, expect stories of me falling on my face, eating strange foods, making questionable decisions, and dealing with the unavoidable realities of bodily functions in combat… including dysentery. Because that was life. It was ordinary and unbelievable at the same time.

Furthermore, the language in those moments was raw and unfiltered, and that has carried into this book. I include it not to shock, but because it's the truth. It's how we actually spoke, and changing it for a cultured audience would be disingenuous.

So please, after everything that has been said, give yourself permission to laugh, especially when the joke is at my expense. I certainly do. I lived through those moments, and they have become stories that are still told to this day. Yes, even the ones where I completely lost control of my bowels.

When you read this, I want you to feel the adrenaline, the fear, the sorrow, the camaraderie, the foolishness, the courage, and the absurdity. I want you to laugh in one paragraph and sit in stunned silence the next. I want something here to stay with you. If you feel that range of emotions, even for a moment, then I have done my job.

Additionally, as you move through these pages, you will notice the citations, the letters, the interviews, and the personal accounts. They are all here for a reason: to record and to share our story while we still can. This book is not an attempt to convert you to a political stance, defend a particular policy, or persuade you to adopt my view of the war. It's simply an effort to show what was seen, what was endured, and what it cost.

No claim is made to having all the answers. The goal is to ask the questions that matter, to lay the truth bare, and to trust you to decide what all that means. I simply lay out what happened and the reasons behind the decisions I personally made, even if they were wrong.

As of this writing, the Defense Casualty Analysis System reports that Operation Iraqi Freedom (OIF) resulted in 4,418 American military deaths and 31,994 service members wounded in action.[4] These figures reflect U.S. forces only. Other coalition nations also endured losses. The Brookings *Iraq Index* reports 316 military deaths among partner forces, not including tens of thousands of post-Saddam Iraqi Security Forces, though some sources indicate the final total may be

---

[4] Defense Casualty Analysis System. "Operation Iraqi Freedom (OIF) Casualty Summary by Casualty Category," November 3, 2025. Accessed November 6, 2025.
https://dcas.dmdc.osd.mil/dcas/app/conflictCasualties/oif/byCategory.

slightly higher.[5] Private military and reconstruction contractors also faced significant casualties. According to U.S. Department of Labor Defense Base Act records, 1,611 contractor deaths have been reported in Iraq, though researchers widely agree that the true number is significantly higher due to underreporting and unfiled claims.[6]

Estimates of enemy combatant deaths, including those of soldiers, insurgents, terrorists, and militia fighters, are substantially higher, though the true numbers remain highly uncertain. These fighters came from Iraq and surrounding regions, often without formal unit structure or record-keeping. Many of those killed were never identified or recorded. Available estimates vary widely, but most place the number of enemy fighters killed in the war somewhere in the tens of thousands, with some assessments suggesting between 20,000 and 50,000 or more.[7] However, no definitive figure exists.

Tragically, civilian losses far exceed all other losses. Verified, documented civilian deaths caused directly by violence are estimated at roughly 187,000 to over 211,000.[8] However, when taking into account the collapse of infrastructure, sanitation, public health systems, and internal displacement, mortality estimates increase substantially. A widely cited medical study published in *The Lancet* estimated around 655,000 excess deaths[9], while a later survey by ORB International suggested the total number of Iraqis who died as a consequence of the war could be as high as one million, though that figure has been contested.[10] The full human toll may never be known.

However, a word of caution is in order. Casualty statistics, particularly those circulated online or repeated without verification, can be wildly inaccurate or politically motivated. Even among formal studies, many figures are shaped by methodological limitations, selective reporting, or incomplete record-keeping. Researchers working in good faith often arrive at different totals simply because they define or measure loss differently. Yet despite the uncertainty, one conclusion remains clear. The war exacted an immense and lasting toll on those who lived through it, both

---

[5] O'Hanlon, Michael, and Ian Livingston. "Iraq Index: Tracking Variables of Reconstruction & Security in Post-Saddam Iraq." *Brookings Institution.* Washington, D.C., September 30, 2010. https://www.brookings.edu/wp-content/uploads/2016/07/index20100930.pdf.

[6] U.S. Department of Labor. "Defense Base Act Case Summary by Nation," September 1, 2001 - June 30, 2014. https://www.dol.gov/agencies/owcp/dlhwc/dbaallnation6-30-14.

[7] Crawford, Neta. "Human Cost of the Post-9/11 Wars: Lethality and the Need for Transparency." Watson Institute for International and Public Affairs. Brown University, November 2018. https://costsofwar.watson.brown.edu/sites/default/files/papers/Crawford-Human-Costs-20181108.pdf.

[8] Iraq Body Count. "Documented Civilian Deaths From Violence." Accessed November 6, 2025. https://www.iraqbodycount.org/.

[9] Burnham, Gilbert, Riyadh Lafta, Shannon Doocy, and Les Roberts. "Mortality After the 2003 Invasion of Iraq: A Cross-sectional Cluster Sample Survey." The Lancet 368, no. 9545 (October 21, 2006): 1421–28. https://doi.org/10.1016/s0140-6736(06)69491-9.

[10] Opinion Research Business. "More Than 1,000,000 Iraqis Murdered." Newsroom, Opinion Research Business, September 2007. https://web.archive.org/web/20071224053055/http://www.opinion.co.uk/Newsroom_details.aspx?NewsId=78.

Iraqi and American. The cost was severe, and its effects continue to reverberate long after the last official shots were fired.

**Triangle of Death (2005) - the apparent calm in this photo is a lie**

That being said, firefights, explosions, and witnessing gruesome deaths were a daily reality for thousands of combat arms soldiers across Iraq, especially for infantrymen. My division, and more specifically my platoon and I, lived through this brutality firsthand. Death and violence were everywhere, so much so that they became woven into the fabric of life. Most readers already grasp this reality without it being emphasized, so I have intentionally chosen to explore dimensions of the story beyond the purely macabre.

In the end, what will always remain with me is the unwavering bond with my brothers-in-arms. The loyalty, companionship, and stories we share are indelibly imprinted forever upon my mind. I wish that for one fleeting moment, we could once again stand together and share the sting of battle as Dogface Soldiers.

I wish I could live it all again.

Jeff Harstad

Sioux City, Iowa

# Part I

The Forge

---

...there is no knowledge that is not power.

Ralph Waldo Emerson
"Old Age," *Society and Solitude* (1870)

---

# CHAPTER I

## You're in the Army Now

For those too young to remember or those not yet born, it is nearly impossible to convey what September 11th felt like in the United States. I have attempted to explain it to my daughters several times, but my words always seem inadequate. How does one summarize a day that altered the course of history? Not only for me but for the entire country. In fact, the world as a whole changed on that day.

All of humanity felt the strange shift. It was as though something had dramatically changed overnight. New and highly controversial legislation emerged in the aftermath. Security measures were drastically tightened, though with questionable outcomes. Our allies offered reassurances, and adversaries issued threats of destruction. The television screens and radio airwaves were filled with a myriad of voices, some calling for unity and peace, others for fiery vengeance and war. Some were flip-flopping between both sides of the issue.

It started as a typical Tuesday, with nothing exciting happening by mid-morning. That quickly changed as the airplanes began hitting their targets. The emotions soon ran high, and everyone affixed to their television screens collectively shared them. We experienced feelings of isolation, fear, rage, the unknown, and, surprisingly, eager anticipation. For me, I knew war was brewing. And I needed to get myself into the middle of it.

Perhaps those feelings were what millions experienced upon learning of John F. Kennedy's assassination or the attack at Pearl Harbor. As a remote observer trying to understand those historical events at a distance, both in time and proximity, I suppose I will never truly know what those days felt like. I can easily read the statistics in books, carefully study the black-and-white photographs of sinking Navy ships, or watch the gruesome Zapruder film that captured the President's final moments. While I can understand the relevance of those events at a distance, I don't truly feel the gut punch. I presume this is what my children experience as they see the haunting photograph of the Falling Man. Sad, yes, but it is only historical to them. However, 9/11 resonates for many of us who lived through it much more profoundly than we can fully articulate.

Other than the day I was born, no date has been more influential in my life than September 11th, 2001. I'm not alone in this sentiment; it is the same story for many Americans. It was the day I called my Army recruiter to enlist. He was already overwhelmed by the massive influx of calls and asked me to postpone signing the paperwork until the following day. He also advised me to carefully discuss it with my parents, as I was only seventeen and needed parental consent. Three separate signatures were required.

Delayed Entry Program ID Card

Though I had plans to enlist in the military before 9/11, the terrorist attacks sparked an urgent need to act and accelerated my decision to join. The very next day, September 12th, my parents and I sat at the dining room table as the Army recruiter passed around the documents. My mom trembled with dread as my dad sat as stoically as possible. They knew I had decided on the matter, and they had little choice. They could either grant permission for me to enlist early, or watch me do it without them as soon as my birthday in March rolled around. With trepidation, they signed the papers next to my name as I gave my life over to the United States Army, with New York and Washington continuing to smolder.

All these years later, I'm still often asked why I chose the Army over the other branches of the military. Others seemed to be a more natural fit for my intellect and personality. The answer, though, is quite simple. I wanted to be a frontline combatant, an infantryman engaged in land warfare in hostile environments. This meant I had to pick between the Army and the Marine Corps, as only they have dedicated conventional infantry units. It can be argued that other branches have specialized combat units similar to infantry, such as Air Force Pararescuemen and Navy SEALs, but these units have small teams that are tremendously hard to join. While those sounded exciting, I knew the odds of success were slim. If I failed a selection course, I didn't want to be washed out and end up serving on a submarine or as a mechanic. I specifically wanted to avoid any non-combat roles. I wanted to fight.

I didn't want to join the military to be a cook, a tank driver, or work in intelligence. Those jobs are essential, and I don't wish to diminish their contributions in any way, but I wanted to test myself in direct combat with other human beings. Could I mentally and physically hone my skills to become a proficient warrior? I suppose many young men feel the same way, with a primal urge to match our forefathers' physical strength, bravery, and determination. Could I live up to the legacy of those at D-Day, Guadalcanal, the Chosin Reservoir, and Khe Sanh?

Upon speaking with a Marine Corps recruiter on 9/11, which I did just before the phone call to the Army, he told me the blunt truth to a direct question. I asked if he could guarantee a combat deployment if I enlisted as an infantryman. He suggested there was a strong possibility of this because of the terrorist attacks. *But*, he added, I could also end up on a Marine Expeditionary Unit ship in the middle of nowhere, pulling security indefinitely. That didn't sound like much fun to me, so I quickly called my Army recruiter. In hindsight, I appreciate the Marines' honesty, even though it cost them a recruit.

The Army had a different pitch, though they were likely stretching the truth. Naïve as I was, I didn't understand that nothing in the military is guaranteed, especially regarding deployments. Yet they reassured me that the 3rd Infantry Division (3rd ID) would be part of any upcoming invasion in the Middle East. In fact, they had an available slot for me in that division if I wished to join. This seemed like my best opportunity for combat, so on September 12th, 2001, I enlisted in the US Army as an infantryman (11X and later 11B). Upon completion of Infantry School at Fort Benning (temporarily renamed Fort Moore), I was to be assigned to the 3rd ID at Fort Stewart, Georgia.

Unfortunately, many military recruiters deceive potential enlistees, which often creates embittered troops. These grudges are frequently permanently held against the recruiter and sometimes even the military in general. For example, I was verbally promised a spot in Airborne School after completing basic training. Allegedly, all I had to do was raise my hand and volunteer before I shipped to my permanent duty station, as all willing infantrymen were readily given Airborne slots. Or so I was told. This was a blatant lie; my drill sergeants even laughed at me when I brought it up. I learned a valuable lesson: get everything in writing. Also, *nothing* is guaranteed.

For clarification, all Army recruits who become infantrymen attend the United States Army Infantry School. While this is roughly interchangeable with the term basic training, key differences must be addressed. The Army has two distinct tracks that recruits must follow, depending on whether they are infantry or not.

Typically, non-infantry recruit training in the military is divided into Basic Combat Training (BCT) and Advanced Individual Training (AIT). BCT, commonly known as "boot camp," covers essential skills that all service members must learn, and traditionally features screaming drill sergeants. AIT, however, provides specialized training for a specific Military Occupational Specialty (MOS), or, in other words, a job. The MOS simply refers to the soldier's primary role in the military, with most MOS fields serving in support roles and not engaging in combat. The BCT and AIT phases may occur at different bases, meaning a soldier's "battle buddies" (or classmates for civilians) may vary.

On the other hand, Infantry School begins on day one for recruit infantrymen, and the company stays together from start to finish. There is no separation between BCT and AIT; the entire training is seamlessly integrated and referred to as Infantry School. Essential skills are ingrained throughout the process, including marksmanship, physical fitness, first aid, and land navigation. Additionally, it emphasizes developing hardened shock troops by implementing the pounding rigors of daily drills, such as carrying heavy loads during long road marches and hand-to-hand combat training. However, I will use "Infantry School" and "basic training" interchangeably in this book.

Before shipping off for training, I did my best to get into shape, though, in hindsight, my preparation was laughable. I played baseball throughout my school years, but I wasn't what one would call physically gifted, so I was inexperienced in putting together a training regimen. I jogged around the block a few times, did some push-ups, and tried to devise other exercise programs that weren't particularly useful. In fact, it was horribly inadequate and did very little to prepare me for what was to come.

Ironically, the most beneficial physical skill that I would bring into the military wasn't from regular exercise or traditional sports. It was backyard wrestling. Professional wrestling was all the rage in the mid-1990s, and my brothers and I were obsessed with World Championship Wrestling and the World Wrestling Federation. We would wrestle for hours, using our giant trampoline as a safety net to avoid serious injury. We even built a wooden platform about fifteen feet high to launch ourselves from, much to our mother's horror, in an attempt to imitate our heroes from television. From the stunts we pulled, how we got out alive I'll never know.

Admittedly, pretending to be Hulk Hogan wasn't exactly battlefield training. Yet it unexpectedly taught me leverage and body control, two invaluable skills that later helped me become my platoon's US Modern Army Combatives instructor. As funny as it is, our childhood antics were a precursor to many fundamentals of mixed martial arts (MMA). It shockingly served a purpose other than simply being fun.

I mention all of this because many people have the wrong impression about the armed forces. Television and Hollywood often shape the public's perception more than reality. John Rambo and Jack Reacher may be interesting and larger-than-life characters, but they aren't indicative of the military as a whole. In my experience, the military was a highly diverse collection of individuals from every walk of life.

I served with some knuckle-dragging mouth-breathers who were incapable of being trusted with even the simplest of tasks. Others were absolutely brilliant in leadership, strategy, communication, and even creative thinking. They possessed sharp analytical minds and could easily thrive in any Fortune 500 company. Some men were physical powerhouses, while others struggled with every run and physical training (PT) session. Some enlisted under a court-ordered ultimatum: go into the Army or an orange jumpsuit. And then there were men who possessed graduate degrees and gave up six-figure careers to sleep in the cold mud. It was indeed a wild assortment of personalities, intellects, talents, and physical capabilities. It reflected America.

Since the military was taking pretty much everyone who wanted to enlist in the early 2000s, mainly because of the terrorist attacks and the need for manpower, it wasn't uncommon to see this interesting cross-section of society. While we all came from very different backgrounds, the military had an effective way of leveling the playing field, providing the training and tools needed for success. It rarely gets the credit it deserves for the personal and professional development it offers to so many young men and women. However, as the old saying goes, you still can't get blood from a stone. Meaning that some people weren't cut out for military life, no matter how much training or help they received.

After enlisting, I wasn't required to report for training immediately because I was part of the Delayed Entry Program (DEP). This allowed me to postpone entry until after I graduated. Loathing high school, which I considered boring, too easy, and a waste of time, I finished a semester early and spent the spring and summer working with my dad at his carpet cleaning business. I tried to pad my bank account while working diligently, knowing that my departure date was drawing ever closer.

After months of anticipation, the day finally arrived, and my family drove me to Omaha, Nebraska, where the departing airport was located. That evening, we all dined at Olive Garden for our last meal together. It was a depressing experience and far more difficult than I had ever imagined. After ordering my food, I found I could hardly eat a bite. My stomach was twisted into knots, and I couldn't even touch it. None of us could. I was scared to go into the world alone, but I knew I had to do it.

Saying our emotional goodbyes, I watched as my family slowly pulled away, their minivan receding into the evening as their faces revealed anguish and dread. I was suddenly alone with my thoughts, standing in a motel parking lot, the fluorescent bulbs flickering. So terribly alone.

**Undated (July 2002) – Note from the Harstad Family**

**We all love you, Jeff! Good luck and work hard! – Mom, Dad, and your brothers**

With nothing else to do and nowhere else to go, I made my way to the room the military had reserved for the night. I was boarded with another recruit as he, too, was flying out of Omaha. However, he was supposedly enlisted to become a Navy SEAL and would ultimately have a different destination. He and I talked for a while, but I didn't get the feeling he was confident with his decision to enlist. His enthusiasm was lacking. Sure, I was afraid, but also excited. I was bound for adventure!

In the middle of the night, I heard the latch flip and the door slowly creaked open. He started to leave, mumbling that there was a family emergency and that he had to go. I knew better than that, though. He didn't have a family emergency. He was running away from his enlistment. It seemed the reality of the situation had finally set in.

I didn't sleep a wink that night, tossing and turning fitfully. That morning, I boarded a flight and arrived in a hot, humid Georgian heat wave. August 1st, just shy of one year after the terrorist attacks, marked my first official day in the United States Army. I was set to become an infantryman.

I found myself surrounded by other wandering recruits at the Atlanta International Airport. We were all rounded up and bused to the front gates of Fort Benning, the legendary home of the infantry. I was prepared to be swarmed by screaming drill sergeants, but none of them were to be found. Instead, everyone in uniform appeared lethargic and moderately annoyed. My first impression was one of confusion and mild disappointment. It was very apparent that wherever we were, the soldiers were not enjoying their jobs of escorting new recruits into the base. Our baptism by fire was lackluster, to say the least.

That evening, I learned we had been deposited squarely into the Army's crusty armpit at the 30th Adjutant General (AG) Battalion. This holding facility served as a pipeline for new rotations of basic training companies. It was the equivalent of purgatory. Recruits would arrive daily, and the barracks would slowly fill up. Once a unit was ready for a new training cycle, the fresh recruits from 30th AG would be marched or bused up to their basic training locations and introduced to their drill sergeants and command staff.

Honestly, the 30th AG was the most depressing and disorganized place I would ever experience in the military. Even surpassing war in the Middle East, the 30th AG drained the life force from every recruit. It was utterly disordered, full of undisciplined and apathetic soldiers, and smelled like the inside of a high school locker room. It wasn't a good first impression and was confusing, as I was expecting a grand gateway to America's premier fighting force.

**6 August 2002 – Letter to the Harstad Family**

There are two washers and two dryers in a barracks of over 100 guys. Doesn't make much sense. A lot of things here don't seem to make sense.

The cadre, or training personnel, consisted of regular Army soldiers and drill sergeants, though the latter were noticeably in the minority. Drill sergeants were readily identifiable by their iconic campaign hats. They tended to be far more aggressive and perpetually irritated than their beret-wearing counterparts, with an overtly hostile demeanor toward recruits. We made a point of avoiding them, but this wasn't always possible.

Each day, someone from either group would march us around and complete a requirements checklist before our cycle could begin. This included vaccinations, obtaining gear, receiving haircuts, and other menial tasks. It was a lot of monotonous, busy work that was mostly a waste of time. It felt like we were on an assembly line, and everyone was confused because we expected a very different scenario to unfold.

While there were many subpar soldiers at 30th AG, several squared-away professionals were beacons of light for those of us eager to work hard. One particular monster of a man was Drill Sergeant (DS) Thomas, who resembled a clean-cut, modern-day Viking. He was a towering figure, likely 6'5", if not taller. His neck appeared to be as thick as my thigh, and he would occasionally stroll up and down our hallway with a scowl on his face. We all wanted to be like him simply because of his presence.

DS Thomas had a legend surrounding him. Allegedly, he was in the 10th Mountain Division as an infantryman when the Black Hawk Down incident took place in Mogadishu, Somalia. Rumor had it that he ran out of ammunition during the battle and was swarmed, subsequently killing at least one enemy combatant with an E-Tool, a small, collapsible shovel. Later, a cadre member told me the story was true, except it was multiple fighters, not just one. I tend to believe it. But whether the legend was factual, embellished memory, or simply battalion folklore, it was clear that he wasn't someone to be trifled with.

Thankfully, my stay at the 30th AG was brief, lasting only a week. During that time, we learned the basics: marching, identifying ranks, polishing boots, making our bunks, and other related skills that would be useful during basic training. We were also issued two large duffel bags, filled to the brim with clothing and miscellaneous gear.

Most importantly, we were quickly familiarized with the painful, horrible slogan *"hurry up and wait."* No matter the task, it seemed as though the military bent over backward to facilitate the best way to rush recruits through a crucially important assignment, only to funnel them into a confusing and boring queue. Quite frequently, there were whispers of puzzlement about what we were doing. More often than not, no one seemed to have any idea other than we were standing in line for... *something.*

While adjusting to my surroundings, I witnessed a bizarre incident involving a disturbed recruit who was threatening to kill himself. The barracks were a series of neighboring buildings connected by an elevated concrete walkway, with only a railing to prevent someone from falling. I wasn't sure if a jump from there would be fatal, but I safely assumed it would probably hurt quite badly.

The recruit, whom I didn't know, had gone out onto the boardwalk with a heavy floor buffer and stood near the railing. He had tied the electrical cord around his neck and was threatening to throw it overboard. After he lifted it onto the barrier, I watched, bewildered, as it teetered back and forth; all the while, he screamed obscenities. I had no idea what he was yelling about or what the act was supposed to accomplish, but a large group of us stood in silence, watching with morbid curiosity, unsure what to do. I assumed his brilliant plan was that the cord would snap his neck as the buffer fell, similar to a reverse hanging. Or perhaps it would pull him over the railing; I'm not sure.

A swarm of drill sergeants suddenly appeared out of nowhere and began yelling at the hysterical private. They were hovering around but maintained a perimeter about twenty feet back, clearly concerned he would throw the buffer off the bridge. I don't know the order of the subsequent events. I'm unsure if he dropped the buffer first or if they charged him, and *then* he dropped it, but the result was the same. The machine tumbled in slow motion through the air and crashed onto the concrete below, twisting into an odd angle of dark metal.

The effect wasn't what the suicidal man had thought, though, as he had tied the noose too far from the buffer, giving himself far too much slack. This meant absolutely nothing happened, other than a loud bang and some dangling electrical cable. Well, not at first.

Just as soon as it crashed below, the drill sergeants tackled him down onto the bridge, causing the now taut cable to violently wrench his head and neck backward. Camouflaged bodies smothered this poor guy as the buffer now jerked up and down each time they pulled him this way and that. They appeared to be fighting over his limp body like a pack of hyenas on a hapless wildebeest. Panicked shouting took over as they quickly removed the cable from his neck. With increasing astonishment, I watched as they carried him away, his face now bright red and purple.

I never saw him again, as I imagine he was quickly discharged due to his apparent mental health issues. In hindsight, it is darkly ironic and a bit amusing that his plan had backfired, though he was nearly killed by the people intending to save him. It was just another strange occurrence at the 30th AG.

Interestingly, my strange memory is not the only one from that awful place. If you ever meet a US Army infantryman, ask him about the 30th AG, especially if he served in the early 2000s. I can almost guarantee a colorful story like this one. It may be even wilder. It wasn't called the "armpit" of the Army for nothing.

**4 August 2002 – Letter to the Harstad Family**

We figured out that threatening to commit suicide isn't effective. They put you on a 24-hour security watch. All the drill sergeants scream at you and even urge you on. "Go ahead! Throw yourself off that bridge, and we will have someone take your place! We have you for four more years! You volunteered! No one pointed a gun at you and said that you have to join! You do your duty, and the Army will do its duty."

Portions of our days were spent doing mild conditioning activities in the parking lot, such as running, pull-ups, sit-ups, push-ups, and stretching. Though I wasn't in great shape, it was much easier than I imagined. To simply qualify for entrance into basic training, we had to be able to pass a minimum physical fitness requirement. To receive a "go," we had to obtain a laughably low standard: an 8:30 mile run, 13 push-ups, and 17 sit-ups. As soon as we had precisely our go number, the cadre would stop us and not allow us to do anything more.

A recruit had allegedly died from an enlarged heart and dehydration the week before I arrived, so everyone was walking on eggshells. There were rumors that the cadre had been ordered to prevent any and all injuries before we had the chance to be assigned to our basic training units. I thought, *Is this the Army or a summer camp?* To this day, I still don't know why they handled us with such kid gloves. They even restrained us from exercising beyond what takes place in a high school gym class. This included banning independent running, throwing a Frisbee, or working out in the barracks bay. All I knew was I wanted out of the 30th AG as soon as possible.

We were on complete lockdown from the moment we arrived at the reception center until graduation from Infantry School. This meant we had no freedom of movement and were restricted to our assigned post 24 hours a day. Some military branches don't enforce this rule during certain training classes, allowing recruits to enjoy their free time after training hours are over. This does not apply to the Army, and certainly not for basic training. The rule was strictly enforced, and severe punishment was meted out to anyone caught in an unauthorized location.

It doesn't take long to understand why recruit soldiers quickly hit a wall of frustration at the 30th AG. They're not free to go anywhere, are forbidden from doing extra physical exercise, cannot read books or magazines, are prohibited from watching television, and are stuck in a barracks bay with hundreds of other testosterone-fueled men with nothing remotely interesting to do.

What is the logical result? *Extreme* boredom. This naturally tends to be bad, especially when there are lengthy periods of no supervision. A bunch of bored, immature guys, most of them just out of high school, often make poor decisions. Sometimes, spectacularly poor decisions.

**6 August 2002 – Letter to the Harstad Family**

Yesterday, a Bravo platoon (which we are a part of) messed up an Alpha platoon's barracks. The drill sergeants punished all of Bravo for this: a half-hour of standing in a continuous squat while holding your hands over your head. All because a few idiots screwed up.

The day mercifully arrived when my group was informed that we would be picked up by our new training unit, Charlie Company, 2nd Battalion - 54th Infantry Regiment. Our excitement was palpable as we nervously prepared for the move. Reality began to sink in, and we didn't know what to expect. Would basic training resemble the nonsense at 30th AG, or would it be like the portrayal in the movies?

We were ordered outside and scrambled into formation, each of us carrying our belongings. There, we met our drill sergeants for the first time, and they carried an intimidating presence, occasionally barking orders at a random private. Yet, they appeared as though they were holding back. The confusion continued for all of us, as we thought our first introductions would be more dramatic than what we received. Sure, they looked mean, a little crazy, and ready to take control, but they also looked restrained. Why?

Like the reception center at 30th AG, our first introductions were a letdown. There was no pandemonium or screaming drill sergeants. They were just standing there, looking as if they wanted to get the hell away from 30th AG as quickly as we did. I finally realized at that moment we were standing in a black hole where time was permanently frozen and recruits went to die. OK, I'm exaggerating a bit on that, but there was definitely an aura of bad vibes around the entire place. Trust me.

Sergeant First Class (SFC) Clements and Staff Sergeant (SSG) Ross, two drill sergeants of 2nd Platoon, the Renegades, led our formation as we marched from the 30th AG to our company barracks on Sand Hill. We would soon learn the misery of that name. *Sand Hill.* It still brings back horrible memories.

The sand there was infested with chiggers and other creepy crawlies, and it was thick and clinging. We would later learn that some of our drill sergeants had a favorite pastime in that sand, turning their recruits into "sugar cookies." Despite the pleasant-sounding name, the process was anything but delightful. Under the direct Georgia sun, we would be ordered to perform nonstop drills in the scorching grit. The worst of them was "on your belly, on your back, on your feet." Over and over, we slammed into the ground, rolled, spun, and tried to keep up with their commands, each one coming faster than the last.

Sweat would soon coat every piece of battle dress uniform (BDU, our camouflage fatigues), and the sand would paste itself to us. It worked into every crack and crevice imaginable, ground into our skin, and congealed in our eyes. It mixed with chigger bites and raw, bleeding rashes. That sand was incredibly uncomfortable, with handfuls of it wedged into our boots and underwear. But we didn't know any of this as we marched innocently up that hill. There was no way we could have known the misery that waited for us.

We soon neared our destination, Charlie Company, and made our way to the compound. The very instant the first private's boot touched Charlie 2-54's turf, the drill sergeants erupted in apoplectic fury. They began screaming obscenities, and their faces were twisted in rage. The infamous knife-hand chop of the drill sergeant sliced through the air toward our faces as spittle rained down as quickly as the insults and threats of murder. We all panicked. But what the hell were we supposed to do? Run? The games had officially begun.

With veins bulging and bright red faces streaming with sweat, they demanded we throw all our duffel bags into the middle of the lawn. Our olive-green bags had been spray-painted with our last names, which would quickly come into play. The scorching sun beat down as the jumbled mountain of green was hastily created; all the while, they pinballed around to scream obscenities into random and terrified faces.

Only moments ago, we had been cocky and lulled into a sense of complacency. Perhaps our griping about the 30th AG and our boastful wishes to meet "real" Army drill sergeants was a foolish thing to desire. We would quickly learn to regret this.

The disorganized group of men and I quickly gathered into a formation near the building about fifty meters away. Then the order came from our first sergeant, the man in charge. We were to sprint to the Mount Everest of duffel bags, retrieve our own, and then return to the formation spot near the building in three minutes or less.

Upon the command "go," the real punishment began. We ran to the bags but soon discovered the task was impossible. The pile was disorganized, and the drill sergeants joined in on the fun as they swarmed like frenzied sharks. Five or six of them, apparently experiencing psychopathic rage, would completely surround a soldier and painfully pound him with the brims of their rounded hats, making it impossible for the poor guy even to move. The more he panicked, the worse it became. They would scream vulgarities and demand to know why the recruit wasn't moving while tightening their circle and restricting his movement. It was a hot box of flying spit, insults, escalating threats of physical violence, and impossibly contradictory orders.

Only a fraction of the soldiers found their duffel bags on the first attempt. I did not. As soon as the allotted time elapsed, we were made to sprint back to the formation area for furious sets of push-ups, our arms trembling like jelly as the drill sergeants circled like vultures. Sweat poured out onto the sand as we tried to follow their orders.

Another three minutes were granted, and we made the mad dash back to the pile, attempting to come up with better strategies each lap. One soldier was inadvertently crushed from behind by dozens of stampeding men, and he took a silver padlock to the forehead. These were fastened through the clasp of each duffel bag to keep recruits from "acquiring" other people's gear. He dripped blood everywhere, which mixed with the sweat and sand paste, though the drill sergeants screamed with no sympathy. It made the already chaotic scene surreal.

This task was repeated until word organically spread throughout our company just to grab *any* random bag and bring it back to formation. We would sort it out later. Unsurprisingly, the drill sergeants immediately spotted the deception as the pile went from a hundred bags to zero. This wasn't their first training cycle, and they knew what we had done. Any private holding a green duffel bag with a last name different from the one emblazoned on his chest was made to bake some "drill sergeant sugar cookies." That meant me. As further punishment, the entire company was forced to return all the bags to the pile and start from scratch.

I'm not sure how long our activity lasted, though time seemed to blur, and minutes became what felt like hours. Once we finally got our act together and started working as a team, everything began to fall into place. We solved the problem by having the soldiers who had already located their property assist their battle buddies. In hindsight, it seems like common sense now, but in the mass confusion and heat, paired with the lunatic drill sergeants constantly harassing everyone, it was overlooked. Stress does funny things to people; it can cause panic or make them say bizarre statements. It can also make them blank on blatantly obvious solutions that armchair quarterbacks later second-guess. And this was just day one, only dealing with a couple of bags!

With our bags in hand, we marched into the barracks and were ordered to quickly put our gear away. Every item of clothing, piece of equipment, and bunk had to look identical. For example, socks had to be folded and rolled precisely, and the beds required crisp, clean hospital corners aligned with a perfect diagonal angle. Indistinguishable living spaces were mandatory, and anyone deviating from that plan was swiftly punished.

That evening, and many more after it, we would return to all our belongings thrown chaotically about the barracks by marauding drill sergeants. Closets were tipped over, contents were emptied, every bunk was undone, and shaving cream was sprayed liberally throughout the latrines (bathrooms). It looked as though a tornado had gone through our barracks. Unfortunately, we couldn't sleep until everything was restored to order and thoroughly cleaned. It was inconvenient at first. Then it quickly became demoralizing as stress and exhaustion ramped up.

It didn't take long for the first weak and unprepared recruits to question their decisions to join the Army. Almost immediately, some soldiers attempted to desert, also known as absence without leave (AWOL). Men began cracking under pressure. The physical torture from the sun and relentless "smoke sessions" (grueling, marathon-length exercises), paired with sleep deprivation and constant screaming, was enough to push some recruits over the edge and make abysmal choices. This combination was simply too much for some, and they made decisions they would regret for the rest of their lives.

**12 August 2002 – Letter to the Harstad Family**

We have already had a few guys try to go AWOL. Dumb idiots. They get caught every time. There was a funny thing this drill sergeant said: "If you want to leave, go ahead. From the reception station, walk through the forest line. You will cross two roads and two streams. If you make it this far, you will come to a road. DON'T go left, otherwise you will be within a few hundred yards of the Rangers, and they will kill you. Go right instead. If you don't want to be here, get the fuck out!"

From the very beginning, the platoons began to fracture. Each individual soldier would be reset into his true nature, and the strain would shatter the pretense of machismo and feigned competence. It became easy to spot the natural leaders, the fools, the workhorses, the loudmouths, the class clowns, and, of course, the hated Blue Falcons. A Blue Falcon is everyday military slang for a "Buddy Fucker," or someone who undermines others and betrays the team for personal gain. No one likes a Blue Falcon; everyone immediately learns to watch for them. They're the ones most likely to backstab the unit for their own benefit.

**12 August 2002 – Letter to the Harstad Family**

Our platoon is getting rough with each other. So many people want to lead, and few want to follow. We have been getting in trouble for this, which results in mad PT! I have never sweat more in my life.

The first week was a blur of activity, and I don't remember most of it, mainly due to a lack of sleep. It was incredibly stressful, though I now realize much of it was due to my inexperience. I had never been physically or mentally pushed or intimidated like that in my life. While I wasn't fearful of the drill sergeants hitting me for no reason, except during our sessions of US Modern Army Combatives (mixed martial arts), they had creative ways of generating environments of perpetual strain. Everything we did was somehow wrong, and it was demanding work. This meant we were punished repeatedly and often. Stress was ingrained into our very existence.

Charlie Company was on what they referred to as "total control." This meant we had no say in anything we did. We couldn't choose what we ate in the chow hall, we had to use the latrines when they told us to, and we had to do everything by their rigid schedule. No matter how minor, every detail was preplanned, and we had to follow precise orders. Any deviation from their rigorous timetable resulted in punishment, regardless of who or what was at fault. Need to take a shit outside the schedule? Punishment. Did you drop your pen in the hallway? Punishment. Did a random bird take a dump on your barracks window? *Punishment.*

The only task they allowed us to complete "unsupervised" was laundry, which had to be finished during our short free time at the very end of the day. This is when I would typically write letters home. Everyone in the platoon was arguing about when they would get to use the washing machines, as our clothes were filthy, and this became a primary source of conflict. The hostilities boiled over into fistfights, and the drill sergeants intervened by dropping the hammer on us. They installed a padlock on the laundry room door, meaning we stayed dirty until we began working as a team. That took quite some time.

**14 August 2002 – Letter to the Harstad Family**

Today we went through crazy PT. The drill sergeants are nailing us for everything.

In PT, we do organized exercises in a dirt field. Anyhow, people were brushing themselves off while at attention (when we were required to stay motionless). So, the drill sergeants made us roll in the dirt and sprayed us with water. We were covered in mud.

To add on to this, some of us haven't been able to do laundry in over a week. We only have 4 washing machines. We have been sleeping in dirty clothes for the past week, and the drill sergeants know this. They want us to work as a team for everything, which is impossible for our platoon.

In the beginning, 2nd Platoon only had two drill sergeants, meaning we were understaffed by one. We thought it was fortuitous to have one less drill sergeant to worry about; at least, we did at first. Occasionally, we would be marched around by a drill sergeant from another platoon, such as to the chow hall or to another non-training event. We were not "their guys," so they treated us exceptionally poorly. They treated us even worse than their own men, and that was saying something!

The drill sergeant we despised most was DS Whiteley from 4th Platoon. A slim man with glasses, he looked more like a member of a college lacrosse team and didn't fit the mold of a traditional disciplinarian. Yet, he was extraordinarily ill-tempered, and we knew we were in deep shit whenever he was near. He seemed to have an almost uncanny ability to find a recruit's biggest insecurities and exploit them mercilessly. For me, it was my size. At 119 pounds, I had joined the Army at the minimum allowed weight for my nearly 6-foot-tall frame. He relentlessly came up with things to harass me about, though he was far worse with many other recruits. It always seemed as though he was making fun of someone's weight, their supposed "mental retardation," weird facial features, or whatever vulnerability he could exploit. He was the absolute master of creating stress and was disturbingly effective at his job. But we didn't like him. No, we *hated* him.

**27 March 2025 – Reflection of Toby Glass, 3rd Platoon, C 2-54**

Learning the psychology of each drill sergeant in the company was key to surviving. They were all different, and each one had his own way of beating you into submission. And they were all extremely effective at what they did.

For example, DS McCrary was an old-school Southerner who valued one thing above all else: going hard. The formula with him was simple. Work until you had nothing left, keep your mouth shut, and don't give him a reason to notice you.

DS Schneider played mind games. He thrived on chaos and loved forcing us to think on our feet. If you stayed sharp and calm while everything around you was falling apart, he mostly left you alone.

DS Whiteley used belittlement as his weapon. He was relentless and on you for every little thing. With him, "good enough" didn't exist. If you wanted even a sliver of approval, if you could call it that, you had to go far beyond the standard.

One day, DS Whiteley marched us to the chow hall around lunchtime. He paced incessantly and screamed as we received whatever the cafeteria workers decided to slop onto our plates. We were forbidden from eating any sweets, including cakes, cookies, slices of pie, or other sugary treats that were so temptingly displayed before us. Overweight recruits naturally struggled with this torment more than I did, but it still bothered me to see that heavenly sugar call my name.

"Give Skeletor two of everything!" DS Whiteley screamed, pointing directly at me.

My eyes probably bulged from my skull as the workers piled massive helpings of meat, potatoes, vegetables, and other food onto my plate. It was a comically large amount of rations, and there was absolutely no way to eat all of it. But I decided to give it my best shot. I had no choice.

Sitting down, we were given two minutes to eat everything on our plates and drink two large glasses of water. As soon as I sat in my chair, DS Whiteley took the chair directly across from me and leaned across the table like a tiger ready to pounce. He stared directly at me, his hands grasping the table's edges, and he began screaming to eat faster.

I stuffed my face as quickly as possible, carefully avoiding eye contact, as he kept hurling the insults. However, there was a finite limit to how much I could put into my small frame, and my stomach quickly ceased cooperating. No matter how much he screamed and threatened my demise, I couldn't take another bite for fear I would puke all over him.

"Get outside, dipshit!" he screamed at me. I tried to remove my tray from the table and return it to the receptacle station, but this just set him off into another vicious tirade. So, I ran outside with hopes of fleeing his immediate wrath.

This didn't work as he swiftly followed me and then proceeded to make me do an ungodly number of sit-ups, flutter kicks, and other exercises until I barfed into the wood chips near the pull-up bars. Afterward, he marched me back inside to get a fresh plate of food. Again, I nearly vomited in the chow hall but kept forcing the food down into my stomach. Although I couldn't finish my second plate, he was a bit more lenient this time and ordered me outside to clean up my vomit. I scooped up large handfuls of wood chips and deposited them into the trash can, with ripe tendrils of mashed potatoes and stomach grease snaking around my arms.

Over time, my stomach slowly adapted to accommodate the massive helpings of nutrient-dense foods shoveled down my throat. Only a handful of times did I vomit from being force-fed enormous quantities of food and water. The method was clearly working, as I was gaining weight, albeit slowly. The mechanisms of Infantry School were coming together, bit by bit.

I wish I could say the first week of Infantry School was easy for me. That would be a lie. I would also like to say I found myself as a natural leader, unfazed by their games. That, too, would be a lie. I was simply doing my best to stay afloat and keep up with everyone else. My hands and feet were blistered, my mind was foggy, and I smelled like a mixture of dirt and vomit. The weeks ahead would progress at a blinding pace, and the difficulty would only intensify. Could I possibly hold on?

# Reflections

Basic training is the quintessential introduction to the military that everyone immediately thinks of, as it has been popularized in movies like *Full Metal Jacket* and *Stripes*. However, neither of those movies is a totally accurate portrayal of Infantry School, at least not from my experience, with one being a war drama and the other a comedy. Yet, both represent some realistic elements, and Infantry School was like a weird mash-up of the two movies. On the one hand, there was a lot of yelling, insults, sleepless nights, grueling physical challenges, and constant running from one place to the next. On the other, there were moments of practical jokes during sleeping hours and time for genuinely enjoyable training exercises.

At its core, basic training strips away individualism and instills the belief that the unit must live or die as a single entity. Something as simple as a buzz cut, which all recruits receive at the beginning of training, reinforces the idea that everyone is on an equal playing field. As each task is tackled, some soldiers naturally rise to the top and stand out. However, they quickly learn that survival is not an individual endeavor. Instead, they must rely upon the collective strength of the group and their fellow soldiers to accomplish the mission. In this way, it is in their best interest to help everyone, including, most importantly, the weakest members of the platoon. Personal achievement becomes a distant secondary concern.

An outside observer would likely notice quite a few things about our training that we didn't, simply because we were too close to the action. For instance, our platoon started as a group of about 55 men with wildly unique mannerisms and personalities. Very few of us worked as a team, and we often clashed in the first couple of weeks, doing exercises and activities with an individualistic mindset to the best of our abilities. As the drill sergeants turned up the pressure and began picking on the weakest links, though, stronger platoon members began stepping up to support their struggling peers. The fastest runners would sprint back to offer encouragement to those lagging behind, while those with sharper minds began helping recruits who had difficulty memorizing information. It was crucially important that the weakest be brought up to standard, or we all paid for it.

The true genius of basic training lies in the drill sergeant's ability to foster an environment of "us versus them." They punish the group as a whole, constantly search for weak links to exploit, and go out of their way to make all recruits miserable. Doing this effectively teaches natural leaders to prioritize building competency in their fellow battle buddies. This is more important than personal glory. It is mandatory for everyone, as those who are struggling in a particular area will be the soldiers who later get people killed in actual combat.

Instructors like DS Whiteley were, without a doubt, assholes. Many civilian readers are likely horrified at his behavior, which only escalated as our training cycle progressed. However, he was an integral part of why we began to band together and work as a team. We shared a mutually acquired hatred for some of our drill sergeants, doing everything in our power to prove them wrong and protect our own. In this way, he and the others were directly responsible for our accomplishments and the progression of civilians into soldiers. It didn't matter if we liked or hated them. They were effective at their jobs.

When I first arrived at Fort Benning, I couldn't understand why our barracks were routinely destroyed at the end of each day. Exhausted to the point of physical collapse, I stared in disbelief at the disorder in front of me and realized that it would take hours to put everything back together. All I wanted to do was fall onto my bunk and sleep. Our bay had looked spotless when we left and was now destroyed. How could we clean everything and get enough sleep? It was during moments like these that fights and frustrations would boil over. It was all done by design.

The military has adopted a saying attributed to the Greek poet Archilochus: "We don't rise to the level of our expectations; we fall to the level of our training."[11] This is a critical lesson to know. We must train our minds and bodies to handle physical strain, stress, anxiety, hunger, thirst, and other primal instincts. It's unrealistic to think we can handle any situation flawlessly without proper training and experience.

Thus, it didn't matter that our barracks bay was spotless. The point of the exercise wasn't cleanliness. They destroyed it to put us through the experience of doing precise work while being hungry, frustrated, and exhausted. It was intended to be difficult.

Exercises can offer invaluable training and enlightening revelations by introducing stressors in a controlled environment. Rather than rising to meet idealized, unrealistic expectations, especially those formed during moments of rest and comfort, the more likely scenario is falling to the base level of training. The military understands this well and argues that it is better to experience as many new challenges in a training environment as possible, rather than the unforgiving reality of combat.

I suspect you have heard someone confidently declare that they would rise to the occasion during a stressful situation at least once in your life. Statements like, "If someone did that to my child, I would go into momma bear mode. Nothing could stop me." While I wish this were true, the fact is that "momma bear mode" would probably be nothing like what they imagine. In fact, it would likely be a disaster unless adequate training and preparation had taken place beforehand.

With that in mind, one piece of advice I give to men and women alike is to learn as many practical skills as possible before a stressful situation arises. After all, it is a matter of when, not if, a rainy day shows up at your doorstep. It is much better to be prepared and to know your mental and physical limitations before being forced to confront them in a potentially unpleasant event. I don't believe you need to be an expert in every field, as that's impossible, but everyone should have a basic level of competency. Additionally, everyone should be aware of how they are likely to react in a generalized stressful environment, preferably before anything serious happens. This can be the difference between success and failure.

---

[11] Feloni, Richard. "Tim Ferriss Lives His Life According to an Ancient Greek Quote That Helps Him Prepare for the Worst." Business Insider, December 1, 2017. https://www.businessinsider.com/tim-ferriss-favorite-quote-greek-philosopher-archilochus-2017-12?op=1.

Looking back on my time in Infantry School, it was a wonderful experience. It was difficult, yes, but vital for my progression into manhood. I suspect most people have not encountered something like this in their entire lives, which is truly unfortunate. It began a profound transformation into a new realm of discipline, leadership, and commitment to excellence. As strange as it seems, it was a change I didn't even notice happening.

**20 August 2002 – Letter from my father, Bill Harstad, Jr.**

Do whatever you need to do. Work hard, then turn around and work even harder – harder than you've EVER worked in your life, and you'll be better off for it. It's called the "rite of passage," where a boy turns into a man. Some men never go through it, some do… and some sooner than others. You are now in your rite of passage. Good luck. I know you will do fine.

The only question mark in my mind is the 30th AG's perplexing nature. It was the antithesis of everything I believed military training to be. It was like standing in a long line at the DMV, where the attendant calls "next" and wants to get you out of the door as quickly as possible, your license and questions be damned.

I still suspect it was simply a poorly run unit staffed by people who hated their jobs. The monotony and inefficiency were staggering, and almost no one seemed to care. But every now and then, I get the nagging feeling that perhaps this was a wolf in sheep's clothing. Was all this done to lull us into boredom and despair, only to throw us into the jaws of Infantry School? I just don't know, but I do have my suspicions.

# The Infantryman's Creed

I am the Infantry.

I am my country's strength in war, her deterrent in peace.

I am the heart of the fight… wherever, whenever.

I carry America's faith and honor against her enemies.

I am the Queen of Battle.

I am what my country expects me to be, the best trained Soldier in the world.

In the race for victory, I am swift, determined, and courageous, armed with a fierce will to win.

Never will I fail my country's trust.

Always I fight on… through the foe, to the objective, to triumph overall.

If necessary, I will fight to my death.

By my steadfast courage, I have won more than 200 years of freedom.

I yield not to weakness, to hunger, to cowardice, to fatigue, to superior odds,

For I am mentally tough, physically strong, and morally straight.

I forsake not my country, my mission, my comrades, my sacred duty.

I am relentless.

I am always there, now and forever.

I AM THE INFANTRY.

FOLLOW ME![12]

---

[12] U.S. Army. "The Infantryman's Creed." U.S. Army Fort Benning and the Maneuver Center of Excellence. Accessed November 7, 2025. https://www.benning.army.mil/infantry/199th/ibolc/content/PDF/The Infantryman's Creed.pdf.

# Chapter II

## Blood Makes the Grass Grow Green

The shock of basic training had subsided, and we were now in the midst of trying to keep up with the grueling demands of our tight schedule. Early morning PT was always on the agenda, with chow and several hours of classroom time set aside to learn our craft before we attempted to put it into practice in the field. Some of these were entertaining and informative, though others were terribly boring. I enjoyed many hands-on courses, such as first aid and land navigation. Some of the others, like sexual harassment prevention, not so much.

The drill sergeants taught most of the lessons, though other military service members occasionally taught a portion. The room was delightfully air-conditioned and gave us a respite from the outdoors, though this was sometimes counterproductive as people would start to nod off, especially after chow. The drill sergeants were always hunting for the slightest hint of fatigue. One drooping head was enough. The recruit who nodded off, along with anyone unfortunate enough to be near him, would be punished on the spot. As such, we routinely slapped our neighbor's arm or back if we perceived a slower-than-normal blinking rate.

A third drill sergeant joined our platoon early in the cycle and quickly made a name for himself by the staggering amount of PT he dished out. DS Nash, a staff sergeant, was difficult to figure out. He was somehow intense and intimidating, yet funny and a bit crazy at the same time.

During one lecture, as DS Nash gave an animated presentation, a private started doing the slow-motion chicken bob in the front row. Our new instructor leaped from the elevated stage in mid-sentence, and Superman tackled the drowsy private out of his desk. They both crashed to the ground, and we looked at them with amazement. He grabbed the dazed private by his collar and lifted him.

With exaggerated concern, he declared, "I thought you were going to fall on the floor! It's a good thing I saved you!"

Yeah, he "saved" him by tackling him straight out of his chair and onto the linoleum floor. A few of us started to chuckle under our breaths from the incident, but we immediately stopped when he shot all of us a cold stare. It was probably not a smart thing to do.

In another section of the classroom training, an outside soldier from a different battalion, likely from the dreaded Human Resources department, was brought in to give a presentation on workplace violence. Yeah, seriously. Wasn't that the *entire* point of our jobs? As our drill sergeants patrolled around, the message being relayed was that our cadre members were not allowed to hit us, threaten us with physical violence, or cause undue stress from harassment. I couldn't believe that we had to listen to that garbage. We sat in silence, feeling the ridiculousness of the wasted hour. Drill sergeants just bored holes of hatred through our skulls as the rambling lecture finally ended. We knew better than what we were being told and definitely understood the unspoken truth. They had creative ways to get around these silly mandates.

**2nd Platoon, Charlie Company, 2-54 INF - "The Renegades"**
**Photo by Visual Touch Photography**

**29 September 2002 – Letter to the Harstad Family**

(We were being too loud) DS Whiteley said it perfectly: "If you don't shut the fuck up, I'm going to break all of your legs! And when I say I'm going to break your legs, I don't mean that literally... *I mean I'm going to break your fucking necks!*"

Anyone who has had the fortune, or misfortune, depending on your perspective, of working for the U.S. government will instantly recognize that many so-called "training classes" are little more than a complete waste of time. It's a shame because some of the topics *could* be valuable. Instead, training is frequently led by incompetent and unintelligent individuals who lack the ability to convey information. Ah, it brings back memories of college!

More often than not, the person leading the class was "voluntold," a portmanteau of "a willing volunteer" and "told to do it." It isn't much fun to sit in a class purportedly about how to be a safe driver of military vehicles, for instance, in which the instructor repeatedly mispronounces words and clearly has no idea what is going on. Unfortunately, this is far too common in government work and is about as painful as it sounds. And don't let anyone tell you this doesn't happen, because they are full of shit if they do.

After all the classes and our strenuous physical activities were finished for the day, we would have some free time in our barracks bays, though it wasn't much at first. Each soldier had been paired up with someone else, with this duo being considered "battle buddies," and they shared a bunk bed. Everywhere one went, the other was required to go. Then again, we called everyone in the platoon a battle buddy, but I'm trying to make the point that we were essentially attached at the hip with another person, 24/7.

Mine was Private (PVT) Harper, a soldier with whom I had almost nothing in common other than we were both white. He spoke with a thick Southern drawl and grew up very differently from me, but we surprisingly became quick friends from the strain of the training. It appeared that very different men were slowly being transformed into soldiers, all because of the Army. Those soldiers then grew close to one another, even when the pairings were wildly bizarre and mismatched.

Across the bay, everyone was being thoroughly entertained by PVT Greenlee, nicknamed "Opie." He had naturally assumed our platoon's class clown role and thrived in that position. The best way to physically describe Opie was as an adult version of Squints Palledorous from *The Sandlot,* a '90s coming-of-age movie. Opie was about as skinny as I was and wore the notoriously ugly prescription lenses the military gives out to recruits during training. They use the dysphemism "Birth Control Glasses" (BCGs) for good reason. The glasses could scare off females from a mile away.

**24 August 2002 – Letter to the Harstad Family**

Our platoon has its moments. We have some really great guys.

There is this one guy we call "Opie," and he is funny. He is loud, but small. He is about my size and talks constantly with a great sense of humor.

Lately, as we're standing in formation and the drill sergeants aren't around, he will call out a private's name (from a different company) that is passing by. We all stand, plain-faced, looking straight forward. The private will look around and only see our formation standing still (and not knowing where it came from). It is so hard to keep a straight face with the odd look on *his* face.

He can also make a cricket noise that sounds perfect. He does it while we are standing in formation. It sounds so real that the drill sergeants think there is a cricket in the barracks somewhere.

PVT Greenlee provided the much-needed comedic relief for the platoon. It kept all of us sane, and I mean that literally. The stress from basic training was overwhelming at times, but one silly remark or bizarre outburst from Opie was enough to twist the knob on the pressure relief valve. It gave us something to laugh about in an unpleasant situation. However, had Opie done almost any of these activities in the civilian world, he would have been immediately fired on the spot, or worse.

### 26 August 2002 – Letter to the Harstad Family

Last night, Opie, the platoon clown, went crazy due to a sugar rush. He was wild as hell. He ran up and down the barracks, screaming in cadence. It was funny. He was goofy the whole night. We asked him how much sugar he had (snuck from the chow hall). From what I understand, he put sugar in his bread, water, main course, and just about everything else.

### 30 August 2002 – Letter to the Harstad Family

Opie is as crazy as ever. He runs around doing crazy stuff. Last night, he snuck up on people sleeping in their bunks and mauled them. We all thought it was hilarious. We rolled around laughing as he would jump from bed to bed, trying to bite people.

We started learning hand-to-hand combat in designated training blocks for US Modern Army Combatives early on. Drill sergeants often used this as an excuse to rough up recruits during martial arts training. Remember those mandates from Human Resources?

Our head instructor, DS Whiteley, who always appeared to have a screw loose, was laser-focused on teaching Combatives proficiently. I didn't know him well enough to determine whether his personality was an act, but it didn't matter. He routinely asked for volunteers to demonstrate moves to practice, such as arm bars, guard passes, and a variety of chokeholds and submissions. I happened to be one of his favorite "volunteers," though I don't recall ever raising my hand to be choked unconscious willingly.

I aggressively fought back to the best of my ability, but more than once, I found my vision dimming away to sparkly stars and then blackness. He was amazingly competent at fighting, too, so it wasn't as though he needed a smaller guy, such as me, to demonstrate the moves. He just liked picking on people. And he was scary good at what he did.

The US Modern Army Combatives program originated in 1995 at the 2nd Ranger Battalion, then expanded to the rest of the Army shortly thereafter.[13] It gained popularity during the early rise of the Ultimate Fighting Championship (UFC), a period when MMA was exploding in the media. The program's goal was straightforward: to equip soldiers with the skills to defeat enemy combatants in hand-to-hand combat. Or, at the very least, pin them down until someone else caved in their skull.

As viewers of the early UFC will remember, MMA was a blossoming full-contact sport initially divided into specialized disciplines, such as Karate, Taekwondo, Judo, and many others. It was enjoyable to see which discipline would win out over the others, and it was the talk of school kids and adults alike. Very quickly, it was identified that experts in only one skill often had glaring holes in their combat approach. Knowing many elements, such as grappling, striking, *and* groundwork, was more critical as the sport evolved. Plus, conditioning became crucial for all fighters.

Using this philosophy, Combatives tried to teach universal practical skills that didn't require a lifetime of training, though this would obviously help. Recruits practiced submission techniques, escaped from dangerous positions, and worked to control the enemy's body to establish dominance. More advanced techniques were learned as the training progressed, though this wasn't like becoming a *Grand Master of Kung Fu*. Simplicity was intentionally built into the formula to foster retention. After all, every service member must be prepared to always defend themselves and should be proficient in self-defense and combat techniques.

Unfortunately, the truth is that many units within the military quickly abandon Combatives and even fundamental weapons training, instead focusing solely on their primary responsibility. That is, whatever their specific job is within the military. This was sadly on full display in Iraq and Afghanistan, particularly when non-combat arms units were ambushed, such as that of Jessica Lynch's company.[14] I don't mean to single them out, as many others were guilty of this neglect, but it became an infamous example of being unprepared in a combat zone. This highlighted an even greater need for easy familiarization and memorization.

Attempting to maneuver my body away from DS Whiteley's clutches, I would soon find myself gasping for air after receiving a cross-collar choke that he had quickly applied. Slipping away and proud of the fight I had put up, I would then find he bypassed that illusion and put me into a rear-naked choke until my vision went dark. I refused to "tap out" at first, simply out of stubborn

---

[13] U.S. Army. "Modern Army Combatives." U.S. Army Fort Benning & the Maneuver Center of Excellence. Accessed November 8, 2025. https://www.benning.army.mil/Armor/316thCav/Combatives/.

[14] Peters, Katherine McIntire. "Hard Lessons." Government Executive, April 1, 2004. https://www.govexec.com/magazine/2004/04/hard-lessons/16406/.

pride. This wasn't a good idea on my part, as we were supposed to be training, not trying to kill each other. Ego got in the way of learning.

Though I initially hated wrestling with him, primarily because he beat the shit out of me every time, I must give DS Whiteley credit for teaching me a lot about many techniques. He was part of the reason I gravitated toward Combatives later in my military career, and I grew to enjoy our sparring sessions. His personality was just so unlikable at the time that it was difficult to appreciate what he was actually doing: teaching us to stay alive.

**19 August 2002 – Letter to the Harstad Family**

Yesterday, I was standing in the food line. A big, tall guy in front of me started swaying back and forth. He started to fall, and we caught him, like battle buddies do. He didn't pass out, but lost all energy.

I helped him outside, where he sat down. Drill sergeants are always there to make sure everyone is ok. I'm standing next to the kid when DS Whiteley walks up and says, "How many fingers am I holding up, shithead?" He just raised his middle finger at him. What a dick!

This guy also threatens to kill us if we do something wrong. We all know he is joking, but some take it seriously. They think he is crazy. He's not, he's just mean.

He also screamed at some kid when he fell over during the road march. *"Get the fuck up! What the hell are you doing, you dipshit? Get your fucking ass up right now! You are a disgrace to the Army!"* He says that kind of stuff all the time.

When he isn't picking on you, it is funny to listen to him. He has a great sense of humor for an evil guy like him.

We frequently found ourselves being marched over to the bayonet course. There, we would affix M9 bayonets onto the ends of our rifles. Full-sized human dummies had been set up on a quasi-obstacle course, dangling from chains or suspended from steel bars. We charged full speed at the dummies and thrust our bayonets deep into their pink bellies. We then pulled the blades free, driving them back into their faces or necks. It was a lot of fun, though it's weird to admit that in hindsight.

"Kill, kill, kill, with cold blue steel!" we shouted in a repeating chant, stabbing and butt-stroking the dummies relentlessly.

"What makes the grass grow green, privates?" the drill sergeants screamed.

"*Blood* makes the grass grow green, drill sergeant!" we roared back.

"Whose blood?"

"The enemy's blood, drill sergeant!" we cried.

"Kill, kill, kill, with cold blue steel! Stab and twist!"

Every activity was reminiscent of this, where violent acts were being normalized and even encouraged. "Brainwashing" is perhaps the word that comes to mind, but we don't typically like to use that term in the West, especially when it refers to what we do to our own people in the infantry. That word is exclusively reserved for places like North Korea and Nazi Germany, where coercion and mental manipulation were overt and malicious. Perhaps "conditioning" the recruits is a softer, less unsettling word, but I suspect the reader will understand what was happening, regardless of which expression is used. To be clear, I'm not equating our training with the brutality or ideological extremism of those regimes. The comparison lies not in intent or morality, but in the shared method of breaking down old identities in order to build new ones.

Desensitization was also deeply woven into our training. Repeated exposures to physical pain, hunger, stress, and drills, such as the bayonet course, were applied repeatedly and applied often. This method may sound barbaric or uncivilized, but this is how the military wins. It wins by killing people faster and more proficiently than the enemy. It wins by unleashing overwhelming, demoralizing violence on the enemy until they break. As Secretary of Defense Pete Hegseth aptly stated in his book, *The War on Warriors*, "When you send Americans to war, their mandate should be to lethally dominate the battlefield. If that makes you uneasy, keep us at home."[15]

All this was done while keeping the general public away from the area. We had no contact with the outside world, except for the occasional letters, when allowed, that trickled in during mail call. Our precious, backlogged letters would pile up, and phone calls were forbidden until later in the cycle when the drill sergeants deemed we had earned them. *Everything* had to be earned, with nothing given for free. However, these privileges, when given, were quickly revoked at the slightest breach of conduct, reinforcing compliance with every little detail. We didn't want to mess up.

Short, five-minute conversations with parents, wives, and other family members were a double-edged sword. They motivated us to complete our training and provided a powerful incentive, fueling our resolve to persevere through the torment. But they could equally cause tremendous emotional stress as people learned of impending divorces, ill children, parents facing health problems, or worse. Still, the training always came first. Everything else was secondary.

**11 August 2002 – Letter from my mother, Joyce Harstad**

Sometimes, when you hear your loved one's voice, it makes you feel better, and you know you can make it another day, or week, or month. I think Dad would like you home, working. I think he is getting tired of Dan (my brother) and I. We don't have that dark sense of humor you two do. Anyhow, remember to go to church and get a little shut-eye. Until we see each other again, Mom.

Another platoon had a family member come up with a truly fantastic idea. He or she sent giant boxes of cookies, allegedly around a thousand, to their loved one during the early weeks of

---

[15] Hegseth, Pete. The War on Warriors: Behind the Betrayal of the Men Who Keep Us Free. eBook edition. Broadside Books, 2024. Chapter 11

Infantry School. Looking out my barracks window one evening, I saw over 50 recruits lying on their backs, doing flutter kicks and sit-ups while eating every single cookie. This spectacle went on for over an hour, with the sounds of vomiting soldiers cascading through the breezeway. While we enjoyed watching, we had to be careful not to be detected. If the drill sergeants had seen us peeping through our windows, they might have forced us to join the fun. Was it an elaborate practical joke played on his son by a former soldier, or was this a sweet gesture from a loving grandmother trying to brighten her grandson's day? I have no idea, but it made me feel warm and fuzzy inside, knowing it wasn't me blowing chunks this time.

### 23 August 2002 – Letter to the Harstad Family

Tell Grandma not to send any cookies unless I tell her it's ok. My drill sergeants will eat them if she does.

We also had been hearing rumors of war brewing in the Middle East beyond the limited special operations already underway in Afghanistan. It was frustrating not knowing what was happening on the outside. It's difficult to explain, but the uncertainty of our fate, with graduation only a few months away, was unnerving. Would we be deployed directly into war as soon as we left? Would everything calm down? The drill sergeants kept telling us that we would be thrown into the "human meat grinder" as soon as we left Infantry School, so we had better shape up quickly. It was impossible to tell if they were serious or just saying that to make us train harder. Perhaps it was both. Regardless of the truth, it motivated me to push myself harder than ever.

### 30 August 2002 – Letter to the Harstad Family

Today, we had a little bit of excitement. Someone caught a glimpse of the TV (in the drill sergeant office) and ran back to tell us. On Fox News, there was a news flash saying, "Iraq Attack?" I saw it for myself. Supposedly, something is happening over there.

### 31 August 2002 – Letter from my father, Bill Harstad, Jr.

Here on the home front, a war with Iraq seems almost imminent. When, I don't know – but soon, I suspect. The media is in an almost frenzy.

### 6 September 2002 – Letter from my mother, Joyce Harstad

Bush is trying to get clearance to take out Saddam for good. Might be before you guys are ready. Who knows? Videos, or other sources, say that Saddam is making nuclear bombs. They have proof. Doesn't sound good to me.

All of this seemed to make the training even more critical. In the meantime, our drill sergeants continued to play many psychological games on us to cultivate an environment of teamwork. We were smoked for well over an hour one evening after DS Ross located a can of chewing tobacco, considered to be contraband, in one of the ceiling tiles. Up and down our drill sergeants strolled as we held ourselves in the push-up position, sweat starting to puddle on the floor. They demanded to know who was responsible for the can of dip. No one would fess up to it.

**A look inside the 2nd Platoon "Renegades" barracks bay (2002)**

They continued to smoke the platoon as one drill sergeant began isolating soldiers, taking them back to the latrine. My turn came, and I was ordered to stand at parade rest, in which we rigidly stood with our hands interlocked behind our backs to speak with non-commissioned officers (NCOs). DS Clements demanded to know who the owner of the chewing tobacco was. I genuinely didn't know and kept repeating it, staring straight ahead and carefully avoiding eye contact. Even if I did know, I wasn't about to be a Blue Falcon and rat someone out. That wasn't my way.

Ultimately, no one confessed to the deed, and no one squealed to our NCOs. They were left with no answers about who owned the can of dip. Growing frustrated, they reprimanded us for contaminating the barracks floor with sweat. Thus, they threw all our beds onto the ground, emptied every locker, and created a chaotic mess of socks, boots, and other random gear. This had almost become routine. One way or another, we would be punished. There was no escape from punishment.

### 31 August 2002 – Letter to the Harstad Family

Still no word on who "smuggled" in the contraband. All suspicions point to the drill sergeants. We'll find out someday...

DS Nash discovered another vulnerability that irked me just as badly as when they made fun of my weight. At eighteen, I looked much younger than I really was, and they were relentless in offering me pacifiers, baby blankets, or a nipple to suckle from. I should have found it funny because I think it is hilarious today, but I didn't at the time, and it got to me. Because of this, they kept doing it and rubbed it in whenever possible. It was an important lesson, but one I struggled to follow. *Never* show weakness and just laugh off the insults.

### 23 August 2002 – Letter to the Harstad Family

The drill sergeants have been making fun of me due to my baby face.

One of them walked up to me and yelled, "Harstad, did you shave this morning?!"

I replied, "Yes, drill sergeant."

He screamed, "Why, Harstad?!"

I looked at him and said, "Because you told me to, drill sergeant."

He looked closely at my face and said, "How old are you, 11, 12? I bet you could neglect shaving this whole week!"

These relentless acts and insults brought the platoon closer and ultimately united us. Whether it was us being smoked together or if it was one of us being picked on individually, it became the mentality of us versus them. We started to resent all their constant harassment and maltreatment. Again, I was incredibly young and naïve then, but in hindsight, this was entirely by plan. Before going into training, I knew they would mess with me, but I didn't anticipate this. It sometimes felt personal. They did all these things to test our resolve. Were we going to crack under pressure and throw our battle buddies under the bus? Or were we going to unite and stand together, refusing to give them any excuse to punish us?

Some recruits still couldn't get their shit together, though. They couldn't figure out how to blend in with the program. It was too much strain, too much exercise, and too much filth to crawl through daily. People continued to contemplate going AWOL, even including my battle buddy. Thankfully, we talked him out of it after considerable persuasion, but it was a stark warning to me that the stress was getting to people. Even those we considered strong and unbreakable some-times had weak moments. They weren't immune, especially when the strain originated at home, as in the case of PVT Harper.

### 26 August 2002 – Letter to the Harstad Family

Yesterday, around 2000 (8:00 PM), as we were doing trash cleanup, I witnessed a soldier do an in-credibly funny and stupid thing.

This soldier (from a different company) has been trying to get kicked out for a long time now. He either went to jail or the Army. He is a small and weak little guy.

Anyhow, a few drill sergeants had platoons marching around the ¼-mile track. This guy, under-wear over his shorts, a set on his head, and a towel tucked into his shirt like a cape, ran around like Superman. The drill sergeants freaked and chased after him. It was brutal what they did to him.

The last time I saw him, he was doing the low crawl around the muddy part of the track. He had grit stuck in his face.

### 27 August 2002 – Letter to the Harstad Family

Today, during PT, a soldier from a different platoon said that he wasn't going to do this anymore. After the drill sergeants screamed at him for a while, they called the MPs (Military Police). They hauled him away to jail. We clapped as they handcuffed him.

Within a few weeks, we had all assigned nicknames to each other. This was often more en-joyable than simply calling each other by our last names, as no one ever used first names. In fact, I don't think anyone ever called me "Jeff" the entire time I was in the military. The unwritten rule was that picking your own nickname was unacceptable, though many desperately wanted to change their given moniker for obvious reasons. Handles such as Hippie, Rez, Ivan, and Hass were some milder nicknames, though they could be considerably cruder.

I was given the nickname "Whispers." This was because I rarely spoke and simply watched and listened to everything around me. I quickly learned to stay out of the spotlight, regardless of whether it was positive or negative attention. Drawing attention to yourself in basic training is not a particularly bright idea. The best tactic is to blend into the scenery so no one even knows you are there. It served me quite well.

Whispers was a lot better than what my previous nickname had been. During the first few days, our entire platoon gathered in a big circle around the drill sergeants, and we were forced to awkwardly introduce ourselves as they glared at us. Like everyone else, I said my name, where I was from, why I had joined the Army, and a few other "interesting" facts. For some reason, don't ask me why, I mentioned that my mom and dad owned a carpet cleaning business back home.

Well, this was somehow lost in translation, because the drill sergeants heard "carp cleaning," as in the fish. So, for the next few days, I was referred to as "Fish Stick," which was funny because of my size. But when they said it, it suspiciously sounded a lot more like "Fish Dick." And mind you, this was years before that South Park episode with Kanye West. When I tried to correct them, explaining that it was "carp-et," they clearly weren't interested.

To add insult to injury, when I was finally able to call home, I told the story to my dad, who thought it was hilarious. Thinking he was being funny, he changed the business logo on an envelope containing a letter he sent me, reflecting the new name "Harstad Carp Cleaning" and featuring a big green fish. The mail, which always passed through drill sergeant hands, was gathered up one day, much to their delight. One look at that logo sent the drill sergeants into a frenzy.

"Fish Dick's got some mail!" And that's how I got my first nickname.

For some reason, though, the nickname had a short shelf life. Maybe they moved on, or maybe the Army's fun police (ahem, those HR buzzkills) paid a visit, but it faded fast and was replaced by Whispers. To be honest, basic training was the funniest place in the world, assuming it wasn't *you* they were picking on, and you weren't the one caught laughing at their insults. The drill sergeants were some of the quickest and funniest people imaginable.

The first few weeks flew by at a lightning pace. They were little more than extended, grueling exercise sessions, interspersed with drills on marching, proper saluting, and attending those miscellaneous classes. In the Army, and I'm confident this applies to the other military branches, we would learn everything through a method known as "crawl, walk, run." For example, during the first phase of Infantry School, we routinely checked out rubber M-16s from the armory, nicknamed "rubber duckies." These practice rifles were dimensionally identical to real M-16s but constructed from solid rubber.

We would go everywhere with these M-16 replica rifles. We marched around with eight pounds of black rubber that we pretended was real. The point was to become used to the weight and feel of the rifle while avoiding the dangers and responsibilities of an actual weapon. Drill sergeants would carefully observe us to ensure we were always correctly carrying them. Great care was taken never to point them at anyone and to keep our fingers out of the trigger area. We had to earn the right to carry an actual weapon, so we started with the rubber duckies. They also became an extension of our bodies, going with us everywhere, even into the latrines and showers. God help you if you left it unattended or misplaced it!

**28 September 2002 – Letter to the Harstad Family**

The drill sergeants have been making us carry these things called "rubber duckies," which are rubber M-16s. They weigh the same and look the same, but they are rubber and don't fire.

Last night, I got up to use the bathroom. I see Opie, in shorts without a shirt, standing by one of the stalls. He has a big smile on his face, and in his hands is his M-16 rubber duckie. I can tell that someone is taking a crap.

Anyhow, Opie kicks in the door and yells, "Freeze! Get down, get down!" He charged the crapper like a SWAT team. The guy in the stall could only scream. You can never tell what Opie will do next.

At first, not much exciting training was happening, and much of our time at Infantry School was spent marching from one location to another. Road marches, in which we walked along a trail or a paved road, quickly became more difficult. The pace was sometimes challenging to manage, falling somewhere between a fast walk and a jog. Recruits were outfitted in full combat fatigues, a 40-pound rucksack, a load-bearing vest (LBV), water, a Kevlar helmet, and always the rubber duckies. In total, this meant I was carrying well over half my body weight, just in gear. Being so skinny had its drawbacks. I needed to put on some weight quickly.

Our first road march was a brisk two-mile trek, which doesn't sound that difficult. However, I quickly learned that much of the gear issued to us was shoddy at best. Honestly, it was shit. The boots were amazingly uncomfortable, and the soles were rock hard, offering little to no shock absorption. Almost immediately, we all developed bleeding blisters that routinely caused much discomfort. These were exacerbated by walking through some of the swampy terrain. Wounds had to be treated with care as they easily became infected, and I don't believe anyone escaped unscathed without severe bouts of athlete's foot, jock itch, feet that looked like raw hamburger, other types of fungus, and painful, pus-filled sores.

**14 August 2002 – Letter to the Harstad Family**

It was about 100 degrees and 100 percent humidity. We got back and ate. We took some courses on radios and stuff like that. We then marched back. During this time, it was one of the lowest and highest points in my life. It was the lowest because it began to downpour. The sand turned into slippery goo. People fell and were covered in orange slime.

It was also the highest point. While moving through the woods, we heard a lot of outgoing fire. The whole company, four platoons consisting of 50 men (apiece), was marching in two files. We heard loud machine guns, mortars, and small arms fire (from the nearby range). We then had a formation of Black Hawk and Apache helicopters go right over our formation. I was literally saying, "Holy shit!" as they flew over. Between the choppers and gunfire and explosions, it was deafening.

After a bit of toughening up, our feet became leathery and resilient as the marches steadily increased in distance. Our equipment was often joked about as being made by the "lowest bidder," and I sincerely believe that this was the case. I often wondered why we were given such inadequate gear, especially for grunts expected to walk everywhere, though I now suspect there were two primary reasons.

First, it was likely a filtering mechanism. While it sounds cruel, it served as a silent and effective way to weed out those who couldn't handle the physical punishment from those who could. Did the Army want wimps serving on the front lines? Probably not. I suspect this was their way of thinning the herd.

Second, it was a method to toughen up the survivors, as combat would likely be far worse than anything we complained about, such as blisters and rashes. They didn't need grunts throwing pity parties because of a bit of discomfort. I don't know if these are the actual causes, or if the government was simply too cheap to buy top-notch gear, so please take my speculations with a grain of salt. I strongly suspect it was a mixture of the two, but I may be wrong.

After returning from a road march or the field, we'd immediately have to shine our boots to a blinding, funeral-home gloss, because they always came back looking like they'd been used to kick rocks for sport. The day after one march, we were standing in formation while the drill sergeants prowled the ranks, doing their little gear inspection safari. One drill sergeant, whose name escapes me, stopped right in front of me and immediately locked onto my boots.

"AHHHHHHH!" he screamed, inches from my face and directly into my ear. Flinching was a felony, and any reaction at all was basically an invitation for every drill sergeant within a five-mile radius to come over and take turns. So, I did what every good private does. I stood there like a statue while he had a full-blown emotional event.

Then, out of nowhere, his voice dropped to this calm, polite tone. "Harstad," he said, almost friendly. "Did you polish your boots last night?"

"Yes, Drill Sergeant!" I yelled.

The second "yes" left my mouth, he detonated again. "WITH A FUCKING HERSHEY CANDY BAR?! PUSH!"

And that was my cue to start doing push-ups until *he* got tired. Not me. He stormed around in circles as he screamed to himself and to no one else in particular. His direction would suddenly change course, looking as though he was having a full-blown schizophrenic attack.

In my defense, I *did* polish my boots. I just didn't do it very well. I was exhausted and figured I could skate by. Big mistake.

At the end of August, we all received our first paychecks from Uncle Sam. The pay was adequate for me since I was single and had no dependents. Many guys already had families, though, and the budget area was a little tight. This was a source of genuine concern for many in the platoon. As is commonly understood, the military isn't exactly known for paying big bucks to enlisted soldiers. In fact, it is a poverty wage, only modestly offset by covered healthcare, food, and housing. If a young man or woman wishes to become rich, the military is not the place to do it.

**29 August 2002 – Letter to the Harstad Family**

Got paid today. $1,089.60. $294.50 went to taxes, college (GI Bill), Social Security, retirement, etc. That gives me $795.10 earned this month! Wow! Poor man!

# Reflections

One of the most valuable lessons I learned in the military came after Infantry School from one of my team leaders at the 3rd ID, Sergeant (SGT) Marcus Brister. Still a young and impressionable enlisted soldier, I impatiently emerged from a class similar to the sexual harassment prevention course that we had to sit through while in basic training. I was vocal about my displeasure at the pointless waste of time. I complained that we had better things to do than sit through a bunch of "hippie crap."

"Hold on," SGT Brister interjected. "Harstad, we must attend all training sessions with an open mind and the curiosity to learn. Even this 'hippie crap.'"

"How so, Sergeant?" I asked, figuring he was just messing with me.

"What did you learn from that class?" he asked, temporarily dodging my question.

I thought about it for a fraction of a second. "Not a damn thing," I replied, confident in my answer.

"That's where you're wrong," he stated bluntly. "You learned something in there. You just aren't thinking about it hard enough."

By this time, I was thoroughly confused. He had just gone through the same dumb class as the rest of us. He *had* to know that it was a waste of precious time and was taught by an idiot. There was no way he thought it was valuable. How could he? I just shook my head because I had no clue what he was getting at.

"In a year or two, you will probably be in my position and will likely have your stripes if you keep working hard," he continued, motioning to the sergeant stripes on his collar. "When you have them, will you teach your guys what you learned here today?"

"Hell no, Sergeant," I replied. "That instructor didn't know a grenade from his elbow."

"Exactly." That's all he said while he let it sink in, as I just stared at him in confused silence. After a lengthy pause, he continued, "You aren't going to teach them that… *because* you learned a valuable lesson our instructor just taught us. He taught us never to teach a class like that to anyone. He taught *you* what *not to do*."

The proverbial light bulb went off. While much of the information we received was garbage, we had to hone our detection skills to sift out the diamonds from the sand. With this mentality, bad training classes were sometimes just as valuable as the good ones. We had to learn to identify poor leaders, flawed information, and inadequate communication skills. By doing this firsthand, we could say, "I'm never doing or saying that."

Thus, I learned by example from countless incompetent leaders and bad training classes while working for the government, swearing that I would never do or say those things to my subordinates in the future. While some may claim this is simply putting a positive spin on a bad scenario, or lipstick on a pig, I disagree. We can learn in any situation, no matter how poor, provided we are appropriately skeptical, willing to push back on incorrect information, and careful not to fall into the dogma of "that is how it has always been."

A related lesson was also taught to me around the same time, though I cannot recall where I first encountered it. I learned so many things early in my career that it is impossible to keep them all straight. It is entirely plausible that this, too, comes from SGT Brister. He was incredibly influential in my professional growth, and it sounds like something he would have said.

The idea is that we should treat our brains like a toolbox. Every place we go, every scenario we encounter, and every leader we meet gifts us with various tools. Some of these tools are universal and may be utilized in multiple applications, such as a hammer. Another may be very specific and highly specialized, like a miniaturized screwdriver designed for precision work. Others are complete trash, such as the soft plastic wrench made in a foreign sweatshop. Our responsibility is to identify when and where these tools should be utilized in the future. It is equally essential for us to keep the rubbish out of the toolbox.

Although I hated some of the courses in basic training and despised some of the people there, I was quickly building up my toolbox. I just didn't realize it at the time. I wouldn't know it until I arrived at the 3rd ID and met SGT Brister, though that revelation was still far off. Plenty of drama had to happen between now and then.

Another observation I'd like to touch on is that I was acutely aware of what the Army was doing to us, especially during desensitization in our training blocks, such as the bayonet course. There wasn't anything malicious about the training. It was quite the opposite, as far as I'm concerned. It was designed to prepare us for the realities of combat, keep us alive, and win wars as quickly as possible. Quick wars usually mean less suffering and fewer deaths overall. That is the point.

The same principle applied to our instructors' actions. They were trying to break us to later rebuild the foundational pieces into proficient warriors capable of ruthlessly killing the enemy and enduring the hardships on the battlefield. Nothing was designed for comfort, nor was it designed to make soldiers feel good about themselves or boost egos. It was intended to smash and defeat the enemy. Was it harsh? Sometimes, it was, but it worked.

Militaries around the globe have learned lessons over thousands of years of experimentation, all with the intent of improving performance. In 1961, psychologist Stanley Milgram conducted a famous and thoroughly disturbing experiment to measure participants' willingness to obey a person in a position of authority. Willing participants were tricked into believing they were taking part in legitimate research and that they would merely be assisting the experimenters by testing other volunteers.

To conduct the fake experiment, the participant took on the role of a "teacher." The teacher would simply assist the authoritative experimenter, dressed in a professional laboratory coat, by testing the influence of physical punishment on a person's ability to memorize specific details. The person whose memory was being tested was given the role of "learner."

All three individuals then entered a room, and the learner was strapped into what appeared to be an electric chair. After he was tightly restrained to prevent escape, the experimenter and teacher went into a nearby room. The strapped-in learner, who was seemingly immobilized in the electric chair, could be heard but not seen.

Next, the authoritative experimenter, again dressed in a professional-looking lab coat, instructed the teacher to administer a remote, progressively stronger electric shock to the learner if a wrong answer was given. The teacher could read the levels as going from "Slight" to "Danger: Severe Shock." To reinforce the apparent authenticity of the procedure, the teacher was first given a mild shock as a demonstration of what the learner would experience.

The ruse was that the supposed "learner" was a paid actor calmly sitting in the chair. When the experimenter instructed the teacher to administer an electric shock because of an incorrect answer, a button was pressed that generated an electrical noise, and the learner faked being zapped by crying out in pain. No actual shock was delivered, though. It was entirely bogus. Later iterations of this experiment intensified the drama, having learners plead for mercy, claim severe pre-existing medical conditions, or even scream in apparent agony.[16] Some versions went so far as to suggest that the maximum shock could potentially be fatal.

Shockingly, every volunteer "teacher" administered shocks of up to 300 volts when instructed to do so by the authoritative experimenter. Even more startling, 65% continued all the way to the maximum level of 450 volts despite believing that the shocks were causing severe pain or possibly serious harm.[17] Videos from the original experiment, widely available online, vividly showcase the participants' internal struggle. Yet, researchers noted, almost none of these volunteers would have independently chosen to commit such acts of violence. Why, then, did seemingly ordinary individuals willingly deliver such dangerous shocks to complete strangers?

The proposed explanation was that human beings are predisposed to obey authority figures, or those they believe are responsible for the situation. The presence of the experimenter in a laboratory coat, who also held a position of authority, was enough to push ordinary people to commit shocking acts against fellow citizens. While many teachers displayed signs of distress, such as sweating, trembling, or nervous laughter, most continued to administer physical punishment up to and including the very end of the scale.

Upon publication of the experiment, much of the public was outraged by this research. Many psychologists questioned Milgram's methodologies and the test results. Regardless of how one feels about this experiment, it suggests that we should, at the very least, dig deeper into the uncomfortable realm of human nature. Only by understanding and confronting these uneasy truths can we begin to understand the military and the processes of conditioning soldiers to kill. It may not be easy to consider, but perhaps it doesn't take as much conditioning as expected. Maybe it simply requires a figure of authority.

Military drills like the bayonet course focused on the horrific act of penetrating human flesh with a sharp instrument. All of this was done while our leaders screamed encouragement. I

---

16 Romm, Cari. "Rethinking One of Psychology's Most Infamous Experiments." *The Atlantic*, January 28, 2015. https://www.theatlantic.com/health/archive/2015/01/rethinking-one-of-psychologys-most-infamous-experiments/384913/.

17 Milgram, Stanley. "Behavioral Study of Obedience." *Journal of Abnormal & Social Psychology* 67, no. 4 (October 1, 1963): 371–78. https://doi.org/10.1037/h0040525.

suspect the reader is likely uncomfortable with this observation, but that is precisely what happened. The military needs recruits who are willing and able to kill on command. By conducting drills and appointing a leader, they can increase the percentage of participants in acts of violence.

However, this does not always translate easily into success. At least one study, while controversial, has found that a sizable percentage of combatants, even those explicitly trained, refuse to fire or will intentionally miss an enemy soldier if given the opportunity.[18] The findings, presented by S.L.A. Marshall, have been debated, with some historians questioning his methodology. However, the core idea that many soldiers hesitate at the moment of killing has influenced military training reforms for decades. This means they are always looking for additional ways to increase the likelihood of active killing. It's not a pleasant thing to consider, but it's what must be done to win. Meaning, Pete Hegseth's statement becomes even more important to heed. "If that makes you uneasy, keep us at home."[19]

---

[18] Marshall, S.L.A. Men Against Fire: The Problem of Battle Command in Future War. University of Oklahoma Press, 1947, 50.

[19] Hegseth, The War on Warriors: Behind the Betrayal of the Men Who Keep Us Free, Chapter 11.

# Chapter III

## Infantry Leads the Way

The Renegade Platoon began working together as a team, and we were finally hitting our stride. Sufficient sleep remained elusive, even though the Army had mandated we obtain a seemingly undefined minimum. As far as the drill sergeants were concerned, there were always ways to fudge the corners.

Fire Watch was a standing order from the very first day, in which we had to post two roving guards in our bay at all hours of the night. The purpose of it was to simulate sentry duty, designed to protect the platoon at nighttime from the enemy or other hazards (e.g., actual fire). In basic training, it served to ensure everyone was accounted for at all times. It was also an excuse to interrupt sleep cycles. This just added another layer of stress to the daily grind.

About fifteen minutes before my shift, I would be roused by one of the previous guards. Quickly changing into my BDUs, I'd grab my flashlight with a red lens to avoid disturbing other soldiers who were sleeping. Quietly, I would then walk up and down the barracks bay and count the number of sleeping recruits. My battle buddy would do the same, and we had to report this number to drill sergeants, who routinely and randomly checked on us at all hours of the night. If a guard fell asleep, someone was missing from his bunk, or we failed to challenge someone entering the bay correctly (usually drill sergeants), there was hell to pay.

Strangely, we also seemed to be suffering from bouts of electrical problems, in which the fire alarms would suddenly go off for no reason, sometimes multiple times a night. Each alarm meant we had to run downstairs and quickly form up in the breezeway. Drill sergeants would then check to make sure everyone was there and that there was no danger, after which the alarms would be reset, and we could make our way back to the bunks.

While they always went through the motions of checking for fire, we had the nagging suspicion that it was all an act. It was a calculated tactic to mess with us, further disrupting our fragile sleep cycles and keeping us in a constant state of readiness.

**28 August 2002 – Letter to the Harstad Family**

We got in trouble today again for moving while at attention. DS Clements is making 7 guys do Fire Watch at a time instead of 2. Yes, I have watch tonight. The thing that sucks is that I'm getting up at 3 AM tomorrow and marching a long distance.

We eventually earned the removal of "total control," in which the drill sergeants micromanaged every minute of our existence. One perk was that we could now choose what foods to eat, though we were still prohibited from eating sugar or sweets. As funny as it sounds, this was like Christmas Day ten times over. Tiny, newfound freedoms became motivating and joyful events.

We also had more free time in the barracks, though we were still strictly prohibited from sleeping, eating, or doing anything remotely fun. The only sanctioned activities were writing letters, exercising, cleaning, or studying our handbooks. That didn't mean we wouldn't try to sneak a little more shut-eye, though at the great peril of being caught.

**27 August 2002 – Letter to the Harstad Family**

Sleep is priceless around here. I've seen guys pay others $5 for relieving them from a Fire Watch duty. Sleep is hard to come by.

Once in a great while, we'll have nothing to do. One guy slides into his little cubicle (storage closet) and sleeps while another guy watches. When the drill sergeant walks in, a whole bunch of guys yell, "At ease!" (This gives everyone a quick warning that he's coming.) Pop up real quick, and you're fine. You have about three seconds before he can see us.

A lot of guys lie under their beds and pretend as though they're tucking in sheets. They're actually sleeping. Same thing applies here.

Once we were released from total control, our leadership designated four recruits as squad leaders and one as the Platoon Guide. Squad leaders were designated first-line supervisors and were assigned several men each, while the Platoon Guide was responsible for everyone. The duties for each position were minimal at first, such as marching us to chow or creating a Fire Watch list. As the cycle continued, however, more critical tasks were assigned.

Unsurprisingly, they tended to put the biggest assholes, screw-ups, Blue Falcons, or people who couldn't keep their mouths shut in these positions. This meant we were often led by our weakest members and people who had said or done terrible things to the platoon, damaging everyone's trust. The result was an uncomfortable feeling of dread. What could possibly go wrong?

**29 August 2002 – Letter from my mother, Joyce Harstad**

I want to tell you, Jeff, that everybody here is so damn proud of you. Keep fighting till the end. You will make it. There is not a doubt in my mind. You are one very special kid of mine. And yes, you learn to forgive people for the bad stuff that they have done. It makes you a better person.

Opie, our buddy who had everyone in stitches from laughing so hard, was unexpectedly given a squad leader position. The drill sergeants just beamed with joy when they announced his name, figuring this would be a lot of fun. I was incredibly worried because he wasn't known for his leadership abilities. In fact, *everyone* was concerned. If he or our other designated soldiers screwed up too much, our drill sergeants would put us back in total control and snatch away the tiniest of our hard-earned freedoms.

Amazingly, against all odds, Opie was superb in his newfound position of power. He rose to the occasion and became a stellar squad leader. The drill sergeants were shocked, and probably secretly proud, that their plan had backfired.

**2 September 2002 – Letter to the Harstad Family**

Opie is doing great as squad leader. He has been calm while in front of the drill sergeants. It's behind their backs that the real fun stuff takes place.

Last night, Opie had another spell. He pulled off his shorts and pulled his underwear up the crack of his butt like a thong. He went running around the barracks, laughing and jumping around.

Believe me, if you were here, you'd find this stuff funny too. When boredom sets in, and total seriousness is preached all day long, stupid stuff like this cracks you up.

The only person who wasn't thoroughly impressed by Opie's miraculous transformation into a superb leader, even including our own drill sergeants in the Renegade Platoon, was DS Whiteley. No matter what Opie did, DS Whiteley was relentless and all over him. Too many jokes and previous wisecracks had made our buddy the target of perpetual harassment and smoke sessions.

Opie became accustomed to doing push-ups whenever 4th Platoon came around. Indeed, there was a point when Opie automatically dropped to the push-up position in preparation as soon as he saw 4th Platoon in the area, even if DS Whiteley hadn't yet been spotted. It was like a reflexive, defensive maneuver ingrained by sheer repetition. However, as you'll soon see, Opie got his own form of revenge. And we would scream about it until our stomachs hurt, and we cried wonderful tears of laughter.

One of the most exciting and nerve-racking days of training came in the form of visiting the notorious gas chamber. Everyone had heard stories of how terrible it was, with recruits vomiting or passing out from holding their breath. Before setting foot inside, though, we had to attend lengthy training on Nuclear, Biological, and Chemical (NBC) warfare.

**Undated – Letter from my uncle, Mark Harstad**

I can't tell you how proud of you I am. I read all the letters you send to your mom and dad, and I'm really happy you're getting along so well. I had my doubts at first, but I can tell you're going to be one hell of a soldier.

P.S. I hear that gas chamber thing really sucks.

We suited up in charcoal-lined BDUs, known as Mission Oriented Protective Posture (MOPP) gear, and practiced putting on our rubber boots and gloves. These items were designed to protect grunts from NBC attacks, but we didn't know how well they would work in combat. They were hot, suffocating, and made every movement feel clumsy and slow. Evan Wright, an embedded journalist who later accompanied Marines during the invasion of Iraq, would summarize it best: "What the MOPP basically does is encase you in your own private panic attack."[20] We also learned how to correctly put on and seal our gas masks before clearing them, a crucial step before breathing.

The gas used in the chamber was CS, short for Corson and Stoughton, the surnames of the chemists who first synthesized it in 1928. Commonly known as tear gas, CS is not technically lethal when used correctly, though it causes intense discomfort, burning, and irritation. While the 1925 Geneva Protocol banned many forms of chemical warfare, it didn't specifically outline which substances were prohibited, leaving a significant amount of gray area in interpretation.[21] I personally think it's darkly funny that you're allowed to shoot and stab enemy soldiers in the face during war, but gassing them is off-limits, apparently. Today, the U.S. military primarily uses CS gas for training purposes, and it retains the right to employ it in certain non-combat situations, such as riot control.

Once we had thoroughly trained and demonstrated our competency by promptly putting on our equipment in the field, our drill sergeants began a lengthy smoke session. We ran sprints and crawled through the sand in our thick suits. It was tough, and the gas masks fogged up with our sweat, making it hard to see. The NBC suits acted like an oven, keeping our heat trapped next to our bodies until we were drenched in perspiration. Finally, they ordered us to remove all our equipment except for our masks. It was the only NBC item worn into the gas chamber.

As I marched into the concrete bunker, my heart raced with anticipation, but I found that my gas mask held its seal and that my breathing remained uninterrupted. It was reassuring to see

---

[20] Wright, Evan. Generation Kill: Devil Dogs, Iceman, Captain America, and the New Face of American War. G.P. Putnam's Sons, 2008, Page 39.

[21] Sadeghi, McKenzie. "Fact Check: It's True Tear Gas Is a Chemical Weapon Banned in War." USA Today, June 6, 2020. https://www.usatoday.com/story/news/factcheck/2020/06/06/fact-check-its-true-tear-gas-chemical-weapon-banned-war/3156448001/.

that our equipment would actually work. However, a handful of soldiers had not gone through the sealing process properly and began coughing violently. Tiny wisps of CS had penetrated the barrier and worked toward exposed mucous membranes. They, not I, began to feel the bite of the gas.

**27 August 2002 – Letter to the Harstad Family**

How did the gas chamber go? It sucked.

We walked in, about 50 of us. You can see this white haze in the room. Immediately, sweaty parts of your body begin to burn. Then it becomes worse and worse until it is almost unbearable. They walk us around the center, where the CS gas is brewing.

They take four at a time. You break the seal of your gas mask. At first, I was fine. For about three seconds, I didn't feel a thing. Then it hits you hard! Your eyes slam shut, unable to open. Tears stream down your face. Snot begins to pour out as the gas attacks your sinuses. You try to breathe, but it feels like taking a breath underwater. Your head feels like it's going to explode from a lack of oxygen. Coughing, gagging, puking, you try to stay focused.

They keep you like this for about 30 seconds, then you leave. You hit the fresh air, and it is like heaven. You can breathe again. You cough like crazy and gag some more. It is difficult to open your eyes.

Inside the gas chamber, I found that there wasn't anything I could do to force my eyes open once my mask had been removed. Nor could I breathe in that horrible gas. It felt like my lungs had been set ablaze, and my skin burned fiercely with an intense, stinging heat. I tried my best to remain calm and was mostly successful, though I was fearful of passing out because I couldn't breathe. Each time I took a breath, even a small one, I began to cough violently with searing pain. This forced out what remaining pure air I had left in my lungs, only to be replaced with acidic needles stabbing me from inside.

I was near panic when my time finally expired, and I was shoved out of the door by a masked drill sergeant, though I couldn't see who it was through the blur of tears and snot smeared across my face. We were supposed to flap our arms like birds upon exiting, but I don't know why, because it didn't help. A few moments later, I squinted and watched as my battle buddies started to stumble their way out. I soon found myself laughing and crying at the same time.

**30 August 2002 – Letter to the Harstad Family**

He (Opie) did a funny thing at the gas chamber. DS Whiteley, a pretty mean guy, has been causing a lot of problems for him.

Anyhow, Opie drinks about two canteens of water just before he goes in the chamber. He gets in there and stands right next to Whiteley. Then he starts puking all over Whiteley's boots. A lot of water! They just pushed him out of the door. He came out with a big smile on his face!

I was relieved when the gas chamber ordeal was finally behind us and even more thankful that we wouldn't have to endure it again. It was time to move on to bigger and better training.

One of my favorite activities, and certainly a lot less painful, was our visit to Eagle's Tower. This 40-foot wooden structure was designed to help recruits conquer acrophobia and build confidence through a series of elevated challenges. We practiced rappelling, navigated rope bridges and cargo nets, and pushed ourselves well outside our comfort zones.

One of the tower's most exhilarating features was a rope that stretched from the top and gradually sloped down to the ground, ending at a distant pole. Our task was to shimmy along the rope on our bellies, using one leg as a stabilizer while pulling ourselves forward without falling off. There were no safety harnesses. It was just you, the rope, and a long way down to the net if you slipped. Looking down was awesome as I was unafraid of heights, though it was clearly terrifying for some. Many recruits struggled to maintain balance, crawling at a snail's pace and clinging tightly to the line. They shook with fear as drill sergeants yelled at them to hurry along.

**With my dad and brothers at Eagle's Tower, much later on graduation day**

The event was easy for me, though. I could zip down the line in mere seconds, never once needing to slow down or even correct my balance. To my surprise and quiet satisfaction, I was the fastest in the company by a wide margin, even beating out our seasoned drill sergeants. Being extremely skinny had some advantages, though not many.

Before doing any of this, however, we had to sit through something called a safety brief. Veterans will roll their eyes at the very mention of those words simply because they are usually filled with obvious information about areas of concern. Safety briefs are like the warning labels on laundry detergent packets that say, "DON'T EAT." Why are they on there? Because some moron ate one of them. So, we had to sit through a bunch of tedious nonsense to do anything that was even remotely dangerous.

Our company marched into the compound and took its place in the bleachers as all the drill sergeants began their presentations. They explained in detail how to ascend the tower, safely rappel to the ground, and follow other safety-related procedures. A volunteer was requested from the bleachers, and one recruit quickly raised his hand. He was told to meet a drill sergeant at the back of the tower for additional instructions. The recruit jumped up, Kevlar helmet on his head and LBV around his body, and sprinted to the back.

I listened with mild interest as we learned how to protect ourselves from falling to our deaths. It wasn't complicated, but they had to review every little detail. As the lecture continued, a sudden commotion broke out at the top of the tower. The drill sergeant and private had presumably already made their way to the summit, though we couldn't see them from where we sat.

"I don't want to go down!" we all heard someone yell. "I'm scared!"

The unmistakable screams from our drill sergeant followed. Obscenities and threats of death became more and more animated, with our view still blocked by the tower. We heard the terrified volunteer's feeble response, saying he no longer wanted to rappel down. This kept going back and forth as we sat fixated on the hidden action and unfolding drama.

Suddenly, a sickening scream emanated high above us as a soldier wearing a helmet and LBV was flung off the tower's summit without a rope, quickly speeding up and slamming face-first into the dirt 40 feet below. An audible gasp emanated from the crowd as I recoiled in horror, his head snapping off and rolling away like a pinball. I quickly looked up and saw someone standing at the tower's edge, looking down, laughing hysterically. We all sat in silent shock as we tried to process the murder that had just taken place.

All the drill sergeants screamed in laughter, but I didn't think it was funny. That is, until I looked at the "body" again and noticed that it was a pink CPR dummy dressed in our BDUs and LBV. They had expertly pranked all of us, with mild chuckles slowly flowing through the recruits as they realized what had happened. What a dark sense of humor grunts and drill sergeants have!

My turn eventually came, and I excitedly climbed the cargo net to rappel down. Our commander, Captain (CPT) McCormick, stood at the tower's top. We rarely interacted with him as he was the company commander, meaning he made the important decisions behind the scenes. As an officer, he naturally had less direct involvement with us enlisted soldiers. It typically went through the drill sergeants. We only saw him occasionally, usually for a motivational speech or another leadership directive.

Despite the distance, he was a good guy and well-liked. He had been to combat as a Ranger and had an impressive stack of medals and qualifications, so we all looked up to him. Everyone respected and almost idolized him.

**27 August 2002 – Letter to the Harstad Family**

CPT McCormick came up to me today while we were on the tower. He says to me, "Private Harstad, did I pronounce that correctly?"

I replied, "Yes, sir!"

He continued, "Do you like me, Private?"

I nodded and said, "Yes, sir!" I completely meant it.

He says, "Liking leads to loving, and loving leads to fucking. I'm not into fucking men. So, do you like me?"

I reply, "No, sir!"

He nods, smiles, and walks away. Nice guy. He's got a goofy sense of humor.

After spending much time getting intimately acquainted with our rubber duckies (did you enjoy that paragraph transition?), our company finally began learning the meat and potatoes of our profession, namely, shooting things. As I mentioned, we always used the "crawl, walk, run" approach for every piece of training, though sometimes it felt like overkill. This was the "walk" portion of marksmanship.

One of the first drills we practiced was the dime drill, done hundreds of times, if not thousands. Lying prone on the ground, I would establish a stable shooting platform and look down through my rifle's iron sights. Despite Hollywood's portrayal, the M-16 has minimal recoil and can easily be fired one-handed or even by a small child, though I obviously don't recommend either. Because of the low recoil, it is common practice for soldiers to place their noses directly on the charging handle of the rifle. This is done as a point of reference, meaning I look down the sights the same way every single time, with no "x, y, z" variation (coordinates in space). By doing this, my round groupings will be consistent. In other words, it ensures my bullets will go where I *think* they will go.

My battle buddy would then carefully balance a dime on the end of the muzzle with our rifles unloaded. I'd inhale and then slowly let air escape from my lungs, gently squeezing the trigger. If I flinched in anticipation of recoil or from the click of the trigger, the dime would fall. If I were "surprised" by the trigger pull, the proper reaction when firing, the dime would stay put.

As weird as all of it may be for those inexperienced with rifles, these foundational skills were crucial for maintaining tight shot patterns and accurately hitting targets 300 meters or more away (over three football fields). For American readers, one yard is approximately 0.91 meters.

Steady breathing, smooth trigger control, and proper body alignment were all important for consistently hitting targets.

Becoming a proficient marksman requires careful attention to several fundamentals. Trigger control is one of the most important: using the pad of my finger keeps the shot steady, while using the joint or the fingertip can pull the rifle off target. Breathing is just as critical. To steady the shot, I try to slow my heart rate if possible and exhale gently as I squeeze the trigger.

Consistency in how I rest my cheek on the rifle, known as a cheek weld, ensures that my sight picture remains the same every time I aim. A solid shooting position also matters. Pressing the softer parts of my body, like my flat forearms and torso, firmly into the ground creates a stable foundation, whereas relying on bones like elbows or knees leads to wobbling and instability. Finally, I stay aware of my profile. Am I flaring my elbow out like a chicken wing and exposing more surface area, or am I keeping a tight, low silhouette that is harder to see and harder to hit?

Now, the reader who doesn't care about shooting or firearms, in general, may have skimmed over the preceding paragraphs. Honestly, that's perfectly fine. The point is that there are many things to be cognizant of to hit targets consistently. An experienced marksman can usually tell what he or she is doing incorrectly simply by looking at the targets downrange. Shot patterns, or lack thereof, offer valuable insight into correcting shooting behaviors.

Practice and proper repetition foster a second-natured ability to perform fundamentally sound marksmanship, which is important while killing human beings. It isn't as simple as clicking the "X" button on the Xbox controller, especially when factoring in adrenaline surges and the enemy shooting back. The best shooters, including elite special operations soldiers, always self-evaluate their skills and fundamentals. *Always*.

The "spray and pray" method is sometimes the default for soldiers in the heat of combat, especially when they are out of control or panicked. Spray and pray simply means shooting as fast as possible in the general direction, praying that a stray round will hit the enemy. However, this is not how the military teaches recruits, even if it feels instinctive in high-stress situations. It leads to wasted ammunition and rarely accomplishes anything productive, aside from occasionally keeping the enemy's head down.

When we weren't practicing dime drills, our drill sergeants had wheeled in a television set equipped with a custom-made Army video game, and we could practice fundamentals during our limited free time. The best way to describe it was a more advanced but less fun version of Nintendo's *Duck Hunt*. A rifle replica was attached to the system, and we would lie in our bay, aiming down the iron sights and lining up targets on the screen. When we pulled the trigger, the system would show where we had hit and whether or not we had killed the target.

However, it wasn't as cool as it sounds, as the technology was still new and left much to be desired. But... they had left us unattended with a television set locked with a passcode. After a bit of tinkering, I figured out how to bypass the lock, and we cautiously watched *Jeopardy!* with giddiness, the one show that reliably came through with no antenna.

**29 September 2002 – Letter to the Harstad Family**

We set up our TV in the front of the bay (even though we almost never get to use it). Opie stands up there one evening and says, "Hey, check out what is on TV!" We all looked, and there he is, standing completely naked in front of the TV. He screams, "I can't believe you all like gay porn! What's wrong with you?!"

On the 2nd of September, we all celebrated Labor Day. Thankfully, we had no work to do, and almost all the drill sergeants were at home. By that time, we were off total control, and they left us mostly unattended, with the condition that we remain in the bay, the laundry room, or the chow hall. A drill sergeant would randomly show up multiple times throughout the day to catch people doing unauthorized activities, such as sleeping. So, we had to remain vigilant.

It also turned out to be painfully boring, especially after I had grown used to the flurry of drills we had been doing each day. Basic training and much of my time in the military, as I would later find out, seemed to involve going 100 miles per hour one day and then being stuck at a standstill the very next. Hurry up and wait.

**2 September 2002 – Letter to the Harstad Family**

Well, I'm sitting here, bored as hell, and I figured I'd write. Today we have absolutely nothing to do. From what I've heard, we sit in the barracks all day long and go to chow three times. That's all!

Can't leave, can't go to the PX, can't use the phones, can't read books, no magazines, just sit and do nothing. Sleeping isn't allowed, but everyone is going to. What a waste of a day. Stupid crap like this really makes me mad. Instead of allowing us to go to the PX or use the phones, they make us stay in the barracks and do nothing. We've already cleaned everything. Hell, we even buffed the floors yesterday! Maybe they'd like us to count how many holes are in the ceiling panels...

Today, I hate the Army.

My battle buddy, PVT Harper, keeps a log of everything we do. I pretty much do the same thing, except in the letters to you guys. Anyhow, he is writing about me going crazy with boredom. He thinks it's absolutely hilarious. At least someone is having fun...

He just brought up a very good point. I'm getting paid to sit on my butt! Didn't even cross my mind...

You have to find something to do! We were so bored a little while ago that we collected a bunch of pens. Pens with caps on the top were designated black, and uncapped pens were red. We laid them down on the tiles to represent a grand checkers board. We played checkers!

As our Basic Rifle Marksmanship (BRM) progressed, we entered the "run" phase and began firing live rounds at paper targets to obtain a "zero." This meant adjusting our sights to hit targets accurately, both close and far away. Like before, we drilled on this over and over again before moving onto actual lane targets or those that moved back and forth. We did it so many times that it became boring. It was simply being drilled into our heads.

**4 September 2002 – Letter to the Harstad Family**

While we were doing BRM, DS Nash inspected our bay (while we weren't there). You can guess what he did to us: extreme PT! He lectured us on how to clean the barracks while we stayed in the push-up position. You may not think it is hard to stay in a push-up position, but try it for 52 minutes. He lectured us the whole time as we held ourselves up in pain. I still can't lift my arms; I'm so tired.

**6 September 2002 – Letter to the Harstad Family**

Friday, during the zeroing of our M-16s, I finished early. Anyhow, the range cadre made all of us who had finished walk through the woodline near the road and pick up every single pinecone, piece of trash, and every single twig. Yes, I said pick up all these things in about a football-field-sized area. It took us forever! Airborne Rangers? Nah! Forest Rangers!

Around the time of BRM, everyone in the company came down with a nasty bout of influenza. People were dehydrated, vomiting and diarrhea were common, and coughing and sneezing attacks were rampant. It was miserable being on a road march with diarrhea, running ahead of the column to shit liquid in the woodline next to the road, only then to use an extra sock or dried leaves as toilet paper. It was essential to do it quickly to avoid being left behind.

Everyone was wretched. I felt I needed several days of lying in bed to recuperate, but there was no way it would happen. Widespread illness didn't stop training; we had to continue even though we were sick. The mission always comes first.

**10 September 2002 – Letter to the Harstad Family**

We've been doing a lot. Today, we left at 0400 for a road march. We marched 7½ miles there. I have never felt so bad in my life. I'm still sick, and I wanted to die. I just pictured you guys at the end of the march. We did pop-up targets. My M-16 jammed so many times, I was fed up. Top that off with a nasty cold and extreme nausea, and it doesn't sit well.

The one silver lining from that time was that our drill sergeants were slowly transitioning from pure disciplinarians and monsters into quasi-normal leaders. I don't want to give the impression that they were suddenly friendly or unwilling to dish out the pain, as they most certainly were.

Still, there were signs that they were beginning to recognize our improvements and our development into becoming soldiers. At this time, we could see their actual leadership qualities and the valuable talents they possessed. They truly cared about their men, even if sometimes it didn't feel that way. I found myself no longer resenting them, but instead wanting to earn their approval. I suppose it was similar to a son looking for validation from his father.

**10 September 2002 – Letter to the Harstad Family**

My nose is so dry due to me blowing it. Well, today we went onto the firing line, and I started shooting. We are supposed to put our noses on the charging handle and look down the sights. Every time I shot, even though it didn't kick hard, it made blood spurt out of my nose (because it was dry and sore). I had a puddle, and DS Ross saw it. He freaked out (thinking I had shot myself) and ran over to me to make sure I was okay. He cleaned me up (my face was a mess). You start to like a guy when he's willing to clean blood off your face.

**31 March 2025 – Reflection of Logan Hastings, 2nd Platoon, C 2-54**

I once had a one-on-one conversation with DS Ross. I confided in him the difficulties of being in basic training, such as mental strain, physical exhaustion, and constant pressure. He made a positive impression on me and recommended reading the Book of Job as motivation to endure and overcome my struggles.

Qualification was easy, and we had to hit at least 23 of 40 pop-up silhouette targets at distances of 50 to 300 meters. The only problem was that they sometimes had holes from the thousands of rounds already fired. Or perhaps they just didn't work correctly. I recall shooting a target at the 50-meter line, which was effortless at that range, and I could see a little piece of it flick off from the impacted rifle round. DS Ross, who was crouched next to me, saw it too. I protested while he laughed and shrugged, saying, "Just make sure you hit the others."

The more enjoyable course was the moving target range, which simulated soldiers jumping up from a trench and running a short distance before dropping back down seconds later. It was so much fun shooting these targets nicknamed "Ivan," a leftover from the Cold War era. Having spent my teenage years hunting in Iowa, I had no trouble knocking all the targets down with glee. I would have gladly practiced all week long on that range had they let me.

After we finished shooting for the day, we had to carefully police the area and pick up spent brass casings and unfired ammunition. All of this would be turned over to the armorer on the range immediately. Once that chore was complete, drill sergeants would inspect our rifles and magazines to make sure we didn't possess any "brass or ammo." We were also required to stand with our arms and legs outstretched, forming a big "X," so they could search us carefully for any of those items. If a recruit had even one spent casing in the bottom of his pocket, there would be swift, overwhelming amounts of PT coming, whether it was there intentionally or not. Further, the searches gave drill sergeants an excellent excuse to rough us up.

**11 September 2002 – Letter to the Harstad Family**

After we shoot (on the range), we get searched by our drill sergeants for brass or ammo. Our platoon's drill sergeants are cool about it and pat us down quick and easy. However, today I had DS Whiteley. He slammed his hands into me hard and knocked me all around while searching, knocking me to the ground. He enjoyed doing it. Not much fun!

**12 September 2002 – Letter to the Harstad Family**

I got roughed up pretty good today during brass and ammo check. DS Lafferty hit me so hard my feet literally left the ground. I have his palm prints still on my chest. I just smiled as I stood back up.

All the rifle shooting took place on the first anniversary of September 11th, poetically foreshadowing what was in my near future. That morning, we all marched to a parade ground and listened to a motivational speech by one of the base commanders. We also watched a stirring video projected onto a massive screen, showing haunting reminders of the attacks. Images, such as the Falling Man, were intermixed with snippets of video taken that day. All of it was overdubbed with emotionally stirring music.

The intent of the video worked. I was reinvigorated to work even harder and push past my struggles. I didn't care that I was sick. Nor did I care that we were still sometimes being treated like maggots. I wanted to find the people and organizations that were responsible for the attacks and kill them. I wasn't scared, and I wasn't hesitant. My battle buddies felt the same way. We wanted blood. It was the motivation that we didn't know we needed but desperately did.

**11 September 2002 – Letter from my mother, Joyce Harstad**

It's been a year that I would like to forget.

I remember after we sat and watched the towers being hit that you came to me and said, "Mom, will you be alright taking care of the kids without me?" I asked you why, and you said that you were going to call and join the Army, but you had to make sure I was going to be alright first.

Of course, I wanted to be selfish, but I couldn't. At that point, you not only made me proud of you, you also made me scared for you. Proud because your family came first, and second, you wanted to defend your country.

Back at home and across the nation, similar memorials were taking place. My family attended one of those events, as did many. At the same location, my dad ran into a longtime customer, Doug Slechta. I looked up to Doug as a teenager, as he was a former Green Beret with experience in Vietnam. He was quiet and stoic, yet carried a presence far greater than his slight frame suggested. Though he was older and no longer of proper fighting age, the attributes he had acquired in the service never left him. As the saying goes, "Once a Marine, always a Marine." This holds true for every branch of the military.

**11 September 2002 – Letter from my father, Bill Harstad, Jr.**

We all stood and watched the fly-bys of the F-16s. As I stood behind Doug, I could kind of see a younger version of him, the "old soldier" in him from way back. He stood up really tall, and I could tell he was moved by it all.

The following day brought another kind of motivation I'll never forget. The pain and suffering endured during Infantry School can sometimes be overwhelming, but the mind is interesting. How quickly I could ignore the bleeding blisters on my feet and the pains in my back. How quickly I could stand with my head held high, as though I were ten feet tall.

**12 September 2002 – Letter to the Harstad Family**

Another battalion had graduation, and their families were outside. We happened to be marching in the 100+ degree weather. They looked at us like we were gods or something. A bunch of older guys, you know, World War II vets most likely, had big grins on their faces as we walked by. You forget the pain for a while.

The weather in Georgia slowly began to change as graduation day inched ever closer. The oppressively hot, humid days gave way to a more bearable, temperate climate. Sometimes, the nights could even be a bit chilly. When the rains came, as they always tend to do in that part of the country, we would be soaked to the core and left shivering. Our regular Field Training Exercises (FTXs) could be miserable; we would sleep outdoors for several days at a time under a tiny poncho stretched over a hole dug in the mud. The small depressions we had scooped away filled with the soupy water, and it wasn't uncommon for us to shiver together in a futile attempt to stay warm like tightly packed sardines.

**15 October 2002 – Letter to the Harstad Family**

DS Nash impressed me today. We were standing in the field, freezing cold and wet, learning squad movements in a forest setting. A guy in our platoon didn't have his wet-weather top (which he was supposed to bring), while DS Nash was wearing his. Without hesitating, DS Nash (saw this guy suffering and) gave him his. That shows his dedication to his troops. Little things like that make us look up to him.

I began to loathe sleeping outdoors, which is comical now as I would have well over three more years to do precisely that. Why the hell did I pick the infantry again? More than once, I seriously questioned my thought process when I had picked my MOS. Couldn't I kill bad guys *and* sleep in a Holiday Inn?

When we spent nights back at the barracks, it was wonderful to clean our gear and our bodies of the filth that had accumulated. Showers took on an entirely new level of pleasure. Evenings at the barracks also meant plenty of continued hilarity when the drill sergeants weren't around.

**1 September 2002 – Letter to the Harstad Family**

Something funny happened a little while ago. A private named Hamilton was trying to burp really loud. He succeeded, but managed to crap his pants at the same time.

**8 September 2002 – Letter to the Harstad Family**

We just bet a guy that he couldn't do the "impossible sit-up," aka the "Ranger Sit-Up."

What we did was find a gullible person (in the platoon). We told him that there was no way he could do it. He agreed (to try), and we tied a towel around his eyes and then told him to keep his eyes shut. We spun him around a whole bunch and then had him lie on the ground. Everyone stood around and cheered him on.

We then had him hold his arms out like a bird.  We all told the guy, "One, two, three!" He then sat up with all of his might. Meanwhile, Opie had pulled his pants down and put his bare ass right where he thought the guy's face would end up when he did the sit-up.

As he sat up full strength, we pulled off the towel, and his face went right into Opie's cheeks. Hilarious! Can't beat that for entertainment.

**12 September 2002 – Letter to the Harstad Family**

Opie has been pretty calm today. They took away (found) his sugar. However, he did manage to get ahold of a CPR dummy from somewhere and ran around with it. He put it in lockers and scared people, dropped it out of the windows onto people, grabbed people's asses with it, stupid crap that was funny. Somehow, he never got caught.

I'd be remiss if I didn't mention the rare occasions, twice, if my memory serves me correctly, that we were required to participate in "mandatory fun days." Yes, the Army has these, even for Infantry School, and they are pretty stupid. It is a bit embarrassing to admit it now, as it really took away from the seriousness of what we were doing. Plus, I'm sure my Marine infantry brothers are laughing hysterically at this. I can't imagine they had any fun days that the Marine Corps mandated, but you never know.

Essentially, it was a few hours off where we were bused to a location on post for an event. On the 28th of September, it was a concert featuring a girl group that felt like a knockoff version of the Spice Girls. I wish I were making this up, but I swear it's entirely true.

**Mandatory Fun Day & the Chicago Honey Bear Dancers**
**Photo courtesy of the Chicago Honey Bear Dancers**

The group was part of the *Chicago Honey Bear Dancers*, an offshoot of the Chicago Bears cheerleading squad in the '70s and '80s. Honestly, none of us were even remotely interested in their music or dance routine. We simply gorged ourselves on junk food and ogled the attractive women on stage. We hadn't seen a female in nearly two months, let alone a pretty one. For us, it was dinner *and* a show. I even had my photograph taken with the girls, which I sent to my parents as evidence of how difficult Infantry School was. Perhaps it made my brothers consider joining, if only for a moment.

**28 September 2002 – Letter to the Harstad Family**

Well, I just got back from the concert. At least we got to get away from basic for four hours. I ate a medium pizza, a hotdog, a BBQ rib sandwich, five Mountain Dews, four fruit punch Gatorades, and nine candy bars. I felt like crap after eating all of that, but it tasted great. The concert consisted of five girls dancing to music and occasionally singing. It was just like dance squad at North High School. DS Whiteley said it was the dumbest shit he had ever seen.

In hindsight, the group was probably trying to be patriotic by putting on a show for the troops. This is pretty admirable. Plenty of acts do this even today, especially with the USO. The girl's dance routine was just out of place for many testosterone-fueled grunts in Infantry School, but that was the Army's fault, not theirs.

After eating and drinking our weight in sugary and fatty foods, the drill sergeants made us pay for our indiscretions. Nothing was free, and everything was done at a cost. The follow-up PT was brutal. They destroyed us and made us pay dearly for every pizza and candy bar we had eaten. But it tasted so good!

On top of everything, it's pretty comical to think that we were throwing grenades one day and then watching a very estrogen-fueled concert the next. The Army, it seems, has a peculiar sense of humor.

As the days continued to pass, we looked forward to the mid-point of October, when Family Day was held. On that single day, families could visit with recruits and energize them to finish the course. In the meantime, we shot machine guns, went on plenty of road marches, ran the obstacle course, slept in the woods, threw grenades, launched rockets, and choked each other unconscious during Combatives. It went by at a blistering pace.

My parents flew down to Georgia and rented a car so they could watch us do early morning PT. At first, they were shocked at my appearance. My dad later joked that I looked like a cancer patient. With my buzzed hair and skinny, gaunt frame, I'm sure I did. With tearful hugs and much catching up, we visited the National Infantry Museum and Soldier Center as they looked on in wonder. This life was utterly foreign to them, my chosen path, but I did my best to represent my new brothers with honor. Their visit was terribly brief, but it was an excellent chance for them to see my transformation in progress. Imagine traveling over a thousand miles to spend a single day with your son at work. Those were my parents. It was much-needed motivation to finish out Infantry School.

**My mother, Joyce, inspecting my bunk during Family Day - and yes, this is me *after* putting on weight**

Unfortunately, several days later, I suffered a significant setback that threatened to derail my budding career permanently. An unforeseen injury had the potential to snatch away the prize I had worked so hard to achieve. I couldn't claim to be an infantryman yet.

# Reflections

When my parents arrived to watch our early morning PT session at Fort Benning, I was already suffering from delusions of grandeur. I suspect many of us, particularly those young and impressionable, were all in the same boat. As budding infantrymen, we were superior to everyone else, or so I thought. We were at the top of the food chain, so to speak.

**With my father, Bill, at Eagle's Tower during Family Day**

This is something that is beaten into the heads of infantrymen everywhere, regardless of whether they are in the Army or the Marines. As we went on long runs with the drill sergeants calling out cadence, it wasn't uncommon to hear and repeat the chant, "If you ain't infantry, you ain't shit!" We looked down on fellow non-infantry soldiers for many reasons. While some were legitimate, it was a genuine "us versus them" rivalry in the making. We had to establish the pecking order in the grand scheme of things, or so we believed.

This type of behavior is not isolated among Army infantrymen. Marines, both infantry and non-infantry Devil Dogs alike, are notorious for thumbing their noses at the other armed forces. They proudly proclaim their allegiance to the Corps, sometimes with comical rapidity and overwhelming derisiveness. Don't believe me? The next time you encounter a Marine Corps veteran, look at his or her vehicle. It will likely have one of the following: a Marine Corps bumper or window sticker, a personalized license plate showing off their branch of service, or perhaps a magnetic Eagle, Globe, and Anchor. Maybe those are surprisingly not on the veteran's automobile. In that case, they likely display a giant Marine Corps flag in their yard, have a service tattoo, or some other signifier to the general public that "I'm a Marine!" Your veteran will likely have all these items, plus a pair of embroidered Marine Corps underwear. OK, that last one was a joke.

Members of the special operations community often take this elitist mindset to a whole new, hilarious level. It's rare to pick up a memoir by someone formerly in special operations, or even watch a video with them in it, without exclusively hearing terms like "operator," "SF," "the Unit," "SEAL," or "MARSOC" thrown around as everyday vocabulary. They are used repeatedly as a badge of superiority over conventional forces and everyone else, and it quickly becomes quite funny once the reader or viewer is aware of the terms.

Many of these guys go out of their way to avoid calling themselves "soldiers," "sailors," or whatever the standard term is, as if the word itself is taboo. Instead, they refer to themselves as "operators," sometimes in a hushed tone or even in the third person. I suppose it is another way to separate themselves from the "less worthy" military community.

I'm not necessarily trying to make fun of this behavior, as I did it, too. I even had a big "Infantry" tattoo permanently inked on my leg. I'm simply trying to shine some light on this interesting tribalism, in which we manufacture hostilities against people who are our allies and who often do excellent work. As I pointed out in Chapter 1, all MOS fields and branches are vital to the war effort. Army and Marine infantrymen, even the elite special operations variety, can survive in combat for a few days, perhaps a few weeks if lucky, without assistance from the rear. That means if they are cut off from resupply, they will die. No matter how amazing the grunts are, they will eventually expire from a lack of food, water, medical supplies, and ammunition. And they will ultimately be overrun.

It took me quite a while to figure this out. My revelation came from Command Sergeant Major (CSM) Clarence "Cap" Stanley, whom you will meet later in the book. He was a tanker, responsible for leading and working with a tank crew on vehicles like the M1 Abrams. Since he and his tanker soldiers weren't infantry, I initially had little hope of learning anything from them. I also figured they would be a liability next to us on the battlefield. How *wrong* I was. More on this later.

These days, I occasionally chat with people about their military service, or sometimes it is their son or daughter who is serving. More than once, I've had someone make a snide comment because I wasn't a Marine, a Ranger, or a member of whatever super-secret club only the "cool kids" get to join. I just smile and shake my head. I used to be that way, too. Unfortunately, some people never mature and grow out of that silly phase.

But perhaps a bit of historical context is needed to understand the roots of tribalism, particularly between infantry and non-infantry units. It was traditionally understood within the services that infantrymen faced the greatest risks of capture, injury, and death, far surpassing those of other MOS fields. After all, they're the ones charging the pillboxes, stabbing and burning people to death, machine-gunning enemy soldiers, and blowing things up with rocket launchers.

In 1943, in the midst of World War II, the Army began awarding the highly coveted Combat Infantryman Badge (CIB) to infantrymen and Special Forces soldiers who engaged the enemy in ground combat. The decoration portrays a cocked musket surrounded by a wreath. It was created to honor and recognize those grunts who were directly in the middle of harm's way and facing a most gruesome death.

Later, in 1952, the Army instituted another way to acknowledge the infantry's unique sacrifices in the face of danger: the Infantryman Shoulder Cord, a light blue fourragere worn on dress uniforms around the soldier's right shoulder. Soldiers who successfully graduated from Infantry School and were assigned to an infantry unit could wear this beautiful and easily identifiable decoration.

In addition, grunts who passed Special Forces selection were awarded the famous green beret, while Rangers wore the black beret. Likewise, Airborne Paratroopers earned a maroon beret to set them apart from other soldiers. It was another way to make grunts stand out from the masses of the Army, most of which held non-combat positions and would never see or hear a shot fired in anger.

Unfortunately, the Army and its leadership began to slowly erode these honors bestowed upon its tip of the spear, seemingly hoping to foster inclusivity and general morale. In 2001, the Army snatched the black beret from the 75th Ranger Regiment. It then unceremoniously distributed it to *all* soldiers, even those who permanently sat behind desks and held non-combat roles.[22] This didn't go over well with the Rangers, unsurprisingly, who had *earned* that coveted cap. In response, they quickly ditched the black beret and began wearing a tan version to set themselves apart from the masses.

In 2005, the Army continued the collectivist march by approving the Combat Action Badge (CAB). These were awarded to non-infantry personnel who served in a combat zone but didn't necessarily fire their weapons or even come close to any threat of injury or death. They looked nearly identical to the CIB, except the musket was removed and replaced with a knife. Slowly, the Army was trying to hand out prizes to everyone uniformly, regardless of the dangers faced.

None of this suggests that our non-infantry brothers and sisters don't contribute mightily to combat successes. Tankers, scouts, forward observers, attack helicopter pilots, and many other MOS fields are often in the middle of the fight, too. The difference I'm trying to highlight is that grunts are expected to close with and destroy the enemy on foot by all means necessary, whether with a rifle or an E-Tool. Grunts are the ones intentionally and repeatedly thrown into the human meat grinder.

To be blunt, the real gripe came from watching that new badge get handed out like candy to support personnel sitting in air-conditioned offices all day. The way the Army rolled it out cheapened what the infantry had already earned. The blue cord, the distinctive berets, and the Combat Infantryman Badge, symbols that once meant you had survived hell, were reduced to just

---

[22] Associated Press. "Army Troops to Wear Rangers' Black Berets." *Los Angeles Times*, February 10, 2001. https://www.latimes.com/archives/la-xpm-2001-feb-10-mn-23783-story.html.

more decorations on a uniform. And the men who fought and bled for them were expected to smile and pretend it didn't matter. But it mattered.

Former Secretary of Defense and Marine Corps General Jim Mattis expressed a similar sentiment in his book, *Call Sign Chaos*. He wrote of his combat Marines returning from Iraq in the early 1990s: "In the months following my return to the United States, I grew increasingly displeased with the lack of individual recognition my men were receiving. Frontline service in the military brings with it the dignity of danger, and recognizing valor is critical."[23]

To suggest that infantrymen have a chip on their shoulders would be an understatement. They are often the underdogs, poorly equipped, tired and hungry, and ordered to perform impossible tasks with limited or no resources. It doesn't take a genius to determine why they look down on the rest of the military. Marine Corps infantry veterans, likewise, will undoubtedly feel the same way. So, while I may have been suffering from delusions of grandeur, it wasn't entirely without merit.

Beyond this, I was also unquestionably suffering from a form of cognitive bias in which I grossly overestimated my abilities. By the time Family Day rolled around, I thought of myself as a master of warfare because of my brief training in Infantry School. It's embarrassing and completely ridiculous now, but I swear it's true. I had quickly gone from skinny high school kid to Viking Berserker, all in about a dozen weeks, or so I believed.

The phenomenon is known as the Dunning-Kruger effect.[24] It is a fascinating psychological term that, when fully understood, reveals people often become overly confident in their newfound abilities. They believe themselves to be miraculously competent in whatever metric is being measured. It doesn't matter if their experience is minimal.

What happens is that a person will start off knowing very little about a topic. This means that he or she will naturally have a low confidence level in performing a task. As an example, I feel this way about brain surgery. I haven't the faintest idea of where to begin, other than to cut the poor guy's head open. Meaning I wouldn't even consider doing it for fear of killing the patient, and rightfully so.

As new information about the topic flows in, however, confidence in performing the task suddenly spikes and is soon considered "high," especially when minor successes are achieved. Some people have described this as the "Peak of Mount Stupid." This was me in basic training. Unfortunately, some never leave this mountain of stupidity and consider themselves beyond reproach. I suspect most readers have had a boss who fits into this category at least once in their lives.

Those who continue to learn and open their minds to the possibility that they know very little about the topic will soon fall into the "Valley of Despair." Their confidence levels drop pre-

---

[23] Mattis, Jim, and Bing West. *Call Sign Chaos: Learning to Lead.* eBook edition. Random House, 2019, Chapter 3.

[24] Dunning, David. "The Dunning–Kruger Effect: On Being Ignorant of One's Own Ignorance." *Advances in Experimental Social Psychology*, Vol. 44, edited by J. M. Olson and M. P. Zanna, 247–296. San Diego: Academic Press, 2011. https://doi.org/10.1016/B978-0-12-385522-0.00005-6

cipitously, even as they become more proficient with each new piece of information. The complexities and staggering tangents become overwhelming, shattering the previous illusion of mastery.

Assuming the person doesn't give up and continues to learn and practice, competence and confidence will rise, though not as quickly as at the beginning of Mount Stupid. This is referred to as the "Slope of Enlightenment." Here, most of the work is focused on achieving mastery.

Finally, once confidence and competency reach high levels of expertise, the "Plateau of Sustainability" takes over. This is where the person understands what he or she "knows" and what he or she "doesn't know." It is here that the person can become realistic in making decisions and future observations.[25]

I earned two of my college degrees later in life than most people do, in my mid-30s. It was comical to see some of my fellow students, young and inexperienced, emerge with their Bachelor of Science or Master of Science degrees, all with the attitude that "I know everything." I just laughed and thought, *I've been there before*. In that moment, I made a conscious decision not to return to the apex of Mount Stupid.

For those who actually learned something in college, and let's be honest, that number is frightfully few these days, it should be glaringly apparent that a degree only briefly introduces a topic. Even having a Master of Science degree, which sounds incredible and really important, is only scratching the very surface. Actual knowledge comes from putting those ideas to work and earning competency through experience. Degrees are only as valuable as the piece of paper they are printed on, nothing more. Remember this when some jackass proclaims his or her superiority because of that magical piece of paper they have, which was probably sprinkled with fairy dust.

I was delusional as an infantryman Zeus atop Mount Stupid, believing I was far more capable than I actually was. No one slapped me across the cheek and said, "Wake up, dumb ass." I was intoxicated by the aura of the infantry and our superiority at home and on the battlefield. Fortunately, this didn't last too long as I realized the error of my ways and promptly went tumbling down into the valley, hitting many boulders along the way. I suppose it was a blessing that the façade shattered early. We were good; there was no question about it. But the enemy could still kill us if we didn't sufficiently prepare.

---

[25] Pontefract, Dan. "The Dark Empath and Dunning-Kruger Effect on Leaders." *Forbes*, March 11, 2025. https://www.forbes.com/sites/danpontefract/2025/03/11/the-dark-empath-and-dunning-kruger-effect-on-leaders/.

# Chapter IV

## This Place is Worse Than Hell

I sat on my bunk and stared at my foot. A cast encased it from my toes all the way up to my right knee, and it smelled like a biology experiment gone wrong. My mood was as sour as the cast, and I was excelling in the art of sulking. I felt utterly drained of energy, and the only impulse I had was to kill one of my violent bunkmates. I didn't *actually* plan to do anything, but let's just say I wouldn't lift a finger if he were to spontaneously combust.

How had I fallen so far in such a short time? Somehow, I'd gone from the familiar chaos of Charlie 2-54 to the spiritual equivalent of Dante's First Circle of Hell. I was no longer with my brothers-in-arms or within the familiar confines of the company. Instead, I found myself stuck in a prison without bars, surrounded by strangers, and feeling more isolated than I ever had in my life. I was back at the 30th AG.

Like Icarus, son of Daedalus in Greek mythology, I had presumably flown too close to the sun. The wax that held those feathers of flight together had started to melt away because of its rays, though I was blissfully unaware of my impending fall from the skies. Unlike Icarus, however, I would not be swallowed by the sea. Instead, I'd be stranded and spend months aimlessly doggy paddling, desperately hoping for rescue.

Several weeks prior, we had been in the field on FTX. Our company was learning how to set up and react to ambushes, cross dangerous obstacles (e.g., barbed wire), and properly dig, occupy, and fight from foxholes and other hasty fighting positions.

The climax of the FTX was a live-fire exercise on 110-yard lanes. In forty-degree weather, we crept over a channel wall and crawled through waterlogged trenches under the cover of night, steam billowing from our mouths. Brilliant flares arced overhead to light the way while strands of concertina wire snagged our gear and helmets. The cadre fired live machine gun rounds above us and set off small dynamite charges in nearby sandbags, each blast showering us with sand. Red tracer rounds, regular bullets tipped with a small pyrotechnic charge, zipped over our heads as we crawled. It looked like a *Star Wars* laser battle and was genuinely fun, even as it left us exhausted and shivering in the cold.

We then completed our final true road march of the cycle, making our way back to Charlie 2-54. Mere weeks remained until graduation. We were so close that the drill sergeants finally let us get high-and-tight haircuts, leaving our buzz cuts in the rearview mirror. We no longer looked like cancer patients; we looked like infantrymen.

The fifteen-mile march, where we nearly had to jog to keep up, marked the culmination of months of punishment for our legs and feet. In the darkness, I followed my battle buddy up and down the winding gravel roads, guided only by the tiny glow-in-the-dark "cat eyes" on the back of his Kevlar helmet. The road through the woods was so black that I couldn't see anything moving around us at all.

The trek itself wasn't anything special; we had done it many times. That was infantry life. You marched everywhere, sometimes you ran, and you carried whatever they loaded onto your back. But it never occurred to me that simply walking with a heavy pack might injure me, but that is precisely what happened.

Within the first hour, a nagging pain started in the middle of my right foot. At first, it was just an annoying throb. I kept walking and focused on my breathing, trying to cue up songs in my head to pass the time. We were forbidden to speak or make noise to stay "tactical," so my internal soundtrack did the talking. I vividly remember John Lennon's "I'm Stepping Out" playing on endless repeat, fitting enough given all the stepping we were doing.

Several hours later, my right foot began to slap the ground with each stride. My ankle had gone stiff, and I couldn't flex it properly. The pain sharpened, and I thought I felt a popping with each contraction. Still, there wasn't anything to do but keep moving.

If I told a drill sergeant, he could throw me on the trailing pickup and send me back to the barracks, which meant humiliation and almost certain "recycling." That close to graduation, it felt like career suicide.

Soldiers who cannot complete training because of injury, poor performance, or other issues are recycled, a word that sends ice through any recruit's veins. It means starting basic training over, sometimes from the very beginning. It happens often enough, but the term is usually used as an insult, implying that someone is weak, undisciplined, or unworthy of graduation. Being a recycled soldier is treated as a mark of failure.

So, I kept my head down and marched. We were moving at a brisk four miles per hour, and I focused on my breathing and the rhythm of my steps. The goal was to find a groove, smooth out my movements, and avoid wasting energy. We all picked up little tricks, like zigzagging up steep hills, and over time, your body quietly learned to shed unnecessary motions to conserve calories.

As time passed, my foot turned into an awkward club, and I knew something was wrong. I walked with it twisted out at a strange angle and could no longer roll my ankle the way you do in a normal stride. It was stiff, and my shin felt like it was about to explode.

By the end of the fifteen miles, I was in severe pain. My uniform was several pounds heavier with sweat, yet I felt cold and shook from the temperature and the adrenaline crash. I was also increasingly worried about what was actually happening inside my leg.

I sat on the edge of my bunk and tore at my gear, dreading what I was about to see. I was convinced my foot was broken and mangled. Even fully unlaced, my boot refused to come off my swollen foot. My battle buddy, PVT Harper, watched through tired eyes and tried to help. We finally managed to force it off, but only by damaging the boot's tongue. I felt enormous relief once that leather shell was gone.

Then, a wave of tiny needles of pain surged across my entire foot as though I had fallen asleep on it hours ago. My throbbing heartbeat surged, making it feel like it was being pumped up like a basketball. Removing my sock, I found that everything was severely swollen and beet red, with small, dark bruises and abrasions on the midfoot. Dark blood under the surface of my skin, a sickly display of black and yellow, also collected around all my toes. The only positive sign was that it wasn't deformed or dislocated.

Harper and I kept quiet, hoping a few hours of sleep would fix it. Instead, I woke to find my foot even more swollen and brutally painful, especially under any pressure.

We were scheduled for a short track run, and I doubted I could make it. I forced my foot into a running shoe and yanked the laces tight to stabilize it. It hurt like hell, but I pushed through and tried to hide the injury. During the run, I felt a sickening pop with each step, and my stomach turned, but I managed to conceal my limp for one more day.

By the following evening, I could barely walk. Two older soldiers in the platoon heard about my injury, likely through Harper, who was just looking out for me. I still refused to say anything, despite the pain and lack of sleep, but they eventually convinced me to show them. I was simply fixated on finishing. In my mind, I could always see a doctor in a few weeks... after I'd earned that blue infantry cord.

**22 October 2002 – Letter to the Harstad Family**

I fractured the third metatarsal bone in my foot during the road march.

Jersey and Fahrenholtz (soldiers in my platoon) saw me limping around the bay, and they made me show them my foot on Sunday night. They looked at it and freaked. Jersey ran down, even though I told him he didn't need to, and got the drill sergeant to take me to the emergency room.

I spent all night there, from 8:30 PM until 3:45 AM. They wrapped my foot, took X-rays, put it in a little restraining boot, and gave me crutches. I'm supposed to do no PT, marching, running, or jumping for one week. The only good news is that the doctor said it wasn't too bad of a break. If I get better soon, like within two weeks, I can graduate normally.

As long as I can persuade the doctors to get me off crutches, I will do everything, whether it hurts like a bitch or not. Right now, I don't care about anything except graduating with C 2-54. The drill sergeants are surprised I made it that far with a broken foot. Crazy, the smallest guy toughed it out for over ten miles with a broken wheel!

**27 October 2002 – Letter to my brother, Mike Harstad**

I'm stuck walking around on crutches. They really suck, and I want to break them. It gets so frustrating not being able to do simple stuff, like carry your own tray in the chow hall.

It was hard to tell my parents because I felt like a failure. Only weeks earlier, I had been proud to show them my progress on Family Day, and now it all felt as if it were being swept away like a fading dream. My platoon did their best to support me, but we all knew the truth: I would be recycled and wouldn't graduate with them. I could feel it in my gut that our paths were about to part for good.

### 19 October 2002 – Letter to the Harstad Family

I just got done talking to you on the phone. Like I said, I hurt my foot.

I don't know how I did it, but about two hours into our 15-mile road march, it started to hurt on top of my foot. It continued to hurt throughout the duration of the march, but I kept going, don't ask me how. It makes me feel depressed just thinking about it.

Opie, probably my best friend here, has been a huge morale boost for me. He's been helping me along pretty well by joking and pushing me along. It's good to have great friends in the Army.

### 22 October 2002 – Letter from my father, Bill Harstad, Jr.

I do remember a bit when we saw you on the 12th; you had a slight limp on your right side, but I just took it as you maybe being sore or whatever. But again, I won't sit here and speculate about what it might be.

I couldn't participate in physical training with my foot locked in a big Aircast boot, so the platoon pushed on without me. It hurt to watch them go. But standing on the sidelines did something strange.

For the first time, I could really see the transformation that had taken place. These were no longer the disorganized, pathetic soldiers I'd met on day one. Had we really changed that much? How had I missed it? The metamorphosis had happened right under our noses.

Training had hit its crescendo, and the recruits had started acting like infantrymen. We weren't getting smoked daily anymore, with beds squared away, the latrines spotless, and drill sergeants rarely needing to bark orders. Discipline and competence had taken over.

My heart ached, but I was also proud of my brothers. All that hard work had paid off for them. Even as the difficulty spiked toward the end, little moments of humor still slipped in, and it was a relief to laugh in the middle of such a bleak stretch.

### 25 October 2002 – Letter to the Harstad Family

The captain came in today for an impromptu inspection. He walked into the latrine, and we heard, "Oh my God! I think I found a hamster!"

There wasn't anything I could do except cheer them on from the sidelines and offer a few words of encouragement. In truth, I was the one who needed support. It felt like I was slowly drowning while my friends marched on just above the surface, unaware I was beneath the water, quietly sinking out of sight.

**24 October 2002 – Letter to the Harstad Family**

I had a great conversation with DS Schneider yesterday while I was sitting on the bleachers. He came over and talked with me for a while, since I had been sitting there alone for about four hours. He basically rejuvenated my pride in the Army and my country.

Unfortunately, the boost didn't last. A follow-up appointment showed the break was worse than they'd thought, and the medical staff put me in a full plaster cast. The doctor was sympathetic but refused to sign the paperwork clearing me to return to duty.

Still, I tried every negotiation tactic I knew. I told him I would walk on a bloody stub if I had to, that I just wanted to graduate. He only shook his head and patted my shoulder. His decision was final. For a moment, I even considered forging the document, but thankfully I thought better of it.

The injury was bad enough that the Army sent me on convalescent leave. They didn't want to babysit me while I healed, so I was given orders to go home, or really just anywhere but there, and to return in a few weeks. I had mixed feelings about this. Seeing my family would be great after so long away, but it also felt like the Army was quietly tossing me aside.

With such short notice, airline tickets were too expensive, and no one volunteered to drive me to the airport. I could have asked my parents for money, but my pride staged a veto. So, I did what any broke, injured private does. I hobbled on crutches into a Greyhound station and began a multi-bus odyssey from Columbus, Georgia, to Sioux City, Iowa. The trip took more than twenty-four hours, and every bus was packed with a rolling parade of humanity that ranged from mildly eccentric to "I hope this guy's not collecting teeth." I'm convinced you haven't truly lived until you have spent several hours on a Greyhound in the middle of the night. That's the *real* crucible, not basic training.

When I got home, everyone was friendly, but I hated being seen in public on crutches. I ditched them as soon as I could and clomped around on my hard cast instead. It was less about pain and more about pride. I was back from Infantry School without actually finishing it, and that fact felt louder than anything I could say.

People asked the usual questions, and conversations drifted toward the old concerns about me being too skinny or too fragile for the infantry. In hindsight, I'm sure I was reading too much into it, but at the time, it felt like nobody was shocked that I had been recycled. It stung, and I didn't exactly rise above it.

Looking back, they probably meant well. But at the time, I took it as a challenge. My anger focused into a single point. I wanted to get back to Fort Benning and finish. Not just wanted, I *needed* to finish.

I did what I could to stay in shape, which is not easy with a cast from toes to knee. I even went pheasant hunting with my dad and brothers, shuffling through endless snowy fields with my cast wrapped in a garbage bag to keep it dry. It wasn't ideal training, but it beat sitting on the couch playing video games and feeling sorry for myself. Still, I could feel my body getting softer and knew the next cycle wouldn't cut me any slack.

Going back to Fort Benning after leave was rough because I knew exactly what I was walking into. I arrived late one evening at the barracks, which felt like a ghost town. Almost all my peers had already graduated and moved on to their units. I had missed everything. The bunks were empty. My friends were gone. Opie and Harper were now somewhere downrange with the rest of my platoon. I realized I wouldn't see them again. That chapter had closed without me.

Perhaps the only silver lining, if you could call it that, was that three soldiers from my platoon were also injured and scheduled to be recycled with me. Privates Hastings, Hansen, and Homme (pronounced like "What's up, homie?") had all suffered similar injuries. The four of us were the only ones from our platoon who didn't make it through the cycle, aside from one early departure who was medically discharged in the first week after suffering seizures.

Injuries were rampant in the platoon thanks to the rigors of Infantry School, but a surprising number of guys still managed to squeeze through. Two other privates had fractures in their feet, and another, who led the company in PT, was marching on a fractured tibia. Plenty of others hid their injuries to avoid being recycled. Toughness or preparation didn't seem to matter much; broken bones and overuse injuries hit everyone without discrimination. The four of us who were left behind simply had damage that couldn't be disguised and hidden. It was a tough break, so to speak.

DS Ross stepped in and took on a big brother role for those of us who stayed. After being smoked by him every day, it was strange to see that side of him. It felt as if he had taken off the drill sergeant hat and gone back to being a regular guy, trying to keep us optimistic about making it through a different cycle. We did odd jobs for him and deep-cleaned the barracks just to stay busy. The company commander and DS Nash would stop by now and then with their own words of encouragement. Their kindness and leadership stuck with me, especially in a place where no one expects compassion.

**29 November 2002 – Letter to the Harstad Family**

I talked awhile with Captain McCormick. He came by our bay in civvies (civilian clothes) and talked to me about my last name. He even knew there was a Harstad, Norway, and could tell me right where it was. I was amazed at how much he knew about the place. He could even pronounce our last name "Norwegian style."

We were still living at Charlie 2-54 when a new cycle started, giving us a front-row seat to watch a fresh batch of disorganized, clueless recruits from 30th AG get absolutely annihilated. A few months earlier, that had been us. CPT McCormick's only order was that we were not to talk to them; they had to figure out the game on their own. That made sense to me, and I watched with genuine interest as they endured the same brutal treatment we'd just survived.

### 30 November 2002 – Letter to the Harstad Family

It is funny watching the new recruits. They are so unorganized and confused. DS Ross snuck around the back way and came up the stairs today. He started screaming, "I caught you! Why are you sleeping?" He smoked them hardcore, and when they left, he smiled at us and said how much fun that was.

### 4 December 2002 – Letter to the Harstad Family

I go out and do PT every morning with DS Ross and DS Nash. They treat us really well and let us do whatever we can (even with light-duty orders). DS Nash was furious at the platoon last night. He told them straight up, "I hate you. In the nine years I've been in the Army, I've never seen a more worthless group of individuals ever!" He absolutely ripped them apart.

One evening, when DS Ross was alone in his office, I quietly asked, "Did you ever figure out who put that can of dip in the ceiling tile?" I jerked my thumb over my shoulder toward the spot in the bay where he had "discovered" it.

A slow grin spread across his face. He practically started glowing. "As a matter of fact, yes," he said. "I put it up there. Don't tell these guys, though. I'm going to do it to them next week."

That was it. Confirmation at last. The drill sergeants had planted the evidence. I kept my mouth shut like he asked, but I still laugh every time I think about a grown man hiding tobacco in ceiling tiles just so he could smoke a platoon.

Unfortunately, our time at Charlie 2-54 ran out, and we were sent back to the 30th AG, the armpit of the Army. It housed a platoon of recruits who couldn't pass their PT tests, called the Fitness Training Unit (FTU), and a rehabilitation platoon for soldiers scheduled to be recycled into another training company once their injuries healed. Mixed in were recruits on their way out of the Army altogether, for medical or mental health reasons.

From the moment we arrived, the four of us were stunned by what we saw in the rehab platoon. Somehow, it was worse than anything we had experienced at the 30th AG reception center. The place was full of shells of men with broken limbs, wrecked backs, and assorted training injuries. On top of that, it seemed to be the collection point for the genuinely unstable. A decent number of our new roommates qualified as full-on nutjobs. It was depressing, unsettling, and a daily challenge to stay motivated.

**Early 2003 – Letter to the Harstad Family**

One guy broke both legs in basic and is in a wheelchair. One of our guys accidentally let him go while pushing him, and he rolled down a big hill. He hit the curb at the end really hard and flew, smacking face-first into the ground. He flew about 15 feet. I laughed about it all day, but he didn't think it was funny.

Even more surprising was the state of the drill sergeants in the rehab unit. Many of them were broken, too. Some had left their units because of injuries; others, according to the rumor mill, were there because they were unstable and no one wanted them back. After spending time there, I tended to agree that this place was a dumping ground for more than a few undesirable soldiers *and* drill sergeants, though there were exceptions. It may be hard for a civilian reader to accept, but not all service members are squared away or worthy of praise. Because of this, for those trying to get back to a real training unit, 30th AG felt like a minimum-security prison.

**10 December 2002 – Letter to the Harstad Family**

It REALLY sucks here. Everyone here is violent, retarded, unwilling to work, or all three. We basically do nothing all day long. We sit, and sit, and sit, and sit, and sit. Did I mention we sit? This is the worst experience of my life. I can't wait to come home! This place is worse than Hell!

Our barracks was one big open room with rows of faded turquoise bunk beds, each covered in some kind of water-resistant material that felt designed for easy hosing after a biohazard incident. Everything was lined up, neat and drab, like a discount prison showroom. The showers rarely had hot water, allegedly due to "maintenance issues," even though no one ever seemed to maintain anything. Mushrooms literally grew out of the tile, and black mold spread across the ceiling.

As before, we were strictly forbidden from any physical activity, yet had endless hours of free time. So, I read a monster stack of books. Reading was one of the few activities not openly discouraged, at least until someone screwed up, which happened almost daily. Then the books would vanish as group punishment, because nothing says "healing environment" like taking away literature from injured soldiers.

Our fellow prisoners (sorry, "soldiers") were perpetually angry and complained about everything. Apathy and boredom ruled the place, and fist fights broke out regularly. The drill sergeants couldn't smoke us because we were technically injured. Still, I would have preferred push-ups over sitting on my bunk for hours staring at a blank wall. Occasionally, we were allowed to watch DVDs, but we only had a handful of movies, so before long, I could quote them line for line, which is its own special form of insanity.

Depressing days at the 30th AG Rehab Platoon (late 2002)

**11 December 2002 – Letter to the Harstad Family**

I swept an entire warehouse today. Private Harstad, master janitor. Not much to write about. Fights are so common that Ultimate Fighting has taken over instead of the Army. By the way, the drill sergeants here don't care. They never come by, so everyone messes around constantly. I want to leave so badly.

**6 January 2003 – Letter to the Harstad Family**

Someone got caught with a porn mag, and we lost the TV. What a shocker. Stupid retards.

One of our regular responsibilities was CQ, Charge of Quarters. In practice, that meant a lowly private got paired with a sergeant or above at the company front desk. It was an easy but mind-numbing job that ran for twenty-four hours with no sleep allowed. We cleaned, answered the phone, and made sure no one wandered in who didn't belong there. It was like being a night-shift mall cop, minus the mall and the authority.

One evening, I was wiping down the desk and talking with the NCO I was paired with. He told me he had been a private when the Persian Gulf War kicked off in the early '90s. He and his best friend were in a mechanized infantry unit, fighting from the Saudi border into Kuwait. His buddy was killed in a friendly fire incident when his Bradley Fighting Vehicle (BFV) was destroyed.

At some point, I realized he was crying as he told the story, and I had absolutely no idea what to do, so I just sat there quietly. It's terrible enough to lose a friend in combat, but I could only imagine the anger of knowing your own side did it. Then, without warning, he exploded, grabbed a box of pens, and hurled it against the wall, sending plastic missiles ricocheting around the office. I sat perfectly still and stayed quiet, partly out of respect and partly because I didn't want to be the next thing he threw. He was clearly carrying a lot, and I suspect the Army parked him at 30th AG not so much as punishment, but in the hope that someone, somewhere, might help him.

A lot of people in the unit were like that, which made us feel as though we were tiptoeing around emotional landmines all day. We were trying to rebuild our own busted bodies while living with people who could go from zero to violent breakdown without warning.

Another routine job we drew was KP, Kitchen Patrol, the Army's version of community service for the crime of existing. Nobody wanted it. We had to get up before everyone else, limp into the chow hall, and start refilling milk machines, hauling filthy trays, and mopping floors that were instantly dirty again. The grand finale was sitting on a stool in the back, peeling an endless parade of white hand-grenade potatoes into a 55-gallon drum. It was mind-numbing work, and most of the civilian staff treated us like we were dogs, barking orders as if they were drill sergeants and we were lucky to be allowed indoors.

Somehow, the worst duty of all was assisting the S1 section, the admin office. That place felt like a hostile foreign country run by bitter female soldiers on power trips. We were "loaned" to them as if we were broken equipment, then spent hours sorting medical files and personnel records, and feeding mountains of paperwork into a shredder that sounded like it was chewing through people's careers. If anything in there was confidential, they didn't seem bothered. While we did all the grunt work, they sat around gossiping in out-of-regulation uniforms, looking annoyed that we were even breathing in their workspace. Their unofficial mission seemed to be making sure we never forgot how worthless we were in their eyes.

**31 March 2025 – Reflection of Logan Hastings, 30th AG Rehab Platoon**

They made us help those ladies in administration. Honestly, they were unpleasant and seemed to go out of their way to make our lives miserable. We had to shred stacks of documents in this huge paper shredder, so we turned it into a little game. Depending on how many pieces of paper we sent through the shredder, it made a different tone. We started coming up with songs to play, like "Eye of the Tiger" by Survivor. That pissed them off, especially when we kept jamming up the machine.

To make matters worse, one soldier in the rehab platoon, PVT X (name redacted), was a constant problem for everyone. I always suspected he was malingering and that most, if not all, of his injuries were imaginary. He was a large, physically imposing man who seemed to hate anyone with a different skin color, and he reserved a special contempt for Asians.

His favorite nickname for a fellow injured soldier, PVT Nguyen, was "Ho Chi Nguyen," and he would pull the corners of his eyes back whenever they spoke. PVT X routinely argued with the drill sergeants, and I never understood why they didn't simply remove him from the unit. Instead, they left him in the rehab platoon, where he pushed everyone around and faced no real consequences. He took out his aggression on all of us and was universally hated.

At night, X would pick someone for a wrestling match, even though we were all recovering from broken bones or other injuries. He would beat his opponent into submission, leaving split lips and bloodied noses as a kind of nightly entertainment and a reminder of who sat at the top of the food chain. Built like a football linebacker, he didn't have much competition from those of us with crutches and wheelchairs, so it was never a fair fight, and he knew it. He used that advantage to pawn off his duties on whoever he disliked that day, which was pretty much everyone.

We also learned to sleep with one eye open. X liked tipping over bunk beds while people were sleeping, stealing from footlockers, and generally making life miserable. He behaved like someone who belonged in an institution, not in an open barracks full of injured recruits. He was reported repeatedly and chewed out more than once, but nothing meaningful ever changed. PVT X stayed, and we had to live with it.

The rehab unit itself was a dumping ground for the broken and the forgotten. Officially, it was where you went to mend and get back to training. In reality, it was where time stalled and motivation died. Anyone who showed up there intending to heal up and return to their unit was quickly worn down by the chaos and neglect. The easiest escape route was a medical discharge, and the Army's handling of the whole situation made no sense to me. The indifference from the command staff, mirrored by the growing apathy of many injured recruits, was hard to believe but impossible to ignore.

My three buddies from Charlie 2-54 were among those who took the exit. Privates Hansen, Hastings, and Homme all received medical discharges for injuries they had sustained at Infantry School. I couldn't blame them. Their leg injuries were far more serious than my fractured foot and would have kept them trapped at 30th AG with the lunatics for months, peeling potatoes and rotting in boredom. It was a miserable setup, and everyone knew it. Most people wanted any way out.

Their departure left me on my own and hit me especially hard because Hastings and I had become close friends. Overnight, I went from having a small tribe to being the lone holdout. Still, I understood why they made the choice they did.

That left me as the sole survivor of the original Renegade platoon. At that point, I made a simple decision: no matter what happened, I was going to finish Infantry School. I wasn't going home again without my blue infantry cord.

The only positive memory I had of the 30th AG was meeting DS Parkins, a man I eventually grew to respect and who helped me a lot. At first, though, he seemed less like a mentor and more like a future entry in my therapy notes. I had a brief but memorable encounter with him at reception in the first few days before Infantry School started. My fellow soldiers and I were standing in line, waiting to shuffle into the chow hall, when DS Parkins spotted PVT Homme. Reading the last name on Homme's BDU top, he asked him how to pronounce it. Everyone did this because of the strange name, and no one ever guessed it was pronounced like "homie."

Parkins stood there, inches away from Homme, deep in his personal space, just staring. He had this weird, almost sinister way of glaring at people, like he could see straight through your skull and rummage around for answers you definitely didn't have. Homme, visibly intimidated by the looming drill sergeant, stammered and quietly got out the correct pronunciation. The rest of us stood frozen, silently praying he would get bored and go terrorize some other poor soul.

He didn't. Parkins kept staring straight through Homme with his beady little eyes, then suddenly let out a wildly exaggerated laugh in response to the pronunciation, the kind of over-the-top cackle the Joker would use in a Batman movie. He did this for at least a full minute, inches from Homme's face, without breaking eye contact. It was incredibly awkward. Everyone felt uneasy and mildly concerned that this man might have just lost his marbles right there in the chow line.

Finally, he slowly turned, still laughing that bizarre laugh, and walked into the chow hall. We heard his cackle echo and fade as he disappeared into wherever drill sergeants go when they are not actively ruining someone's day.

So, when I later realized he was one of the drill sergeants at the rehabilitation unit, I was thoroughly concerned. To make matters worse, the soldiers there swore he was crazy and said he did all kinds of off-the-wall things. That sounded like a bright red flag telling me to stay as far away from him as possible.

**31 March 2025 – Reflection of Logan Hastings, 30th AG Rehab Platoon**

I remember DS Parkins really laid into Homme at the 30th AG reception before basic training. Homme wore these thick Coke-bottle glasses (BCGs). DS Parkins kept going at him, saying, "Homme, your glasses are so thick, I bet you could burn ants with those things! Can you see into the future?!"

After about a month at 30th AG, I pulled a routine overnight CQ. It was the usual schedule: sit at the front desk, answer the phone, keep an eye on the building, and point lost, half-panicked recruits in the right direction. That night, DS Parkins was in charge and sat right next to me.

At first, I was on edge, waiting for him to launch into one of his bizarre performances, especially with all the green recruits shuffling around, but he didn't. He talked to me for hours like a normal human being. He asked about my life, how I ended up in the 30th AG, and, for some reason, I laid it all out for him. I didn't hold back, and I expressed how dysfunctional this place was.

For whatever reason, he seemed to take a liking to me. I think he respected that I still wanted to finish basic training, rather than opting out like so many others. It had to be demoralizing watching a steady stream of soldiers decide they were done with the Army while higher leadership did almost nothing to fix the system. He was a drill sergeant, but there was only so much he could do when the real direction of the unit was decided far above his pay grade.

**31 March 2025 – Reflection of Logan Hastings, 30th AG Rehab Platoon**

DS Parkins really helped me out as I was leaving. I talked to him about how depressing it was at 30th AG, and he offered support and guidance. My memory is that he did this for several people. It was my impression that he didn't want to be there any more than we did.

DS Parkins asked me what my goal was, and I told him I wanted to get back to Charlie 2-54, hopefully into 2nd Platoon again. He just nodded and moved on, keeping up the façade of a loose cannon. The next day, I was in a training program to get back into shape, and I'm pretty sure he was bending more than a few Army medical guidelines to make it happen.

I was ecstatic about working out again. My foot no longer hurt, and he and I regularly went to the cadre-only gym. While I sweated on the machines, he quietly pulled strings with my old company. Without telling me, he arranged for Charlie 2-54 to take me back into the same platoon, with the same drill sergeants, assuming I could pass their reintegration standards. He kept that part tucked away until he was sure I was ready.

One day, I was running on the treadmill when a random sergeant walked in, saw me, and lost his mind about a trainee being in the NCO gym. He didn't believe I had permission. After he smoked me and called me every name in the book, I went to Parkins and explained what happened. Parkins just laughed, scribbled something on a scrap of paper, and told me to go right back.

When I walked through the gym door again, the same sergeant nearly exploded. I handed him the note as if it were a peace treaty. He snatched it, read it, and his face tightened with anger. It said, "Leave the kid alone. He's trying to return to his company." It was signed by DS Parkins and added that if there were any problems, the sergeant could come find him. Nobody ever said a word to me about the gym again. Parkins had basically handed me diplomatic immunity on a Post-it.

Under his direction, I ended up in charge of the FTU, a herd (pun intended) of overweight or out-of-shape recruits who couldn't pass the initial, easy PT test just to get in. Some of them, as unbelievable as it sounds, couldn't run for more than a minute at a time. A few were pushing a hundred pounds over what anyone would consider a reasonable fighting weight. Others had mystery ailments or conditions that made you wonder how the Army had ever cleared them in the first place.

I led their PT sessions and tried to give them a small taste of Infantry School and what to expect. At first, I did what had been done to me. I ran them, smoked them, and pushed them the way our drill sergeants had pushed us. The problem was that even when I thought I was going easy, I was smoking them into the ground. The drill sergeants eventually pulled me aside and told me to dial it back because they were genuinely worried someone might have a heart attack on my watch.

It felt strange leading anyone when I hadn't even graduated yet, and stranger still to be told I was too aggressive by the same kind of people who once treated me like a chew toy. Still, once I adjusted my approach, it gave me a real sense of purpose. I was no longer just killing time at 30th AG. I was getting my first real taste of leadership, trying to drag this motley group a little closer to the standard instead of watching them fall further behind.

At some point in all this, Parkins pinned my new PV2 (E-2) rank on my collar. It was just a slight bump in pay, and it was still very much the bottom of the food chain. However, he turned it into a little ceremony instead of treating it like regular paperwork. I could tell he genuinely wanted me to succeed and was doing everything he could to push me forward.

When I wasn't leading PT or hiding in the barracks with a book while PVT X made life miserable, Parkins had me driving the company van and running errands. He sent me to driver's training so I could get my Army license. Most of my runs involved shuttling injured soldiers to appointments and making fast-food runs for the cadre. It didn't seem like much, yet it felt good to do something useful instead of peeling potatoes or feeding paper into a shredder.

Some days, he took me out to places like Eagle's Tower, the rifle range, or the obstacle course. A different company would be training there, and Parkins would get permission for me to tag along, observe, and help where I could. I mostly cleaned weapons and watched, but it kept the drills fresh in my mind so I wouldn't return to Infantry School completely rusty.

In all seriousness, Parkins deserves most of the credit for getting me out of the 30th AG. He saw me at my lowest, and instead of writing me off, he decided I was worth the effort. After timing my runs and watching my PT scores, he finally decided I was ready to go back. About a week before it happened, he told me I was returning to Charlie 2-54. I was stunned. I had no idea if I'd ever get back, let alone to my original platoon.

On the day he drove me over, he asked me to write to him when I knew my graduation date. He wanted to be there. The least I could do was promise that I would.

I was so grateful I could barely keep from crying, and I told him thank you for helping me. He raised a hand to stop me and said, "Just get in there and finish it." That was it. So, I climbed out and made a beeline back to my home.

As I walked into my old barracks, I saw him outside doing his trademark stare-down routine with a group of privates who were painting thousands of little rocks blue. He was in full performance mode, looking half insane and scaring them into standing a little straighter. I had to bite back a laugh. They didn't know him like I did. I knew now that it was mostly an act, a masterful art form of controlled intimidation.

All I knew was that it felt good to be home. I had been given a second chance, thanks mainly to one drill sergeant who decided not to give up on me. Now I just needed to finish what I had started.

# Reflections

I hesitated to write this chapter. It covers the most depressing stretch of the book, at least in terms of sheer duration. I even skipped it for a while and thought about cutting it entirely, pretending those months in the rehab platoon never happened. The problem is that it would leave a several-month hole in the story, and anyone paying attention would wonder where I disappeared to and why I came back looking more bitter and sleep deprived.

Basic training misery is easy enough to explain. The disaster that was the 30th AG is not. It was less a unit and more a slow-moving train wreck that somehow got a mailing address.

Even with everything I have written about it, I still cannot fully capture the apathy and dysfunction there. It was a black eye on the Army, and I include it here for a straightforward reason. It happened, I watched it, and I lived in it. And it still does not make sense.

Later in life, as I mentioned in the Introduction, I became a police officer and spent plenty of time in our local jail and in a nearby mental health facility. The rehab unit at the 30th AG was almost identical to those places. We were locked down, forced to mingle with people who had no business in uniform, and were treated like we were somewhere between inmates and lab rats. There were no bars on the windows, but the threats of Article 15s, harsher punishments, or simply being left to rot at the 30th AG indefinitely did the job just fine.

Seen from leadership's perspective, there was at least a partial explanation for why things went so wrong. The rehab platoon, FTU, and reception center were full of people who were lazy, malingering, or clearly unstable. Some absolutely needed to be shown the door. That part made sense to all of us.

What didn't make sense was how the system handled everyone else caught in that dragnet. Mixed in with the chaos was a sizeable group of men who simply wanted to heal up and get back to their units. Instead of separating the poison from the medicine, the Army leadership tossed us all into one big pot and turned up the heat. The solution was a blanket policy. Everyone was treated as a troublemaker, a lost cause, or a future paperwork problem.

Only a few people, like DS Parkins, bothered to see the difference. He stood between us and the system and tried, in his own strange way, to pull a few of us back from the edge. Leadership at higher levels, on the other hand, turned that place into a one-way jagged pipeline out of the Army. I cannot imagine any responsible leader looking at the results and calling it a success. It was a monumental failure, and I doubt it was unique to the 30th AG. If you feel like ruining your day, search for "30th AG Fort Benning" and enjoy the parade of horror stories. Even worse, I suspect there are many more units just like it. Meaning, that dysfunction likely isn't just a Benning problem.

All this feeds into a larger truth about the military itself. It's a fascinating place and really is a cross-section of America. You get every kind of person, from future CEOs to guys who are one bad decision away from starring on a true crime documentary. Since the early 1990s, it has been encouraging to see civilians show strong support for veterans, especially after the grotesque treat ment Vietnam vets received. But sometimes that support turns into blanket hero worship, and that's where things get a little distorted.

To an outsider, veterans can sometimes look like superheroes. The popular image is of noble warriors who selflessly sacrifice everything to protect freedom and democracy, preferably in slow motion, with an American flag waving in the background. That's true for *some* people. I have known those men and women.

For others, the reality is less inspiring. Some joined for reasons unrelated to service or sacrifice. A few used the military as training to become better criminals, victimizing civilians and fellow veterans during and after their time in uniform. Others were slugs or classic Blue Falcons, contributing nothing beyond a body count on a roster and a cautionary tale at safety briefings.

The point is not to tear down veterans as a whole. It is to be honest about what places like the 30th AG produced and what they revealed. Not everyone in uniform is a hero, and not everyone who struggled there was a lost cause. The tragedy is that the system often treated them the same.

As such, I strongly disagree with former Minnesota Governor Jesse Ventura, a Navy veteran, when he said in an interview with Dan Abrams, "You don't criticize anybody about their service. If they were honorably discharged, they did the job they were asked to do."[26] This is a supremely hyperbolic statement and doesn't hold up in the real world. He should know better than to say something like that. I still enjoy Mr. Ventura's books and his pro-wrestling persona, and I understand why he said what he did: it plays well in politics. But that doesn't make it true, and it isn't an excuse.

Blanket statements like that help no one. Veterans are not a single, uniform block of moral excellence. Each one of us has a record, a personality, and a history that should be judged on its own merits. An honorable discharge is a piece of administrative paperwork, not a certificate of sainthood. Nor does wearing a specific tab, trident, rocker, or branch insignia automatically mean someone is squared away. Every community, no matter how elite it claims to be, has its share of people who shouldn't be there.

Having witnessed this firsthand, I know of some former colleagues who received honorable discharges and who were, frankly, awful human beings and awful soldiers. Take your pick of adjectives: lazy, incompetent, untrustworthy, unmotivated, thieving, narcissistic. Plenty of them slipped cleanly through the sieve that should have caught them. So, when we automatically assign honor or heroism to everyone with "honorable" stamped on their paperwork, we don't just flatter the unworthy. We also cheapen the sacrifices of those who actually earned the respect.

The reverse problem exists, too. I served alongside several outstanding soldiers who received less-than-honorable discharges for failing drug tests, most often for marijuana. Some of them were tremendous in the field, in combat, and in how they treated the people around them. Of course, none of this excuses their behavior. They made a serious error by taking drugs during off-

---

[26] NewsNation. "I Want a Woman President Before I Die: Jesse Ventura | Dan Abrams Live," 3:02, August 12, 2024. https://www.youtube.com/watch?v=D97U0_AYlWA.

duty hours and ultimately paid the price. But, while I have stayed away from illegal substances my entire life, and I believe discipline and consequences matter, it still turns my stomach to know that some of those soldiers carry a permanent black mark on their records, while a few "honorably discharged heroes" I mentioned earlier skate by untarnished.

All of this is why I'm wary of treating titles, veteran status, or discharge ratings as moral shortcuts. "Veteran" means we signed up and served in some capacity. Sometimes it means we went to a combat zone, sometimes it doesn't. The definition is fuzzy, and the reality behind it is even messier. What matters is the *person*, not the stamp on their DD-214.

At the 30th AG, I met soldiers whose behavior would horrify most civilians. PVT X was a prime example. Once I left, I never found out what happened to him, though I suspect he was eventually discharged before finishing training. I wouldn't be shocked if his paperwork ended up saying "honorable," because the military can be surprisingly casual with administrative details when it's convenient. Again, the label tells you very little. The person tells you everything.

Every so often, a kind stranger has offered to buy my groceries just because I had an Army T-shirt on in the supermarket, especially in the mid-2000s. I always politely declined and thanked them for the gesture. Part of me wants to launch into everything I have written here, but the checkout line is not the place for a dissertation on veteran mythology. Instead, I smile and silently hope they never cross paths with one of the professional "veteran heroes" who make a living milking unearned gratitude.

To some readers, all of this may sound harsh or cynical. For most veterans, though, it is just common knowledge. I'm not asking anyone to stop being generous toward the military community. Most of the men and women I served with truly deserve every bit of respect they receive. I'm simply suggesting that people pay attention.

The best advice I can offer is to trust your instincts. If something feels off about a story, a claim, or a person, it's probably bogus. No one is obligated to hand over money, favors, or special treatment just because someone mentions military service. Treat veterans the way you would treat anyone else. Be kind, be respectful, and if something doesn't add up, feel free to keep your wallet and your hero worship firmly in your pocket.

Additionally, I have noticed an odd inflation of self-importance among some veterans across various communities. During my time working as an Alcohol Safety Action Program (ASAP) officer, in which I was responsible for apprehending alcohol- and drug-impaired drivers, I arrested at least a dozen veterans who self-identified as previously being in the Navy. As a follow-up, I almost always received the same answer when asking what they did in the service. It became so comical that I kept a tally sheet in our office.

The most popular response was, "I was a SEAL," usually spoken in the hushed tones of a secret agent on a mission, trying to save the world from impending doom.

"Man, you guys must hire nothing but criminals," I would reply. "You're the third SEAL I've arrested this year!" Apparently, SEAL Teams must recruit almost exclusively from Sioux City's drunk driving population.

Of course, I knew they were lying. I always had a bit of fun poking holes in their stories, though. The problem I have with it is that many people probably bought the snake oil these fake SEALs were selling. Perhaps they said it in the hope of being released before being criminally charged. They just ran into the wrong veteran to try their bullshit on.

I have never understood this weird inflation of self-importance, but it's a sadly familiar story. I mentioned this earlier, and it is worth saying again. Everyone who serves, from frontline combat troops to the quiet admin clerk in the back office, matters to the overall effort. The machine doesn't run without all the unglamorous pieces.

Instead of padding their résumés with imaginary battle stories, these folks should own what they actually did and wear it as a badge of honor. True, movies are not usually made about cooks, unless Steven Seagal is involved, but they're the ones who fed us and kept us moving. Take them away, and the whole operation falls apart in about two days.

According to a 2023 Pew Research Center analysis of Census data, just over 6% of U.S. adults are veterans.[27] Only approximately 60% of those who join the military ever deploy, and of those, only about 10–20% serve in a combat zone.[28] Further siphoning that figure down, a mere 10% of veterans can claim to have been in combat.[29] These numbers are approximate and vary by era, branch, and data source.

"Combat" itself is a slippery word. It can mean being shot in the head, or it can mean being on a large base where a grenade goes off somewhere on the perimeter hundreds of meters away. Both situations are technically "combat," but they're not remotely the same experience. When people talk about having "seen combat," they may be referring to very different realities.

The broader point is that only a tiny fraction of troops see sustained, close-up combat. An even smaller number spend most of their time actively engaging the enemy as their primary job. It is a genuinely small slice of an already small population.

To my knowledge, no one has formally applied a Pareto analysis to combat within the U.S. military, but I suspect the pattern would look familiar. The Pareto Principle, or "80/20 rule," is the idea that roughly 20% of a given group is responsible for about 80% of the outcome.[30] It shows up everywhere in nature and human activity. In this context, it isn't hard to imagine that a relatively small share of combat-arms soldiers accounts for a disproportionate share of actual fighting, while many others can truthfully say they were "there" without being in the thick of it as often.

---

[27] Schaeffer, Katherine. "The Changing Face of America's Veteran Population." *Pew Research Center*, November 8, 2023. https://www.pewresearch.org/short-reads/2023/11/08/the-changing-face-of-americas-veteran-population/.

[28] Midwest Disability LLC. "What Percentage of Soldiers See Combat?" Midwest Disability, December 13, 2019. https://www.midwestdisability.com/blog/2019/12/what-percentage-of-soldiers-see-combat/.

[29] Bledsoe, Everett. "Answering: What Percentage of the Military Sees Combat?" The Soldiers Project, January 29, 2025. https://www.thesoldiersproject.org/what-percentage-of-the-military-sees-combat/.

[30] Guy-Evans, Olivia. "Pareto Principle (the 80-20 Rule): Examples & More." Simply Psychology, September 21, 2023. https://www.simplypsychology.org/pareto-principle.html.

Now, I'm not claiming an exact 80/20 statistical match. I'm simply using the Pareto Principle heuristically, as a rough way to describe how unevenly combat exposure is distributed. I also realize this drifts into percentages and power-law curves, and some of you are rolling your eyes and saying, "Get back to the story Jeff." But the bottom line is simple. Very few people fight. Almost everyone else supports those who do. That's part of why it was so funny to run into an endless stream of supposed SEAL Team operators getting arrested for drunk driving in Sioux City, Iowa. *How far are we from the ocean again?*

Of course, this is a joke, but it illustrates that we must use common sense when talking to veterans. They are people, too. Most are good, upstanding citizens, but some will lie, cheat, and steal to obtain unearned benefits. It's *your* job to watch out for them.

My final red flag word of caution is to watch out for the following statements: "My official job in the military was (fill in the blank), but I was basically infantry." An even funnier one is, "I was in Special Forces, but I can't talk about it or show you proof." No, you weren't. Those lines are absurd, but they still work on people who don't know better.

People who earn their place in combat arms and special operations are usually very clear about what they did and proud of the path they took. A legitimately earned blue cord, green beret, or SEAL Trident represents training and hardship that exceed the basic pipeline. Now, while I'm proud to have been an infantryman, I never went to Special Forces selection or Ranger School, so I don't pretend otherwise. I was a lowly grunt, nothing more. Their berets represent an entirely different level of accomplishment on the battlefield, and I respect them for that. Trying to borrow someone else's honors is, to use the technical term, a dick move.

So, don't assume unearned honors for yourself, and don't let others slide by unchallenged when they spin tall tales in your presence. You don't have to be confrontational, but you also don't have to nod along politely when your instincts are screaming at you.

The final topic I wish to highlight is the lasting impact of DS Parkins. I've wondered why he came along when he did, helping push me in the right direction as I went through the struggles of the 30th AG. I believe it's exactly like the famous saying: "When the student is ready, the teacher will appear. When the student is *truly* ready, the teacher will disappear." This has been attributed to many sources, including the Buddha and the Chinese philosopher Lao Tzu. Perhaps it is a universal truth that manifests everywhere, regardless of who originally wrote it.

Whatever its origin, it fits. When I was flailing, Parkins stepped in, pushed, coached, and occasionally dragged me forward. He didn't coddle me, but he threw me a life raft before I slipped under. Finding people like that and gravitating toward them in hard times is crucial. The real trick in life is spotting the ones who are actually reaching out a hand versus the ones who are just along for the ride.

Then, he disappeared. I eventually lost contact with him after leaving Fort Benning, and I still don't know how his life or career turned out. I don't even know his first name. Every so often, though, I think about him and hope he's doing well. That is another truth about military life. You cross paths with remarkable people, your lives overlap for a brief and intense stretch, and then you scatter. The connection is temporary, but the impact sticks around for a lifetime.

# Chapter V

## Turning Blue

My drill sergeants greeted me with muted smiles that never broke their stern façade as I walked back into the barracks bay, bags in hand, rejoining 2nd Platoon. I suspected they had assumed I would be swallowed for good by the black hole of the 30th AG rehab unit. But that fate wasn't meant for me. I was the only one who made it out of there.

Instead of starting over at the very beginning of basic training, which would have been brutal, I was allowed to jump back into the cycle. I had to repeat a few of the weeks and some of the training I had already completed, but that didn't bother me. I was just happy to be back and relieved I didn't have to start from day one. In fact, I welcomed the chance to brush up on my skills before our final FTX, when everything we had learned would be put to the test. My time at the 30th AG felt like wasted weeks, and it left me feeling unprepared. Fortunately, DS Parkins had taken me along to those training events during my rehab stay, which made the transition back to C 2-54 a little smoother.

As anyone who has spent time in a gym can tell you, training on a treadmill indoors is very different from moving out in the field. The pounding of the road and gravel on my feet hurt, and I was quietly terrified of injuring my foot again. Months of idleness had taken a noticeable toll on my conditioning.

Even so, I settled back into a groove quickly and threw myself into the work, determined to catch up and improve. As an unexpected bonus, none of the drill sergeants gave me any extra grief. Even DS Whiteley left me alone, though he was still on a rampage with everyone else. I still got smoked whenever the platoon messed up, as expected, but they never singled me out or even raised their voices at me. I never knew exactly why, but I assumed it was because I had come back when no one else did.

**27 March 2025 – Reflection of Toby Glass, 3rd Platoon, C 2-54**

DS Whiteley picked on everyone. As stupid as it was, he put his finger over the "GL" on my uniform, leaving only "ASS" for my last name, and then said a smart comment about it. He did little things like this to everyone, just to keep up constant levels of harassment and stress.

**I beat a former Marine during one of our grappling sessions. Whiteley was so impressed, and still riding the whole "ASS" joke, he shook his head and said, "That's a good piece of ass right there!"**

One evening back at the barracks, I sat on my bunk writing a letter to my family. Across the open bay, a group of guys were wrestling around, burning off energy like caged animals. Then I heard a heavy thump, a muffled scream, and a confused rumble of voices from their direction.

I looked up from my letter and saw one of the recruits holding his hand out in front of his face. For a second, my brain refused to cooperate. He just stood there, pale and glassy-eyed, staring at his own hand. A small semi-circle of soldiers had formed around him, all frozen in place, saying nothing. His ring finger looked white. Not "bad circulation" white, but white as bone. Because it *was* bone.

His name escapes me, but he jumped onto the back of another soldier during their little wrestling match. His buddy, though not trying to hurt him, reacted on instinct and flipped him over his shoulders. As the recruit went airborne, he reached out to catch himself. The ring on his finger snagged perfectly on the metal frame of the bunk. His whole body weight followed. The ring held, but his skin didn't.

The result was ring avulsion, a lovely little phenomenon where the skin, veins, fat, fingernail, muscle, and everything else gets peeled off the finger like a banana, and the bone is left standing there on its own. The rest of his finger lay crumpled next to the bunk like a discarded sausage casing. From where I sat, there wasn't even much blood, which somehow made it worse.

Someone sprinted off to find a drill sergeant, while we helped our battle buddy into a pickup truck and loaded him up, finger parts and all. He was clearly in shock, staring straight ahead and barely speaking. His finger rode along in a separate bag, but the hospital staff couldn't reattach it. The doctors amputated what was left of the skeletal stub while we sat in the bay debating whether he would be recycled or medically discharged. It was a bad time to lose a finger. We were so close to graduating.

To our surprise, he returned to the barracks a day or two later, one finger short but in bizarrely good spirits. Painkillers probably helped, but he was joking with us as we tried to guess what they had done with the missing finger. In classic infantry fashion, we decided it was perfectly good meat going to waste.

Someone suggested preserving it in formaldehyde and using it as a training aid for future cycles, with a permanent warning label against barracks wrestling. We pictured a drill sergeant roaring, "This is what happens when you wrestle in the barracks," and then whipping a mummified finger out of his pocket, waving it in front of horrified recruits like some grotesque show-and-tell item.

Aside from that isolated incident, everything else went surprisingly smoothly. One of the last remaining requirements was Eagle Run, a 5-mile brigade run that had to be completed in 45 minutes. The pace itself wasn't difficult. It was basically a jog, and running five miles wasn't a big deal. The real problem was the sheer number of bodies packed into a single giant formation, creating a massive slinky effect.

Our company was somewhere in the middle of the pack, which was one of the worst places to be. We would slow to a shuffle or even run in place for no apparent reason. Then, out of nowhere, a huge gap opened in front of us, and we would have to sprint like lunatics for a minute just to catch up. It was less like a steady run and more like interval training designed by a sadist with a stopwatch. Awkward or not, we got it done.

I also passed my final PT test, the last box I needed to check before the week-long FTX. I wasn't a PT stud and didn't blow anyone away, but I posted solid scores in every event. That alone felt like a victory. I had been most worried about it after spending several months mostly idle, watching my fitness leak away in slow motion.

Earlier in the training cycle, we had been routinely hauled around in dreaded "cattle cars," and nobody missed them when they started to disappear. In Army basic training, cattle cars are long, enclosed metal trailers attached to trucks used to transport large groups of recruits to secluded ranges outside marching distance. They look exactly like what they are named after: metal boxes built for livestock. Only the Army swaps cows for privates.

There were no benches inside, at least none that were operable. We had to stand packed shoulder to shoulder, bodies and rucksacks mashed together so tightly it felt like we were shrink-wrapped. Every bump on the rutted back roads of Fort Benning sent a fresh ripple of pain through knees, backs, kidneys, and overloaded shoulders. The air inside was hot, stale, and barely moving, and guys sometimes started to hyperventilate from being squeezed in so hard. Somewhere on the other side of the world, Gulag guards probably used similar contraptions for prisoner transport. It was a dark comparison, but the nightmare of cattle cars seemed to be universal.

By the time we reached the end of the training cycle, those cattle cars were mostly gone from our lives because we were marching everywhere instead. As crazy as it sounds, I preferred walking. I would rather carry the load and go on foot than be jammed into one of those mobile torture crates again. I'm also reasonably sure the drivers took pride in hitting every pothole on post. I can't prove it, but it certainly felt like part of their approved training plan.

The final FTX, known as The Bayonet, was designed to replicate combat conditions and allowed us to practice our newly acquired skills. We had to march everywhere tactically, sleep in the woods, conduct patrols, dig foxholes, set up ambushes, and operate like we were in actual combat. Difficult and strenuous drills were commonplace, such as when certain platoon members were designated as "casualties." We then had to carry all their equipment, in addition to our own, while also evacuating them on stretchers through the woods. These team drills were designed to exhaust everyone.

On the final day of the FTX, our parents and family members, following instructions sent earlier in letters, illuminated a blue light on their front porches all evening. The intent was to send us positive "vibes." If we made it, we would officially be infantrymen. If not, there was always a subsequent cycle.

We had to march 25 miles with rucksacks and heavy gear while occasionally stopping to conduct raids or other drills. The road march was broken into smaller pieces but was even more difficult and exhausting than the 15-mile nonstop trek. We would drop our packs at a designated spot during the raids and then conduct the operations. Once complete, we would run back to our gear and continue the march out of there. We hadn't slept in nearly two days, which was mentally and physically punishing.

Marching in the middle of the night along a darkened road, I was so exhausted I literally fell asleep on my feet. While I was trudging forward in that half-dream state, a drill sergeant threw a mortar simulator into the middle of the road, and I never even registered it. These training grenades emit a whistle, then explode with a sharp blast and a blinding flash. The next thing I remember, I was in the ditch, hugging the ground and pulling security. My body had already reacted before my brain woke up. We then sprinted down the road to break contact and outrun any follow-on attacks. As strange as it sounds, it was reassuring. Even half asleep, the training was taking over, and all those hours of drills were finally sinking in.

A wave of emotions washed over me as we finally marched out of the woods, with Sand Hill property and urban terrain surrounding me. We all experienced a surge of adrenaline as we approached the barracks. As was customary, other soldiers' companies were standing along the roadway for the final stretch. All the recruits were clapping and cheering us on, motivating us to finish out strongly. Though we were all beaten up and exhausted, our backs suddenly straightened and our pace quickened. It was one of the greatest moments of my life. All those months of difficult training, punishment, and time away from home were coming to an end.

We were allowed to sleep and recover the next day, though a cleanup process was immediately necessary to prepare for our "Turning Blue" ceremony. This is where we were officially christened as infantrymen, with the blue infantry cord affixed to our uniforms. Of course, this meant we had to deep-clean everything to a level that would make a cleanroom technician envious. Every inch had to be spotless.

**30 March 2025 – Reflection of Toby Glass, 3rd Platoon, C 2-54**

I was working on Battalion Staff Duty one evening with a few other privates and DS Schneider. He kept telling us, "Guys, the general is coming! This place needs to be spotless!" So, the other privates and I cleaned like mad, scrubbing everything in sight to get it perfect for this mysterious general.

About 45 minutes later, a short Asian guy in civilian clothes walked through the front door carrying a paper sack. DS Schneider bellowed, "Battalion, attention!" We snapped to, standing as straight and rigid as fence posts, hearts pounding, waiting to impress the high-ranking officer.

The man just looked confused and held out the bag. Without missing a beat, DS Schneider snatched it from him and shouted, "General Tso, sir!" The delivery driver shook his head and walked away.

He had us convinced some important general was on his way, not a bag of General Tso's chicken for dinner.

My family drove all the way from Iowa to Georgia for the ceremony. My three brothers, Dan, Mike, and Brad, were excited to see what a military base looked like. None of them had ever been to Georgia or anywhere near a military post before, so everything about it was new. It was all especially thrilling for my youngest brother.

**27 October 2002 – Letter to my brother, Brad Harstad (then 9 years old)**

By the way, yes, there are Pokémon cards down here in Georgia. The only thing is that they are guarded by four Army Rangers. You have to beat all four of them up in order to have the cards.

With CPT McCormick, who always led the way

After a short, rousing speech from the command staff and cadre, my dad had the honor of presenting my blue cord. I stood motionless as he looped it through the epaulet on my right shoulder. We all cried. In that moment, it felt like my adult life was finally headed in the right direction, and all I wanted was to make my parents proud.

I cannot speak for my family, but for me, the greatest source of pride was simple. I didn't quit. There were so many moments when it would have been easier to give up, to let the injury and the setbacks win. But I didn't. I made it.

True to his word, DS Parkins arrived before the ceremony began. He brought several soldiers from the 30th AG rehab platoon with him, and they congratulated me on finishing. I remember one of them in a wheelchair, the same soldier who had gone airborne down the hill in the previous chapter, drawing worried stares from the families gathered around us. I shook Parkins' hand and tried to tell him how much his support during those dark months had meant to me, but the words came out smaller than the gratitude I felt. It was a wonderful day and one I will never forget. Unfortunately, it was also the last time I ever saw him. He drove away, and our paths never crossed again.

**Shaking my dad's hand as a brand-new infantryman, with DS Parkins watching**

After the ceremony, everything felt strangely off balance. We were no longer treated like complete garbage. I don't mean we were suddenly treated like royalty, but the open hostility disappeared almost overnight. For months, we had been ordered to stand inches away from the man in front of us in the chow line, "nut to butt," mechanically shuffling forward in lockstep. As we received food from the lunch ladies, we moved sideways along the line, never actually walking forward until it was time to head to a table. It was all very mechanical, and it became second nature.

Once we received our infantry cords, though, that process changed, and it felt almost wrong. It was like skipping school while knowing your classmates were stuck taking a test. There was this lingering sense that we were about to get caught, even though we weren't doing anything wrong. We now stood in line like regular soldiers. We were officially no longer recruits and were expected to carry ourselves like professionals, which included walking normally in the chow line.

On March 6, 2003, I proudly stood beside CPT McCormick, DS Nash, and DS Ross as we posed for photographs together. Like DS Parkins, those men were a major reason I made it through. As a gesture of recognition, they presented me with a pair of green jungle boots they had purchased specifically for me, simply because I had come back to Charlie 2-54 and finished what I started. Holding those boots, my diploma, and the blue infantry cord was an overwhelming feeling of accomplishment.

**With DS Ross: hiding chewing tobacco better than we hid in the field**

I went around the company and thanked each drill sergeant, something I'm sure they didn't hear very often. I was genuinely grateful for everything they had done and all they had taught me.

Even DS Whiteley, the toughest and most demanding of them all, was included in my thanks. Every one of them had a hand in making me a better soldier.

**With DS Nash, who turned push-ups into an Olympic sport**

**30 March 2025 – Reflection of Toby Glass, 3rd Platoon, C 2-54**

The infantry is a place that either makes you or breaks you. You find out who you are and what you are capable of, fast. It demands things you didn't even think your body could do. It can be miserable, but with the right mindset, it becomes a good misery, the kind that leaves you stronger.

If you decide it isn't going to beat you, you can always take one more step, and then one more after that. Somewhere along the way, you realize you can push through almost anything.

Everyone knows the physical demands of being a grunt, but the mental demands outweigh them by far. If fear, pain, or the weight of authority gets into your head, you will fall hard.

In basic training, plenty of soldiers were miserable, and some tried to hurt themselves just to get out. With at least one soldier, I had to talk him back from suicide. But if you make it through, something shifts. You carry a quiet confidence afterward, like life can throw its worst at you and you will still stand. And when you survive that with your buddies, you build a permanent bond .

Jeff Harstad, the author of this book, has become the most trustworthy and supportive friend I could ask for. We survived it together. We still fight life's battles together. No matter what comes our way, neither of us has to go through it alone. That's what comes from being in the infantry.

My only regret was that my original platoon couldn't be there to celebrate with me. I missed Harper, Opie, Hastings, and the rest of the guys. Still, I finally had my prize, that blue cord. I was a grunt.

# Reflections

**Graduation Day with the Harstad family**

**Photo by Visual Touch Photography**

Infantry School was a defining chapter in my life that revealed I was capable of far more than I previously believed. I had plenty of assistance from my family, drill sergeants, the commander, and my platoon. But in the end, I had to be the one to cross the finish line. They couldn't do it for me. I had to be willing to endure the suffering of growth.

When I was a young child, I suffered a terrible injury and was left partially paralyzed. Physical therapy and the constant support from my parents brought me back to a mostly everyday existence, though I had to relearn many simple actions that children perform without thought. My dexterity was a mess, certain muscles didn't develop symmetrically, I walked a bit funny, and one of my fingers didn't work (it still doesn't). Lingering but minor effects of the paralysis continue to this day, though I have carefully crafted workarounds for everything.

Only my closest friends knew about this injury. I suspect this information will shock many of my previous co-workers and acquaintances. I hid my injury because I was embarrassed about it and didn't wish to be treated differently. I also didn't want it to define who I was. The paralysis was simply the hand of cards I was dealt, but it was up to me to play them to the best of my ability. Given the circumstances, I believe I've done quite well.

Why do I mention this at the end of Part I? Would it have been better to lay this information at the feet of the reader before Infantry School began? I don't believe so, because it would have set the message that it was ok to fail, or worse, not even try, and blame it all on my previous injury. There would always be an escape route should things not go my way, at least in the eyes of the reader. This is *not* how I think, then or now. To me, there were no excuses.

Moreover, I hope that this message inspires a young man or woman somewhere to become more than they previously believed possible. It can be done. It won't be easy, though, but that is what makes it worthwhile.

I don't know of anyone else in basic training who overcame something quite like this, though there are surely some out there with similar stories. There are likely better ones, too. We will probably just never know about them. In my opinion, those fighting true adversity don't need to speak of it. Instead, they adapt and overcome, not whine. This ties into my belief that it was best for my condition to be unknown in the story until now.

This invariably leads to the question, "How did the Army permit you to join in the first place?" When speaking with my recruiters, I was honest and explained that I had suffered some serious setbacks as a kid, but remained confident in my ability to work. I could make it, given the opportunity. They were concerned and even suggested I rethink my plans to join the Army.

In response, I immediately dropped to the ground and started doing push-ups, then demonstrated that I could run without falling over. Earning their confidence by regularly joining them for early-morning PT when I could have been sleeping, they offered me some advice for my medical examination.

"Tell the truth, but don't volunteer any information unless specifically asked." I did precisely that.

Should the medical staff have rejected me during in-processing? Perhaps. I like to think that they saw my passion for joining the military and decided to allow me through the doors. This may be just a fantasy, but I like to think that it was the story. More likely, they were probably just overwhelmed by the number of recruits and overlooked it. Or maybe they legitimately didn't consider it to be a problem. I don't know. Regardless of the reason, I passed and was determined to do my best, even with those tribulations.

Before I left for training, my dad offered words of encouragement. He said, "You probably aren't going to be the fastest, the strongest, or the best. That's fine. You have something more important: work ethic. Don't ever let anyone outwork you. Ever. Work harder than everyone else, and don't quit." I heeded his advice.

The mountains we all must climb in life are often seemingly insurmountable from the base. It is here that the separation takes place, the wheat from the chaff. Those determined will begin the trek to the summit, while most will turn back and give in before they even try.

For those willing to go on the adventure of a lifetime, though, a curious thing happens along the way. When we briefly stop to look around, we are momentarily amazed at just how far we have come. It is the path itself that is the true reward. It is the reward of progress.

Infantry School was my forge. I arrived at Sand Hill chasing glory and a war before an injury knocked me off the path. But I left with something less glamorous and far more valuable: the knowledge that I could endure.

Find a mountain and climb it.

# Part II

## Rock of the Marne

---

Certainly, there is no hunting like the hunting of man and those who have hunted armed men long enough and liked it, never really care for anything else thereafter.

Ernest Hemingway
"On the Blue Water: A Gulf Stream Letter," *Esquire* (1936)

---

# Chapter VI

## Out of the Frying Pan, Into the Fire

An infantryman from 3rd Platoon named Private First Class (PFC) Kuntz and I boarded a Greyhound bus and rolled across the state of Georgia. His last name had always been a constant source of entertainment for the drill sergeants, who took great joy in "mispronouncing" it in ways that would make a chaplain blush. When we arrived at Fort Stewart and the 3rd Infantry Division, famously known as the "Rock of the Marne," we were herded into a reception area while someone decided where to put us. We waited there, anxious and curious, looking forward to meeting our leadership and the soldiers we would be serving with.

During the Second Battle of the Marne in World War I, the 3rd ID held its ground near the Marne River in France against a massive German onslaught. They refused to give up any territory and repelled the attack, which turned out to be the last major offensive the Germans could mount in the entire war. Division legend holds that Major General Joseph Dickman shouted, "Nous resterons là," which means "We shall remain here."[31] From that moment on, the division carried the nickname "Rock of the Marne," or simply "the Rock."

The 3rd ID is classified as a mechanized infantry division, meaning it relies heavily on armored vehicles to move and support its infantry. Its primary workhorse is the Bradley Fighting Vehicle (BFV), a powerful and versatile asset on the modern battlefield.

The BFV is crewed by three soldiers: a driver, a commander, and a gunner. It is armed with a 25mm M242 Bushmaster chain gun, a TOW anti-tank missile system, and a 7.62mm coaxial machine gun.[32] It looks like a smaller tank, with a tracked chassis and armored hull, but its most important feature is hidden in the back.

In the rear compartment, six or seven fully equipped infantrymen ride in relative safety until it is time to fight. When the ramp drops, they emerge like angered bees, spreading out and pour-

---

[31] Stark, Dustin. "How 3rd Infantry Division Became the 'Rock of the Marne.'" U.S. Army, December 5, 2023. https://www.army.mil/article/272207/how_3rd_infantry_division_became_the_rock_of_the_marne.

[32] Military.com. "M2/M3 Bradley Fighting Vehicle." Accessed November 19, 2025. https://www.military.com/equipment/m2-m3-bradley-fighting-vehicle.

ing firepower into the fight. This combination of light infantry and heavily armored vehicles can smash into an enemy position and overwhelm it before the enemy has much time to react.

Because of their mobility, mechanized units are best suited for open terrain such as deserts, rolling hills, or plains. They don't perform as well in thick vegetation or tight urban mazes. That made mechanized units, and the 3rd ID in particular, especially valuable in Iraq, which offers large expanses of desert and flat ground.

Kuntz and I eventually linked up with PFC Toby Glass, another soldier who had graduated with us but had driven his own vehicle to Fort Stewart. Toby and I would later become the best of friends, but our first meeting was uneventful. For the moment, we were all just three new guys waiting to see where we would land.

We were temporarily housed in outdated, slightly dilapidated holding barracks that looked like they had been thrown together in the 1950s. Asbestos and lead paint seemed like the only construction materials anyone had bothered to use. We joked about it and quietly hoped the rest of the base was a little more modern.

Getting assigned to a line infantry unit within the 3rd ID took time because the entire division had already deployed to the Middle East. The invasion of Iraq hadn't yet begun, so they were still staged in Kuwait while we started our in-processing at Fort Stewart. We hoped to catch flights overseas as quickly as possible so we could join our unit and be part of the initial push into the country.

Even though we were mostly cut off from everyday life, it was impossible to ignore what was coming. Every radio station, news broadcast, newspaper, and magazine seemed to be focused on the looming war. Hundreds of thousands of American and coalition troops were massed along the Kuwaiti border, and the news occasionally showed tense scenes from Iraqi cities bracing for the storm of men and death machines that was clearly on its way.

Meanwhile, back in the States, we did everything we could to speed things up, only to face resistance at every turn. Reception wasn't as dysfunctional as 30th AG, but there was a familiar, uncomfortable haze of apathy hanging over everything. No one seemed in a hurry to do much of anything, especially when it came to getting us assigned to a real unit.

After several days of waiting, we were finally picked up by the 3rd Battalion, 7th Infantry Regiment, nicknamed the "Cottonbalers." Because the battalion was already in Kuwait, with only a tiny rear element left at Fort Stewart, we were dropped into Headquarters and Headquarters Company (HHC), the garrisoned catch-all.

In a battalion of 500–1,000 men, HHC is the umbrella that covers everyone who doesn't fit neatly into a line company: administrative staff, drivers, support personnel, and assorted odds and ends. It also happened to be where they parked all the late arrivals and stragglers until we could be properly assigned to a fighting company. We found ourselves mixed into a loosely assembled pool of people, most of whom we would probably never see again once the war machine spun back up.

There weren't many people left around the division; the place had a hollow, ghost-town feel. Unfortunately, most of the ones who were still at Fort Stewart reminded me uncomfortably of the crew at 30th AG. It was a ragtag mix of questionable leaders, injured soldiers, and misfits, sprinkled with random civilian employees and new arrivals like us. *Out of the frying pan, into the fire*, I thought.

**My new home, 3-7 Infantry: Cottonbalers**

The official word was that we were now part of Rear Detachment, "Rear D" for short. New soldiers from basic and transfers from other divisions were trickling in, swelling the ranks of Rear D. Like us, they arrived confused, trying to figure out what the mission was, when we might deploy, and who was actually in charge. It was a mess, and communication was essentially nonexistent. I'd estimate there were maybe thirty to fifty soldiers when I first showed up, most of them the people the division had chosen to leave behind. We were outnumbered and outranked by a wide margin.

Most of us were inexperienced and naïve, so we didn't fully grasp what was happening. The division had left this small contingent behind to maintain the post and monitor incoming personnel. What we didn't realize at first was that they had also parked many of their problem children there, likely because no one wanted them anywhere near combat. They were seen as liabilities. So, when we arrived, our first impression was that the entire 3rd ID looked like this circus we had fallen into. Thankfully, we later learned that wasn't the case.

Some of our "exemplary" NCO corps included a compulsive gambler who borrowed money from lower-enlisted soldiers and never paid a cent back, a man who seemed terrified of his own shadow and regularly had panic attacks, and several people who simply vanished for the day the moment roll call ended. There was even a notorious thief. This NCO, posing as a leader, would pop open his trunk, ask if anyone wanted to buy some "bling bling," and show off random items that looked suspiciously like government property. He also sent brand-new soldiers on errands to pick up items from around the post, and we strongly suspected he was using them to lift supplies he'd later sell from that same trunk. As schemes go, it was clever, but it was also blatantly corrupt.

Only a handful of NCOs were truly competent. One of them, who arrived early, was Staff Sergeant (SSG) Hassel, a transfer from another division. He was squared away, a military term meaning prepared, professional, and good at his job. When he showed up, I imagine he was just as stunned as we were by the level of disorganization and incompetence in HHC. He wasn't especially friendly toward us at first, and I suspect he initially assumed we were all part of that original pack of misfits.

Eventually, he realized many of us were just unlucky, having transferred in a few weeks too late to deploy with the main body. But it didn't take long to sort out who the true dirtbags were. Once he understood that, his attitude toward us shifted. That change in demeanor marked a turning point for me, and he became someone I looked to for guidance. He was a good leader and one of the few NCOs we respected in Rear D.

SSG Hassel did what he could: he led us in PT, tried to instill some discipline, and pushed to fix as many of the company's problems as possible. But he wasn't the senior NCO, and there were clearly power struggles playing out behind the scenes. I'm sure it felt like a no-win situation. Anyone with real authority was deployed, and some of the troublemakers left behind still wore enough rank to cause problems for anyone who challenged them. In that environment, even good leaders had to pick their battles carefully.

**30 March 2025 – Reflection of Toby Glass, 3-7 Infantry, Rear Detachment**

We all went to the range with SSG Hassel to shoot and qualify. A bunch of deer were roaming around out there, and he told us not to shoot them. Everyone wanted to, but we listened. While I was shooting at the targets, I couldn't get a zero, and the range cadre started yelling at me, demanding to know how I'd made it through basic training when my rounds were nowhere to be found. They were really laying into me.

I'd been a good shot in basic, too, so I was confused. SSG Hassel came over and asked what I had previously scored. I told him expert. He snatched my rifle away and tried it himself, but even he wasn't hitting anything. It turned out they had issued me a rifle with a slightly bent barrel. He told the armorer, "I wouldn't send Glass into battle with this weapon. Fix it." No one could hit anything with it.

That just about summed up Rear D. But at least Hassel was there. He was a great guy and took ownership of making sure we had what we needed.

One of our first tasks was to get caught up on immunizations. That included the smallpox vaccine, because the military was worried Saddam Hussein might use NBC weapons against American troops.[33] It wasn't excruciating, but it was a strange and irritating process. The medics carried what looked like a Petri dish and scraped a bifurcated needle along the bottom to collect some of the virus. The needle looked like a miniature flathead screwdriver.[34]

Once it was loaded, the medic would repeatedly jab the needle into your shoulder in a tight little circle. After a dozen or more quick punctures, they slapped a loose bandage over it and sent you on your way.

The problem with the smallpox vaccine was that it was contagious. We had to be careful not to touch the site and accidentally smear the virus into our eyes, mouth, or any other sensitive place we cared about. When the scab finally formed and later crusted over, we were supposed to dispose of it like a tiny biohazard to avoid spreading it to other people. The shot was mandatory, which made it darkly amusing to watch people try, and fail, to duck out of the vaccination line.

If smallpox was weird, anthrax was just mean. The anthrax series was a regular visitor, and those shots were given directly into the triceps. Almost immediately, it felt like my arm had caught fire. It burned deep in the muscle, as if someone had injected battery acid straight into the back of my arm.[35] Those were mandatory as well, and after a while, we all started to feel like human pincushions.

The main contingent at 3-7 Infantry Rear D wasn't prepared or inclined to do much of anything productive, so lower-enlisted soldiers like us were usually handed whatever unglamorous jobs were available: garbage collection, cafeteria duty (KP), working in the motor pool, CQ, yard maintenance, gate guard, painting buildings, and any other task remotely within reach. It was tedious, and the NCOs seemed to enjoy stretching these duties into long, dragging shifts.

On top of that, we were held to a one-hour recall. At any time, if they called our cell phones, home phones, or even pagers, we were supposed to be in formation with all our gear within an hour. Our "imminent" departure to join our deployed unit was the reason this leash stayed tight around our necks.

At first, we were genuinely excited. We believed our flights could arrive any day and that we would soon link up with our platoons in the Middle East. As days turned into weeks, that optimism started to fade. Then someone decided it would be entertaining to pull the recall alarm every few days.

---

[33] Barry, John. "Boots, Bytes and Bombs." *Newsweek*, February 16, 2003. https://www.newsweek.com/boots-bytes-and-bombs-140495.

[34] World Health Organization. "How to Use a Bifurcated Needle to Perform Multiple Puncture Vaccination Techniques," November 2022. https://cdn.who.int/media/docs/default-source/immunization/supply-chain/jobaid_bifurcated_needle.pdf.

[35] Vedantam, Shankar. "Lingering Worries Over Vaccine." The Washington Post, December 19, 2001. https://www.washingtonpost.com/archive/politics/2001/12/20/lingering-worries-over-vaccine/d05c4188-5672-4821-b68d-b979964a3ba4/.

The routine became maddening. We would toss everything from the barracks refrigerators, scramble to pack our gear, and sprint to the formation area, hearts racing and bags in hand. Every time, we arrived breathless and ready, only to be told it was a "false alarm." After enough of these drills, it was hard to see them as anything but cruel games. Our hopes spiked, then crashed, over and over. Eventually, very few of us believed any recall was real. It became a live-action version of the story of the boy who cried wolf.

In the middle of all this nonsense at 3-7 Infantry, my friendship with PFC Glass grew stronger. Shared frustration has a way of bonding people. He was about ten years older than me and nearing thirty, which meant he had a decade of life experience I didn't. So, I'm sure I pestered him with endless questions about everything. He was a steady source of information and came from a kind, grounded family in southern Illinois.

When we could get away with it, we took short excursions into nearby Savannah and explored the historic city, fully aware we were technically breaking the one-hour recall rule. If a genuine emergency deployment order had come down while we were out, we would have missed it unless he drove like a maniac on the way back. That possibility was always in the back of my mind, but, fortunately, we were never burned by it.

Glass had struggled with severe plantar fasciitis during our basic training cycle. Like shin splints and other lower-leg issues, it is a common infantry injury that causes inflammation, pain, and swelling along the bottom of the foot and heel. His case was so bad that he was nearly incapacitated and needed regular trips to the doctor for steroid injections and other treatments. Like my broken foot, his injury came from the constant pounding our feet took during road marches, runs, and daily training. I doubt the cheap, poorly fitted boots we were issued at Infantry School did us any favors, either.

**30 March 2025 – Reflection of Toby Glass, 3-7 Infantry, Rear Detachment**

On a long march, one of the privates in my platoon was about to fall out under the weight of the M-249 SAW. I swapped rifles with him to lighten his load. That helped him, but with all the gear we were already carrying, that trade was the straw that broke the camel's back.

I'd started the march with a back injury. Once that flared up, I began compensating without even realizing it, changing my stride just enough to spread the pain into everything else. Before long, my feet were screaming. The boots and the weight did the rest. I kept pushing, but the more I forced it, the more my body fought back. By the time it was over, I'd destroyed the ligaments in my arches and jacked up my knee and my neck, too.

That's just how it was. You pressed on, and sometimes your body paid the bill later.

I remember watching him limp back into the company area during our final FTX at Fort Benning, the same march where all the recruits clapped as we arrived. Later, he told me that was the moment everything finally caught up with him. He had been pushing through a series of minor injuries throughout basic training, and that last FTX was his breaking point.

Those symptoms followed him into the 3rd ID. It was apparent he was in pain and struggling, especially during any kind of lower-body PT. His body was wrecked, and he never had time to properly recover. The jokers at Rear D didn't care that he was hurting and kept putting him on physically demanding assignments when he should have been easing off and healing. I could relate to his suffering, and it reminded me of how hard it had been just to make it back to the barracks on that march to avoid being recycled.

On top of this, Glass had a stutter that developed after a childhood case of meningitis. Those two issues together were a perfect storm, and he became a regular target for our inept and sometimes cruel leadership at Rear D. They were constantly irritated that he couldn't stand or walk for long periods. As the pounding on his feet continued, he became more and more limited in what he could physically do.

At least once, an NCO openly mocked him by stuttering in front of him, imitating his speech for laughs. Glass usually tried to brush it off. Still, it was another example of someone being ridiculed for something completely outside his control. Incidents like that made many of us question the Army and whether we had made a mistake enlisting. We didn't yet understand that the Rear D crowd wasn't representative of the Army as a whole. Combined with my fresh memories of 30th AG, it felt, for a while, like I had made a very costly decision.

Away from the base, though, Glass was charismatic and outgoing. He had a knack for talking to almost anyone, even if his speech could be choppy at times. He was generous with his time, money, and advice. He acted like a big brother to many younger soldiers and was widely respected. That contrast was confusing for some of us, given how often certain leaders targeted him. I'm convinced it was because he was a bit hard-headed, and I mean that in the best way. He had enough life experience by that time to recognize that what we were seeing in Rear D wasn't normal, and he didn't like it. He spoke up about the nonsense more than most, and he caught a lot of heat for it.

Early in our time at Rear D, he and I met a new transfer to the unit, Specialist (SPC) Donald Cyr. Cyr quickly became part of our small circle, and the three of us were almost always together. He had previously served in the United States Marine Corps infantry and had deployment experience, which gave him a certain aura in our eyes. After his initial enlistment, he switched branches and joined the Army, hoping for a combat deployment and a potentially better long-term career path.

### 30 March 2025 – Reflection of Donald Cyr, 3-7 Infantry, Rear Detachment

After my first enlistment was over, I saw people on TV talking about the 3rd ID deploying to Iraq, and I figured that was where I needed to go if I was going to get combat experience. I enlisted, and the Army sent me to Fort Stewart. When I arrived, I couldn't believe what I had gotten myself into. It was a total culture shock. There were so many dirtbags in Rear D.

But there were a lot of really great guys, too. A lot of hard workers and good soldiers. They just needed some guidance and direction.

Born on one of the islands in Southeast Asia, he spent most of his childhood in the great outdoors, hunting and fishing for food, and living a simpler, more impoverished lifestyle than most Americans can imagine. As he described it, it was "island life." That background translated well to being a grunt. He was constantly asked for advice, sometimes even by seasoned NCOs.

We talked with him about his experiences in the Marines and did our best to absorb as much information as possible. Cyr had been in the Pacific in 2000 with a Marine Expeditionary Unit when the USS Cole was bombed by a small boat laden with explosives. The suicide attack, carried out by Al-Qaeda, killed seventeen United States Navy sailors and injured thirty-nine more.[36] The guided missile destroyer was nearly sunk while harbored in Yemen. He and his fellow Marines, a nearby quick reaction force (QRF), were tasked with providing security against further attacks.

Because of that, we constantly pestered him with questions, trying to learn from someone with real-world experience who had deployed before. Cyr hoped to use this knowledge in the Army for the upcoming war, and it would ultimately prove to be beneficial.

Reflecting on his previous experience, Cyr had the unique opportunity to experience both branches, carefully weighing their pros and cons. Both services had distinct advantages over the other, while serving as the primary ground forces of the United States. Though he wasn't initially in a formal leadership role, he drilled into us the importance of being proficient riflemen above all else. Everything else was secondary, a lesson he carried over from the Marine Corps.

### 30 March 2025 – Reflection of Donald Cyr, 3-7 Infantry, Rear Detachment

I was trained to see the individual as the weapon. At 3rd ID, some leaders seemed to believe the weapon was the Bradley. It's a great tool, but the trade-off is that some soldiers became medio-cre. The best weapon is a well-trained soldier: a rifleman first.

While he spoke with genuine pride about his previous enlistment, Cyr was quick to point out that it, like the Army, was often miserable work, redeemed only by the camaraderie of the grunts. He and his platoon regularly trained off the coast, launching from naval ships in the middle of the night in small boats. As they neared shore, they would cut the engine, and scout swimmers quietly slipped into the water, disappearing into the cold darkness. Once the scouts signaled, every-one else followed, dropping into the water and swimming alongside the boat until they were soaked from head to toe. The water was so cold that some of the unlucky men ended up with hypother-mia.

---

[36] Britannica Editors. "USS Cole Attack." Encyclopaedia Britannica, November 12, 2025. https://www.britannica.com/event/USS-Cole-attack.

In the pitch black, they would sometimes feel dolphins brushing against their legs, curious about these strange shapes in camouflage drifting through their world. Cyr would laugh and say, "We all tried to make our legs as small as possible. What if they weren't dolphins?"

**30 March 2025 – Reflection of Donald Cyr, 3-7 Infantry, Rear Detachment**

It ruined swimming for me. For the longest time, I didn't want to get into any body of water at all.

Many people outside the military may not realize it, but much of the armed forces' training is deliberately designed to be uncomfortable. There is a real purpose behind it. It prepares soldiers for combat environments where creature comforts are missing and nothing feels easy. Luxuries like air conditioning, real beds, decent food and water, proper clothing, and other basic amenities are often nowhere to be found. Sometimes, though, the pendulum swings too far, and the misery outweighs any real training value.

For Cyr, spending long nights swimming in frigid water while running grueling drills was the opposite of how he grew up. Experiences like that can slowly turn things you once loved into things you can't stand. I agreed with him. I can no longer enjoy camping or going to the beach. Walking in sand or sleeping in a tent drags me back to memories of curling up in dirt foxholes, coated in mud and sweat, with bugs and snakes. On top of that, weeks of grime and human filth built up in places you didn't know could chafe, and the whole experience crossed from "toughening" into "miserable."

On a typical day in the rear, we were required to be in formation for Reveille at 0600 hours. This daily song, played over a loudspeaker, signaled the beginning of the workday for troops in garrison. Soldiers saluted as the American flag was raised, with vehicle traffic stopping and all activity briefly halted. Immediately afterward, the "Dogface Soldier" song played, and soldiers sang along to the slightly revised, sanitized version made famous in the 1955 Audie Murphy film "To Hell and Back."[37] "Dogface soldier" is slang for an infantryman and is explicitly used for soldiers of the 3rd Infantry Division. Once the formalities were over, roll call was taken to make sure everyone was present and accounted for.

PT came next. The platoon was broken up into squads, each led by a different NCO. Most of our physical training, especially early on, focused on running and simple calisthenics. There wasn't much creativity or variety. Compared to basic training, it was easy, at least in the beginning.

After PT, we were released to shower, clean up, and head to the chow hall for breakfast. The food was better than what we had in basic, and no one stood over us, screaming for us to hurry. With formation not resuming until 0900, we usually had plenty of time to eat. There was still hell to pay for anyone who showed up late, so no one pushed their luck.

---

[37] U.S. Army. "Dog Face Soldier Song." U.S. Army Fort Stewart / Hunter AAF / Kelley Hill. Accessed November 19, 2025. https://web.archive.org/web/20151122063311/http://www.stewart.army.mil/info/?id=464&p=2.

Once we formed up again, the NCOs handed out various duties around the post. The most dreaded was gate guard, and the main gate was the worst assignment because it saw a constant stream of vehicles. We were essentially loaned out to the MP units responsible for post security. Some MPs treated us poorly because we were grunts, and they made us do most of the boring work while they stayed inside air-conditioned shacks. A handful were squared away, so we did our best to work with those and avoid the lazy ones.

**L-R: Toby Glass, Donald Cyr, and the author, trading Rear D misery for Florida sunshine**

The job itself was simple. We greeted everyone driving in, checked identification cards, kept unauthorized people out, and notified the MPs of anything suspicious. Every so often, we would randomly search a vehicle for contraband or weapons. The simplicity was part of the problem. We did the same thing for eight-hour shifts, standing in the sun, repeating the same lines, and trying not to get our feet run over. It was necessary work, but it was also mind-numbing and not what any of us had signed up for.

Cyr and I often volunteered to work at the Ammo Supply Point (ASP) just to dodge gate guard duty. The ASP sat out in the middle of the woods like a forgotten outpost. If we were lucky, they would let us team up. An MP would issue each of us an M-16A2, one lonely 20-round maga-zine, and a radio "for emergencies." With that level of firepower, I assume our real mission was to give the illusion that someone was in charge, not actually to stop a determined trespasser.

The ASP was a good distance from the main part of the base and was wrapped in thick vegetation. A tall chain-link fence surrounded the compound, topped with concertina wire, and just outside sat what can only be described as a knee-deep, swampy moat of mystery water.

As the name suggests, the ASP stored ammunition for the entire base in heavy concrete bunkers. Units would roll out there with trucks to pick up rifle ammo, machine gun belts, and eve-rything else that could launch, explode, or otherwise ruin someone's day.

Our orders were simple. Sit in a small wooden hut outside the gate, look official, and make sure no one tried to enter without authorization. Every so often, we'd walk the concrete paths be-tween the bunkers, daydreaming about actually being overseas with our units instead of guarding a glorified concrete farm. The job was boring, but it gave us plenty of time to talk, swap stories, and, best of all, stay away from the MPs and the rotten NCOs back at HHC.

One day, I was strolling along the raised concrete path when I saw what looked like some-one trying to climb the chain-link fence. My heart jumped. I took off in a sprint toward the "in-truder," ready to confront and hopefully shoot this daring saboteur. As I closed the distance, my terrorist morphed into a large, dark green alligator.

I froze. For a split second, I actually considered shooting it, then realized starting a gun-fight with a protected lizard next to a field full of explosives was probably not a stellar career move. So, I just watched.

The gator casually hooked its claws into the fence and started climbing, like this was some-thing it did every Tuesday. It made it a surprising distance up before losing its grip and flopping onto its back with an undignified thud. After a moment of reptilian embarrassment, it rolled over, waddled off, and disappeared into the swamp like nothing had happened.

Unsure of the proper protocol for an enemy alligator breaching the perimeter, likely of the Al-Qaeda variety, I got on the radio and reported that a wild beast had just attempted to sabotage the ASP, but that I had heroically survived. My joke didn't land and went right over the MP's head.

"Yeah, they do that sometimes," the MP replied, completely unfazed. "A couple of small ones are in the ditches around the bunkers. Don't go in there." That was it: no further guidance, no concern about our reptile insurgency, and zero appreciation for my comedic report.

Naturally, I did the opposite and went all in on alligator reconnaissance. I started carefully checking the ditches and, sure enough, found several small gators lounging around the bunkers like they were on vacation. Before long, I started scooping up the smaller ones like Steve Irwin and flinging them back over the fence into the swamp where they belonged.

Later, I started noticing chain-link fences in the southern United States that arched back-ward near the top at almost a 90-degree angle. I learned they were designed specifically to keep alli-gators out. Once the fence angled back, gravity would do its work, and the gators would fall off.

For a kid from Iowa, where the buffalo and absolutely no alligators roam, this was all new and exciting. Every day seemed to present an odd little lesson, and this time the teacher was a fence-climbing swamp dinosaur. I was fascinated and wanted to see what else the world had to offer.

**Stuck at the ASP, jungle boots on, ready for gator counterinsurgency operations**

Once our shift was over, we were usually driven back to the barracks and signed in at the headquarters front desk. The cycle repeated day after day, with very little variation. At 1700 hours, "Retreat" and "To the Colors" played over the loudspeakers, signaling the supposed end of the workday. I say "end" loosely, because plenty of days dragged on well past that point.

To our supreme disappointment, the invasion of Iraq began on the 19th and 20th of March, 2003. We watched our own division storm across the berm into Iraq on television. While we stood guard at Fort Stewart and played with gators, our unit was pushing forward in real time. Targeted airstrikes hammered Iraqi cities and military bases. Special operations units and the CIA had already slipped into the country, and the 3rd ID was driving west, then turning north toward Baghdad.[38] All we could do was sit on the sidelines, follow their progress on the news like everyone else, and quietly hope for a miracle flight overseas that would let us catch up.

Several days after the invasion kicked off, I was stuck working the main gate, checking ID cards. I was in a foul mood, standing there with my little reflective belt, guarding Georgia while the rest of my unit was storming across the berm into Iraq.

A nicely dressed middle-aged woman pulled into my lane and started digging through the black hole that was her purse. After a full archaeological excavation, she finally found her ID, rolled down the window, and handed it over.

"My goodness!" she said, peering at me over her dark sunglasses. "Are you old enough to be in the Army?"

By this point, I had heard that question from just about everyone: drill sergeants, my recruiter, fellow soldiers, and now even random civilians. I had already been told I looked like a middle schooler stuffed into a baggy Army uniform, so it didn't exactly improve my mood. The joke had gotten old for me, so I decided to make it fresh for myself.

"No, ma'am," I said, keeping a straight face. "I'm actually a high school student in JROTC. I guess they are losing so many people overseas, they called us up to do guard…"

"*Oh my God!*" she screamed, cutting me off. "My son is in JROTC! Is he going to be deployed to Iraq?"

She was now hysterical and on the verge of tears. That was when it hit me that she wasn't teasing me about my age. She honestly thought she was talking to a young teenager in uniform and was picturing her own kid replacing me at the gate, or worse.

My brain lit up with sirens. Abort mission. Bad joke. *Very* bad joke.

"Uh, ma'am, I was just kidding," I blurted out. "I'm actually active duty with 3–7. I'm a real soldier."

Great, Jeff. Perfect. Not only did you terrify a mother, you also told her exactly where to find you and your chain of command. I handed her ID back. I didn't even bother looking at it.

"That is not funny," she said, ice cold, then drove off.

For the rest of the day, I waited for the hammer to drop. I was sure she had to be an officer's wife, already speed-dialing her husband to report the idiot comedian infantryman at the gate. In my head, I could see my future: a thousand years of extra PT, permanent gate guard, and maybe a public execution in front of the division.

To my complete surprise, nothing happened. No one said a word and I never saw her again. Somehow, I had dodged that bullet.

---

[38] Synovitz, Ron. "Iraq: U.S. Army's 3rd Infantry Division Crosses Border, Begins Northward Advance." *RadioFreeEurope / RadioLiberty*, March 21, 2003. https://www.rferl.org/a/1102615.html.

It turned out to be a useful lesson. A lot of the civilians coming onto the post were stretched to the breaking point. Husbands and wives were deployed. Friends, neighbors, and family members were in danger. Nearly everyone had at least one loved one downrange, and they lived in a constant state of low-grade panic. They didn't know where their soldier was, what they were doing, when they would come home, or if they ever *would* come home. On top of that, they lived with the quiet terror of seeing an Army chaplain and a casualty notification officer walking up to their door in the middle of the night.

I still think the joke was objectively funny. But I also learned there are times you keep the material in your head inside and just say, "Yes, ma'am, I'm old enough."

A short time after my gate guard incident, I was assigned, along with several other soldiers, to attend funerals in my Class A uniform. These were for service members killed in the early stages of the war. It was a heart-wrenching duty and one of the most difficult things I have ever had to do. During one procession, a fallen soldier's mother broke down hysterically and screamed while Taps played. I didn't personally know the soldier or his family, but the grief in that moment was overwhelming. It has stayed with me for over twenty years.

I can still see her desperately clutching the flag and the casket, repeating his name over and over while tears poured down her face. Even worse, there was absolutely nothing we could do or say to ease the family's nightmare. Our role was simply to stand there, remain professional, and give them space to grieve in their own way.

I'm not an emotional person. Some have even described me as a "robot" when it comes to showing sadness or fear. But that day left a massive lump in my throat and brought me to the edge of tears. No book or movie can truly capture the essence of a military funeral. It must be witnessed in person. It's one of the most moving and haunting experiences imaginable.

To the 3rd ID's credit, they did an exceptional job honoring their fallen soldiers, at least at the ceremonies I attended. The utmost respect was shown to the families and friends of the dead. Even so, it was clear that no amount of precision, symbolism, or ritual could ever fill the permanent void in their hearts.

Before long, and much to my surprise, my wish to deploy to Iraq was granted. Now it was my own family's turn to live in a constant state of anxiety. I could only hope that they would not be the ones standing beside a flag-draped casket, living through the heartache of losing a son and brother to war.

# Reflections

Like Benning, my introduction to the 3rd ID was lackluster. But it was still an improvement over 30th AG, if only because I finally had some freedom during my off time. I was no longer trapped with the same lunatics twenty-four hours a day with no hope of escape. We were still subject to recall, though, so there were limits, but at least there was breathing room.

By the time we arrived at Fort Stewart, the main body had already deployed. In hindsight, it would have been far better if we had simply been shipped overseas and told to catch up. The logistics would have been a nightmare, but most of us would have adapted quickly enough.

To make matters worse, the people left behind at Rear D represented the division's worst elements. I don't think that outcome was entirely intentional, but the effect was disastrous. It was a rude awakening for those of us new to the Army.

The division understandably didn't want to take injured or problematic soldiers into a combat zone. There was a war to fight, after all, and it wasn't going to be won by dragging along people who could not or would not pull their weight. In a perfect world, the Army would have separated the dirtbags outright. In reality, that process is slow and paperwork-heavy, and the chain of command had more pressing concerns. The next best option was to park them on post, keep them busy with menial tasks, and hope they stayed out of the way.

The flaw in that plan was putting the wrong people in charge. Many of the soldiers left behind, and some of the leaders over them, were terrible examples to set in front of young, impressionable infantrymen. Their attitude and behavior spread through the formation like a plague. Morale was abysmal. New soldiers learned quickly that you could either sham your way out of work or blend in with the general climate of cutting corners and flirting with unprofessional, sometimes outright unlawful, behavior.

While writing this book, I spoke with Donald Cyr and Toby Glass about their memories of 3rd ID Rear Detachment. They also described it as one of the most depressing and unprincipled times in their lives. None of us looks back on it fondly.

In Chapter 4, I challenged the idea that every service member is automatically honorable simply because they raised their right hand. Without belaboring the point, I hope this chapter reinforces the reality that soldiers are human beings. They bring the same vices, weaknesses, and destructive tendencies into uniform that exist outside of it. Treating military service as an instant badge of virtue is, in my opinion, incredibly naïve. The reality is more complicated, and this period at Rear D is a clear example. Each veteran is an individual and should be evaluated as such. Doing your homework before buying into the aura of automatic righteousness is wise. OK, I'll quit now.

But none of this means the 3rd ID was devoid of good people, far from it. Fort Stewart was full of young men, hardworking soldiers, who were eager to do their jobs. We had just graduated from Infantry School and were hungry to put our new skills to use in modern combat. A lot of that came from youthful inexperience, but it also came from a genuine desire to serve a higher purpose. The attacks of 9/11 were still a fresh wound, and many of us believed we were heading to the

Middle East to stop future attacks before they reached our shores. Twenty years later, the logic behind that belief is open to serious debate, but at the time it felt unquestionably right.

To this day, I remain best friends with Toby Glass and Donald Cyr. In fact, they are godfathers to my two daughters. I also keep close ties with several other men from my Rear D days, and we still shake our heads at the stories from that period. Maybe we were meant to go through that mess together. Difficult times have a way of welding people together that nothing else can.

The absurdity of Rear D made arriving in Iraq feel almost like a relief, as insane as that sounds. Imagine despising a place so much that you would rather go to a country where your odds of being violently killed were exponentially higher, just to escape the nonsense back home. That is about the only honest way to sum up Rear D.

Reflecting on that difficult time, the most significant realization is that none of it would have happened had I not broken my foot several months prior. Had I graduated with my first basic training cycle, I would have arrived at Fort Stewart with a different set of guys. I may have ended up in another battalion altogether, crossed the border with the main body, and had a wholly different experience. I almost certainly wouldn't have become friends with those I now consider family. They would be utterly unknown to me. How different it could have been.

I don't particularly subscribe to the concept of fate, though I recognize that it occasionally presents itself and sometimes appears real enough. I find it difficult to believe in preordained destiny, especially when random chance is a viable alternative. However, I will say that everything seemed to fall into place in just the right way. I still ponder the possibilities had it not happened this way, though those paths are perpetually darkened and clouded with speculation.

I remember standing at C 2-54, speaking with my dad on the phone. My leg was in the new Airboot, and I poured out my heartache and disappointment. It appeared then that all doors of possibility were closing. My dreams were permanently shattered because of a weak bone in my foot. Wrapped up in the immediacy of the moment, I couldn't fathom that this was merely a detour off the main road in my plan. At the time, it felt like a permanent roadblock.

The benefit of hindsight is that it allows us to see how flawed our interpretations can be in those moments, especially when emotions are involved. Unfortunately, we are rarely able to project our story arc in the opposite direction accurately. Had I been able to see into the future, perhaps even six months, my outlook on life would have changed drastically. I would have seen my broken metatarsal as an annoyance, possibly even a mildly "good" thing, not the dreaded dream killer I believed it was.

During times of self-reflection, I have found other stories of this same pattern in my lifetime. Often, positive things can emerge from seemingly terrible incidents. The passing of a loved one or the loss of a career opportunity are usually traumatic events. Yet I now try to avoid falling into despair when tragedy strikes, though I certainly mourn the loss in the heat of the moment. Once that mourning phase is over, I open my heart and mind to see what emerges from the chrysalis.

This outlook may be a product of my generally optimistic nature, and I entertain that possibility. But I have always tried to find meaning in adversity. What can I learn from it? What can I do to conquer the next challenge? Am I just complaining, or am I working to overcome? The lesson also applies, however, to those who are far more pessimistic. For them, it may be even more beneficial.

Self-reflection is a vital tool I have learned to use daily, and it can be a powerful mechanism to identify historical patterns. It helps us understand what happened and, maybe even more importantly, reveals why the event matters to us. Once the patterns have been identified, it becomes easier to navigate similar events in the future. In this way, we can more accurately see how our choices, unforeseen circumstances, and even misfortune can shape the trajectory of who we become.

Because of this, I regularly advise on how to use self-reflection after a significant life event, whether good or bad. Look back on it a month, a year, or a decade later, and I suspect you will find, with increasing clarity, outcomes that couldn't possibly have been anticipated. Once you identify the routes, it becomes easier to pinpoint the forks in the road. It also becomes easier to respond in the future, not just react. It isn't perfect and cannot guarantee success, but it offers endless perspective. In many ways, perspective is what turns life's chaos into meaning.

I must admit that my broken foot was a bitter disappointment in 2002. In 2025, however, I can confidently say it was one of the best things that ever happened to me. It set the stage for everything that was to come, and taught me that what looks like an ending can actually be the beginning of a better path.

# Chapter VII

## The Cradle of Civilization in Flames

The lands of ancient Mesopotamia have witnessed the rise and fall of countless civilizations over thousands of years, with each distinct people leaving their mark on the region known as the Fertile Crescent.[39] Long before recorded history, Neanderthals and early Homo sapiens overlapped in the Middle East.[40] Only much later would recorded history document the ebb and flow of the Sumerians, Assyrians, Akkadians, Persians, Babylonians, Greeks, Romans, and even the British, all living and spilling blood within these lands.[41] It's as if every significant power of the era was required to set foot in what is known today as Iraq.

In what is perhaps the oldest surviving literature from antiquity, *The Epic of Gilgamesh* tells of a remarkable odyssey undertaken by a demigod hero in the regions around Mesopotamia. Written over many years, possibly as early as 2100 BC, the epic follows Gilgamesh and was partially preserved on clay tablets.[42] Like Hercules, who was also mythologized as a demigod, Gilgamesh possessed superhuman strength and established what is often described as a Jungian archetype. That is, a universal or collective unconscious pattern of thought that persists through many generations, as defined by Swiss psychiatrist Carl Jung.[43] It also describes a Great Flood, sharing many striking similarities with the biblical account of Noah and the Flood in the Book of Genesis.

Beyond literature, the region was home to one of the Seven Wonders of the Ancient World, the Hanging Gardens of Babylon. Said to have been built in or around present-day Hillah

---

[39] Mark, Joshua J. "Fertile Crescent." World History Encyclopedia, March 28, 2018. https://www.worldhistory.org/Fertile_Crescent/.

[40] Balter, Michael. "Humans and Neandertals Likely Interbred in Middle East." Science, January 28, 2015. https://www.science.org/content/article/humans-and-neandertals-likely-interbred-middle-east.

[41] Khadduri, Majid. "Iraq." Encyclopaedia Britannica, November 19, 2025. https://www.britannica.com/place/Iraq.

[42] Archaeologist. "The Epic of Gilgamesh: The World's Oldest Known Literature." The Archaeologist, September 9, 2024. https://www.thearchaeologist.org/blog/the-epic-of-gilgamesh-the-worlds-oldest-known-literature.

[43] Britannica Editors. "Archetype." Encyclopaedia Britannica, September 25, 2025. https://www.britannica.com/topic/archetype.

(just south of Baghdad), or possibly in Nineveh, some have attributed it to King Nebuchadnezzar II, who may have built it for his wife.[44] This has been disputed, and there is no definitive information about who was ultimately responsible for its construction. What is generally accepted, however, is that it was an overwhelmingly beautiful garden, with various trees, vines, shrubs, flowers, and other vegetation carefully layered across terraces. Irrigation canals reportedly provided fresh water to the flora, fauna, and those Babylonians lounging within the confines of the paradisiacal grounds.

In addition to great literature and world wonders, Iraq also boasts a significant first in the annals of human history. Sargon of Akkad (circa 2334-2279 BC), king of the Akkadian Empire of Mesopotamia, was the ruler of the first multinational empire in recorded history. He unified several kingdoms under one banner and reportedly lived somewhere near the confluence of the Tigris and Euphrates Rivers.[45] However, the exact location has been lost to the fog of time. The empire stretched from the Persian Gulf through Iraq, including portions of Turkey and Iran.

Much later, after the fall of the Akkadians, Babylon would arise near present-day Baghdad. It first flourished under Hammurabi (circa 1792-1750 BC) and is well known for originating one of the earliest sets of codified laws in history, the Code of Hammurabi.[46] Babylon also made significant contributions to literature, mathematics, and astronomy.

However, not all aspects of that era were positive. It was referenced negatively several times across many books of the Bible, including Genesis, Daniel, Jeremiah, Isaiah, and even Revelation. In the Book of Genesis, the people of Babylon reportedly began constructing the Tower of Babel, which reached upward to the heavens. Symbolism within the book highlights the dangers of hubris. It also explicitly details the consequences of disobedience to God.

Iraq's Golden Age came later, in what is commonly called the Islamic Golden Age (circa 750-1258 AD). Caliph al-Manṣūr established the capital city of Baghdad around the year 762, and it became one of the world's largest and most advanced cities.[47] Like Babylon, it greatly contributed to mathematics, literature, science, and theology. Reportedly, it allowed a relatively peaceful coexistence between Muslims, Christians, Jews, and other religious groups.

In the 16th century, Iraq fell under Ottoman control. It remained this way until World War I, when the defeated Ottoman Empire collapsed, temporarily leaving the region with undefined borders and a patchwork of countries. European powers, particularly the British, had plans for the

---

[44] Cartwright, Mark. "Hanging Gardens of Babylon." World History Encyclopedia, July 27, 2018. https://www.worldhistory.org/Hanging_Gardens_of_Babylon/.

[45] Baird Rattini, Kristin. "Meet the World's First Emperor." *National Geographic*, May 3, 2021. https://www.nationalgeographic.com/culture/article/king-sargon-akkad.

[46] History.com Editors. "Code of Hammurabi." HISTORY, May 28, 2025. https://www.history.com/articles/hammurabi.

[47] Hawting, G.R. "al-Manṣūr." Encyclopaedia Britannica, October 3, 2025. https://www.britannica.com/biography/al-Mansur-Abbasid-caliph.

area and began dividing it into specific countries. Only in 1932 did modern-day Iraq declare independence, even though its history goes back further than almost any other country.[48]

After gaining independence, Iraq went through decades of turbulence, revolutions, and authoritarian rule. In 1979, Saddam Hussein took control of the country and became president.[49] Within a year, he launched a massive war against neighboring Iran, which had just undergone its own revolution and emerged as an Islamic theocracy under Ayatollah Ruhollah Khomeini.[50]

That new Iranian leadership moved quickly to break with the West, especially the United States, and adopted an openly hostile posture that contrasted sharply with the Western-aligned monarchy it had replaced. The rupture exploded into the Iran hostage crisis of 1979, when 66 Americans were seized and 53 were held for 444 days. Eight U.S. servicemen were killed during the failed rescue attempt, Operation Eagle Claw, in April 1980.[51] That history is complicated and outside the scope of this book, but it helped shape everything that followed between Iraq, Iran, and the United States.

Holding to the saying, *the enemy of my enemy is my friend*, the United States began assisting Saddam Hussein and the Iraqi government in its fight against neighboring Iran, despite warning signs of his authoritarian and often brutal ways. This included economic aid, intelligence, training, and even the sale of military technologies.[52] Details emerged decades later that the US even turned a blind eye when Iraq gassed Iranians with chemical weapons.[53] While technically pro-Iraqi in the conflict, the United States grudgingly formed the uneasy alliance, with former Secretary of State Henry Kissinger reportedly remarking, "It's a pity both sides can't lose."[54] In a way, they both did.

The eight-year-long Iran-Iraq War ended in a stalemate. Hundreds of thousands were dead, both countries suffered tremendous financial loss, and concern over the proliferation of WMDs

---

[48] History.com Editors, "Iraq Wins Independence," HISTORY, January 24, 2025, https://www.history.com/this-day-in-history/october-3/iraq-wins-independence.

[49] Britannica Editors, "Saddam Hussein," Encyclopaedia Britannica, October 25, 2025, https://www.britannica.com/biography/Saddam-Hussein.

[50] Britannica Editors, "Iran-Iraq War," Encyclopaedia Britannica, September 28, 2025, https://www.britannica.com/event/Iran-Iraq-War.

[51] ASOMF. "Operation Eagle Claw." U.S. Army Airborne & Special Operations Museum. Accessed November 20, 2025. https://www.asomf.org/operation-eagle-claw/.

[52] Battle, Joyce, ed. "Shaking Hands With Saddam Hussein: The U.S. Tilts Toward Iraq, 1980-1984." The National Security Archive (George Washington University), February 25, 2003. https://nsarchive2.gwu.edu/NSAEBB/NSAEBB82/.

[53] Harris, Shane, and Matthew Aid. "Exclusive: CIA Files Prove America Helped Saddam as He Gassed Iran." *Foreign Policy*, August 26, 2013. https://foreignpolicy.com/2013/08/26/exclusive-cia-files-prove-america-helped-saddam-as-he-gassed-iran/.

[54] O'Grady, Sean. "We Should Look Back on Kissinger Not Just as a 'War Criminal' – but as a Man Who Sought Peace." *The Independent*, November 30, 2023. https://www.independent.co.uk/voices/henry-kissinger-death-obituary-nixon-vietnam-b2456059.html.

became a global issue.[55] In Iraq, Saddam found that morale had decreased, while tensions ramped up to potentially unsustainable levels within his land.

On August 2, 1990, only about two years after the end of his devastating war with Iran, Saddam Hussein again ordered an invasion. This time, he directed Iraq's massive military to seize its tiny, oil-rich neighbor, Kuwait. Historians point to several reasons behind his decision, but a key factor was the billions of dollars Iraq owed in outstanding war loans to Gulf neighbors, including Kuwait. Combined with longstanding territorial claims, accusations that Kuwait was overproducing oil and driving prices down, and allegations that it was siphoning Iraqi oil from the Al-Rumaylah field, Saddam had a ready list of justifications for the attack.[56]

Immediately, international outcry began to demand the Iraqi withdrawal from Kuwait. Saddam and his forces were given a deadline by President George H.W. Bush, but they remained steadfast and refused to comply. Everything then quickly began to unravel for the Iraqi troops.

A coalition of more than thirty nations rallied behind the leadership of the United States, devastating the Iraqi military in a 100-hour, one-sided slaughter. They ejected the Iraqi army from Kuwait and pursued fleeing forces temporarily, inflicting heavy casualties. Instead of overthrowing the Iraqi government and its dictator, though, the coalition withdrew its troops back from Iraq. At the same time, the United States maintained a significant military presence in the area.

As a young child, I sat glued to the television, watching CNN with my dad. We watched in real-time as our military crushed the occupying Iraqis. The images of anti-aircraft fire, Abrams tanks, Stealth Fighters, and night vision are permanently burned into my brain. I suspect this is where my fascination with becoming a soldier began.

**23 April 2025 – Reflection of my father, Bill Harstad, Jr.**

I remember sitting down to dinner and watching the developments unfold live on television. I thought, *Oh God, here it goes*. It was the first time we could watch a war as it happened, all from the comfort of our homes. It captivated me, and you seemed to be captivated by it too.

The Iraqi armed forces suffered an embarrassing defeat, with estimates of battle deaths in the tens of thousands. By comparison, *combined* coalition forces suffered nearly 400 deaths and less than 1000 wounded.[57] This turned into a tumultuous ceasefire and included crippling sanctions, though minor clashes sporadically took place between American and Iraqi units from the early 1990s to 2003.

---

[55] History.com Editors, "Iran-Iraq War," HISTORY, October 2, 2025, https://www.history.com/articles/iran-iraq-war.

[56] Britannica Editors, "Persian Gulf War," Encyclopaedia Britannica, November 4, 2025, https://www.britannica.com/event/Persian-Gulf-War.

[57] Imperial War Museums. "What Was the Gulf War?" Accessed November 21, 2025. https://www.iwm.org.uk/history/cold-war/gulf-war/what-was-the-gulf-war.

To provide a brief insight and different perspective of Iraq, it's essential to consider the perspective of one of our Iraqi interpreters who assisted us much later in 2005. She, like many other honorable Iraqis, made great sacrifices to help her country and people.

### 1 August 2025 – Reflection of Noor "Nora" Aljassim, Interpreter for the U.S. Military

When I was in high school, the regime made us sign these pieces of paper saying we would be loyal to the Ba'ath Party. It didn't really mean anything, but if we didn't sign, there was always the chance our family would be killed. So, we all signed.

Then some officers from Saddam's Army came to our high school and took everyone out to a big field. They taught us how to shoot an AK.

We also had to swear that we would fight to free Palestine and make ourselves martyrs if needed. Everyone just went along with it because if you didn't, that would be very dangerous for you and your family. That's just how it was.

Following 9/11, President George W. Bush declared Iraq to be a part of the "Axis of Evil" in a State of the Union address in 2002.[58] Much of the run-up to the invasion involved the expulsion of United Nations weapons inspectors, supposed terrorist sponsorship, and Iraq's alleged possession of WMDs. All these explanations for war would be highly controversial in the coming years, to say the very least.

### 23 April 2025 – Reflection of my father, Bill Harstad, Jr.

I remember the coverage in the early '90s and how horrible Saddam Hussein was. We saw images of gassed children, and the media portrayed him as a Hitler figure.

So, when we heard after 9/11 that he was developing weapons of mass destruction, we decided it was a really big deal and that he needed to be taken out. Knowing what we know now, I'm not sure it was the right call.

I still think he was a terrible man, but I don't know about the explanations for war. It seems like they greatly exaggerated the WMD thing.

In early 2003, it was apparent that war was inevitable as the two regimes, Washington and Baghdad, had irreconcilable differences. The Bush administration in the United States took a very blunt position in that its goal was regime change. This approach stood in stark contrast to the limited, narrowly defined goals of the Gulf War in the early 1990s. Unlike that earlier conflict, the

---

58 Bush, George. "President Delivers State of the Union Address." The White House Archives, January 29, 2002. https://georgewbush-whitehouse.archives.gov/news/releases/2002/01/20020129-11.html.

United States and its planned overthrow of the Iraqi government received a lukewarm response from much of the international community.

As an example, Turkey would not allow US forces to open a second front of attack in the northern part of Iraq from its territory.[59] Yet, Turkey was previously a strong ally during the original Gulf War in the early '90s. This indicated a very mixed response to the latter conflict, even before it had begun. There were many warning signs of the difficulties ahead.

Special operations groups and the Central Intelligence Agency (CIA) worked in the area to begin preparations for the invasion. Around the 21st of March, a "shock and awe" campaign began, in which massive airstrikes targeted military and government installations.[60] Shortly thereafter, President Bush gave the order, and American and British troops swarmed across the border in a mad dash to capture Baghdad. Everyone blitzed through the ill-prepared Iraqi forces as they were pummeled in another lopsided massacre.

Thousands of American paratroopers landed in the northern territories of Iraq as the bulk of coalition forces invaded from Kuwait. The US 1st Marine Expeditionary Force plowed through the defenses along Highway 1, directly in the heart of the country. The United Kingdom's 1 Armoured Division proceeded north, while the US 3rd Infantry Division initially went west. All units were essentially spreading out before being used as a massive pincer on Baghdad.[61]

Like the earlier campaign in Afghanistan, the war immediately proved deeply unpopular with much of the corporate media landscape. In his book *Call Sign Chaos*, General Jim Mattis described how a seemingly regular interaction with the press nearly sparked an international incident. Apparently, the shenanigans were well underway in *both* Iraq and Afghanistan.

Mattis writes: "…I was directed by higher headquarters to speak with reporters on board ship. Without thinking much about it, I told them, 'The Marines have landed. We now own a piece of Afghanistan and are going to give it back to the Afghan people.'"[62]

The problem was that the quote was seriously, and to me suspiciously, truncated when it was published, drastically changing the meaning. It simply appeared, "We now own a piece of Afghanistan."[63] Was it an accident, or was it intentional? I'm not sure, but many of us kept noticing the same kinds of distortions afterward.

---

[59] Boudreaux, Richard, and Amberin Zaman. "Turkey Rejects U.S. Troop Deployment." *Los Angeles Times*, March 2, 2003. https://www.latimes.com/archives/la-xpm-2003-mar-02-fg-iraq2-story.html.

[60] CNN. "A Day of Sirens, Bombs, Smoke and Fires," March 21, 2003. https://www.cnn.com/2003/WORLD/meast/03/21/sprj.irq.aday/.

[61] Associated Press, "U.S. Forces Within 35 Miles of Baghdad," *KFVS*, April 2, 2003, https://www.kfvs12.com/story/1211870/us-forces-within-35-miles-of-baghdad/.

[62] Mattis and West, *Call Sign Chaos: Learning to Lead*. eBook edition. Chapter 5.

[63] The Guardian. "Feet on the Ground." *The Guardian*, November 26, 2001. https://www.theguardian.com/world/2001/nov/27/afghanistan.guardianleaders.

Jim Lacey, an embedded reporter for *Time*, accompanied the 101st Airborne Division during the 2003 invasion of Iraq. He detailed two similar events in his fantastic book *Takedown*. By far, it is the most accurate portrayal of the war that I have read. It even describes the conflict he had with his employer over the framing of the war.

The first incident occurred during the ongoing push toward Baghdad. Lacey powered up his satellite gear, checked his email, and found a message from *Time* announcing the next cover in blunt red letters: "WHY ARE WE LOSING?" He was then told to find stories that would support the title. After speaking with the commander he was traveling with, Lacey replied back that the premise was wrong and that the 3rd ID was on the verge of taking the capital within days. He warned the magazine that if it ran the cover as planned, it would make itself look foolish, bizarrely announcing defeat just as the military was storming the capital. *Time* backed off after a major editorial fight, and replaced the headline with the more open question, "WHAT WILL IT TAKE TO WIN?"[64][65]

What strikes me wasn't simply the pessimism of the original headline, but the method. The editors were not asking what was happening. They were asking for confirmation of what they had already decided to print.

Lacey later tells a story from Baghdad International Airport through Colonel William Grimsley, commander of the 1st Brigade, 3rd ID. Early on the morning of April 4th, Grimsley stood beside his Bradley when a Fox News crew member asked to go live. He agreed, and the timing was perfect. At that very moment, Iraq's information minister, hilariously nicknamed "Baghdad Bob," was once again on television insisting that American forces were nowhere near the airport. He had made the same claim the day before, even escorting the international press out to the airfield to "prove" it.

This time, the denial didn't just ring hollow. It detonated on camera. As Fox widened the shot, U.S. forces were sitting in plain view at the airport. The minister's performance kept rolling, but the picture had already killed the script. Grimsley remembered it as a rare moment when the battlefield and the information war collided before a global audience. "We told the world we were winning," he said, "though the BBC still seemed to doubt it."[66]

My question is, did the BBC doubt it, or was this wishful thinking on their part?

To be fair, the distortions didn't come from only one direction of the political aisle. While some outlets seemed determined to declare defeat before Baghdad had even fallen and sometimes painted regular military joes as villains, others were just as eager to declare vindication. I remember watching pro-war coverage that treated every rumor of a chemical find as a clinching revelation. In

---

[64] Lacey, Jim. *Takedown: The 3rd Infantry Division's Twenty-One Day Assault on Baghdad.* eBook edition. Naval Institute Press, 2013, Chapter 14.

[65] Time. "What Will It Take to Win?" TIME, April 7, 2003.
https://content.time.com/time/covers/0%2C16641%2C20030407%2C00.html.

[66] Lacey, *Takedown: The 3rd Infantry Division's Twenty-One Day Assault on Baghdad.* eBook edition. Chapter 21.

late April 2003, preliminary field tests on a cache near Baiji triggered breathless reports that nerve agents like cyclosarin had been discovered.[67] A year later, when an old Iraqi artillery shell containing sarin was found rigged as a roadside bomb, it was presented by some as proof that the central WMD story had been right all along.[68] Those finds were real, but they were remnants from an earlier era, not the active stockpiles the country had been promised.[69] The headlines still sprinted far ahead of the truth.

Looking back, that was the pattern on both sides. One camp tried to force the war into a narrative of catastrophe and quagmire before the first act had finished, while demonizing those troops just doing their jobs. The other tried to force it into a narrative of clean justification, where any scrap of ambiguous evidence became a trophy. In both cases, the story was chosen first and the facts were recruited later. That is why my contempt is not partisan. If the press starts with the conclusion, it doesn't matter whether the conclusion is "we are losing" or "we were right all along." Either way, the reader was being sold certainty where there should have been none at that stage.

Back at Fort Stewart, I kept seeing the same kind of gloomy news coverage, holding out hope I could soon join the fight. Some of it went so far as to bizarrely imply we were in serious trouble, with inevitable military defeat at hand. That didn't square with what our chain of command was telling us, and our guys had a far better pipeline to the truth than any news desk anchor.

I asked a young NCO I trusted, "What's going on over there. Why doesn't what we're hearing match what's on the news?"

An older NCO nearby caught the question and answered without hesitation. "They're lying. They're trying to distort the war. Get used to it. They aren't your friend."

And that was it. There was no debate and no explanation needed. It was a stern, simple warning, and it has stuck with me the rest of my life.

Upon arriving at Baghdad's doorstep, the 3rd ID, under the leadership of Major General Buford Blount III, implemented a daring tactic called "Thunder Runs." He utilized armored vehicles, such as the BFV and the M1 Abrams tank, and he sent a massive convoy of troops racing through the city to probe for weaknesses and rapidly seize crucial objectives.[70] The 3rd ID came under relentless attack from thousands of soldiers, tanks, armored personnel carriers, and other defending forces, but shattered their way through everything like a rampaging bull in a China shop.

---

[67] Holguin, Jaime. "Blair Believes Weapons Will Be Found." CBS News, April 30, 2003. https://www.cbsnews.com/news/blair-believes-weapons-will-be-found/.

[68] Tully, Andrew. "The Sarin-Laced Shell: Major Discovery or a Curious Relic?" *RadioFreeEurope / RadioLiberty*, May 18, 2004. https://www.rferl.org/a/1052831.html.

[69] Fox News. "Report: Hundreds of WMDs Found in Iraq," June 22, 2006. https://www.foxnews.com/story/report-hundreds-of-wmds-found-in-iraq.

[70] Pfeffer, Alexander. "The Gamble for Baghdad - an Account of the 2003 Thunder Runs," June 1, 2025. https://www.americangrit.com/post/the-gamble-for-baghdad---an-account-of-the-2003-thunder-runs.

In a matter of weeks, the Saddam Hussein regime and the Ba'athist Iraqi government collapsed. Hussein, his family, and government members went into hiding.[71] A US propaganda tool then appeared to assist in finding the fugitives, the famous "personality identification playing cards."[72] Officially produced by the Department of Defense, this deck resembled standard playing cards. However, the center of each card had a different photograph of a most-wanted Iraqi, the fugitive's job, and sometimes additional information to assist troops in identifying suspects.

A massive manhunt was underway to round up all of these wanted individuals. Almost all were killed, captured, or later released, with only a small handful ever remaining at large.[73] The Ace of Spades, Saddam Hussein, was finally captured on December 13, 2003, in Operation Red Dawn by Task Force 121 (including Delta Force) with support from the 4th Infantry Division.[74]

All this meant even though the government had technically been defeated, combat was still ongoing everywhere due to the mad dash to capture the capital city. Entire cities, filled to the brim with fighters, had been bypassed while thousands of foreign insurgents poured into the battlegrounds from neighboring countries. Ambushes, sniper attacks, and a terror campaign replaced traditional forms of warfare. The government had fallen, but combat would continue for years.

**1 August 2025 – Reflection of Noor "Nora" Aljassim, Interpreter for the U.S. Military**

I became a linguist by accident.

After the initial war calmed down in 2003, I was shopping for groceries in Baghdad. I saw American soldiers standing on a corner yelling at an old man. He didn't understand what they were saying. I knew English, so I asked, "Do you need some help?"

They said, "Finally! Someone who speaks English. Just tell him he's getting too close and he needs to stay back."

In Iraqi culture, we use our hands a lot and stand very close when we talk. Of course, that made the American soldiers nervous. I explained it to the old man. He apologized, and everything was fine.

---

71 Reuters. "Timeline: Invasion, Surge, Withdrawal; U.S. Forces in Iraq," December 15, 2011. https://www.reuters.com/article/world/timeline-invasion-surge-withdrawal-us-forces-in-iraq-idUSTRE7BE0EL/.

72 GWOT.org. "Iraq's Most Wanted Deck of Playing Cards," May 22, 2023. https://gwot.org/docs/iraqs-most-wanted-deck-of-playing-cards/.

73 Davis, Clint. "What Happened to the 52 Most-Wanted Iraqis From the U.S. Military Card Deck?" *WPTV News Channel 5 West Palm*, April 22, 2015. https://www.wptv.com/news/national/what-happened-to-the-52-most-wanted-iraqis-from-the-us-military-card-deck.

74 DIA Public Affairs. "Our Place in History: 'We Got Him!' the Anniversary of the Capture of Saddam Hussein." Defense Intelligence Agency, December 13, 2013. https://www.dia.mil/News-Features/Articles/Article-View/Article/566928/our-place-in-history-we-got-him-the-anniversary-of-the-capture-of-saddam-hussein/.

After that, the soldiers asked if I would help them translate. They said they didn't have any money to pay me. I agreed anyway because I thought it was a good idea, and maybe it would help keep things calmer on my street. I volunteered with them for about six months.

Later, I was hired by a private company and paid to translate for the military. Because I worked with them, I became close to many American soldiers. I had them over to our house. They met my family, and sometimes we ate dinner together. I just wanted to help.

Reflecting on the invasion, it is fairly clear that the United States government had well-defined goals associated with overthrowing the regime. The military properly executed this goal, and the ruling powers quickly fell under the weight of the US war machine. It is equally apparent that there was little to no preparation for what to do after the inevitable happened. Once the government collapsed, what was the plan? It was as though our civilian leaders just assumed that democracy would suddenly spring forth in utopian harmony.

Instead, massive civil unrest broke out, leaving the military in a serious predicament. Rival groups within the country almost immediately began to fight one another, with sectarian violence between Sunni and Shia militias commonplace. The façade of a blossoming democracy gave way to chaos. Once units seized Baghdad, it was as though the map's limits had been reached, even though well over half the expedition remained. For many grunts on the ground, there was little in the way of definitive instruction from the government. At times, contradictory orders filled the void, and armchair quarterbacking slowly began creeping in. The head-scratching continued as soldiers asked, "Alright, now what do they want us to do?"

### 1 August 2025 – Reflection of Noor "Nora" Aljassim, Interpreter for the U.S. Military

The media does not tell it truthfully. Saddam wasn't a good guy, I want you to know this, but under his regime, there were strict standards. As women, we could walk down the streets without fear. The Army was always out there, so a man would not dare attack us. We went to clubs and parties. We even celebrated Christmas, which was a national holiday. It was a multicultural nation, with Muslims, Jews, Christians, Yazidis, and Sabians. But you had to know the rules, not go against the regime, and watch out for his crazy son, Uday. If you did those things, you didn't have a problem.

After the war, we thought things would quickly be much better. But where was the democracy? Instead of getting better, there were no standards, and people didn't know how to handle it. Builders would put up buildings that just fell. Drugs came into the country, and the hospitals were terrible. The internet arrived, but everyone went crazy with the technology, and now there is pornography everywhere. People started getting attacked on the streets, and it was no longer safe to walk alone. Then everyone began fighting with each other, Shias and Sunnis. The only things that really improved were for businessmen and for traveling, but most people didn't have the money to travel.

In hindsight, perhaps the United States' worst decision of the entire war came on May 23, 2003. Paul Bremer and the Coalition Provisional Authority dissolved the remaining Iraqi Army and began a process of de-Ba'athification.[75] Hundreds of thousands of Iraqi military men, many of whom had been used to maintain an uncomfortable semblance of stability in the country, suddenly found themselves unemployed.[76] Many took their weapons and went home, only to later be recruited by insurgencies and terror groups. Often, this was because they needed money to support themselves and their families.

**1 August 2025 – Reflection of Noor "Nora" Aljassim, Interpreter for the U.S. Military**

When the Saddam statues fell, we were all cheering, but my dad kept saying, "This is going to be very bad." He believed it because the dictator was being replaced by people who were crooks. They fought each other and stole money, and nothing seemed to improve.

Then came what we saw as the biggest mistake. After the war, Bremer fired the entire Army. Military service had been mandatory, so we immediately wondered what would happen to all those soldiers with no jobs. They still needed money to live. The mood changed fast. Waving hands turned into rocks being thrown, and soon, people were shooting at the Americans. Instead of fighting alongside them, many Iraqis were now paid by insurgent groups to fight against them.

The United States' plan, or lack thereof, can be debated endlessly and will hopefully serve as a lesson for future generations. But in the meantime, my friends and I found ourselves in what could only be described as a powder keg waiting to ignite. Everything around us felt volatile, as if one spark could set off a chain reaction of violence.

---

[75] Slevin, Peter. "Wrong Turn at a Postwar Crossroads?" The Washington Post, November 20, 2003. https://www.cs.cornell.edu/gries/howbushoperates/disband.html.

[76] Al Jazeera. "400,000 Lose Jobs in Iraq Disband Order." *Al Jazeera*, May 23, 2003. https://www.aljazeera.com/news/2003/5/23/400000-lose-jobs-in-iraq-disband-order.

# Reflections

The history of Iraq and the Middle East could fill library shelves with tomes, and my attempt to cover even a few highlights is woefully inadequate. Still, a brief synopsis felt necessary. These lands would become the center of American attention for decades, and one could argue that the reverberations are still felt in the United States today.

As a voracious reader, I knew I was about to step into a region steeped in layers of human history. As General Jim Mattis once wrote, "If you haven't read hundreds of books, learning from others who went before you, you are functionally illiterate."[77] I was nowhere near General Mattis's level of awareness, but I understood the spirit of what he meant. I couldn't wait to see the remnants of ancient Babylon with my own eyes.

Looking back, I still marvel at how improbable it all seemed during my childhood. I could never have imagined that one day I would walk in the same exotic places I used to watch on CNN. If my dad had known that those images would lead me there in 2003, would he have forced me to turn off the TV? Again, self-reflection has a way of revealing the forks in the road we don't recognize at the time. The Gulf War in the early 1990s was one of those forks, and it almost certainly set me on the path to Iraq.

The media would later play another significant role in my life, but in a darker and more unsettling way. While I didn't see it early in my military career, Iraq forced my eyes open. The longer I served, the more I learned to be wary of the press, not because cameras were dangerous, but because the story often felt decided before anyone asked a question. Complex events were routinely being compressed into simple stories, and too many reporters were not trying to understand what was actually happening. Instead, many were trying to prove what they already believed. Neutrality could be treated like a nuisance, and facts were welcomed only if they fit the script.

Still, it is rarely wise to condemn an entire profession. To be fair, I have met capable, fair journalists who work hard to get it right, and I have read reporting that captured more truth than official press releases. But their presence didn't erase the obvious and concerning pattern I could see, and the lesson I carried forward was simple. Speak carefully, be skeptical, assume your words will be trimmed into someone else's narrative, and never forget that the headline isn't always looking for the truth.

As podcaster and writer Michael Malice once wrote, "The battle is won when the average American regards a corporate journalist exactly as they regard a tobacco executive."[78] In other words, they're not necessarily serving you the truth. They're selling you a product, and you're the target market, not the beneficiary. It's worth remembering that every time a polished voice on a

---

[77] Mattis and West, *Call Sign Chaos: Learning to Lead.* eBook edition. Chapter 7.

[78] Malice, Michael. "The battle is won when the average American regards a corporate journalist exactly as they regard a tobacco executive." *X (Twitter)*, October 6, 2018. https://x.com/michaelmalice/status/1550548097452969984.

screen tells you what to think. Do your own research, read primary documents, and seek firsthand accounts.

Growing up in a somewhat sheltered environment, I had believed the nightly news was simply a transparent window into daily life around the world. It took me time to come to my senses. I didn't realize how often it can be used as a vehicle for propaganda until I watched the gap expand between what I was living and what people back home were being told. This only escalated when I arrived on scene to see it for myself. Unfortunately, some people never notice that gap. They accept what they are given without scrutiny, even when the contradiction is right in front of them. On the bright side, I think more people see it now than they did in the early 2000s, partly because of the internet, but the problem certainly hasn't vanished.

When I later became a police officer, the pattern intensified to an almost absurd level. Distortions, selective editing, and outright lies weren't rare. They were routine, and it often felt like a two-front war. In one direction, we dealt with criminals and violence. In the other direction, we defended ourselves against a hostile storyline that often portrayed us as the enemy. So, while Iraq was where I first learned the shape of that fight, law enforcement was where I learned how normalized it had become.

For a while, I naïvely thought this was an isolated American problem. Then I traveled throughout Europe and realized it wasn't at all. The same machinery exists there, too, and sometimes operates even more aggressively. That realization left me suspicious in ways I never expected as a young man. Unfortunately, I don't trust much at face value anymore, and I'm sure it comes out in my writings. I wish I did, but honesty requires me to say I rarely do. Maybe I'll let my guard down one day, but I doubt it.

My late uncle Mark had a crude but fitting expression for this kind of thing: "Don't piss on my leg and tell me it's raining." It's funny in a blunt sort of way, but it cuts to the heart of something important. Disinformation isn't just a lie. It's the insult of trying to convince you that what you can plainly see in front of your face isn't happening at all. When deception becomes a currency of power, staying alert and skeptical isn't cynicism. It's survival.

And that circles back to the larger lesson I began to learn in Iraq, even before I had words for it. Military success does not guarantee political stability. It's a lesson that should have been settled in Vietnam, yet we keep relearning it at significant cost. We had a long way to go before anyone could claim victory in Iraq. Toppling a tyrant didn't summon a utopian wave of Western values, and it unshackled forces that had been held down for decades. The fighting might have won the capital, but the story was only beginning, and it would demand patience and sacrifice far beyond the initial triumph.

# Chapter VIII

## Rustamiyah

In sharp contrast to much of the media's portrayal, we felt tremendous support from the American public. Another Vietnam never materialized, at least in terms of how most of us were being treated. The war was divisive, sure, but most people still rallied behind the men and women overseas. Thankfully, the troops and the war's politics were mostly kept separate.

Just before I was deployed, several local schools sent me letters and cards, wishing me luck and offering their support. Those small gestures meant a lot to me. They also brought my terrified parents some comfort, knowing how many people cared. I shared the notes with my fellow soldiers, and they did more than you might think to brighten our days.

**21 May 2003 – Letter from Kim Linafelter, Teacher at Bryant Elementary School**

Your dad came to school today. We read our letters to him. He had tears in his eyes. He is very proud of you, and so is your little brother (Brad, who was in her class). Stay safe and God bless you!

**25 March 2003 – Letter from Casey Riewaldt, Teacher at Sunny Slope School**

We are writing from Sunny Slope School in Omaha, Nebraska. Your cousin, Nick Grage, is one of my students. He and his classmates have written letters and made cards for you. I hope you enjoy them.

I personally would like to thank you for your service and dedication to our country. You have chosen a noble profession, and I wish you the best in your career. Take care and remember that you are in the thoughts and prayers of many.

Like only young children can manage, a lot of those letters were unintentionally hilarious. Some were packed with earnest advice on how we should fight and win the war, complete with tactical tips straight from a child's imagination. We passed them around and laughed hard, not at them, but at the innocent and heartfelt way they tried to help.

**21 May 2003 – Letter from Seth, Student at Bryant Elementary School**

I am proud to be an American. Thank you for going against Iraq and trying to save the USA. Make sure you hit Saddam Hussein or capture him. Bring some goggles because it is sandy. The world is in your hands. Bring a first aid kit to the war. I hope you don't get injured. Keep fighting. Here's a clue for you: to destroy a tank, aim at the bottom of the tank. It's your choice, save the USA or lose and die! Be safe and go win! P.S. Go USA!

**21 May 2003 – Letter from Drew, Student at Bryant Elementary School**

Thank you Jeff for protecting our country. It was thoughtful. Thank you for being in the Army. I hope you come back to home. I am sorry if you throw up. I think you will make it back. I hope you will blow up Saddam's head with your M2 Bradley tank. Go USA!

**25 March 2003 – Letter from Megan, Student at Kluckhohn Elementary School**

Hi Jeff! Are you scared to go to Iraq? I would be scared. Do you have to wear those Army tanks on your back?

Some of the letters expressed a real hope that the people of Iraq would be freed from Saddam Hussein's oppressive rule. Even at that age, those kids seemed to understand the larger purpose behind what we were doing. Their words were simple, but they hit hard. They reminded us why we were there, and of the ideals we believed we were fighting for.

**21 May 2003 – Letter from Hannah, Student at Bryant Elementary School**

Thank you so, so, so, so much for going out there and fighting for our country. Stay safe! I know you and everyone else in the Army can save the world. I know you can! You can free the Iraqi people. Peace on Earth!

**25 March 2003 – Letter from Eric, Student at Sunny Slope School**

I hope Iraqi Freedom works. I also hope you get home safe. And I don't want to make you nervous. Just be good and be safe.

For teachers nationwide, the wars and September 11th were complex topics. How is a teacher supposed to inform their young students about what is happening in the country while protecting them from unnecessary fear and anxiety? Small children often have little to no concept of time, distance, or context. It must have been a terribly difficult minefield for many of our admirable teachers to navigate with care and compassion.

**25 March 2003 – Letter from my aunt, Jane Snyder, Teacher at Kluckhohn Elementary School**

I talked to your dad last night. He said you could be deployed soon. I can't imagine what you are feeling. I know we are all scared to death for you. Just do me a favor and KEEP YOUR HEAD DOWN!

We had a long talk about the war in my classroom today. It's kind of tricky talking to kids because you don't want to scare them. They were very interested in what you are learning and what you'll be doing.

We love you, Jeff! Keep up the good work!

Even with caring teachers, many young students at the time were constantly bombarded with doom and gloom on TV and the radio. One particular letter filled my eyes with tears more than twenty years after it was written.

**21 May 2003 – Letter from Peter, Student at Bryant Elementary School**

Thank you for taking part in the Army and trying to save us all. Please try your hardest to save us because if you don't, well, I'm too young to die and so are you. I believe in you.

In the sweltering Georgia summer of 2003, I stood in formation, sweating through my uniform and weighed down by a rucksack and duffel bag. I wasn't particularly excited. We had done this routine so many times before, and it never led anywhere. This time was different, though, and you could feel the shift the moment the word started to spread. New intelligence said the 3rd ID needed extra manpower to cover injuries, combat deaths, and simple exhaustion. The plan was to split Rear Detachment into three groups. Two would fly overseas in staggered waves to rejoin their units on the front lines, while the last group stayed behind to keep Rear D running.

Cyr and I, along with another close friend, SPC Casey Enos, were told we were in the first group. We would leave immediately, with the second set of grunts following at some unknown date. There was no ceremony and no fanfare. We just shouldered our gear, loaded up, and got bused to Savannah, nearly an hour from where we'd been standing. We didn't know it then, but we would be the only 3-7 Rear D group to make it out of the United States. Everyone else was left behind and missed Operation Iraqi Freedom I entirely.

We assumed we'd be flying over on a military cargo plane, something rugged like a C-130. Instead, they marched us straight onto a civilian airliner. We stopped in Frankfurt for a short layover, and some beautiful German women approached us with packages of candy, toiletries, and other little comforts. It was generous and a little overwhelming. While we were already overloaded with gear, we tried to cram those gifts wherever we could.

It was a surreal sight, carrying machine guns and assault rifles onto a commercial flight while attendants did their best to steer snack carts around our massive piles of equipment. Most of them treated us like sons or little brothers, trying to stuff us full of food. They asked where we were headed, whether we were nervous, and whether we had people waiting for us back home. At the time, it just felt like another small kindness on a long road. I didn't yet realize how much it would matter. The moment we departed, my parents fixated on the phone, afraid it would ring with word that the war had already claimed me.

**Summer 2003 – Letter from my father, Bill Harstad, Jr.**

I was out tending my gardens, and Mom came to the back patio door, holding the phone in her hand and in tears. Needless to say, my heart dropped.

I tentatively said, "Hello?", and that is when the magic began. Ann (Pelmear) started by saying that she was your flight attendant out of Savannah. She also said those incredible words you gave her to tell us, "Tell my Mom and Dad and brothers that I love them."

Both Mom and I were reduced to tears by the time the ten-minute phone call ended, but it was an unbelievable gift from a woman we've never met. I think the reality is finally sinking in about you being there, and frankly, it's a tough pill to swallow. Be safe and get your ass home ASAP.

**Summer 2003 – Letter from my mother, Joyce Harstad**

I got a phone call from the flight attendant, Ann, who flew over with you guys to Kuwait. She said that the flight went fine, that you are in her prayers, and that it is an honor to have these kinds of flights, as they all fight over them. She said (your son) wanted to tell you that he loves you very much and, of course, I started to cry.

Only later did I understand what had truly happened on that flight. A few of the attendants quietly wrote down our names and numbers from back home, as they moved through the cabin. None of us thought much of it. To them, it was a way to reach across the distance and steady the people waiting at home. To my parents, that unexpected voice from across the world landed like a heart attack and then, in the same breath, a miracle.

To flight attendant Ann Pelmear, wherever you may be, and to every flight attendant who has ever done something like this for our troops, thank you. Your kindness will *never* be forgotten.

Hours after departing Europe, we landed in Kuwait City. When the airline doors opened, an intense wave of heat slammed into us as we squeezed our overladen bodies outside. We shuffled across the tarmac toward a line of military buses, and one sergeant standing next to me exasperatedly blurted, "Where is all of that jet wash coming from?" It felt like we were being pelted with tiny grains of sand inside an oven.

A nearby soldier working at the airport overheard his undirected inquiry and laughed. "Welcome to Kuwait!" he said.

It was dark as we drove through the city, weaving along the highways and around oil refineries that spewed massive flames from their flare stacks. The sounds of construction and moving machinery rumbled in the background, though not readily visible in the night. Kuwait City felt busy and alive at all hours. Humvees with machine-gun turrets ran to our front and rear, ensuring a smooth passage. The ride was quiet and uneventful, and I stared out the window, mesmerized by the realization that I had made it. It was the first time I had ever been outside the country, and it happened during a momentous time in history, in a distant and exotic land.

The drive itself was surreal. Our convoy blended into regular traffic, with high-priced BMWs and other luxury cars whipping past us, their headlights slicing the dark. That speed and polish was starkly contrasted with our drab, aggressive-looking military vehicles, loaded to the teeth with weapons and gruff faces. Mixed into the flow were hoopties, cars literally falling apart, missing doors, lights, and even seats. The juxtaposition was jarring. Worlds ran parallel, yet they seemed to move in opposite directions. It was the privileged and the hardened, the affluent and the exploited.

A significant percentage of the Kuwaiti population was foreign-born at the time, and that is still true today, with many jobs available for those willing to leave their home countries.[79] Oil was booming then, and the government was rolling in dough, even amid the usual price swings. The country was brimming with expensive, flashy buildings, cars, and jewelry. Yet amid the extravagance, the faces of dirty and impoverished Bedoons appeared along the roadsides. They were intermixed with migrant workers, some of whom endured cruel mistreatment at the hands of their employers.[80] Their stark appearances kept snapping the illusion of an oasis paradise back into reality. It was unmistakably a country of haves and have-nots.

Our base sat just outside Kuwait City, nestled in the rolling sands of the Arabian Desert. Not a single tree or trace of vegetation could be found. The landscape was a drab, muted tan, and giant dunes lazily rolled as far as the eye could see, forming a terrifyingly hostile environment that looked uniform in every direction. Landmarks were nonexistent in that shifting sea of sand, and we stayed inside the confines of the base for fear of getting lost. The sky was filled with fine particles

---

[79] Huda Ata. "Kuwait's Population Hits 4.9 Million, With Expats Making up More Than Two-thirds." Gulf News, May 10, 2025. https://gulfnews.com/world/gulf/kuwait/kuwaits-population-hits-49-million-with-expats-making-up-more-than-two-thirds-1.500122608.

[80] Human Rights Watch. "Kuwait." Accessed November 24, 2025. https://www.hrw.org/world-report/2020/country-chapters/kuwait.

of dust, sand, and smoke, creating a haze that turned the sun an eerie shade of orange. Even through the haze, the sun's bite was merciless on any exposed skin.

Before arrival, we were prescribed the antibiotic Doxycycline as a preventative measure against malaria, and we were ordered to take it throughout our stay. You would think mosquitoes would hate the desert climate, but they still seemed to thrive wherever we went. Later in my deployment, I found them as thick in Baghdad as I had ever seen, rivaling even the lakes of Minnesota. Doxycycline reportedly protected us from infectious diseases, including malaria, but a noticeable side effect was a drastic reduction in the body's ability to protect against sunburn.[81] That meant we had to stay covered head to toe, always. Easier said than done in a desert where temperatures sat well above 105° Fahrenheit (40° Celsius) and often climbed far higher.

**Summer 2003 – Letter from my grandmother, Toni Harstad**

The temp is 90° here (Sioux City, Iowa), but I suppose that would be chilly for you guys. According to the TV, it is very, very hot over there. We have had a number of hot days here, but nothing like your place.

During the daytime, everything absorbed the sun's energy and turned brutally hot to the touch. I took to wearing black Mechanix gloves, one of my most prized possessions, everywhere I went. They were valuable because they prevented contact burns while still being flexible enough for precise weapon manipulation. Our weapons were jet black and radiated heat like space heaters. They were absolutely miserable to handle. Putting the stock of the rifle against my cheek when aiming was about the last thing I wanted to do for painfully obvious reasons.

Worse, we might as well have been the main course of a turkey dinner inside our vehicles. None of them had air conditioning, and many didn't even have working fans. When the fans did function, they just pushed hot, humid, stuffy air around, offering no real relief. In the back of one Bradley, a temperature gauge read a staggering 156° Fahrenheit (69° Celsius). With the ramp closed, we were literally being slow-roasted, fighting to stay conscious without tipping over into heat exhaustion, and my head constantly pounded from dehydration. Mind you, this was just from sitting there, with no exertion required. We chugged water from Camelbaks and liter bottles, but it poured out of our bodies nearly as quickly as we put it in. It was a miserable place.

At night, the temperatures dipped into the 80° Fahrenheit (26° Celsius) range. Strangely, that often felt cold to us after the oppressive heat of the day. It was almost comical watching soldiers bundle up in cold-weather gear on what any sane person would call a warm evening. The desert is a strange, unforgiving place, and you don't really understand it until you've lived in it.

---

[81] Mayo Clinic. "Doxycycline (Oral Route)." Accessed November 24, 2025. https://www.mayoclinic.org/drugs-supplements/doxycycline-oral-route/description/drg-20068229.

Besides suffering through the stifling heat, our stay in Kuwait was mostly uneventful, though we were always getting handed random duties around the camp. I suspect most people picture a massive base with huge walls and layered defenses, but our location was anything but a fortress. A single rolled strand of concertina wire separated us from the outside world, with power generators equipped with work lights stretched along the boundary like a thin, buzzing lifeline. Some nights, we pulled patrol duty, walking the perimeter wire with loaded weapons. It was creepy. The diesel generators hummed constantly and cast a dead fluorescent glow. The light revealed rolling desert sands that stretched out into the night, and then, without warning, a wall of blackness simply took over. It looked less like darkness and more like a portal into another world. On a map, it would have been labeled, "Here be dragons."

There wasn't much fear that we'd be attacked in force inside Kuwait, since combat was mostly isolated across the border, a little over 50 miles away (85 kilometers). Still, there was mild concern that rogue insurgent elements could slip into the country and take a potshot at a lightly defended base. For that reason, patrols and fire watches were set up to keep the compound safe. Every armed patrol I pulled was boring and monotonous, with no contact and no sign of anyone out there. I didn't even see goat herders or roaming Bedouins. The only life we ever spotted in that subdued landscape was the occasional camel and a venomous horned viper half-buried in the sand.

Undoubtedly, the worst attack on U.S. soldiers in Kuwait happened a few months before I arrived. In the early hours of March 23, 2003, SGT Hasan Karim Akbar, a combat engineer with the 101st Airborne Division, sabotaged the lights at Camp Pennsylvania by shutting down a generator. Under cover of that sudden darkness, he hurled multiple M67 fragmentation grenades into tents where his fellow soldiers were asleep, then opened fire with his M4 as people scrambled in confusion. Two officers were killed, Army Captain Christopher Seifert and Air Force Major Gregory Stone, and fourteen others were wounded in the blast and gunfire.[82]

Akbar was tackled and captured on the spot, and the Army later tried him for premeditated murder and attempted murder. In 2005, a court-martial found him guilty and sentenced him to death, a verdict that has been upheld through the military appeals process.[83] As of this writing, he remains on military death row at the U.S. Disciplinary Barracks at Fort Leavenworth, waiting in that legal limbo where military capital cases can sit for years.[84]

Like most military camps in the region, the place was a sea of enormous tents. Dozens of them held troops, cots, equipment, electronics, and makeshift offices. Each of us was assigned an Army cot and a small spot to stack our gear. Someone had to be posted in the tent at all times to avoid theft. Or, as it's often phrased in the military, to keep things from being "acquired" by some-

---

[82] Barrouquere, Brett. "Appeal for Soldier Convicted in '03 Grenade Attack." AP News, August 27, 2014. https://apnews.com/domestic-news-general-news-dbf3b894977b42d8b3bf2818a3fc1f42.

[83] "United States v. Akbar." 74 M.J. 364, *United States Court of Appeals for the Armed Forces*, August 19, 2015. https://www.armfor.uscourts.gov/newcaaf/opinions/2014SepTerm/137001.pdf.

[84] Death Penalty Information Center. "Soldier Sentenced to Death for Iraq War Murder." *Death Penalty Information Center*, March 14, 2025. https://deathpenaltyinfo.org/soldier-sentenced-to-death-for-iraq-war-murder.

one else. We didn't get too comfortable, though. We knew we were headed to Iraq in a matter of days.

Once the logistical hurdles were cleared, we boarded a C-130 as the sun started to slide down the sky. Its red hue reflected off the ground, creating odd mirages in the sand, with twisting shimmers of heat haze playing tricks on the eye. Our brief stay in Kuwait was over, and Baghdad waited just over the horizon.

The cross-country flight was brief, and before long, we were slipping into Iraqi airspace. The crew killed every light inside the aircraft to avoid attracting unwanted attention. The cabin went nearly pitch black, the kind of darkness that presses against your eyes. There was only the raw whine of the engines and the occasional floating red glow of an aircrew flashlight, like tiny embers drifting through a cave. We sat shoulder to shoulder in that sealed metal tube, listening to the turboprops chew the air, knowing we were finally crossing the line into the war raging below.

Not long after, a message circulated that the landing would be rough. The crew explained that the area around the former Saddam International Airport was still hot, meaning it was too dangerous for a normal approach. Small arms or rockets from scattered insurgent pockets could easily take down a slow, obvious target. So instead of a standard descent, we were going to make a corkscrew landing, a spiraling drop that no one with a weak stomach would enjoy.

The C-130, four roaring turboprop engines vibrating through the fuselage, entered the recently renamed Baghdad International Airport (BIAP) airspace at a normal cruising elevation.[85] It slowed, then tipped hard into a steep bank. The floor seemed to tilt under us as the aircraft began to spiral downward, corkscrewing in a fast, stomach-lifting plunge. The pilot had to keep us inside the invisible safety envelope of secured airspace, so the turns were dramatic and relentless as we dropped in altitude. Everything felt heavier than it should have, especially our heads under weighty Kevlar helmets. We sank and spun with no view of the outside world, trapped in blackness, our equilibrium chasing something it couldn't find. Each time the aircraft seemed to level out, and gravity felt normal for a heartbeat, it lurched again into another hard bank, pinning us to our seats. Our bodies kept bracing for the end, but the spiral just kept going and going, engines screaming, darkness swallowing everything, like we were riding a roller coaster inside a black sarcophagus.

The wheels suddenly slammed into the runway, and the impact hit like a punch, a thunderous welcome to the dusty soils of Iraq. As quickly as it started, it was over. The pilot quickly taxied to an unloading area, and we piled out into the night. A group of 3-7 Infantry soldiers was waiting with a convoy of Humvees and a Light Medium Tactical Vehicle, an LMTV, a 2.5-ton truck with a bed like an oversized pickup. We loaded our gear and weapons into the back and waited for instructions, the adrenaline from spiraling out of the sky still buzzing in our blood.

1st Lieutenant (1LT) Michael MacKinnon, a 1997 West Point graduate, led the convoy and became our first real introduction to Iraq. Calm, squared away, and already carrying himself like someone you wanted in charge, he told us we'd be moving from the airport to Camp Muleskinner,

---

[85] Garamone, Jim. "Renamed Airport Gateway to Iraq's Future." U.S. Air Force, April 4, 2003. https://www.af.mil/News/Article-Display/Article/139574/renamed-airport-gateway-to-iraqs-future/.

later known as Camp Cuervo. Today it's more widely known as the Iraqi Military Academy Rustamiyah, the site of Iraq's oldest military academy, about six miles southeast of Sadr City.[86] To get there, we would have to drive straight through some particularly rough neighborhoods that hadn't been pacified yet, with active fighting still ongoing. He gave us a quick, motivating rundown of what to do if we ran into an ambush. The gist was simple. We were going to smash through it and keep rolling, while everyone unloaded on anything that moved. Then he went back to prepping like this was just another day at the office. There was just one problem, though, and a moment later, he stopped cold.

"What do you mean you don't have any ammo or plates?" 1LT MacKinnon asked, disbelief spreading across his face. But he wasn't angry at us. He was genuinely trying to understand what he'd just heard when someone spoke up.

"Sir, they sent us out here with no ammunition, and none of us have any armor," one of our guys said, explaining for the second time that we had nothing, absolutely nothing, to defend ourselves with. We had no ceramic plates, no rounds, no grenades, and no radios or phones. We had absolutely no way to protect ourselves, except maybe by throwing our empty M-16s and swinging M-249 SAWs like baseball bats. We were standing in the middle of an active war zone, completely exposed, and the paperwork apparently just said, "Good luck, and don't die."

This glaring logistical failure had been clumsily overlooked in the scramble to push us out of Kuwait and into Iraq. Back in Kuwait, leadership assured us we would draw ceramic plates (the heavy ballistic inserts for our body armor), ammunition, grenades, and the rest of our combat load from units at the airport as soon as we landed. But when we touched down at BIAP, the leadership there flatly disagreed. They insisted we should have been issued all that in Kuwait, long before we ever foolishly stepped into a war zone. The result was instant finger-pointing over who had screwed up badly enough to send us forward completely combat ineffective.

So, that was that. With nothing available at the airport, we now had to draw everything from our unit. The only problem was… our unit was dozens of miles away, on the other side of hostile territory, and the route there ran through those neighborhoods we'd just been warned were still very much hot. Clearly, no one had thought through what "reinforcing the unit" was supposed to look like once real bullets were involved. We were left in a precarious spot, and this wasn't how any of us imagined our combat tour beginning.

1LT MacKinnon stared at us, still trying to process the insanity of it. The LMTV we were about to ride in was unarmored, basically a troop truck with thin steel sides meant to keep rucks from bouncing out. It wasn't built to stop fragmentation, let alone bullets. Up front, there was a single mounted machine gun in a turret. *One gunner.* That was the whole security plan for a truck full of unarmed, unarmored soldiers rolling into Baghdad.

But to his credit, he didn't freeze or panic. He did what good officers do. He made an improvised plan and, halfway through, started to make us believe it would work.

---

[86] Global Security. "Camp Rustamiyah." Accessed November 25, 2025.
https://www.globalsecurity.org/military/world/iraq/camp-rustamiyah.htm.

"OK, here's what we're gonna do," 1LT MacKinnon said. "If we have contact along the way, all of you in the back hunker down and stay out of sight. Those of us with ammunition will react to the ambush and kill them all."

He ran through a quick, stripped-down version of Battle Drill 2, which infantrymen know as "React to Contact." In practice, it was going to be ugly. None of us could return fire or offer any help. Our main jobs would be to not get in the way, not die, and not go running off screaming like a four-year-old who just saw a monster under the bed. But still, we were liabilities sitting in an un-armored metal box. It would be like shooting fish in a barrel if something happened.

At least he had a plan, something no one else seemed to offer. More importantly, 1LT MacKinnon exuded confidence and competence, and in that moment, it mattered more than armor ever could. We all knew what capture by the enemy meant. It wasn't just about becoming a prisoner. It meant torture, sexual assault, and execution. We would learn later, as the war dragged on, just how far the insurgents were willing to take it. The 2004 murder of American contractor Nicholas Berg, broadcast to the world, made that horrifyingly clear.[87] We also discovered that some of them used power tools to torture people to death.[88] Meaning, surrender wasn't an option, not for any of us.

"What do we do if *they* get killed?" someone behind me muttered. The unease in his voice echoed what we were all thinking. Almost everyone was troubled by the idea of relying on this small group of unknown soldiers for our protection. After all, we had no means of returning fire, no communication devices, and absolutely no idea where we were or where we were going.

"Run and don't get caught," Cyr said, his response muted and flat.

That was it. That was the plan forced on us by poor preparation. Just run away and don't get caught. With a bit of luck, the good guys show up before anyone grabs you and saws your head off on TV. And hopefully you have been doing that extra PT, because outrunning Al-Qaeda on foot was suddenly on the short list of required survival skills. *You're gonna need it.*

Before we rolled out, the machine gunner on the LMTV leaned down and asked if we actually had ammunition. He sounded like he expected us to laugh and admit it was all a joke. Like any second, someone was going to pop out and yell, "Smile, you're on Candid Camera!" When nobody did, he just shook his head and handed out a few of his extra magazines to the M-16 riflemen. It was kind, and it was something, but it felt like slapping a Band-Aid on a gunshot wound. At best, a couple of soldiers had thirty rounds to their names. That was what we were going into Baghdad with.

---

[87] Lathem, Niles. "U.S. Hostage Screams in Horror as He Is Beheaded." *New York Post*, May 12, 2004. https://nypost.com/2004/05/12/u-s-hostage-screams-in-horror-as-he-is-beheaded/.

[88] Ligabo, Amebyi. "Summary of Cases Transmitted to Governments and Replies Received, Addendum to Report of the Special Rapporteur on the Promotion and Protection of the Right to Freedom of Opinion and Expression." United Nations, March 26, 2007. https://docs.un.org/en/A/HRC/4/27/Add.1%26Lang%3DE.

**22 April 2025 – Reflection of Donald Cyr, HHC 3-7 Infantry**

We landed at Baghdad International Airport, got out of the C-130, and had a quick review on battle drills.

My first memory of Iraq was encountering Lieutenant MacKinnon. He told us that if we made contact with the enemy and were attacked, we were to "kick ass and take names." He was awesome, and I was glad he was the first person we met.

When he realized we were there with no ammo, he was pissed. So, they passed one magazine around to everyone, maybe just enough to get us out of a firefight. Still, I thought, *Yes, I'm in the shit now! This is great!*

I watched *Lawrence of Arabia* when I was a kid, but I never expected to be fighting in the same region.

The convoy pulled out of the airport and into the city, cruising briskly along streets littered with rubbish, burned-out automobiles, and the occasional wrecked armored personnel carrier (APC). Iraqi tanks and unexploded landmines still dotted the shoulders, sometimes roped off for later disposal by combat engineers. A thick, putrid haze of burning trash and diesel hung in the night air, and sporadic bursts of machine gun fire cracked in different directions with no warning and no pattern.

Heaps of rotting garbage piled high on street corners, proof of how completely public services had collapsed. Sewage ran openly through the gutters. Rats the size of house cats and feral dogs moved through the filth like they owned it. Somewhere ahead, for reasons I still don't understand, a machine gunner was hosing rounds into the sky. Red tracers arced upward, bright and lazy, before burning out in the dark. Every so often, a distant *whump* rolled across the city as mortars or something heavier detonated out of sight. We rode in silence, eyes straining into the black, waiting for the first sign that the chaos outside had decided to notice us.

**22 April 2025 – Reflection of Donald Cyr, HHC 3-7 Infantry**

Everywhere smelled like a mixture of shit, fuel, and burning trash. It all just hung in the air.

From BIAP to Rustamiyah, it was about a twenty-five-mile (40-kilometer) drive, crossing the Tigris River and skirting through Al-Dura, Al-Saydiya, and Karrada. I leaned over the LMTV's railing, taking in my surroundings with a mix of awe and nerves. Almost no one was outside. A curfew had been imposed and violating it at that time was dangerous for Iraqi civilians. Simply being seen near a passing American convoy could get someone marked as a potential threat. Everyone seemed to be on edge, military and civilian alike, locked in a relentless state of hyper-vigilance.

That said, as I soon learned, great care was taken by everyone to avoid harming noncombatants and creating pointless casualties. We were here to liberate the country, and no one wanted the death of an innocent person on their conscience. At the same time, we had to learn to react to any perceived threat while under enormous pressure.

Even more dangerous for the civilian population was the city's open anarchy. Roving insurgents and other armed militias were targeting coalition forces and each other, often in crowded areas, indifferent to civilian casualties. In many cases, they actually wanted stray American fire to hit someone innocent, hoping it would turn the public against us. So even when the streets looked empty, we could never relax. We had to expect an attack at any second. It was a complete mess.

Everything seemed to be on fire, and smoke was everywhere, making it hard to see. Gunfire continued erupting in the distance in different directions. The whole ride felt like a balancing act that could spiral out of control at any moment.

A few times, our small convoy slowed to a crawl, not because of dense traffic, but because the road ahead simply stopped being a road. Asphalt gave way to broken slabs, cratered patches, and piles of debris that forced us to weave around them like a drunken snake. The city looked bombed out and somewhat abandoned, lit only by scattered fires and the occasional working streetlamp. Broken buildings loomed on both sides like rotting teeth. Their windows were shattered, their insides exposed, and in the flicker of flames, you could see furniture, curtains, and personal junk hanging in midair as if the people had just evaporated mid-sentence.

Every intersection felt like a decision point between life and death. We would roll up, the machine gunner would sweep rooftops and alleys, and then we'd punch through fast, as if the whole city might wake up and lunge at us if we hesitated. The problem was that you couldn't yet tell what a threat looked like. At least I couldn't, not then. That would come later. A dark doorway, a parked car, or a shadow could be nothing. Or it could be a man with a rocket-propelled grenade (RPG) waiting for the perfect angle on an unarmored truck full of fresh meat. Nonetheless, my eyes stayed glued to the corners and windows, but with haze in the air and darkness swallowing everything, it felt futile.

At one point, a burst of gunfire cracked somewhere nearby. It wasn't aimed at us, at least not that we could tell, but the sound was close enough to make everyone tighten up. The turret gunner swung his weapon toward it, and the entire truck went dead quiet except for the roaring engine. I remember feeling the urge to duck, then laughing at myself because ducking inside an LMTV with no armor was basically just choosing which part of your body you wanted to be hit first. We stayed hunched anyway, because that's what your instincts tell you to do.

Rolling on, Baghdad kept throwing little shocks at us. A burned-out bus sat sideways across a lane, still blackened and warped, like it had died right there in protest. A couple of blocks later, we passed a tank half-sunken into rubble, its turret twisted at an unnatural angle. Some streets were so dark that the only things you could see were the red glow of the tracer arcs overhead. Others were lit by open trash fires that painted the walls orange and made the smoke look thick enough to grab hold of. It felt less like driving through a capital city and more like slipping through a ruined set from a dystopian movie, except there were no cameras or directors to be found.

When we finally pulled through the gates of the Iraqi War College, we breathed a giant sigh of relief. We had taken no direct fire along the route, and somehow, we were all still upright. As we piled out of the LMTV, we hurried to pass equipment down to the soldiers waiting below. With my rucksack on my back, I jumped, preparing to land several feet down. The cargo bed sat at close to six feet off the ground and, like a complete rookie, which I was, a loose strap from my pack caught on a piece of metal. Instead of dropping straight down, it spun me sideways, and I pancaked into the ground, flat on my stomach, knocking the wind out of me. I wasn't hurt, but I was painfully embarrassed. I had just face-planted in front of the soldiers who were about to be my new unit. Not the best first impression. For every military reader out there, this is *precisely* why drill sergeants in basic training lose their minds over unsecured straps.

**Baghdad, summer 2003: The author (left) with SPC Corey Gilley, standing in front of a Saddam sign that reads: "One Arab nation with an eternal message. Saddam Hussein, a great leader of a great people."**

I spent the next couple of weeks traveling between the War College and Camp Baler, where the brigade headquarters sat. We had been told we were there to take the load off soldiers who had already been in country for months, and that we should volunteer for every duty. So, I did. I was up and ready any time a convoy was heading out or an errand needed to be run, no matter how small. It was fun just being there and helping out. It beat doing gate guard at Fort Stewart, and in my mind, I was finally where I was supposed to be.

**22 April 2025 – Reflection of Donald Cyr, HHC 3-7 Infantry**

We were going out all the time in soft-skin Humvees, the unarmored ones. By chance, luck, or divine intervention, we never got hit. We just kept rolling wherever and whenever they needed us.

Trying to figure out the lay of the land was confusing at first. Baghdad was a huge city, and roadblocks and wrecked vehicles seemed to be everywhere. Cyr was far more experienced, and he was always bent over a map, studying routes and landmarks, while I typically used my spare moments to hunt down cold Fantas or orange Ranis somewhere on base. In hindsight, I should have been glued to his shoulder a bit more, paying close attention to everything he was trying to teach me.

**22 April 2025 – Reflection of Donald Cyr, HHC 3-7 Infantry**

I always studied the map because I was afraid of getting lost if we were hit. I knew a Bradley or a tank was usually on or near every bridge. So, if anything went wrong and we ended up on foot, I understood that the smartest move was to reach the next bridge as quickly as possible.

Learning from my fellow soldiers, it became clear right away that warfare was a moving target. As the weeks passed, they took me out past the wire more and more, either on convoys or patrols. We would use tactics that had worked yesterday, only to find the enemy had already adjusted to them. Then we would change what we did, and they would counter with something new of their own. It was a constant back-and-forth that often chipped away at the advantages we were supposed to have in technology, weapons, and training. It felt like the rules were changing every day.

One example was how we handled overpasses. For a convoy, an overpass was a serious danger point. Insurgents would sometimes hide at an empty bridge and carefully peek over the concrete barrier. When the convoy went underneath, they could fire straight down onto the roofs of Humvees or into the exposed bodies of turret gunners. They could also drop a grenade or another explosive, hoping to disable a vehicle and trap the rest of the convoy. To counter that, our gunners trained their weapons on the exact spot where a head could appear, ready for contact at any second. As our vehicle passed under the bridge, the gunner had to spin quickly around and then cover the far side. From the outside observer's perspective, it probably looked strange, with every gunner in the convoy pirouetting in near-perfect unison as we rolled through.

The insurgents noticed the tactics. They watched, learned, and adapted. Soon, they started stretching fine wire from pillar to pillar across the roadway under bridges. At 30 miles per hour or more, that wire could cut a turret gunner badly, or even kill him if his head and shoulders were exposed above the armor shield. Even when it didn't take someone out, a wire to the face or neck was probably not something you walked away from lightly. We adapted again, and in response, we welded tall metal poles to the front of the Humvees. The poles stuck up higher than a gunner's head and would slice the wire before it reached the turret.

Then the enemy adjusted again. They began attaching grenades or other explosives to those wires. When the pole severed the wire, it also armed the explosive, sometimes detonating right away and sometimes a moment later. We countered by adding sandbags around key points of the turret to soak up some of the blast. And so it went. Every bloody lesson was answered with a new trick, which forced another change on our end. That back-and-forth cycle continued throughout my deployment and was even more pronounced when I returned in 2005. Staying alive meant staying alert, because the enemy was paying attention too.

Early in my stay, I attended a week-long Combat Lifesaver course at Camp Baler. They wanted us to get quick, hands-on medical training before we saw real combat, so most of us new guys were in attendance. Overall, it was a good class. We learned how to triage wounds in the middle of a skirmish, apply tourniquets and pressure bandages, treat sucking chest wounds, stabilize soldiers in shock, and a dozen other skills that could mean the difference between life and death on the battlefield.

The part I enjoyed the least was practicing intravenous fluid administration, even though I found it pretty easy. The rule was that we had to give each other IVs, so we were routinely poked by newbies who had never done it before. That was the painful and annoying part. Needles got jammed through missed veins again and again. By the end of it, my bruised, needle-tracked arm was a clear reminder of why I had no desire to ever repeat that portion of the class. I walked out of there with an arm that screamed "heroin addict" to anyone who didn't know better, which wasn't the look I was going for with the ladies. Not that there were any around.

One evening, I crashed at the brigade's Forward Operating Base (FOB) and met PVT Derek Hauger, another guy who would later become a close friend. He was about my age and had been parked on what probably felt like semi-permanent KP, supposedly for disciplinary problems. I never witnessed any of that myself. What I did witness was a relentlessly funny character who lived to joke, stir the pot, and poke the bear, especially with our NCOs, in the same spirit Opie had. His humor was so quick and so fearless that I figured it had to be connected to why he was getting temporarily blacklisted. At the time, I just knew he made the place lighter. I had no idea yet how much more there would be to his story, or how clearly he would eventually show what he was truly made of.

My temporary cot was inside a packed building with dozens of other soldiers. It was hot and humid, and it smelled like feet that needed washing. All the grunts in there were in desperate need of a bath. I decided to go outside and find other accommodations in the cooler nighttime air. Cyr and Enos were already out there, stretched under an awning. I climbed onto the top of a Humvee near the motor pool and used it as a makeshift bed.

Several hours later, I slowly opened my eyes and stared into the night sky. Half asleep, I drowsily processed what looked like red fireflies streaking overhead. Then I heard indecipherable yelling and snapped out of that dreamlike state. The "fireflies" weren't fireflies at all. They were tracer rounds. Insurgents were attacking the perimeter of the FOB, only a short distance from where I had been sleeping. I rolled off the Humvee with my weapon in hand and hit the ground, still groggy and trying to figure out which way was up.

The commotion had roused Cyr and Enos, too. I found a concrete barrier and dropped behind it for cover, but I could barely see what was happening up ahead. So many tracers tore through the air that, for a second, it looked like anti-aircraft fire. Chaotic bursts of rifles and machine guns cracked back and forth, but in the dark, it was impossible to tell if the moving figures were friend or foe. Power generators and random lights lit up my immediate area, yet they cast hard shadows, making it even harder to see what was going on near the perimeter. I was staring into a mess of noise, streaks, and silhouettes, momentarily unsure of what I should even be doing.

A few minutes later, a swarm of Bradleys came barreling out of nowhere toward the area of attack. Someone fired a flare high into the sky, and it drifted down under a parachute, casting a harsh, eerie glow that made everything look unreal and too sharp at the same time. Sporadic small arms fire crackled somewhere ahead while I tried to weigh my options. I wasn't doing any good, hunkered down where I was. I figured it was best to get back to the building where I was supposed to be sleeping. They would probably be doing headcounts and maybe need bodies for a counterattack.

Reinforcing the perimeter on our own would have been reckless, though SPC Enos seemed eager to jump into the fight. But the grunts on the wire would have itchy trigger fingers, and I had no idea about the layout. Meaning, we would have been charging blindly into the night if we went up there, and I certainly didn't want to get shot by my own guys because they thought I was the enemy.

Enos and Cyr, who had been assigned to a different platoon, stayed near the awning as I moved alone toward the barracks. Ducking low, I stuck close to the cover of several concrete T-walls around the motor pool. The Bradleys' engines whined just outside my field of view. Their tracks dug into sand and concrete, making that heavy chunking sound as they pivoted and hunted for something none of us could see. I paused for a second and watched more machine gun fire erupt along the perimeter. Red tracers punched up into the night and fizzled out high overhead. I still couldn't see what they were shooting at, but I could spot the barracks a short distance away.

Then everything went still. Even the tracks stopped turning, leaving only the faint hum of diesel engines hanging in the background. The flare's parachute kept swaying overhead, throwing sharp, shifting shadows across the sand and debris. The hair on my neck stood up and my heart thudded in my chest as I crept forward. It was too quiet, the kind of quiet that feels wrong. It felt like something was about to step out of the darkness at any second, and I was already bracing for it.

Suddenly, a Bradley roared and opened up with its machine gun. The blast made me jump, and I dropped low again, even though the rounds were nowhere near me. Someone screamed in the background, but I couldn't tell if it was them or us. More volleys of automatic fire snapped back and forth, tracers ricocheting into the sky. Then a soldier came sprinting out of the blackness past me with a handheld radio, so close and so fast that it scared the living hell out of me. For half a heartbeat, I imagined a suicide bomber had broken through, and I was about to get tackled into the dirt. Instead, it was just one of ours, running hard with frantic, indecipherable chatter spilling from his radio. I'm not proud of it, but I think my soul temporarily left my body because of that guy.

I crouched low and ran the rest of the way to the barracks with my weapon in hand. No one was outside, and when I reached the door, I fumbled with it in the dark, my fingers clumsy and slick. I yanked it open and stepped back into the heat and smell, instantly sweating like I had been dropped into a sauna. My eyes fought to focus in the dim light, with just a small glowing bulb on the wall. For a split second, my heart stopped as an impossible thought grabbed me. *I had run in the opposite direction and ended up in the wrong building.* I knew it was a stupid thought, but I couldn't shake it. Had I somehow made it past the perimeter in that foggy, disoriented state? Was I now outside the FOB, in hostile territory? Where the hell was I?

But no, that wasn't the case. My overactive imagination and sleep-deprived mind were playing tricks on me. My eyes slowly adjusted to the familiar layout as my heartrate slowed. Everyone else was shockingly still asleep inside, sprawled on their cots, seemingly unaware that anything was happening outside. I stood there a moment, breathing hard, then shook my head and forced those ridiculous, terrifying ideas out of my brain.

Outside, everything quieted down and went back to normal, just as I was trying to rouse my buddies. We were never summoned for a counterattack, and it was soon as though nothing had happened at all. That was extremely odd to me, because it was the first time I had witnessed a dangerously close firefight, and I had no real grasp of what was happening while it was going on. It was a learning lesson that combat is frequently confusing. You often have no idea where the enemy is, who is firing, where your people are, or what exactly is unfolding around you. The movies usually skip over that part. Once everything calms down and an after-action review (AAR) takes place, countless unanswerable questions often remain floating in the air.

### 22 April 2025 – Reflection of Donald Cyr, HHC 3-7 Infantry

That first firefight was crazy. It looked like it was raining tracer rounds in our direction. We were sleeping across the street from the brigade building, near the Humvees under an awning, when it kicked off. We ended up being more spectators than participants, but Enos was completely jacked up and ready to fight. He and I ducked under a concrete structure and waited it out, while you headed back to your building. I kept thinking that one of those rounds arcing into the air was eventually going to come back down and find me.

When everything finally calmed down, I left my vest and helmet on the rest of the night. They were miserable to sleep in, but I wasn't taking them off. I still have no idea what all the shooting was about.

That was the first time we met SFC Gilpin, who later became a command sergeant major. He was another quiet professional that we highly respected.

As infantrymen, our primary job was to secure the FOB and the surrounding territory using static defenses and mounted patrols in Humvees and Bradley Fighting Vehicles. A lot of it amounted to a "show of force." We were trying to maintain order simply by being seen. That meant standing or walking around in full combat gear, talking to locals, and watching for anything that felt off. It sounds simple on paper, but the area was in such rough shape that it was exhausting work.

The landscape was nothing like the rolling dunes of Kuwait. Here, the ground was a hard mix of yellow dirt and sand, packed down by thousands of years of human traffic and then chewed up by war. Fine, golden dust coated everything and lifted into the air with the slightest breeze. Trees, shrubs, and scraps of vegetation were always caked in it. Every bit of greenery looked like it was barely hanging on, sucking whatever moisture it could out of the barren ground and gritty air.

Many of the neighborhoods were battered and broken in a way that is hard to describe. Even the more modern homes, built from stone or concrete, were often cracked and crumbling. In poorer sections, many people lived in mud houses pasted together like antiquated gingerbread walls, soft around the edges and scarred with age. Power lines drooped haphazardly from building to building. Bullet holes stared out from almost every wall. The people looked as worn down as the streets, showing signs of illness and fatigue, and almost always dressed in tattered, dirty clothing.

Once you stepped outside the relative safety of the FOB, you were fully exposed. We were open targets in a dangerous, unpredictable environment, and it was impossible to tell who was hostile just by looking at them. None of the fighters wore uniforms. Instead, they dressed in traditional garb and blended into the civilians moving through the neighborhoods. That meant the enemy could strike first and then disappear into the crowd. It wasn't a traditional battlefield anymore. It was a deadly rhythm of ambush and counter-ambush.

Another problem was trust, and you couldn't assume anything about anyone. A person who smiled and waved could turn on you a moment later, using friendliness to lull you into letting your guard down. And any bystanders who witnessed an attack suddenly seemed to develop amnesia when questioned through interpreters. They always had "no idea" how insurgents had set up an ambush right under their noses, even when it was apparent they were lying. Sometimes they were terrified to talk. Other times, they were complicit, helping foreign fighters melt into their streets and alleys.

The insurgents also exploited civilians as shields, where they used women and children to protect themselves. Combatants would carry babies or walk inches behind women while crossing locked-down areas. We found explosives and weapons tucked under children's beds, and little kids were used as lookouts, passing word of troop movement to hidden fighters. They knew we were unwilling to shoot noncombatants, especially women and children. That was precisely why they did all these things. Instead, we fired warning shots overhead, which only bought a few seconds before they slipped away and tried again.

The enemy was often very smart and creative because they knew they couldn't survive in head-to-head combat with us. As a substitute, they relied on tactics that are considered dishonorable and reprehensible in the West, but they often worked. Within minutes, the cycle of using human shields would start again, building until a full-blown firefight became unavoidable.

**1 August 2025 – Reflection of Noor "Nora" Aljassim, Interpreter for the U.S. Military**

Insurgents started using little kids against Americans. They'd pay a child fifty dollars and slip him a grenade to hide in his pocket. Then they'd tell him, "Walk up to the Americans and ask for chocolate. When they give it to you, throw this next to them. They will like it." Some of the kids actually did it, not necessarily because they wanted to hurt anyone, but because they were tricked into it.

After I learned that was happening, I started asking kids at patrols and checkpoints, "What's in your pocket?" If I found a grenade, I took it from them.

Insurgents didn't stop at human shields. They also hid weapons in mosques, hospitals, and even ambulances. It was clever, in a sick sort of way. If American forces refused to search or strike those places, the weapons would be pulled out later and used against us. But if we did hit a mosque or an ambulance, propaganda videos would spread instantly, showing the "infidels" desecrating untouchable targets. Either choice fed their story, so it was a no-win situation.

Garbage, boxes, furniture, and loose junk were everywhere, piled in streets and left in random spots. Any of it could hide explosives or serve as a distraction, setting up a perfect kill zone. There were too many variables in the city to count. Every patrol felt like trying to solve a puzzle while someone kept swapping the pieces. The number of potential threats was overwhelming, and it took every ounce of focus just to track suspicious people and suspicious objects at the same time.

To complicate it further, unexploded military ordnance was littered around and repurposed by insurgents into improvised explosive devices (IEDs). Grenades, mortars, mines, and makeshift bombs could be readily found on streets and in random buildings. Adding to the volatility, Iraqi civilians were allowed to keep a single weapon, such as a pistol or a fully-automatic AK-47, at home or work for self-defense.[89] With that many weapons floating around in a city already fraying at the seams, people sometimes called it the Wild West, and it didn't feel like an exaggeration.

Making matters worse, incidents like the Abu Ghraib prisoner abuses, carried out by a small number of American soldiers, eroded the image we were trying to project to Iraqis and to civilians back home.[90] Most of us could only shake our heads in disappointment. It didn't represent what the vast majority were trying to accomplish, but it still gave us a black eye that the military spent years trying to heal. Atrocities like that also became an excuse by some in the media to paint

---

[89] Coalition Provisional Authority. "Coalition Provisional Authority Order Number 3 (Revised) (Amended): Weapons Control." *Cyber Cemetery*, December 31, 2003. https://govinfo.library.unt.edu/cpa-iraq/regulations/20031231_CPAORD3_REV__AMD_.pdf.

[90] Taguba, Antonio. "AR 15-6 Investigation of the 800th Military Police Brigade." *Internet Archive*, May 27, 2004. https://ia601301.us.archive.org/29/items/TagubaTortureReport/Taguba-Torture-Report.pdf.

all soldiers with a broad brush, the same kind of over-generalizing that veterans from Vietnam knew all too well.

Eventually, I was assigned to 1st Platoon of Bravo Company, 3rd Battalion, 7th Infantry Regiment, while Cyr and Enos stayed back in HHC. I didn't see much of them for the rest of my shortened tour. Bravo Company went by the nickname "Bandits," and it became my new and permanent home. I was glad to join their ranks, even if I was starting over again, surrounded by strangers.

**1st Platoon quarters at the War College, a concrete sauna we called home**

I was formally introduced to the platoon, led by 2nd Lieutenant (2LT) Johnston and SFC Page. Our living quarters were on the ground floor of an Iraqi military barracks in an open-bay setup that reminded me of basic training. The temperature stayed over 90° Fahrenheit day and night, and there were no air conditioners. The upside was that I finally drew a pair of night vision goggles and the rest of the gear I would need for the duration of my stay.

**15 August 2003 – Letter to the Harstad Family**

It isn't too hot right now (8:45 AM), probably 85–90 degrees. Around 11 is when it gets really hot, 120 degrees. You want to avoid going outside at all costs when it gets past 1100. Around 2200 (10 PM) is when it starts cooling off. The weird thing is that the wind is blisteringly hot. Once in a while, it gets hot enough that it is hard to open your eyes because of the heat. Imagine wearing an Interceptor vest with plates, a Kevlar helmet, web gear with ammunition and grenades, a rucksack, and more. It sucks!

I quickly hit it off with my bunk neighbor, one of the platoon's M-249 SAW gunners, PFC Noel Mata. We bonded over South Park, music, and our quirky senses of humor. He had one of those infectious personalities that made it impossible not to like him, and he did his best to welcome me into the group, even as I still felt out of place among all the new faces.

At first, I felt like an outsider, an observer looking in through a window. Over time, though, the platoon slowly pulled me into its orbit. I began to feel like I truly belonged to a group of soldiers I could trust and admire. I looked up to my fellow enlisted grunts as they carried themselves like seasoned warriors, and I did my best to emulate them. They took me in and gave me purpose, and I will always be grateful to those veterans who had suffered and endured through the early days of the war.

They schooled me in the unit's history and the daring missions they had recently survived. I was envious of their stories and desperately wanted to be part of future operations. They had been integral in the Karbala Gap, Objective Peach, BIAP, and supported the Thunder Runs. Despite the intensity of those famous operations, they minimized friendly casualties while destroying the enemy. They were the true heroes of the initial invasion, and all I wanted was to follow in their footsteps.

Not all my first impressions were positive, though. Looking back, I know I ruffled a few feathers because I constantly volunteered for duties and convoys. None of those guys knew about my frustrations with the 30th AG or the incompetence I had seen in Rear D, nor would they have cared much at that level of exhaustion. From my perspective, I was just trying to get involved in as much as possible and make up for lost time. From their point of view, though, I was showing them up in front of the NCOs by jumping on every task. I should have handled it differently, and that was my mistake. A more measured approach would have served me better, and I took some grief for going overboard. Live and learn.

My first memorable assignment at Bravo Company was an eye-opening experience. I walked to the motor pool and was shown a heavily damaged LMTV, the same kind we had arrived in, that had been knocked out of the war. A soldier from our battalion had recently been killed in it, with several more wounded, and the truck looked every bit the part. There were holes punched through the metal, patches of blackened and charred paint, and a jagged cavity where an RPG had apparently entered. Dark, dried blood stained the headrest and dashboard. It was a pretty gruesome scene.

**Two BFVs and an Abrams tank in position among war-scarred buildings in Baghdad**

I was ordered to clean the LMTV because it was reportedly being returned overseas. For several hours, I separated live from expended ammunition, cleared out trash, and scrubbed the bloodstains as best I could. I understood exactly why they had given me this job. The soldiers of 3-7 didn't want any of their own cleaning up the site of their friend's death. They chose me to do it because I hadn't known the deceased, and because it was the perfect reminder for a new grunt of what combat could do. As strange as it may sound, I felt honored to clean the place where he had died, and I did it with as much respect and care as I could.

**Bravo Company BFVs and LMTVs ready to roll at the Iraqi War College, Rustamiyah, summer 2003**

Several other vehicles needed the same treatment, and I spent a few more days working through mangled pieces of metal. Not all were as gory as the first, but each bore unmistakable signs of combat. Many had drained themselves dry of oil and other fluids, as if they had bled out into the hard-packed sand where they sat. Seeing so many vehicles bound for the scrapyard was a somber sight. They were the ghosts of battles recently fought just beyond the wire.

My family knew none of this. From the moment I arrived overseas, my connection to people back in the United States had all but vanished. We didn't have access to phones or computers, and mail wasn't a guaranteed luxury. While I stayed busy working, my family struggled with the sudden communication blackout. The same torment must have been playing out for thousands of other families across the United States.

At some later point during my rotation, a colonel stopped by and asked if the platoon wanted to use a satellite phone to call home. Naturally, everyone jumped at the chance. The only way to get a signal was to go up on the roof, and even then, it was hard to keep a connection. I climbed partway up a metal tower perched atop the building, hanging on with one hand while holding the phone to my ear with the other, listening for any response.

**On the roof where I often slept to escape the heat — right next to a helicopter pad that ran 24/7**

My dad picked up, but there was a noticeable delay on the line. He exasperatedly shouted for my mom to join the call, and they were both ecstatic to hear my voice. It felt good to talk to them, but the fear in their voices was unmistakable. I tried to explain that I was safe where I was, yet they saw through that lie immediately. As if on cue, gunfire erupted in the background, and they could hear it, which only deepened their dread.

I was forced to climb back down the tower to safety, but the connection broke up so severely that I could no longer make out what they were saying. That was the end of my only phone call during the deployment. Even so, I was grateful for those few moments of contact. More than twenty years later, my dad still talks about that call and the heart-stopping fear he and my mom felt while they listened to the war raging in the background.

**July 2003 – Letter from my father, Bill Harstad, Jr.**

I'm falling into this funk again, sitting here wondering what it is you're doing and what you have been doing. I can't even imagine, kid. I think, from my standpoint, these letters help me deal with all of this as well. I could take basic training, no problem. But "My kid is in Iraq fighting" is a difficult time for me emotionally. Be safe!

When I did write home, my letters rarely contained anything of genuine interest. At Fort Stewart, we had been instructed to censor our own mail, removing details such as our location, daily activities, or anything else that insurgents could use if intercepted. Maybe it was paranoia, but I made a point of keeping everything deliberately vague.

**7 August 2003 – Letter to the Harstad Family**

Everything is good here. I have been mainly doing small bullshit work here and there, like burning the shitters, serving chow, pulling guard, etc.[91] Once in a while, I go out on convoys, and that is where it is fun.

Even the most unglamorous jobs had their moments of dark comedy. The previous letter detailed one particular duty I was stuck with, and I didn't like it much for obvious reasons. I was the man in charge of burning refuse, which typically meant giant tubs of feces and urine. The indoor plumbing didn't work, so we couldn't use the showers or squat toilets in the barracks.

For reference, squat toilets are the norm in that region of the world. They're essentially a porcelain-lined hole in the ground that you squat over to do your business. The drain hole is small, so toilet paper isn't used. Instead, Iraqis typically use a splash of water from their hand to clean their backside, somewhat like a much less powerful bidet. I had always heard that the left hand was used for hygienic purposes, which is why using it for anything else carried an implied insult. For example, offering to shake hands with your left hand is considered deeply disrespectful.

Outside our building, several makeshift outhouses had been pieced together from plywood. Under the plywood seats, large metal buckets served as a repository for all the waste. Every day, since I was low man on the totem pole, I would open the hatches built into the bottoms of the outhouses and pull out the disgusting pails. Flies and shiny dung beetles swarmed in droves as we tried not to gag on the stench.

We would douse the contents with diesel fuel, though it often refused to catch. The cure for this was to mix in some gasoline, which soon created a raging inferno of human waste and the unsettling popping sounds of beetles. After the flames died down, we used metal garden poles to stir the witch's brew and added more gasoline until it was completely charred. Then it was buried. For lack of a better phrase, it was truly a shitty job.

During one of these delightful and aromatic sessions, SPC Tangi from another platoon was tasked with helping burn the latest batch. It was always risky to pour gasoline onto the metal tubs a second time, as they would often spontaneously combust from residual heat. We didn't want to be caught in an explosion, especially when one of us was holding a gallon of gasoline. Tangi thought he had outsmarted the danger. He decided to stand back, place one end of the metal garden pole into the bin, and pour gasoline down the other end at an angle.

---

[91] Lutheran Military Veterans & Family Ministries, Inc. "Military Burn Sites Iraq." Accessed November 25, 2025. https://lmvfm.org/media/2020/07/Military_Burn_Pit_Locations.pdf.

It was an excellent idea… until it wasn't. In a flash faster than any Hollywood special effect, the flames shot up the pole and scorched every bit of hair off Tangi's face. He was uninjured but looked like a cancer patient undergoing chemotherapy. We laughed about it for years afterward, and he was indeed lucky only to lose his eyebrows and eyelashes.

On another occasion, just before leaving Iraq, PVT Barnes and I were assigned burning duties for any extra gear or supplies we didn't plan on bringing back to the United States. This meant that people were discarding almost everything that wasn't tied down. Toothbrushes, beds, batteries, books, clothes, broken electronics, garbage, and hundreds of odds and ends were unceremoniously heaped into a massive pit about twenty yards from our barracks. All of it, too, was set ablaze.

As we circled the smoldering crater, making sure the pile burned completely, a loud "pop" suddenly rang out. It sounded exactly like a rifle shot, and Barnes screamed in surprise. Convinced we were under sniper fire and standing there without our gear, I instinctively sprinted toward the barracks to get help. Out of my peripheral vision, I saw Barnes wildly flailing at something white on his face and chest.

A discarded can of shaving cream in the pile, apparently still full, had built up enough pressure and exploded, though to me it sounded like a muzzle blast. The result was Barnes taking a Stay-Puft Marshmallow Man amount of sizzling shaving cream to the face and body. We laughed about it hysterically afterward, but for a few seconds, I was certain we were goners.

Even when you were ankle-deep in human waste or babysitting a burning trash pit, there was still room for a good laugh. Sometimes, that was the only way to get through those tasks.

Between odd jobs around camp and the occasional convoy, I tried to eat as much as I could because I was dropping weight fast. We rarely got any hot chow, and when we did, it wasn't exactly appetizing. Trays of shelf-stable, gelatinous eggs, bacon, and cheese came mixed together in aluminum pans that were hard to choke down, often resembling yellow slime. Still, it beat going hungry, the way so many people outside our wire were at the time.

When those trays ran out, we lived on Meals Ready to Eat (MREs). They weren't awful, but they were a poor replacement for real food. One day, I managed to pack away nine full MREs. I even mixed the condiments, coffee grounds, and Tabasco sauce into a disgusting slurry just to squeeze out every last bit of food. Each meal averages around 1,200 calories, yet with all the digging, errands, bullshit work, and heat, I still grew thinner by the day and dropped back down to 120 pounds.

I wish I could say the rest of my tour in Iraq was more exciting, but most of my days were spent doing menial tasks and picking up whatever I could from my fellow soldiers. Much of that deployment was uneventful and monotonous for me. I was only there for part of the year and was mainly given jobs of limited importance. The original 3-7 soldiers deserve the credit for the big moments and successes of OIF I. I was simply an extra body, filling an open slot in a platoon that had already carried the heavy load. Still, I was grateful to be there, living out my dream of contributing to the team and being part of a special group of soldiers, even if my primary job was to pull guard and burn their shitters.

However, there is one memorable event that stands out in my brain. On August 19, 2003, an explosion rocked our side of Baghdad. Late in the afternoon, a deep, rolling blast hit our FOB hard enough that you could feel it in your chest.[92] The building shuddered, loose gear rattled, and for a few seconds, most of us assumed something on our own base had just blown up. I stepped outside and saw a thick column of smoke rising from across the city in the direction of the Canal Hotel, where the United Nations had its headquarters.

Details came in slowly, in bits and pieces. A suicide bomber had driven a large flatbed truck packed with old military munitions right up to the UN building and detonated it. The blast tore through the structure and the surrounding area. Twenty-two people were killed, including the UN's Special Representative in Iraq, Sérgio Vieira de Mello, and well over a hundred others were wounded.[93] The sirens started almost immediately and didn't stop for a long time. From our position, there wasn't really anything we could do but listen to the chaos and watch the distant smoke drift over the city.

Later, we learned the attack had been carried out by Abu Musab al-Zarqawi's group, Jama'at al-Tawhid wal-Jihad. At the time, his name was just starting to show up in briefings, another foreign fighter leader in a growing stack of threats. Before long, his organization would rebrand as Al-Qaeda in Iraq (AQI) and, eventually, help spawn what became known as the Islamic State of Iraq and Syria (ISIS).[94] He made his reputation with car bombs, suicide attacks, and filmed the horrifying beheadings of Westerners that dominated the news. That bombing at the Canal Hotel was our first brush with him, even if we were only on the edge of the blast radius.

Not long after that, my shortened tour came to an end. At the end of summer, we rotated back to the United States and received a warm welcome at Fort Stewart. It felt strange to go from concussive blasts and burn pits to handshakes, hot chow, and clean uniforms. I took leave and went home, surprising my family in the process. In my hometown, I was overwhelmed with kind letters, questions, and people wanting to know what Iraq was really like. I did my best to explain that most Iraqis I met were just trying to survive after decades of war and dictatorship. Many were cautiously hopeful that something better might finally be on the way, while others clearly didn't trust us and expected the worst. I also expressed my conviction that if we helped break the country, we had a moral responsibility not to walk away when things got hard.

When I left the Middle East that first time, the situation felt tense but not entirely out of control. Car bombs like the Canal Hotel attack were still rare enough to shock people. By the time I came back in 2005, though, that shock had worn off. As I discovered, much of the public sentiment had drastically shifted, the violence had escalated, and whatever thin sense of stability existed

---

[92] United Nations. "Terrorist Attack 20 Years Later: Survivor of UN Iraq Mission Bombing | Footage Before, During, After," August 28, 2023. https://www.youtube.com/watch?v=F9ZyacwOlMg.

[93] United Nations. "Truck Bomb Terror Attack Survivor Tells All." Accessed November 25, 2025. https://www.un.org/en/video/truck-bomb-terror-attack-survivor-tells-all.

[94] United Nations. "AL-QAIDA IN IRAQ." Accessed November 25, 2025. https://main.un.org/securitycouncil/en/sanctions/1267/aq_sanctions_list/summaries/entity/al-qaida-in-iraq.

in 2003 was utterly gone. My first deployment had been about doing the thankless jobs and earning a place on the team. The next time I returned to Iraq, there would be nothing uneventful about it. Many of us would not come home.

**One of many destroyed Saddam Hussein portraits around Baghdad**

# Reflections

At this point in the story, my tour in Iraq marked the time when everything started to make sense. I had left the dysfunction of the 30th AG and the Rear D permanently behind me. In many ways, entering the warzone salvaged my outlook on military service. I had seriously started to question my decision to join, and, to be quite honest, I was perplexed by how the US military could remain on top of the proverbial food chain with the ineptitude displayed.

By a twist of fate, I made the roster for the only replacement wave that filled the ranks of injured and deceased soldiers for OIF I. Many of the guys left behind in the rear never recovered their careers in the military, especially after being constantly exposed to the immorality and laziness of the Rear D. That is a true shame, as some of those young men were genuinely good people. When I rotated back to the States with the rest of my new platoon, I found many irreparably jaded with the knowledge that they had been left behind. They, too, assumed that the entire military must be similar to what they had been experiencing at Rear D for months on end.

As a lower enlisted soldier at the time, I wasn't privy to much of the inner machinations that took place upon our return. However, it was readily apparent that many of the Rear D scumbags who had spread their pestilence were quickly run out of the Army. Unfortunately, this also included some people caught up in the shenanigans who didn't keep their noses clean.

Nonetheless, I was finally proud to serve alongside a solid group of 3rd ID soldiers. We were awarded the prestigious Combat Infantryman Badge, with the citation reading, "…engaged in combat which involved either Indirect or Direct Fire with Enemy Forces and returned fire in support of Operation Iraqi Freedom during the Liberation of the Country of Iraq." I was only nineteen and had already earned my CIB, an honor that some infantrymen go their entire careers without ever receiving.

The division also received one of the highest awards for heroism, the Presidential Unit Citation, which President Bush authorized. All of us received it and were permitted to wear the bold blue ribbon permanently on our uniforms, with the citation reading, "…sustaining few casualties, the 3rd Infantry Division achieved one of the most stunning victories in military history."[95] It was an excellent recognition of the sacrifices of early OIF veterans.

I greatly appreciated the awards at the time, and still do, but the Army had a weird way of rewarding soldiers. Everyone was given the same recognition, from the grunts who were there at the very beginning to those of us sent as replacements. While outside my control, this caused a bit of friction between the two groups of soldiers. It was kind of immature, but I understand their gripe. They had a much harder time than we did, but the Army responded in typical fashion: just issuing blanket awards to cover all bases. Still, it wasn't like we could turn the awards down. It didn't work that way.

------------------------

[95] Department of the Army. "Permanent Orders 110-15, Presidential Unit Citation, 3rd Infantry Division et al." Army Writer, April 20, 2009. https://www.armywriter.com/army-presidential-unit-citation-3dinfdiv.pdf.

More than twenty years later, one troubling aspect of that first deployment has come back to haunt me, along with thousands of other OIF veterans. We were regularly exposed to toxic burn pits, the standard way of getting rid of trash and feces across Iraq. As I mentioned before, units burned everything in open holes in the ground. The smoke was thick and acrid. I was young and never gave it much thought, but those fumes were undeniably toxic, and everyone on base was exposed to them.

In 2014, the Department of Veterans Affairs created the Airborne Hazards and Open Burn Pit Registry to track these exposures and help link them to long-term health problems. Since then, the VA has added a list of "presumptive" conditions for veterans who served in specific locations during Operation Iraqi Freedom or Operation Enduring Freedom.[96] In plain English, if you meet those criteria and later develop one of those illnesses, the VA starts from the assumption that your service probably caused it. You don't have to prove which exact plume of burning plastic or diesel smoke made you sick.

Like many veterans, I developed severe sinus problems and migraine headaches after coming home, issues I had never dealt with before the war. There's no way to prove beyond all doubt that burn pits did it, but enough of us have reported similar symptoms that the pattern is hard to ignore.

Looking back, it's easy to say we should have done more to avoid breathing that smoke. In reality, public services were nonexistent, trash collection didn't exist, and we couldn't leave uniforms, packaging, or broken equipment lying around for insurgents to scavenge. Hauling waste far away from the FOB would have meant sending soldiers outside the wire just to burn garbage and human waste, which wasn't a realistic option when ambushes were a constant threat. So, we burned it where we lived, and because the FOB was small enough, the smoke settled over everything.

Over time, burn pit exposure has started to be mentioned in the same breath as Agent Orange from Vietnam. Higher rates of asthma, chronic cough, sinus problems, respiratory illnesses, and various cancers have been widely reported among veterans who served near those pits. I'm not a medical doctor, but I strongly encourage any veteran who spent time around burn pits to register with the VA and get evaluated. If you have one of the listed conditions, the presumptive rules mean the VA will generally treat your illness as service-connected, instead of leaving you to suffer in silence and fight alone for proof.

One of the most common questions I still get is whether I was afraid during my deployment. As strange as it may sound, I honestly can't say that I was. My awareness definitely sharpened, and there were moments of genuine concern, like our unarmed ride through Baghdad or stumbling around trying to find my barracks during that first firefight. But even then, I wouldn't describe what I felt as fear. It was something different. It was exhilarating and, in a way, strangely intoxicating.

---

[96] Department of Veterans Affairs. "Exposure to Burn Pits and Other Specific Environmental Hazards." U.S. Department of Veterans Affairs, April 9, 2025. https://www.va.gov/disability/eligibility/hazardous-materials-exposure/specific-environmental-hazards/.

Like many young men, I had found parts of my childhood a bit suffocating. It often felt as though any hint of budding masculinity was being restrained, especially in school. So, when the Army handed me a rifle and those shackles came off, it felt like freedom for the first time in my life. Something primal and long dormant inside me woke up, as if it had been biding its time. I wasn't just allowed but ordered to be strong, brave, and aggressive in killing the enemy. That was a powerful message for a nineteen-year-old, and I quickly became addicted to the feeling.

Instead of avoiding danger, I started looking for ways to insert myself into it. If there was a patrol, a mission, or anything that carried even a hint of risk, I wanted my name on the list. When my buddies told stories about the battles they had fought while I had been stuck doing gate guard back in the States, I never once thought, "Good thing I missed that. That would have been terrifying." My reaction was the exact opposite, and I felt a deep, gnawing envy, alongside a need to make up for lost time.

The only "fear" I carried with me, though that word doesn't quite capture it, was the fear of failing the men beside me. I was terrified that, when it mattered most, I wouldn't live up to their expectations. In Lieutenant Colonel (LTC) Dave Grossman's masterpiece *On Killing*, drawing on Richard Holmes's *Acts of War*, he notes that one of a soldier's first emotional concerns is whether, at the moment of truth, he will be able to kill the enemy or will "freeze up" and "let his buddies down."[97][98] My own experience mirrored that exactly.

According to E.B. Sledge, a World War II mortar infantryman in the Pacific, he and his fellow Marines carried a very similar fear. In his powerful memoir *With the Old Breed*, Sledge writes, "The only thing that we seemed to be truly concerned about was that we might be too afraid to do our jobs under fire. An apprehension nagged at each of us that he might appear to be 'yellow' if he were afraid."[99]

I fully agree with all three men. The internal pressure and my desperate desire not to fail the men beside me drove almost every decision I made. But it wasn't about peer pressure in the usual sense. I never felt anyone pushing or coercing me from the outside. The urge was already wired into my own head, quietly steering me toward danger, insisting that I stand shoulder to shoulder with my new brothers no matter the risk.

Modern neuroscience adds an interesting layer to this. The prefrontal cortex, the part of the brain heavily involved in behavior, personality, judgment, and long-term decision-making, continues to develop into the mid-20s for many people. Because of that, teenagers and young adults tend to lean more on the amygdala, the brain's emotional center, especially under stress.[100] That

---

[97] Grossman, Dave. On Killing: The Psychological Cost of Learning to Kill in War and Society. eBook edition. Open Road Media, 2014, Pages 232-233.

[98] Holmes, Richard. Acts of War: Behavior of Men in Battle. Free Press, 1989.

[99] Sledge, E.B. With the Old Breed: At Peleliu and Okinawa. Presidio Press, 1981, Page 19.

[100] Raising Children Network. "Brain Development: Pre-teens and Teenagers," October 2, 2024. https://raisingchildren.net.au/pre-teens/development/understanding-your-pre-teen/brain-development-teens.

doesn't mean they're incapable of rational thought. It simply underscores that, in youth, the brain is still shifting from a mostly emotional framework toward one that better supports careful, long-term decision-making.

The military is packed with people barely out of high school. Recent Pentagon data show that nearly half of active-duty service members are 25 or younger, and roughly one-third of the Selected Reserve falls into that same age bracket.[101] With that in mind, it's not surprising that so many of us were still riding a mix of bravado and half-finished wiring. Our brains were literally still under construction, especially the parts that handle judgment and impulse control. Looking back in my forties, with a wife and kids, I sometimes shudder at the risks I took and the situations I walked into without a second thought. I was truly young and dumb.

Still, I have often wondered if those of us who chose to enlist, especially into frontline combat roles, are wired a little differently than the average person anyway. I suspect there may be some truth to this. More than once, I have told a story to a non-military friend and watched their face shift into open shock, as if they had been dropped into the middle of Hell and couldn't get out. They imagine themselves in my place and hate every second of it. I usually try to reassure them that, at the time, it felt exhilarating or even darkly funny, but their horror is unmistakable. For them, simply hearing the story is traumatic. For me, it was just Tuesday morning, and I wanted more of it.

As I later discovered, there is a peculiar problem that follows anyone who has lived on the raw mix of dopamine and testosterone that war provides. Once you have tasted that, ordinary life can feel thin. Like an addict chasing a high, you start hunting for the next hit after the guns go quiet. Some go into law enforcement, while others chase women, numb themselves with booze or drugs, buy fast motorcycles, jump out of airplanes, or throw themselves into anything that might bring back even a fraction of that charge. I tried more than a few things on that list myself. The problem is that nothing quite climbs to the same altitude. Even the loudest, brightest adrenaline rush feels a bit muffled, like hearing the world through a wall. The colors drain, and life looks a little gray and washed out.

At this same time in my life, many of my civilian relationships started to fade. It wasn't that I stopped liking those people. I just suddenly found it hard to relate to anyone who had never stood in the dark with goosebumps on their arms, listening to the sounds of battle, and breathing in the thick, hazy air of war. For reasons that go beyond words, the men beside me in combat, no matter their skin color or background, felt more like brothers than much of my own family. When I called someone my brother, I meant it in the most literal way I could without sharing blood. Our lives had been permanently fused by what we had survived together.

Circling back to whether I was afraid or not, I wasn't, but I certainly should have been, especially during our woefully unprepared ride through the streets of Baghdad. The British infantry like to cite a crude but insightful rule of thumb known as the "7 Ps," a phrase with many variants:

---

[101] U.S. Department of Defense. "2023 Demographics: Profile of the Military Community." *Military One Source*, 2022. https://download.militaryonesource.mil/12038/MOS/Reports/2023-demographics-report.pdf.

"Proper Planning and Preparation Prevents Piss Poor Performance."[102] It's a reminder that you need to anticipate obstacles and have a plan in place, even if it's imperfect. Of all the lessons I carried away from that moment in time, this one has stayed with me the most: prepare properly, think ahead about what is coming, decide how you will respond, and act with intention.

Yet even the best-laid plans can unravel. Mike Tyson famously quipped, "Everybody has a plan until they get punched in the mouth."[103] The spirit of that line goes back long before Tyson, and it has been echoed in different ways throughout history, but his version stands out for its blunt honesty. A perfect plan is a myth unless it is so vague as to be useless. Real plans, grounded in reality, are always vulnerable to disruption. That first "punch," whether physical, mental, or figurative, is often the clearest signal that it is time to adapt.

When that disruption hits, the unprepared may freeze or abandon course altogether. Panic and emotion hijack reason, and small mistakes can snowball into disaster. Tyson's line about everyone having a plan until they get punched in the mouth is a reminder that confidence, especially when it has never been tested, is fragile. Fear shows up fast when a plan collapses or when there was never any real preparation in the first place. In those moments, you find out whether you were just hoping things would work out, or you had actually prepared for the day they didn't.

Good leadership, in the military or in civilian life, demands more than blind optimism. It assumes people are trained for contingencies and can react almost automatically when the script gets shredded. As I wrote earlier, a line often attributed to the Greek poet Archilochus captures it well: "We don't rise to the level of our expectations; we fall to the level of our training." The point is not that training makes you fearless, but that it carves responses so deep they come out even when you are scared. That instinctive reaction doesn't fix everything, but it buys time. Time to adjust the plan, time to regain your balance, time to steady your mind, and act from training and clarity, rather than panic.

Fear itself is not a moral failure. Instead, it's simply a warning light that tells you the plan has broken and it is time to improvise. The objective measure of preparedness is not the absence of fear, but the ability to function despite it. Fortunately for all of us unprepared newbies, Mike MacKinnon understood both the "7 Ps" and how to take that metaphorical punch in the mouth, then calmly figure out what to do next. He had done his homework, and he led us out safely.

After I returned home from deployment, I learned from Toby Glass that a member of our basic training cycle had died as a result of heat stroke he suffered in Iraq. PV2 Robert McKinley, just 23, had been in Toby's platoon. He collapsed from heat injuries in Mosul and died on July 8, 2003, in a hospital in Hamburg, Germany. The Army posthumously awarded him the Bronze Star, a small recognition of a life cut short.

---

[102] McNab, Andy. *Bravo Two Zero: The Harrowing True Story of a Special Forces Patrol Behind the Lines in Iraq.* Island Books, 1994, Chapter 3.

[103] Nag, Anwesha. "'Everybody Has a Plan Until They Get Punched in the Mouth.' - How Did the Famous Mike Tyson Quote Originate?" *Sportskeeda*, January 5, 2021. https://www.sportskeeda.com/mma/news-everybody-plan-get-punched-mouth-how-famous-mike-tyson-quote-originate.

On that first campaign with 3-7 Infantry, two soldiers from the battalion were killed in combat operations, though I never met either of them. They were spoken of with genuine respect and affection by men in my platoon. PFC Marlin Rockhold, 23, was directing traffic on a bridge in Baghdad when a sniper shot and killed him on May 8, 2003. SGT Michael Crockett, 27, died on July 14, 2003, while on patrol in Baghdad when his element came under rocket-propelled grenade fire. All three were far too young and are missed by more people than they will ever know.

In war, the loss of comrades is expected in the abstract, but that never dulls the shock when a familiar voice goes silent. The battlefield is dangerous in every direction, not only from bullets and bombs, but from heat, illness, accidents, and plain bad luck. After things quiet down, weeks or months or years later, every surviving veteran I know eventually runs headlong into the same question: "Why did my friends die, but I'm still here?"

**Warriors Walk at Fort Stewart**
**Each tree represented a fallen 3rd ID soldier from Operation Iraqi Freedom and Operation Enduring Freedom**
**It would soon have hundreds**

# In Memoriam

<hr>

PFC Marlin Tyrone Rockhold[104]
1 July 1979 (Hamilton, Ohio) – 8 May 2003 (Baghdad, Iraq)


PV2 Robert Leon McKinley[105]
4 November 1979 (Peru, Indiana) – 7 July 2003 (Hamburg, Germany)


SGT Michael Tyrone Crockett[106]
2 January 1976 (Soperton, Georgia) – 14 July 2003 (Baghdad, Iraq)

<hr>

[104] Military Times. "Army PFC Marlin T. Rockhold." Honor the Fallen, https://thefallen.militarytimes.com/army-pfc-marlin-t-rockhold/256607.

[105] Military Times, "Army PVT 2 Robert L. McKinley," Honor the Fallen, https://thefallen.militarytimes.com/army pvt 2 robert-l-mckinley/256789.

[106] Military Times, "Army SGT Michael T. Crockett," Honor the Fallen, https://thefallen.militarytimes.com/army-sgt-michael-t-crockett/256782.

# Chapter IX

## Land of the Morning Calm

After returning home from deployment, one might expect that our unit would have been given months of rest and relaxation (R&R). In reality, we received only a short block leave at the end of 2003. Upon our departure, the base turned into a ghost town again as almost everyone disappeared for a few weeks. Most soldiers instinctively headed back to their family homes in different states, while a few took their loved ones on long-overdue vacations to more exotic destinations.

That downtime was short-lived and passed far too quickly. Before long, we were right back to the daily grind of cleaning, training, and preparing for war. Business as usual.

3-7 Infantry was housed in a long, continuous building, like an outdoor strip mall where all the businesses are connected but still separate. Alpha, Bravo, Charlie, and Delta companies all occupied this extended structure, while HHC and the battalion leadership were located in an adjacent building. At the front of each company area were the offices for the company commander and first sergeant. The rear held platoon spaces, storage areas, a large table for weapons cleaning, and a reinforced concrete bunker where all weapons and sensitive equipment were secured. That dark, windowless armory would become my office for quite some time.

Almost immediately, I was selected to serve as an armorer, responsible for storing, cataloging, maintaining, and repairing all the weapons and sensitive equipment for 1st Platoon. This was in addition to my regular duties as an infantryman. Still, I welcomed the extra responsibility. I believe my NCOs recognized my organizational skills, because they then sent me to Unit Armorer School, where I quickly learned the job.

The school was a course on post that provided concise, hands-on instruction in identifying damage and repairing weapons, NVGs, and other battlefield equipment. I thought of it as a kind of triage, similar to the Combat Lifesaver course, except it focused entirely on military hardware. It was a solid class and even a bit of fun, though it could have gone much deeper than it did. I had always enjoyed taking weapons apart, studying how they worked, and putting them back together, so the assignment suited me perfectly. I embraced the role and became invested in making the job as efficient as possible.

At the time, our armory housed several million dollars' worth of equipment, with every single item individually serialized. Reinforced steel cages and heavy weapons racks lined the concrete walls, each secured with a padlock. Keeping everything organized and in working order was a constant challenge, especially since we relied on a single Excel spreadsheet to track all equipment for roughly 200 soldiers in the company. Every time the armory door was opened or closed, we had to physically touch each serial number and read it off to verify accountability. The process was tedious and consumed an enormous amount of time.

Occasionally, one of the armorers would misplace a pair of goggles or a rifle, shoving it into the wrong rack. Mistakes happened; there were hundreds of nearly identical items, some with tiny identification marks that were easy to misread. Other times, a repair record would be missing from the spreadsheet, failing to note that an item had been shipped off post. This was the equivalent of a red-alert situation, and the entire company would be "locked down." Everyone hated this because it meant no one, not even the commander, could leave. We were stuck there indefinitely until that piece of equipment was found and accounted for, either in our company area or at a repair facility. These lockdowns could drag on for hours, sometimes well into the early morning.

To streamline the process, I wrote a simple computer program that we ran on the desktop PC in the armory. It was much faster and easier to use than the old Excel spreadsheet, which meant lockdowns became far less frequent. No one wanted to be stuck at work longer than necessary, and my little invention earned me a fair bit of goodwill across the company. Very few soldiers, if any at the time, knew how to write software, so to some of them my program looked like digital sorcery. In reality, it was primitive, but it worked, and it made a huge difference.

While I made it a point to improve efficiency wherever possible, my time as an armorer also opened my eyes to the staggering levels of waste that seemed to permeate not only our company but the broader military system as well. Working behind the scenes gave me a front-row seat. Some of us in the armory even joked about the infamous six-hundred-dollar toilet seats the Pentagon was said to have purchased.[107] The punchline was that we were watching that kind of absurdity play out right in front of us. I suspect this squandering of resources has only grown since I left the military, though that is merely conjecture. On top of that, there was a gross mismanagement of equipment, which I will discuss later in the chapters about our deployment to Iraq in 2005.

As an example, I routinely ordered replacement parts for broken weapons. On an M-16 alone, dozens of tiny pieces are needed to keep it in proper working order. Takedown pins, bolt catch roll pins, selector detents, disconnector springs, magazine catch buttons, and other components are virtually unknown to the casual user. Yet, they are crucial for the weapon to function. Each ordered part would arrive in its own plastic baggie, along with a Technical Manual or a Field Manual (TM and FM, respectively).

---

[107] Hiatt, Fred. "Now, the $600 Toilet Seat." The Washington Post, February 4, 1985. https://www.washingtonpost.com/archive/politics/1985/02/05/now-the-600-toilet-seat/917c98b4-c2fc-40a5-808b-87ff4c5884c8/.

These manuals didn't just explain how to repair the specific component. They included detailed instructions on maintaining and operating the *entire* weapon system. That meant we received a several-hundred-page book for every tiny spring, pin, or plastic piece that we ordered. Multiply that by dozens or hundreds of parts at a time, and it becomes easy to picture the amount of wasted paper and ink, along with the grossly inflated cost of producing each manual.

To make matters worse, the manuals contained sensitive military information and couldn't be recycled or tossed in the regular trash. So, we kept every single one of them. Before long, thousands, or more likely tens of thousands, of redundant TMs and FMs piled up in a back storage area of our company, stacked in Jenga-like towers. Other than the armorers, no one ever read them, and even among us, a copy or two was plenty. We definitely didn't need a thousand.

Eventually, someone decided that all the unnecessary manuals had to go. I spent days tearing apart stapled books and feeding them into a shredder that was far too small for the job, reminiscent of my days with the S1 witches at 30th AG. The shredded paper was put into a dumpster and hauled to the landfill. It was ridiculous we were doing all this extra work, while somewhere a contractor was still churning out endless reams of manuals that almost no one would ever open. Wasteful policies like this seemed to be everywhere.

I never understood why a full manual was included with every single part ordered by a unit, but that was the standard operating procedure. At one point, our lead armorer called the distribution center to ask that no more manuals be sent. His request was denied. I don't believe there was any malicious intent, and I don't imagine someone cackling in a back room, trying to waste as many resources as possible. More likely, it was the inevitable result of a mass-production system that had long since gone off the rails. Government bureaucracy had taken hold, and no one seemed willing, or perhaps able, to fix it.

Around this time, I met a new addition to our platoon, SGT Brister, the man who would teach me several lessons, including the "toolbox." He transferred in from another post and became my team leader. Very quickly, many of the other enlisted guys grew envious because he was an outstanding NCO, and they wanted to be on his team. We had plenty of good NCOs, but he was a cut above most. He could relate to lower enlisted soldiers while still carrying out his leadership duties. Instead of defaulting to "Because I told you so," like so many others, he led with purpose and offered genuine motivation.

I'm reasonably certain I shadowed him like an overly eager pup, but I naturally gravitated toward anyone I sensed had something valuable to teach me. He had attended the U.S. Modern Army Combatives course and became our platoon's hand-to-hand combat expert. I, too, had some experience with the program, mostly from getting choked unconscious by DS Whiteley in basic training. Not quite the same, but it was a start. So, I worked alone with Brister quite often and, to my surprise, became pretty good at it. I finally learned how *not* to black out.

**Undated – Letter from my uncle, Mark Harstad**

Tell all your hand-to-hand combat buddies that if I get the Figure Four Leglock on you, you'll be tapping out real quick. Whoo!

Thanks to SGT Brister's recommendation, the Combatives course became another feather in my cap, right alongside my duties as the platoon armorer. Every additional responsibility my platoon trusted me with felt like I was contributing more to our overall success. I was genuinely grateful for the opportunity, especially since outside training slots were rare at the time. With another Iraq rotation already on the calendar for early 2005, command had little appetite for sending people to schools like Ranger or Sniper, or much of anything else for that matter.

**27 May 2025 – Reflection of Derek Hauger, Bravo 3-7 Infantry**

Brister was great. He was probably my favorite NCO of all time. Johnson, Swears, and some others were good, too, but Brister was almost godlike. He could do no wrong and always seemed to have the right answer.

When we weren't performing tasks in garrison, we routinely went out into the woods for training, which included live-fire exercises, vehicle driving, and constant practice maneuvering with the Bradleys. This accounted for a large portion of our time, as the Brads offered invaluable support to dismounted infantrymen. They could also be highly hazardous for obvious reasons, and we had to know what we were doing to avoid serious injury or death. On one of those early FTXs, I was "officially" initiated into the platoon, though some might call it hazing.

During some downtime in the woods, a number of my fellow platoon members suddenly jumped me all at once with no warning. I was held down and hit repeatedly, mostly hard spankings, and then unceremoniously duct-taped upside down to a tree. Now, I don't want to give the impression I was being beaten into a gang or something extreme like that; it was just a bit of roughhousing and having fun at my expense. I wasn't injured and soon worked my way out of the bindings to the laughter of everyone watching.

While this initiation may sound a bit much to some, I didn't see it that way then or now. Too often today, any roughhousing is labeled as bullying, but I genuinely think there is value in certain initiation rituals. I firmly believe it serves a legitimate purpose when done correctly, provided it doesn't cross a line. The key is knowing when the hazing has gone far enough. Thankfully, my teammates, my brothers in the making, did.

In professions like the Army, daily business has an inherent cost. That cost is danger, with the very real possibility of a gruesome death. As such, new soldiers are heavily scrutinized and sometimes subjected to rough treatment to see how they react to the situation. It isn't done out of malice, but as a test. Does he run away or hide? Does he cry to the platoon leader or throw a fit and get bent out of shape? Or does he stand his ground and fight back to the best of his ability? Sometimes, defusing the situation with humor works even better.

**Initiation complete: I was officially part of 1st Platoon and unofficially part of the forestry program**

In my case, I could tell my fellow soldiers weren't doing it out of cruelty. They were sending a message: *you are one of us now.* Their acceptance came with a price, and they still needed to see how I would react under pressure. As I took the hits, I did my best to fight back, even with the odds stacked against me. When I finally worked myself free, I didn't hold a grudge. Instead, I cracked a few jokes at my own expense, and it was over. I was a little bruised and sore, but I had earned my place. I was now officially part of 1st Platoon, at least in their eyes.

Because I took this beating like a man, I was immediately reintegrated back into the platoon. This was later done to every other new soldier who joined us, though some had different forms of initiation, whether physical or mental. All were teased, tested, and thoroughly probed for weakness. If the enlisted soldiers couldn't cope, the platoon got rid of them in one way or another. It was as simple as that. We needed people who wouldn't melt away when the fire rained from the sky. And it was surely coming.

For those outside professions like ours, it may be difficult to accept that some bullying and hazing are acceptable, or, dare I say it, even *needed*. It's hard to imagine doing this in most civilian occupations. I mean, when was the last time you saw an accountant duct-taped to a tree and spanked outside the office? But the Army is a different beast altogether. The fact is that the military is in the business of killing, and there is no room for anything other than pure meritocracy and teamwork. It is far too serious a profession to coddle emotions and reinforce sensitive feelings. Sometimes that translates into a dirty business that is neither glamorous, polished, nor pretty.

This filtering process didn't apply solely to junior enlisted soldiers. It applied to new NCOs as well. Hell, it even applied to officers and the company commander, who would be initiated a few months later. As an example, after switching from the National Guard to active duty, one sergeant transferred into our platoon. He was likable, though he wasn't particularly bright and had a serious hygiene issue. His lack of cleanliness was so severe that he developed a raging fungal infection in the woods, which weakened his immune system, forcing hospitalization. It became apparent he was unfit for active-duty life, and he soon quietly disappeared, weeded out by our NCO corps.

Another sergeant joined the platoon and immediately set a negative tone. He carried a visible chip on his shoulder and constantly saw racism in the most trivial or unrelated remarks. He was from Puerto Rico and insisted his ethnicity was the reason everyone was out to get him, but his claims didn't match the reality of our platoon. We were already an incredibly diverse group, spread across birthplace, language, religion, skin color, education, and socioeconomic status. We even already had a well-respected Puerto Rican staff sergeant, Juan "Papi" Serrano, whose steady, fatherly demeanor made him one of the most trusted NCOs in the unit. Until the new sergeant arrived, race simply wasn't an issue.

For some reason, this sergeant always complained that people in the Army, especially those with pasty white skin like me, were out to get the ""Cans." As a fairly clueless teenager from Iowa, I had no idea what he meant. I had never heard the term used that way, not realizing it was a pejorative. So, I finally just blurted out, "Sergeant, what's a 'Can' and why is it so mad at me?"

"Of course you don't know," he scoffed with venom, rolling his eyes. "You white boys never seem to know anything or give a shit about Puerto Ricans, Mexicans, Dominicans, or any other 'Cans, for that matter." He really leaned on the "cans" at the end of each word.

That's when it clicked. But before I could respond, Hauger, right next to me and never one to pass up a comedic opening or a chance to deliberately stir things up, was already on his feet, threw his fist into the air, and screamed, "Yeah!" He then proudly declared, "I fucking *hate* those goddamn 'Ans!" He very deliberately dropped the "C" from the beginning of the word, turning it into something completely ambiguous.

Predictably, this new sergeant went nuclear because he heard what he wanted to hear, *not* what was actually said. I hadn't uttered a word, but I couldn't choke back a laugh before it slipped out. So, we were both smoked relentlessly, forced to low crawl over rocks and through pine needles until our hands bled while he lectured us about what terrible people we were. He kept demanding to know why we were so racist against Hispanics. Hauger's purposely provocative line had done its job perfectly and was now drawing an audience from some of the platoon, who watched with mild amusement as we were punished for being alleged bigots.

"Sergeant, what do you mean?" Hauger breathlessly puffed between crawling and push-ups, his voice rising, feigning outrage at the accusation. "I've got no problem at all with Hispanics! I just hate those goddamn '*Ans*. You know, lions, Brians, Ryans… Well, maybe the Mayans. Are they Hispanic? *Oh shit*, I think they are."

As usual, Hauger turned the whole thing into a farce. We still got smoked, and the sergeant held a permanent grudge, but his overreaction had an interesting side effect. It unified the enlisted soldiers against a common enemy: the race-baiting NCO. Everyone saw what he was doing, and it didn't sit well, especially with those of us who had to work with him every day.

Later, I learned that Papi pulled him aside and tried to talk sense into him. He explained that I was close friends with another Puerto Rican from a different platoon, SGT Brian Torres, and that neither Hauger nor I was a racist. Hauger had just twisted his words for a laugh. But this sergeant wasn't interested. He kept insisting the platoon was full of racists and made no effort to adjust his attitude or work cordially with anyone who wasn't Hispanic.

### 22 July 2025 – Reflection of Juan Serrano, Bravo 3-7 Infantry

I saw that he was causing more problems in the platoon than he was helping people. We couldn't have that. I tried talking with him, but it didn't work. After that, I went to the platoon sergeant and we decided to move him. You need guys who are going to show up, do their jobs, and not make life harder for everyone else.

After a while, the platoon leader and platoon sergeant got tired of his conduct and booted him, transferring him out of the line company and over to HHC. Many of us cheered when we heard the news, happy to be rid of him. It was a clear reminder that no one was untouchable. Those who tried to stir up conflict with manufactured complaints were not beyond removal. The message was simple: adapt, contribute, and be a team player, or you wouldn't last long in 1st Platoon.

### 27 May 2025 – Reflection of Derek Hauger, Bravo 3-7 Infantry

I had no racism in my body. I don't think any of us did. We were all one color, in my book. He (the Puerto Rican sergeant) just didn't last long in the platoon because of his behavior. Nobody liked him. When he screwed up, it was always because we were "racists."

When not in the field crawling over rocks and being called a racist, I was shacked up in the barracks with two roommates who treated cleanliness like a myth. Both personally and professionally, I was meticulous about keeping everything spotless and organized, so the platoon's leadership naturally hoped my habits would rub off on these guys. They didn't. *At all.*

They were both nice guys, but our room looked like a crime scene from a hoarder reality show. The only exception was my bed and tiny personal area, which sparkled and shone like it had been staged for a recruiting brochure. We'd regularly get a pounding at the door at 0200, an angry NCO storming in and ordering us to deep-clean the entire room. I grew increasingly bitter about the lack of sleep and submitted multiple room transfer requests. Every one of them was denied.

The final straw came during a scheduled inspection. You'd think, with days of forewarning, that tidying up wouldn't be too difficult. You would be wrong. One of my roommates left a ham-and-cheese sandwich sitting on the counter, like it was part of the décor. The inspecting NCO, with the gleeful mischief of a fifth grader who just discovered glue, took the uneaten portion of the sandwich and quietly stuffed it into this guy's pillowcase. None of us had any idea. Hilarious, right?

Fast-forward two weeks. I started to genuinely wonder if something had died in our room. Either that, or my roommate's natural body odor had evolved into a higher life form. Both theories proved incorrect. One evening, we finally discovered the truth: our roommate had been nuzzling up to the maggot-infested sandwich every night, pressed lovingly against his face like a meaty lavender sachet.

That was the moment I realized I had two options: throw an absolute hissy fit until they transferred me, or fake my own death, go AWOL, and start a new life somewhere far away from ham-scented pillows.

Fortunately, I didn't need to feign a heart attack. SSG Brust, my squad leader, must have seen the desperation in my eyes and took pity on me. He permitted me to move off post, though it would be unofficial since I was still lower enlisted. That meant I had to maintain my bunk in the barracks and would not receive a dime in housing compensation. Everything came straight out of my already light pocket.

Even so, I jumped at the opportunity and moved into a dilapidated mobile home with Toby Glass. It was the kind of place where you could feel the wind blowing *inside* the living room. Also, I'm fairly certain it had doubled as a meth lab in a previous life. We were broke and spent what little we had on enough food to keep us alive. Money was so tight that we couldn't afford to turn the natural gas on, and we slept on air mattresses that were usually half-deflated by morning.

But it was still better than staying with the ham-and-cheese twins. To me, it was paradise. I finally had the freedom to do what I wanted, eat what I wanted, and not clean up after dirty coworkers. Sure, it was a little breezy indoors and the neighbors were sketchy, but I was having fun.

My stay with Toby in the resort mobile home villa was brief, though, because we were scheduled for the first of three training and wargame deployments. Word came down that much of 3-7 Infantry was rotating to Korea for the annual Operation Foal Eagle exercise. Beyond the name and the location, we knew very little about what we would actually be doing. Much like the buildup to the Iraq War, we were put on recall and had to be ready to leave within an hour of notification.

Operation Foal Eagle, the modern successor to earlier "Team Spirit" exercises, originated as a recurring combined field training exercise overseen by the United Nations Command, though the primary participants were the United States and the Republic of Korea (ROK). Beginning in the late 1990s, it became an annual exercise meant to give the South Korean and U.S. militaries a chance to train and work together. At its core, the mission was designed to prepare for a North Korean invasion, with mock battles fought at key locations across the country to rehearse pushing communist forces out of the southern half of the peninsula.

Unofficially, though it was evident to anyone in the area, the FTX also served as an overt show of force against the North Korean regime. The ironically named Democratic People's Republic of Korea (DPRK) consistently denounced Foal Eagle and called it a provocation of war, while the United States routinely denied that charge. It is hard to completely dismiss the DPRK's logic when thousands of troops, aircraft, tanks, armored personnel carriers, and ships are storming through the region, many of them operating just miles from the 38th parallel north and the infamous Demilitarized Zone (DMZ), the thin strip of land separating the two countries still locked in a frozen war.

I'm not arguing here about whether the United States should conduct these exercises. I only want to record what I saw, whether the reader views it positively or negatively. It was apparent that there was deep animosity between the North Korean and South Korean governments. But I would also argue that much of that tension was shared by military personnel and even by regular citizens on the street. There was a noticeable edge between the two sides.

Much of this hostility seemed rooted in the legacy of the original Korean War, the attempted assassinations orchestrated by the North Korean regime, and the ongoing covert operations carried out during the decades-long armistice. The nonstop propaganda machines on both sides also fueled it. To me, it looked as though both halves of the peninsula were ready to duke it out again. They just needed an excuse, like two pit bulls snapping at the ends of their leashes.

Eventually, we were given details about our role in Operation Foal Eagle. We learned that we would deploy to the southernmost part of South Korea, to the country's second-most populous city, Busan (sometimes written Pusan), and gradually work our way north during a month-long field training exercise. The operation was scheduled for March, just before the onset of summer rains and typhoon season, which could have complicated or derailed the mission. That timing, however, brought its own challenge: the cold.

Having acclimated to the heat of the southern U.S. and Iraq, we weren't exactly prepared for Korea's harsh, snowy, and often rainy early spring weather. Georgia occasionally dipped below freezing, but that was nothing compared to the biting cold we encountered on the Korean Peninsula. It was a very different animal.

From Hunter Army Airfield, we flew to Alaska for a brief layover, then continued to South Korea. We landed at a busy airport tucked among rolling hills and the rugged peaks of the so-called Yeongnam Alps. Many of the mountains were dusted with bright, clean snow, and the skies remained a constant, heavy gray, casting a somber tone over everything. Even so, the landscape was striking. Dense forests climbed the slopes, and deep valleys cut through the terrain in every direction. It could have been a fantastic tourist destination, but I was there as a soldier, not a sightseer.

**Soldiers from 1st Platoon, Bravo 3-7 Infantry along the Nakdong River, near Busan**

**From top row left, clockwise: the author, Mata, Velasquez, Root, Sturm, Bowman, Johnson, Lovato**

One of the first things that hit me, and something I still remember vividly, was the smell of the place. Walking around the area, an odd, pungent odor drifted out of residential neighborhoods and restaurants. The source was kimchi, a mix of vegetables salted and fermented in special clay vessels. The clay pots looked like artifacts dug out of an ancient tomb and were sometimes buried in the ground or left out on terraces, which explained why the smell could be so strong. Later, I tried kimchi and ended up liking it, though many Americans treated it as an acquired taste. To each their own, I suppose, but I found it to be a sharp, tangy kick that made almost any meal better.

As we drove through the area, I was stunned to see a large black swastika on the side of a hill next to the highway. For a moment, I was utterly baffled. *Surely* there weren't modern-day Nazis in South Korea. That initial confusion highlighted just how little I knew about Asian customs and symbolism at the time. For most Westerners, the swastika is tied almost exclusively to Adolf Hitler and the National Socialist German Workers' Party. In reality, the symbol on that hillside had nothing to do with either of them.

The swastika has been used for thousands of years by various cultures and religions, including Buddhism, Hinduism, and Jainism. In those traditions, it symbolizes peace, prosperity, and spiritual power, and it often marks the path to a temple.[108] Its use in Asia reflects a much older and far more benign heritage than the one most Americans think of. In other words, I still had a lot to learn about the world.

Driving around to take in the sights of South Korea was enjoyable, but we also had a lot of work to do and couldn't spend all our time sightseeing. We were put to work immediately, fighting an Opposing Force (OpFor) with MILES (Multiple Integrated Laser Engagement System) gear. MILES is the military's version of laser tag, far more sophisticated and realistic than anything you would find at a mall in the U.S., at least when it functions properly. Weapons are equipped with a laser, blank ammunition, and a blank firing adapter. When fired, the system emits coded laser pulses that travel across the battlefield and are picked up by sensors worn on the enemy's helmets and vests.

It's a solid system, but it definitely has flaws. Simply hiding behind vegetation was often enough to block incoming beams from striking a sensor. That wouldn't be the case with actual bullets, which could obviously pass through leaves and twigs. There were also ways to cheat. By repeatedly tapping the top of the laser device, a soldier could simulate firing blanks without making any noise, allowing them to shoot without revealing their position. The punishment was severe for anyone caught doing this, which made it a risky temptation. Still, many secretly tapped their lasers, especially when the OpFor bent the rules and did it to us first.

Under normal conditions, this kind of "laser tag" might have been fun, but the weather and terrain worked against us for much of the training. The skies alternated between flurries of snow and frigid rain, both of which turned the ground into a muddy, slippery mess. In the hills and mountains, this made movement especially dangerous. Our fingers and limbs were stiff from the cold, making precise maneuvers nearly impossible. Combined with the steep elevation and muck, it was treacherous simply to walk, let alone move stealthily through the woods. One wrong step or patch of loose dirt could easily spell disaster.

**20 May 2025 – Reflection of Donald Cyr, HHC 3-7 Infantry**

As part of HHC, we deployed about a week ahead of the rest of 3-7 to recon the training sites in South Korea. I vividly remember watching your unit arrive in the field, everyone soaked, shivering, and slogging through the cold, miserable terrain. Sitting in the relative comfort of a heated Humvee, I watched you all stumble past and thought to myself, *That has got to suck.*

---

[108] Harold Goldmeier, "Swastika," EBSCO, 2023, https://www.ebsco.com/research-starters/communication-and-mass-media/swastika.

As my squad crept along a narrow trail cut into the side of a steep ridge, gunfire erupted from an unseen OpFor across the valley. In the initial confusion and scramble for cover, SGT Wiseman, a team leader in our squad, lost his footing and tumbled down the incline. As luck would have it, he slid right past me. Instinct kicked in, and I grabbed onto his LBV, but his momentum dragged both of us toward the edge. My grip finally slipped, and he disappeared over the lip of the ridge. Somehow, his gear snagged on a small, bent tree that looked like it had no business holding a squirrel, let alone a full-grown infantryman, but there he was, dangling from it.

Hanging above what looked like a hundred-foot drop, Wiseman turned his head, eyes wide and unblinking with pure terror. I slung my weapon and inched toward him on my back, heels digging into the slick mud as I clawed at the ground for leverage. Every pull felt like a losing argument with gravity. His gear kept catching on rocks and roots, jerking him sideways each time I freed it, and nearly ripping him from my hands. My boots slipped, my arms shook, and within seconds, I was drenched in sweat and shaking with exhaustion, even though we had barely moved. For a moment, it felt like the mountain wanted both of us dead and was just waiting for one more bad decision.

While all this was going on, the OpFor kept happily hammering away from across the valley, never once bothering to see if the two dangling idiots were ok. Our MILES gear shrieked non-stop, beeping to let us know we were very heroically "about to die" in the simulation. It was incredibly annoying and distracting, especially with the speakers blaring in our ears. SSG Brust grew visibly furious that they were lighting us up while we were trying to keep a man from plummeting to his doom.

Without hesitation, he climbed to the highest point on the ridge, fully exposed, and started scanning for the enemy. I don't know if he was trying to draw their fire, or if he was attempting long-distance diplomacy with a series of animated hand gestures that translated to "knock it off, dickheads." I lean toward the second option. Either way, they "killed" him immediately, and he unleashed a string of expletives that should probably be classified as psychological warfare.

For the rest of us, the nightmare was still in progress. Running on whatever fumes I had left, I somehow managed to haul Wiseman far enough up the hill to safety. The very second he flopped onto solid ground, my brand-new flashlight, which I had just bought at Ranger Joe's for well over a hundred bucks, popped loose from my vest and cartwheeled all the way to the bottom with a faint, heartbreaking smash. I was too exhausted to be mad in the moment, but later I mourned that flashlight like a lost buddy. That thing represented a serious percentage of my paycheck.

We clawed our way up the ridge on hands and knees, grabbing at rocks and roots to keep from sliding backward. Every inch cost us shredded uniforms and screaming muscles. By the time we reached the summit, we were soaked in sweat, caked in freezing mud, and breathing as if we had just sprinted a marathon uphill. There was no time to admire the view, though. We immediately had to descend our way down the other side, bracing against trees to avoid repeating Wiseman's Cirque du Soleil stunt. Jagged rocks tore at our clothing, and cold mud forced itself into every cut and scrape, making everything sting and burn.

Despite it all, I was deeply grateful this was only a simulated battle. It was grueling enough as training. Had the gunfire been real, it would have been terrifying. Being pinned down on the side of a mountain, exposed to fire with no easy way out, wasn't my idea of fun. I couldn't help but think of the grunts who fought there during the Korean War. They endured this kind of terrain under real fire. My respect for them grew tenfold that day.

Would SGT Wiseman have been killed had he fallen? I'm not sure, but it certainly would have ruined the rest of his day. We all managed to escape the ledge with our lives and limbs, though SSG Brust had to stay in place until a range guide could mark him down as deceased. He continued shouting at the enemy on the other side of the ravine as we made our way out of the area, still incensed at their behavior. Apparently, the OpFor didn't share his idea of playing fair.

**27 May 2025 – Reflection of Derek Hauger, Bravo 3-7 Infantry**

It was very cold in the mornings, and everyone was freezing. Then we were made to climb up hills and mountains that were almost vertical. What a time to be alive! I can't believe we did that.

After each mock battle, we would slog back to our campsite for debriefing, usually caked in mud, smelling like wet boots, and thoroughly exhausted. Despite the chaos, working alongside the ROK soldiers was genuinely rewarding. Many spoke only broken English, which led to frequent miscommunications that were more humorous than hazardous. Our attempts at hand gestures, charades, and creatively butchered Korean often got more laughs than results. Still, their professionalism and good humor bridged any language gap, and we eventually settled into a shared rhythm in the field.

Between training exercises, we were occasionally granted short excursions to explore the areas around whatever military installation we were operating from that day, slowly leapfrogging our way north through the peninsula. That part was fun, too. We tried to take in as many sights and experiences as possible. In today's heavily regulated world, it's remarkable to think that we did all of this without passports or any real civilian documentation, just our military ID cards in our pockets. We were basically turned loose in a foreign country, with our command reminding us in no uncertain terms to stay out of trouble. For the most part, we followed that directive. *Mostly.*

The cities were starkly different from anything I was used to in the Midwest. Narrow, winding pedestrian alleys twisted between buildings, lined with bright, colorful artwork and neon signs. Street vendors cooked food in carts on corners, a haze of delicious smoke drifting through the tight passageways. Strange aromas and unfamiliar spices hung in the air, with skewers sizzling on open grills and steaming dumplings seemingly around every corner.

In some back streets and alleyways, black-market vendors sold everything from bootleg DVDs to knockoff Nike shoes and everything in between. If there was a demand for it, you could find it in those gritty alleys, for the right price. In those pockets, it felt like a surreal mix of beauty, desperation, raw human nature, and survival. The people working those stalls and side streets seemed to be hustling for every possible sale, especially those without the safety net of a steady paycheck. For me, it was an unforgettable education in how people adapt when life backs them into a corner.

One day, a group of us ducked into a small, intriguing restaurant tucked away from the main street. It was exactly the kind of place we hoped would serve authentic local food. The menu was entirely in Korean, which we took as a perfect sign. No English, no cartoon mascots, no "American-style" anything. This was the real deal. Using a combination of wildly exaggerated gestures, hopeful pointing, and loudly enunciated English (as if volume alone could bridge the language barrier), we somehow managed to place an order.

I ordered bosintang, a traditional stew built from a hearty mixture of vegetables, sesame oil, red chili paste, fermented shrimp, a handful of bold, unfamiliar herbs and spices, and gaegogi. Gaegogi is Korean for, *gasp*, dog meat. While that's probably jarring to Western sensibilities, it has been part of Korean culinary tradition for centuries.

I absolutely consider dogs to be man's best friend, and I grew up with plenty of them, but I tried the dish to satisfy my curiosity and to experience the culture on its own terms. The stew was rich and flavorful, though the meat itself was a bit stringy. I won't pretend it was easy to reconcile what I was eating with the image of a family pet in my mind, and I felt more than a little weird about it, but it was undeniably delicious. I figured that if karma keeps a menu, my next life is probably going to be as a fire hydrant in a busy dog park.

In 2024, South Korea passed a law to phase out and ban the breeding, slaughter, and sale of dogs for meat, effectively ending the legal dog-meat industry after a transition period.[109] More South Koreans are becoming pet owners, and many are choosing to forgo the consumption of domesticated animals altogether. That is probably for the best. Still, at the time, I lived by the old adage, "When in Rome, do as the Romans do," and in Korea, for one meal, I did.

After spending some time in and around Seoul, which lies only about 35 miles (56 kilometers) from the DMZ, we pushed farther north to Camp Casey for our final FTX of the deployment. This culminating event was a live-fire scenario where we patrolled through the woods and engaged pop-up targets that appeared without warning. We advanced through winding, narrow trenches and got a taste of the brutal chaos of close-quarters trench warfare. The training was outstanding. We practiced bounding overwatch, flanking maneuvers, shift- and lift-fire protocols, and more. It was an infantryman's dream, and I relished every minute of it.

As night fell, we came down off a large hill and moved into a massive concrete bunker. From the parapet, we engaged an array of targets at different distances, all popping up and disappearing in rapid succession. It felt like Christmas morning, and I went to work with my weapon like a kid at an arcade cabinet with unlimited quarters. It was the kind of fun that is hard to explain to anyone who has never been allowed to unleash that much firepower legally.

The M-240B medium machine gun in our squad belched close to a thousand rounds per minute, sending 7.62 millimeter projectiles ripping into the Korean landscape. Everyone else ham-

---

109 Kim, Hyung-Jin. "Production and Sales of Dog Meat Banned by South Korea's Parliament." PBS News, January 9, 2024. https://www.pbs.org/newshour/world/production-and-sales-of-dog-meat-banned-by-south-koreas-parliament.

mered away with M-16s and SAWs, their tracer rounds carving red lines through the darkness. Geysers of dirt kicked up in time with the streaking tracers that zigzagged unpredictably, all under the eerie glow of an illumination round drifting lazily overhead.

Inside that cramped pillbox packed with soldiers, the noise was almost unbearable. The gunfire pounded so hard that my eyes watered constantly, the concussive shockwaves making everything look slightly distorted, as if we were underwater. I had earplugs in, thankfully, but it was still so loud that it physically hurt each time hundreds of rounds erupted at once when a target appeared. Everyone was grinning from ear to ear, even as the pressure built up behind our temples. Tinnitus and headaches were common souvenirs for weeks afterward, but most of us agreed it had been worth every ringing second.

**The author eats an MRE in the dirt near the DMZ, with his trusty SAW by his side**

In what felt like the blink of an eye, our formal mission had come to an end. Before we left, though, we made one final, unforgettable stop at the DMZ and the Joint Security Area (JSA). We left behind the color and energy of Korea's cities and stepped into one of the strangest and most hostile landscapes I had ever seen. Endless rows of razor wire, surveillance cameras, tank traps, and warning signs about hidden landmines dominated the terrain. Elevated guard posts and checkpoints dotted the area around the JSA, and the entire region hummed with military activity.[110] It felt completely inhospitable, locked in a constant state of tension, like a place that existed solely to remind the world that the Korean War had never really ended.

**20 May 2025 – Reflection of Donald Cyr, HHC 3-7 Infantry**

I was with the command staff as we toured near the DMZ. They showed us tunnels wired with explosives, ready to be blown if a North Korean invasion ever kicked off. Landmines were scattered everywhere. I struck up a conversation with an artilleryman who told us about where the U.S. military previously had nuclear-equipped artillery shells sitting on standby, just in case the DPRK ever got any funny ideas.

From our vantage point on the South Korean side, we scanned "no-man's land" with binoculars and picked out the infamous town of Kijong-dong, which, somewhat hilariously, translates to "Peace Village." A massive North Korean flagpole with a six-hundred-pound flag dominated the center of town. From a distance, it looked picturesque and prosperous, a proud showcase of the regime's supposed success. But it was all theater. The buildings were hollow shells, some with windows painted on or missing entirely, and no one actually lived there.[111] It was a propaganda set piece. Convincing from afar, maybe, but the illusion fell apart the moment you looked closer.

The binoculars and viewfinders confirmed every rumor we had heard. We could see crude openings where windows should have been, yet most had no glass, just empty gaps. Some buildings were painted in bright, cheerful colors and looked inviting at first glance. Then you peered through the openings and realized there were no interior walls, no furniture, and no signs that anyone had ever called these places home. Aside from the occasional street sweeper or a passing military formation, the town was eerily still. It was a counterfeit stage, propping up the regime's image and fooling only the truly uninformed.

Years later, I would read about Michael Malice, a Soviet-born American who emigrated to the United States as a child, and his decision to visit Pyongyang on a tightly controlled, guided tour lasting about a week. He wanted, in part, to see firsthand the kind of totalitarian oppression his family had known in the Soviet Union. In describing the trip in a speech at Bucknell University, he later said that being in North Korea was "like going to another planet, back in time."[112] From what I could see from the South Korean side, I would have to agree wholeheartedly. I can only imagine what it must have looked like up close.

---

[110] Sheposh, Richard. "Korean Demilitarized Zone (DMZ)." EBSCO, 2025. https://www.ebsco.com/research-starters/history/korean-demilitarized-zone-dmz.

111 Ryan, Rowena. "The Mysterious Fake Town on North Korea's Border." New York Post, July 11, 2014. https://nypost.com/2014/07/11/the-mysterious-fake-town-on-north-koreas-border/.

112 Malice, Michael. "Michael Malice on North Korea | Bucknell University - 2017." Uploaded by Bucknell Conservatives, March 9, 2017, 7:25. https://www.youtube.com/watch?v=4UuY8XdXHjg.

Across the border, North Korean patrols were clearly visible. Some of the higher-ranking officials could be observed watching us with their own pairs of binoculars, like mirrored surveillance. Audio propaganda, both in Korean and comically poor English, could be heard from massive loudspeakers that were stationed in North Korea and faced south. They featured derogatory audio clips that called for the destruction of the United States and the "puppet regime" in South Korea. Strange music, the rhythmic sounds of soldiers marching, and other surreal noises could be heard through these audio broadcasts.[113] It was beyond unsettling, as though I was in the middle of a dystopian novel. It served as a chilling reminder that the Korean War never officially ended. It merely settled into a fragile ceasefire that could once again erupt into hostilities at any moment.

**"Peace Village" with the giant flagpole just barely visible in the fuzzy distance**

We then proceeded to the House of Peace and its surrounding area, a site designated for diplomatic meetings between North and South Korea and for international delegations. Nearby, there is a small concrete barrier that bisects the North from the South, but it is not intended to keep anyone out. It can easily be stepped over, though almost no one dares to do so for obvious

[113] Associated Press, "North Korea Takes Down Propaganda Loudspeakers From Tense Border," CNN, August 9, 2025, https://www.cnn.com/2025/08/09/asia/north-korea-removes-speakers-border-intl-hnk.

reasons. Situated directly on top of this concrete divider is a building that is "cut in half," with each half being "owned" by the respective nation in which it sits.

**Republic of Korea military police standing guard in the JSA conference building**

Guards from both North and South Korea were found throughout the area, wearing strange military outfits. The South Koreans typically came equipped with oversized helmets and aviator sunglasses, whereas their northern rivals sported bulky coats and furry hats that looked far too big for their heads. They stood motionless, facing each other in a pose that could only be described as something from a cheesy '80s karate flick. Their fists were clenched, elbows bent tightly at their sides as if paused mid-bicep curl. They maintained a half-squat stance, seemingly ready to spring into action, with faces locked in a permanent scowl.

Almost all the South Korean guards I observed were tall and muscular, resembling professional wrestlers more than soldiers. This was starkly contrasted with the communists, who were much smaller and looked a bit underfed. The entire spectacle felt choreographed for maximum psychological effect, serving not only as security but also as a form of visual propaganda and power projection.

Had I not seen all the strangeness for myself, I wouldn't have believed it. It was as though these people were living in a completely different reality, from the way they dressed to how they acted with their counterparts across the invisible divide. It was a throwback to the repressive days of the Soviet Union, and I was there to witness the same echoes in Korea personally. Unfortunately, this perplexing reality persists well into the 2020s, with no signs that it will end anytime soon.

The most unforgettable moment came when we entered the building straddling the concrete border. A thick black line ran through the center, signifying that even in this room, definitive lines of demarcation had been established. With permission from the South Korean guards, some of us crossed the black line and quickly went into North Korea.[114] We posed for photographs and took in the magnitude of the situation. Standing on that side of the room was electrifying.

Nothing dramatic occurred, but we had been sternly warned that North Korean guards could be unpredictable and sometimes extremely aggressive. The ROK soldiers cautioned us to remain vigilant for anything out of the ordinary. For a few brief moments, we could say that we had crossed into enemy territory as military servicemen, though the act was largely symbolic. I briefly considered doing something theatrical for the camera, then decided I didn't want to be the guy who started World War III as a joke.

One infamous example of North Korean hostility we learned about during our visit to the DMZ was the 1976 "axe murder incident," which took place in the JSA, essentially right where we were standing. The story was recounted to us by personnel on site, and it was a stark reminder of how quickly things along that border could go from routine to lethal.

On August 18th, 1976, CPT Arthur Bonifas and 1LT Mark Barrett were in the JSA with a small work party near an infamous site called the "Bridge of No Return." That bridge had last been used for a prisoner exchange in 1968, when North Korea released the crew of the USS *Pueblo*, an American intelligence-gathering ship whose crew had been seized, beaten, and tortured.[115] Bonifas and Barrett had a simple mission: trim a poplar tree that blocked the line of sight between United

---

[114] Bogart, Peter. "Tour of the Joint Security Area Truly Memorable." U.S. Army, May 9, 2016. https://www.army.mil/article/167461/tour_of_the_joint_security_area_truly_memorable.

[115] Cheevers, Jack. *Act of War: Lyndon Johnson, North Korea, and the Capture of the Spy Ship Pueblo*. eBook edition. Penguin, 2013, Chapter 14.

Nations Command (UNC) checkpoints, a clear security risk.[116] As the South Koreans in the crew began cutting branches, a group of North Korean soldiers approached and demanded that they stop and leave the area. Bonifas ignored them and ordered the team to continue. After a brief standoff, more North Korean troops arrived, and the officer in charge suddenly gave the order to attack.

**The author's view of the "Bridge of No Return"**

What followed was short and brutal. North Korean soldiers seized the tools the workers had been using on the tree and set on the Americans and South Koreans with axes and clubs. Bonifas and Barrett were bludgeoned to death, and all but one of the UNC guards were wounded before the survivors were able to break contact and withdraw to safety.

North Korea's official story flipped the script. State media claimed that it was the Americans who had been the aggressors, that U.S. soldiers had launched an unprovoked attack, and that the killings were an act of self-defense. They even insisted the tree was sacred, allegedly planted by

[116] Fitzgerald, Clare. "The Axe Murders of Two US Army Officers Almost Sparked a Second Korean War." War History Online, January 31, 2023. https://www.warhistoryonline.com/featured/korean-axe-murder-incident.html.

Kim Il-Sung himself. In reality, photographs and film from UNC observation posts and aerial surveillance told a very different story, showing a planned, one-sided assault on an unarmed work detail. Even so, at a Non-Aligned Nations meeting shortly afterward, Kim Jong-Il presented the incident as a grave U.S. provocation, and the conference passed a resolution condemning the United States and calling for the dissolution of the United Nations Command.[117]

The American and South Korean response came a few days later in the form of Operation Paul Bunyan. In simple terms, a large joint force rolled back into the JSA with engineers carrying chainsaws and axes, backed by infantry, special operations troops, artillery, attack helicopters, and nuclear bombers overhead, all on high alert while the work crew finished what Bonifas and Barrett had started.[118]

Accounts of the operation describe South Korean special forces loudly taunting the North Korean guards across the line. Some reportedly ripped open their shirts to show Claymore mines strapped to their chests, holding the detonators in their hands and daring the North Koreans to come across the bridge.[119] It was a theatrical, almost suicidal show of resolve meant to send a clear message: if the North wanted another fight, they were ready to take a lot of enemies with them.

None of the North Koreans took the bait. The poplar tree was cut down to a stump and, eventually, replaced by a memorial to the two officers who had been killed.[120] The crisis cooled, but the lesson was obvious. The entire area was a powder keg, and it didn't take much more than a tree branch in the wrong place to light the fuse.

After nearly a month in country, we flew back to the United States and slid right back into our usual training routine. One thing stuck in a lot of our craws, though. We left just shy of the 30-day threshold required to earn the Korean Defense Service Medal.[121] Had we stayed even one more day, the medal would have kicked in automatically. Instead, we punched out at twenty-nine. According to the rumor mill, someone at battalion didn't want us getting medals for that deployment, so they scheduled our return to land exactly one day on the wrong side of the requirement.

Sadly, that kind of thing wasn't exactly rare in 3rd ID, or in the rest of the military, for that matter. Bureaucratic pettiness had a way of creeping into awards and promotions. Plenty of officers and senior NCOs sometimes seemed to walk away with Bronze Stars, Silver Stars, or other impres-

---

[117] Dean, Josh. "Axes of Evil: Four Days, Two Murders, and One Poplar Tree That Almost Ignited World War III." The Atavist Magazine, July 2018. https://magazine.atavist.com/2018/axes-of-evil-north-korea-dmz-tree-murders.

[118] Sander, Gordon. "When the U.S. Almost Went to War With North Korea." Politico Magazine, September 14, 2017. https://www.politico.com/magazine/story/2017/09/14/north-korea-1976-axe-murder-incident-215605/.

[119] Peck, Michael. "Meet the Panmunjom Axe Murder Incident That Nearly Caused a Second Korean War." The National Interest, September 7, 2020. https://nationalinterest.org/blog/reboot/meet-panmunjom-axe-murder-incident-nearly-caused-second-korean-war-168516.

[120] Wight, William. "Memorial Service Honors Victims of Axe Incident." U.S. Army, August 28, 2017. https://www.army.mil/article/193004/memorial_service_honors_victims_of_axe_incident.

[121] U.S. Air Force. "Korean Defense Service Medal." U.S. Air Force's Personnel Center. Accessed November 29, 2025. https://www.afpc.af.mil/Fact-Sheets/Display/Article/421913/korean-defense-service-medal/.

sive decorations for "heroic feats" that didn't quite match what actually happened. "Creative writing" was an open secret in more than a few award citations. Down in the ranks, a lot of us believed the system favored certain cliques, quietly fast-tracking the careers of a chosen few. It bred a bit of resentment toward some leaders, though certainly not all. Or, as George Carlin once put it when he was talking about America's political and economic elites, "It's a big club, and you ain't in it!"[122]

Korea, cold and mud-soaked - our Bradleys lined up and idling, waiting for the command to move out

To be fair, processing a wave of Korean medals probably meant a mountain of extra paperwork at the battalion level. And no one should serve just to chase medals or promotions. The duty itself should be the reward. The problem was that someone explicitly told us we were leaving early to avoid triggering the award. We had frozen in the fields, slogged through mud, shredded our hearing (and maybe given ourselves drain brammage), and come close to falling off more than one mountain. While the training was excellent, the way it ended felt like a minor betrayal by the people in charge, if the rumors were true.

---

[122] Carlin, George. *Life Is Worth Losing.* HBO stand-up special recorded at the Beacon Theatre, New York, 2005.

**Painted in dancheong, this coiled dragon rides a storm of clouds like a built-in guardian, watching the room from the ceiling the way Korea's old builders believed power and protection should always come from above**

In the end, Korea turned out to be a beautiful, fascinating country that exceeded every expectation I had. It would make an excellent destination for anyone interested in Asia, with outstanding food, rich cultural traditions, and a deep, complicated history. If you ever get the chance, a tour of the DMZ is worth the trouble, while it still exists. Its surreal, almost dystopian atmosphere is hard to capture on paper and leaves a lasting mark on anyone who sees it up close. Walking away from that border, I silently thanked God I had been born somewhere else and not trapped on the wrong side of that line.

# Reflections

I have an imprint of a memory in my brain, like a fuzzy photograph that has faded over time, losing much of its color. In this snapshot, I'm looking into a small, subdivided room in the back of Bravo Company, in which a thick metal cage, similar to a chain-link fence on steroids, surrounds it. Standing next to me is a supply sergeant, though his name, like the fading photograph, has withered away from the passage of time. He is pointing to the mountain of TMs and FMs, all piled into massive pillars of yellowing paper. And just like that, I'm launched into a brain-numbing quest to destroy what looked like an entire forest's worth of government documents.

When I was younger, I spoke of the incident with contempt, using it as a prime example of government waste and inefficiency. I couldn't fathom how so many books, supposedly filled with valuable and secretive information, could have ended up collecting dust in a hoarder's cage, never to be opened once. My assumptions led me to believe it was, by design, a scheme intended to maximize profits at the taxpayer's expense. It *had* to be part of some bloated, corrupt system, all designed to burn money and reward the elites.

But age has a way of sanding down the jagged edges of indignation. Though I still remember many such incidents with a fair amount of scorn, my perspective has shifted a bit. I'm now inclined to utilize Hanlon's Razor.

Hanlon's Razor, like its better-known cousin, Occam's Razor, is a general philosophical rule of thumb that aids in critical thinking and helps reduce the tendency to claim by default, "It's a conspiracy!" Instead, the Razor goes like this: "Never attribute to malice that which is adequately explained by stupidity."[123]

Interestingly, one can replace "stupidity" with several other words and come to a similar conclusion. In other words, your boss *probably* isn't trying to sabotage your career by failing to send you the vital memo of the day. An evil genius? Perhaps. But not likely. It's more probable that he or she's simply incompetent, ignorant, overwhelmed, or perhaps merely forgetful.

So, while I may have initially thought there was a conspiracy afoot, I now tend to suspect that the mountain of paper was simply government bureaucracy at its finest. There were, and still are, a lot of cogs in the machine that weren't paying attention to anything outside their microscopic piece of the pie. Disturbing? For sure, but I have somewhat muted my stance of calling them rotten names and making accusations of debauchery.

In the end, that mountain of paperwork didn't reveal secrets or sinister government plots. What it did reveal was more mundane, but conceivably just as troubling. Sometimes, the greatest threats to efficiency and order aren't rooted in malice, but in something considerably more common: stupidity and indifference.

---

[123] Payne, Laura. "Hanlon's Razor." Encyclopaedia Britannica. Accessed November 30, 2025. https://www.britannica.com/topic/Hanlons-razor.

Beyond government waste, I later came to understand another important lesson from this chapter, one that had completely escaped me at the time. Years later, I came across a story in Dr. Jordan Peterson's book *12 Rules for Life* that hit amazingly close to my own experience of being initiated by my platoon. In that book, he describes his time on a rail crew in Saskatchewan, where a bad day at work can maim or kill people.

In Peterson's story, a new guy joins the crew and shows up with his lunch carefully packed in a nice container from his mother, instead of the usual crumpled brown bag. That one small detail is all it takes. The crew latches onto it, slaps him with the nickname "Lunchbucket," and starts giving him a hard time.[124] Rather than rolling with it or firing back a joke of his own, Lunchbucket gets visibly offended and defensive. The more he bristles, the more they needle him, and the whole thing ramps up.

Peterson's point is that this reaction is not just a personality quirk. It's a warning sign. Lunchbucket's inability to handle some low-grade teasing tells the rest of the crew something about how he might react when things get genuinely dangerous. If he cannot keep his head over a dumb nickname, what happens when someone misses a signal, a load shifts suddenly, or a train is bearing down on them at full speed? In that world, like the Army, emotional stability isn't optional. It's part of the mandatory safety kit.

As Peterson puts it, "The harassment that is part of acceptance on a working crew is a test: are you tough, entertaining, competent and reliable? If not, go away."[125] In other words, the harassment is a crude filter. They aren't just asking if you can do the job when everything is calm. They're finding out if you can keep your sense of humor, pull your weight, and avoid becoming a liability when things go sideways.

From where I stand now, my hazing in the woods and the fate of the two NCOs I mentioned earlier look different through that lens. I passed their test, though at the time I had no idea that was what they were doing. The two NCOs didn't. Was it harsh to boot them from our company? They may disagree, but I don't think so. They needed to go somewhere else, and I'm glad they were reassigned before we deployed back to Iraq.

Finally, one of my most enduring reflections from this time centers on the culture shock I experienced in Korea. To my surprise, it was even more jarring than Iraq. The difference was in how deeply I immersed myself in the local scene. I ate authentic food, explored the countryside, and visited places of global significance. Sure, the war in the Middle East had been intense and disorienting, but standing face to face with the reality of the North Korean regime left me genuinely stunned. I was still fairly naïve, and this trip drove home just how alive and well authoritarianism remained in parts of the world.

---

[124] Jocko Podcast. "The Importance of Having Thick Skin - Jocko Willink and Jordan Peterson," February 23, 2018. https://www.youtube.com/watch?v=futHL4W4E4s.

[125] Peterson, Jordan B. *12 Rules for Life: An Antidote to Chaos*. eBook edition, Chapter 11. Random House Canada, 2018.

Along the way, I ran into cultural symbols that exposed the limits of my worldview. One moment that stands out is when I stumbled across that swastika on the side of a hill. Until then, I thought I had a reasonable grasp of world affairs. Within seconds, I realized how little I knew beyond the narrow confines of an American high school education. That single symbol, stripped of the context I thought I understood, lit a fire under me. I needed to start educating myself properly and move beyond the simplistic, and sometimes flat-out wrong, narratives I had previously been taught.

This period of growth also taught me an invaluable lesson. All of us carry preconceived notions about what we will find in other countries and how the world really works. Most of our opinions rest on a thin base of experience, shaped by where we live, the headlines we skim, and the stories we hear from others. Even today, I meet folks who talk with great confidence about Iraq, its people, its government, and the war, despite never having set foot there, never reading a serious book on the subject, and never speaking with an Iraqi. Now, I'm not saying people should never have opinions. What I am saying is that it is perfectly acceptable, and in my view genuinely wise, to admit when we don't know enough. We don't need to plant a flag on every issue, especially the ones we have never personally encountered.

That moment with the swastika drove home how easy it is to misread symbols. The same thing can happen to an unprepared American visiting Spain during Holy Week. Religious processions there include participants in pointed hoods and long robes that look, at first glance, uncomfortably similar to the outfits worn by the Ku Klux Klan. To a first-time visitor, it can seem like white supremacy on parade when, in reality, it's a centuries-old Catholic tradition.[126] Encounters like that, whether on a Korean hillside or a Spanish street, are precisely when it is most important to hit pause, ask questions, and seek local context before deciding what we are seeing and getting angry.

Speaking as an American, I have come to believe that traveling with an open mind is one of the best antidotes to the quiet ignorance many of us carry without realizing it. This isn't because Americans are uniquely ignorant. No, every country has its blind spots. However, the United States is vast and relatively isolated, so a lot of us grow up with limited exposure to foreign cultures, ideas, and cuisines. Travel doesn't make anyone an expert, but it does have a curious way of shrinking our certainty and expanding our humility, which is not a bad trade.

In his 1869 travelogue, *The Innocents Abroad*, Mark Twain famously wrote: "Travel is fatal to prejudice, bigotry and narrow-mindedness, and many of our people need it sorely on these accounts. Broad, wholesome, charitable views of men and things cannot be acquired by vegetating in one little corner of the earth all one's lifetime."[127]

Amen, Mr. Clemens.

---

[126] Steves, Rick. "European Easter." Rick Steves' Europe, 2016, 16.00. https://www.ricksteves.com/watch-read-listen/video/tv-show/european-easter.

[127] Twain, Mark. *The Innocents Abroad*. eBook edition, Conclusion. American Publishing Company, 1869.

Let's look at food, for instance. The swastika may have shocked me visually, but dog meat gave me pause in a far more personal way. Like many Westerners, I found the idea of eating a domesticated animal uncomfortable, even taboo. I tried it anyway. I cannot claim it unlocked any deep cosmic insight, but it did force me to admit that cultural food practices vary widely and that mine are not the default setting for humanity.

Half a world away, in western Germany, horse meat is, for many, a delicacy. It is lean, rich, and delicious, and it was one of the best steaks I have ever eaten in my life. To a Midwestern farmer in the United States, though, eating a horse might feel a bit like eating one of their kids. The same pattern shows up elsewhere with entirely different animals. In many Asian and African countries, insects are prized as high-protein snacks, while for many Americans, the word *insect* alone is enough to trigger a gag reflex.

The takeaway is simple. One person's food can literally be another's pet or pest. Before passing judgment, it is worth understanding the historical and cultural context that produced a dish. And let's not pretend that "strange" foods exist only overseas. Every country has something the rest of the world finds off-putting or downright disgusting.

In some cases, the foods we mock or avoid are all that people have to eat. In the United States, we have the luxury of wandering through a supermarket packed with chow (pun intended), everything from macaroni and cheese to lobster and caviar. Yet many non-Westerners are baffled by our devotion to ultra-processed snacks, sugary cereals, and hot dogs. Honestly, have you ever seen how hot dogs are made? I still love those things, but it's fair to say the process raises questions. The point is that dietary choices vary, and taste is shaped by far more than ingredients. It's formed by culture, history, and necessity.

So, if you get the chance, travel the world. It won't make you an expert, but it will push you to be a little less judgmental and a little more humble, which is a pretty solid return on investment. It's how you grow as a person.

# Chapter X

## Bengal Tiger on the Bayou

By mid-2004, several turning points were already reshaping the rest of my career and, in some ways, my personal life. Bravo Company came under new leadership, and the two men at the top made such a mark on me that their lessons would last a lifetime. At the same time, 3rd ID began an experimental reorganization in which battalions were no longer "pure" but were restructured into mixed formations called Units of Action. And then there was Toby Glass's medical discharge from the Army because of those persistent injuries, which sent both of us on an emotional rollercoaster.

Taking command of Bravo Company was the newly promoted CPT Michael MacKinnon, call sign Bandit 6, the same officer I had first met in Iraq months earlier when he delivered us safely to our FOB. I was immediately struck by his leadership style: energetic, fiercely competitive, and fully engaged with his soldiers. The first time we officially crossed paths, I was stunned that he not only knew my name but also details about my life that even some people in my own platoon didn't. More than anything, when he spoke to us one-on-one, it was obvious he *genuinely* cared and wasn't just checking a box as part of his job.

At his side was First Sergeant (1SG) Mark Barnes, who embodied everything I had come to expect from a senior NCO. First sergeants have a tough job. They are responsible for leading and mentoring junior NCOs and lower enlisted soldiers, many of whom are inexperienced or still growing up. Just as important, the first sergeant serves as a key advisor to the company commander and often brings far more years of experience to the table. It's a role that requires wearing several hats at once: disciplinarian, mentor, tactician, and sometimes mediator between the officer and enlisted ranks. In my view, 1SG Barnes excelled at all of them.

Together, the two of them brought a remarkable depth of experience and expertise to our company. Their résumés included West Point, Ranger School, Airborne, Air Assault, Pathfinder, and a staggering list of other specialized training. To me, it felt like a dream team had taken the helm. But their presence also came with high expectations. They demanded excellence from everyone under their command, and we did our best to meet the peak standards they set.

Not only had the upper leadership of Bravo Company changed, but our platoon looked quite different as well. Many of the soldiers who had gone overseas to Iraq and Korea were leaving the military or were being permanently reassigned to new duty stations. Adding to the shift, SFC Harold Hill, a steady and capable leader, took control of the NCOs in 1st Platoon. As a result, I found myself advancing rapidly through the lower enlisted ranks, rising from one of the least experienced soldiers in the platoon to a more prominent position in the hierarchy. I was no longer the FNG (fucking new guy).

As was the norm, we spent much of our time not in garrison but in the field, where we conducted numerous unit exercises in the woods and were perpetually filthy. It drove me crazy, but grime was just part of the job. Showers were nonexistent on FTX, so we tried to clean ourselves using baby wipes and napkins from our MRE packs, though this method only went so far. Every so often, we would head to a nearby stream and do our best to bathe. I didn't really like doing this, though. The water was home to some big alligators, and while they never approached us and usually just floated lazily nearby, I couldn't shake the paranoia that one might snatch me up. The idea of stepping barefoot onto one hidden beneath the murky water always unnerved me.

There were no toilets either. We simply had to go far enough away from the sleeping and eating areas to avoid contamination. For defecation, the process involved digging a hole, preferably next to a tree, and then leaning against it. Napkins were often hoarded from the MRE packs for these situations. Once our bowel movement was finished, it was mandatory to bury the waste. The burial served two purposes: to reduce the risk of disease and to conceal evidence of our presence from the enemy. Everything we did was treated as combat training, even the most primitive tasks.

Additionally, it was a strict rule that we had to travel with a fellow soldier at all times, regardless of the location. Even for something as personal as relieving yourself, your battle buddy had to come along, standing a few feet away to keep watch for "enemy activity" while you did your business. Privacy didn't exist, and trying to take a dump with someone standing next to you was kind of awkward. Someone was always within arm's reach, and personal space was a luxury we simply didn't have. The infantry was definitely not a profession for the squeamish or modest.

Sleeping in the mud at night, too, was always an adventure. Because tents could give away our position to the enemy, we relied on the classic "Ranger hooch" for shelter, which is a poncho stretched low to the ground, rigged to keep out the rain and remain hidden from view.[128] But it never worked quite right. I always ended up wet for one reason or another.

Beyond the rain, fire ants, giant spiders, bugs, snakes, and other wildlife routinely crawled in and around our area as we tried to sleep. But the worst were the wild boars, drawn in by the scent of our MREs. More than once, I woke up nose-to-snout with a tusked boar rummaging through my gear in search of food.

---

[128] Savoy, Pat. "What Is a GI Poncho Hooch and Why Should I Know How to Make One?" *Survival Gear Book,* May 12, 2022. https://survivalgearbook.com/blog/what-is-a-gi-poncho-hooch-and-why-should-i-know-how-to-make-one.

During downtime, like prehistoric cavemen, we would affix knives to tree limbs to make spears, then would dig shallow traps to catch one of the boars, intent on turning it into a feast. It really wasn't about survival. It was just something to do for fun, though the thought of roasted pork chops sounded quite delicious after days of MREs.

Disgusting smells of body odor, or what we jokingly called "body cheese," perpetually built up to epic proportions after weeks in the field. I often felt sorry for the wives and girlfriends back at home who had to deal with the aftermath of rancid uniforms and gear crusted in dirt, sweat, or worse. Upon returning home, I developed a ritualistic process of cleaning myself. Removing my boots at the door, I would strip naked in the front entryway and tiptoe down the hallway to avoid contaminating the carpet with my funk. Immediately, I'd toss every filthy piece of clothing straight into the washing machine on the hottest cycle. Then came a long and hot shower, with layers of dirt and filth swirling down the drain like a muddy whirlpool. After life in the field, that first shower always felt like a little piece of heaven.

During one exercise, we were pitted against an OpFor in the woods with MILES gear, much like we had in Korea. As luck would have it, Mata, Hauger, and I ended up working together on the same motley crew, most days under the watchful eye of our unofficial ringleader, SPC Velasquez. Individually, we were all solid soldiers who worked hard and typically took our duties seriously. But together, probably thanks to our age and the way we fed off each other, we had a kind of chemistry that practically guaranteed something ridiculous would happen. Our personalities just clicked, and no matter how serious the mission was, we always managed to find humor in it.

For reasons I still don't understand, SGT Brister left the four of us alone with a walkie-talkie and orders to defend the shack, a small outpost in a makeshift village deep in the woods, so it wouldn't fall into enemy hands. We sat near the open windows, watching excitedly for any activity. When none came after several hours, we naturally grew bored and figured that they had bypassed our immediate area. We listened to the radio traffic, which had excited chatter amongst our unit. They were being engaged nearby, but our shack was depressingly untouched.

Suddenly, our MILES gear emitted high-pitched beeping sounds, accompanied by the snaps of blank ammunition coming from near our position. Shockingly, none of us had been taken out in the initial barrage, suffering only near misses. Peeking out the windows, we could see an element of at least two squads of about twenty men in the woodline. It didn't take a genius to know that we were outnumbered and in serious trouble.

Murphy's Law kicked in as we tried to raise our side on the radio for help, but no one responded. We were all alone and had to fight back on our own, and we certainly couldn't go down without a fight. But knowing we didn't stand a chance in a straight-up firefight against that number of soldiers, we needed a new plan, and fast! I'm not sure who came up with the following idea, but with the four of us alone in the woods, almost any absurd scheme was feasible.

We carefully maneuvered around and hid behind the walls, trying to avoid being shot by the incoming laser fire. At any point, we knew that they would stack up outside and breach the safety of our building. Speed was crucial, or our genius inspiration would fail. That's when Hauger and I launched into a full-blown, staged brawl.

We started by screaming obscenities at each other as loudly as possible, hurling insulting epithets back and forth like bitter enemies. We then began flailing, punching, and kicking in an exaggerated fashion within the darkened room, making a realistic spectacle for the advancing OpFor. They stopped shooting and simply watched with amused disbelief, unsure what they were seeing.

I grabbed Hauger by his BDU top and threw him against the wall, knocking a framed picture to the ground. He responded by shoving me out the open window, still cursing like a man possessed. I landed flat on my back, the air punched from my lungs. Was it stupid? Yes. But against all odds, the act had worked. We'd completely thrown the enemy off; they stood around us, convinced they'd walked into a real fistfight between two young soldiers, and were unsure how to respond.

**Wearing MILES gear during OpFor training at Fort Stewart**
**From top left, clockwise: Mata, Velasquez, the author, Hauger**

"What the hell are you idiots doing?" one of the infiltrating NCOs yelled, his face twisted in rage as he stomped toward us. "Knock it off and start doing push-ups right now!"

Behind him, a squad-sized element clustered in the open, distracted by the commotion and oblivious to their exposed position. In the blink of an eye, Mata, who had been silently hiding behind the door with his M-249 SAW, burst from his hiding position and opened fire, hosing them down with automatic gunfire.

The screams of MILES gear were deafening, as every single one of them was now hilariously dead. Their faces shifted instantly, from amusement to pure rage. We could barely contain our laughter as we quickly bounded back inside and slammed the door behind us. Our ridiculous "ambush" had somehow worked to perfection, and all of us were still alive, while over a dozen enemy troops were now KIA (Killed in Action) outside. We looked at each other in disbelief; there was *no way* that had just worked!

The door exploded open as the NCO kicked his way inside, his MILES gear screeching like a dying rabbit. He was red-faced, livid, and absolutely seething with anger. All of us were ordered to do furious push-ups as he accused us of "making a mockery of the training." According to him, we hadn't just disrespected the exercise, we had *cheated,* and he was going to make sure our first sergeant heard about it. That didn't sound too good to me. 1SG Barnes could squash us like annoying bugs if he wanted to.

Well, luckily, that never happened. Instead, the range cadre, who ran the exercise, couldn't stop laughing, having witnessed the incident firsthand. While the furious squad leader still fumed, they patted him on the back, reassured him that justice would be served, and quietly made a note to return us to the platoon once the exercise ended. When we were finally handed off to our team leader, SGT Brister, it was with a sheepish grin and a knowing nod.

### 27 May 2025 – Reflection of Derek Hauger, Bravo 3-7 Infantry

We were surrounded, so we did whatever we could to adapt and overcome. Getting tossed out the window was the best sell ever. They bought it. You hit the ground flat on your back in a big puddle, and then Mata cut them down with the SAW. They were PISSED!

Naturally, 1st Platoon immediately blamed Hauger for the antics, who loudly protested that it was *my* idea. I didn't admit to it then, and I'm not about to admit to it now. But let's just say… it was a damn clever plan. And even we were a little stunned it had actually worked. For a brief moment, we were legends, at least in our eyes. But in the field, glory doesn't last long.

We nervously stood by as the range sergeant spoke to SGT Brister in hushed tones. I could just barely overhear him say, "…they killed about a quarter of their platoon," followed by muffled laughter. With that, we were never punished, and it was quietly swept under the rug.

Later, while practicing bounding overwatch on lanes, I was reminded that the OpFor wasn't the only threat lurking out there. Bounding overwatch was a fun but exhausting activity, where we worked as a team to advance forward and engage the distant enemy. During this particular drill, we were practicing with blank ammunition. It added to the realism without the danger, and we usually ran the lanes several times before progressing to live ammo.

The concept was simple: I'd lie prone in the grass, weapon ready, providing cover as my buddy sprinted ahead. Before the hiding enemy could "fire" on him, he'd hit the ground and return the favor, covering me as I jumped up and sprinted past. We repeated this leapfrogging pattern, alternating roles until we closed in on our objective.[129]

As I bounded past Hauger and dove onto a small berm like I was storming Normandy Beach, I heard a distinct, ominous rattling off to my left. There it was, a massive rattlesnake, looking like a spawn of Satan himself. It was coiled up and ready to strike. Somehow, I hadn't seen it, and I was miraculously fortunate not to land right on top of it.

I immediately rolled away in a panic as it was right next to my head, letting out a high-pitched shriek in the process. Firing at it wasn't an option, as I wasn't about to intimidate it to death with my blanks. The snake, apparently unimpressed by my quick and evasive maneuvers, lunged at me and somehow latched onto my pant leg. In a panicked response, I resorted to the age-old infantry tactic of flailing and running in zigzags, like a deranged goose in combat boots.

"What are you doing?!" someone yelled, though I was still too busy thrashing about to respond immediately.

It was 1LT Johnston, our platoon leader, who had been quietly observing our bounding overwatch from the rear. He looked confused and moderately annoyed, as this wasn't part of the training manual. Additionally, the company commander was evaluating our platoon, and my dancing was unlikely to be well-received during the AAR.

"Lieutenant Johnston, a BIG snake!" I screamed, still performing my one-man show of Riverdance. I looked down, relieved to see the serpent was no longer hooked onto my BDUs. It now sat a few feet away as though nothing had ever happened.

Still annoyed, 1LT Johnston channeled his inner Bear Grylls, grabbed his E-Tool, and beat the snake to death like it had insulted his mother. Somewhere, there exists photographic evidence of him proudly holding the lifeless snake next to a Humvee like it was a trophy buck. It was a monster.

I walked away with only a small hole in my pants near the top of my boot, narrowly escaping what would have been a very unfortunate bite. Hauger, of course, still gives me hell to this day, not about the snake, but the high-pitched screaming and flailing. According to him, I briefly hit notes only dogs could hear, and he thought it was the funniest thing he'd ever seen.

To our dismay, our team was short-lived. The laughs we shared were countless, but like all good things in the Army, it came to an end. They broke us up, not because of trouble but because of the usual shuffle of platoon reorganization and the identification of individual strengths. Mata was assigned as a Bradley driver, Hauger stayed on an assault team, Velasquez left the Army, and I went with SGT Brister to the Heavy Weapons Squad. It all worked out for the best, but those days together were some of the most fun I've ever had in uniform.

While I wasn't aware of it at the time, CPT MacKinnon and 1SG Barnes were already looking for new opportunities as soon as the Korean deployment was over. One of their most ambitious initiatives was to arrange a joint training session with the Federal Law Enforcement Training

---

[129] TRADOC G2 OE Enterprise G&V. "Squad Movement Formations & Techniques," December 22, 2011. https://www.youtube.com/watch?v=OKRues4Fwrk.

Center (FLETC), located approximately 60 miles south of the base in Glynco, Georgia. Rumor had it that the last combat units to train there were some of the guys who'd ended up in Mogadishu during the Black Hawk Down incident, as the facility is traditionally reserved for federal law enforcement training. The partnership was an ideal match for our unit. FLETC was nearby, well-funded, and staffed by a diverse range of instructors eager to share their expertise with grunts.

To prepare for the mission, we were all sent through a crash-course version of Air Assault training, tailored and led by our competent command team. Although the 3rd ID was traditionally a mechanized unit that operated Bradleys and other tracked vehicles, CPT MacKinnon and 1SG Barnes wanted us to be cross-trained in a range of infantry tactics. Their goal was to make us more lethal, mobile, and adaptable than ever before. As part of that vision, they coordinated intensive training with UH-60 Black Hawk and CH-47 Chinook helicopters.

Out in the field, we flew in both aircraft around the base while the pilots, clearly enjoying themselves, executed wide, looping maneuvers designed to disorient us. They buzzed treetops, spun tight circles, and did their best to make the ride as wild as possible. It became something of a game to see who could keep their lunch down. I always requested a door seat on the Black Hawks so I could dangle my legs out the window, and it quickly became my favorite spot.

On the ground, we trained on calling in close air support and extractions over field radios, properly securing loads for potential sling operations, and, although we never had to use it in combat, the basics of fast-roping. This is sliding down a thick rope from a hovering helicopter into a landing zone (LZ). It was all fun and exciting, and we were eager to show off our new skills.

Just before sundown, we loaded up in a swarm of Black Hawk helicopters, their rotors kicking up clouds of dust as we sprinted aboard. Moments later, we lifted off and soared low over the darkening Georgian treetops. I sat at the open door, cradling a machine gun in my arms, legs dangling in the open air, the fading sunlight streaking the horizon behind us. The rush of wind pummeled my face, wild and smelling of exhaust, while the roar of the engines drowned out everything except the deep, rhythmic *whump* of the rotors overhead. That sound, steady and hypnotic, became a pulse, lulling us into a strange, Zen-like calm. Just below, headlights dotted the winding roads, and I imagined stunned drivers glancing skyward, wondering what shadowy force had just thundered overhead.

There's no easy way to explain it. It was the meditative stillness that crept in just before a mission. It was a kind of mental dimmer switch as if my body were quietly powering down to preserve fuel for what came next. I wasn't asleep, but it was as though I wasn't fully present either. I was detached, just waiting.

As night fell, we flipped down our NVGs, and the world turned to a sea of glowing green. In the distance, the faint outlines of a neighborhood began to emerge. It was our objective. The pilot gave the signal: one minute to the LZ. We started to drop fast.

Then, without warning, my inner dimmer switch would flip at the moment it was time to move. My stomach fluttered with giddy anticipation and every nerve came alive. My senses would sharpen into razor clarity, my vision tunneled, sounds snapped into crisp detail, and the pure electric thrill of adrenaline flooded every inch of me. It was intoxicating and addictive.

I gripped my weapon tighter as the ground rushed toward us. Then, we slammed into the dirt, and the backwash of the rotors filled the air. In a blur of motion, we surged out in a dead sprint, fanning into defensive positions. My boots barely touched the ground as I dashed to cover, heart hammering, every nerve aware.

All around us, more Black Hawks swooped in, grunts spilling out in waves like clockwork. Within seconds, the LZ was cleared and secure. We reestablished comms with the rest of the element and began our assault on the nearby neighborhood. Enemy contact came quickly, with gunmen holed up in the houses, waiting for our arrival. Room by room, we swept through, precise and aggressive, clearing the path to our final objective.

Watching from a distance were dozens of law enforcement trainees from FLETC, wide-eyed and grinning as we gave them one hell of a show. For us, it was just another day at work, but for them, it was an action movie come to life.

The initial mission of taking down the village wrapped up quickly, and soon after, our platoon was called over to meet with clusters of excited federal agents nearby. Some were in training, and others were seasoned instructors. We mingled with them, showcasing our equipment and weaponry while they asked a wide range of questions. It was a strange experience as they treated us like rock stars and genuinely seemed happy that we were there. I suspect they knew we were headed to Iraq in 2005 and that we were taking in every possible training opportunity.

What stood out most was the instructors' eagerness to share their time and knowledge with us. It was, in all honesty, quite odd to me, as they treated us so professionally and politely. I think I can safely speak for my fellow soldiers on this: we were used to being yelled at and talked down to by typical military instructors who seemed more interested in asserting dominance than actually teaching. FLETC was refreshing in this way, and we soaked it in. We didn't want to let these people down after the reception that they gave us.

The following morning, long before the sun rose, CPT MacKinnon had the entire company up and moving. We ran in company formation around the darkened compound while screaming out cadence at the tops of our lungs, waking everyone up. It was the loudest and most aggressive formation that I had ever been a part of, simply because we wanted to make our captain proud. We wanted to show off our colors as much as he did, and I recall a big smile on his face as we zigzagged across campus, carrying our guidon (company flag) at the front of the pack.

He wanted everyone to know that Bravo Company was in the compound and that these were *his* men. The memory still gives me goosebumps twenty years later as I recall sleepy faces looking out of barracks windows, with momentary confusion. A few pointed, and others just stared in awe. All we were doing was running in a tight formation while screaming battle cries and singing cadence, all of it loudly reverberating around the buildings. But something in the air was electric that day, and the amount of noise we were making was unprecedented. It was the closest thing I've ever felt to what Viking berserkers must have done before charging into battle. And I've never been prouder to be part of CPT MacKinnon's company than I was on that day. From his giant smile, I know he felt the same.

The FLETC compound in Georgia boasted a large, fully furnished neighborhood ideal for close-quarters battle (CQB), with dozens of houses, buildings, and streets.[130] We were in awe of the complex. In the practical exercise area, we refined our urban warfare skills and reviewed playback footage of our movements and decisions. It was the finest possible training scenario for an infantry company, likely surpassed only by the elite facilities used exclusively by the special operations community, which were beyond our reach. As far as second options go, this was top-tier.

For several of the CQB exercises, our weapons were equipped with Simunitions, a portmanteau of "simulated munitions." These non-lethal rounds, filled with brightly colored wax and fitted with plastic tips, mimicked the size and appearance of real ammunition.[131] Unlike traditional paintballs, Simunitions were more accurate and far more painful, especially on exposed skin. Force-on-force scenarios required full combat gear, including helmets, gloves, body armor, and gas masks. Even with all that protection, it was impossible to shield every inch of flesh. But that was the point. The sting of a sim round was a physical reminder of a tactical mistake. Pain became the instructor for our failures.

We had several minor injuries from the Simunitions around the company, which included shattered fingernails, large welts in sensitive areas, and wax painfully lodged under the skin. One round smashed through a fellow soldier's reinforced plastic eye lens on his gas mask and impacted his face, though he somehow avoided any serious injury.

During one Simunition exercise, we fought against a local SWAT team and had a lot of fun blasting them in the process. I suspect the learning was mutual. It was clear that they enjoyed it, as our tactics differed significantly from what they were accustomed to or had previously experienced as a police unit. Their methods, geared for law enforcement operations, focused on restraint and civilian safety, and contrasted sharply with our approach. We emphasized overwhelming force and rapid domination. Their approach was to preserve life. Ours was to smash it, closer to controlled chaos, like a tactical demolition derby.

While Simunitions and force-on-force training offered valuable lessons and plenty of adrenaline, there was no mistaking them for real combat. Anyone can pull up videos of paintball battles on the internet, with combatants running around, doing all sorts of crazy things. This doesn't translate all that well to an actual battleground, where fear, anxiety, confusion, pain, and suffering reign supreme. Those reckless paintball maneuvers, while entertaining, would result in premature death. Still, we needed to practice our fundamentals under stress against a live, thinking opponent, with the added consequence of pain when we got it wrong. Simunitions were more than training rounds. They were accountability.

---

[130] Federal Law Enforcement Training Center. "Glynco, Georgia." Accessed December 1, 2025. https://www.fletc.gov/glynco-georgia.

[131] Simunition. "Simunition Overview Brochure," 2024. https://Simunition.com/wp-content/uploads/2024/07/400000143-Simunition-Overview-Brochure_202402.pdf.

FLETC also provided several instructors who taught us a wide range of skills for detecting and avoiding booby traps encountered in the United States and overseas.[132] We watched hours of gory video footage showing people being maimed and killed by a variety of devices. While the content was difficult to watch, it provided an excellent opportunity to learn and avoid making the same mistakes.

We then applied our new knowledge in practical exercises at the CQB village and on the demolition range. Among the most eye-opening segments of our training was our introduction to a deadly and increasingly common threat: car bombs.

IEDs were being embedded into vehicles, known as VBIEDs (Vehicle-Borne Improvised Explosive Device). They were fast becoming a preferred tactic among terrorists in Iraq and Afghanistan. Piloted by suicidal drivers, they looked for targets of opportunity that included patrols, checkpoints, or even FOBs. The most alarming aspect was that it was nearly impossible to detect or defend against them because the explosives were usually well concealed. Additionally, VBIEDs were often indistinguishable from regular traffic and frequently didn't show any apparent signs of danger until it was too late. All they had to do was get relatively close and then detonate a massive explosion.

Coalition forces developed tactics to address the threat, such as funneling vehicles into chokepoints. But even then, the carnage was horrifying. Soldiers and civilians were still being blown apart in staggering numbers, and there were no easy answers.

We trained with various explosive detection tools, the most common being a simple chemical test strip. It required swabbing a surface and placing the strip into a reactive kit that would display a positive or negative result for explosive residue. The downside? It meant getting dangerously close to the suspected vehicle, sometimes crawling beneath it to reach likely bomb placement zones. No robots were available for this type of testing at the time, not for us lowly grunts anyway. That job fell to us. It was up close, personal, and terrifying.

Though the training on car bombs was excellent, I was a bit leery about crawling underneath a car to check for explosives. Yet, this is precisely what I would find myself doing in Iraq within a few months. My rationalization at the time was that if it were an actual bomb, I probably wouldn't even know it. I just hoped it would be big enough that I wouldn't feel a thing... but small enough that the guys around me would survive.

When we wrapped up the FLETC training, I realized it was the most valuable instruction I'd received in the military. Ironically, it didn't come from the military at all, but from a law enforcement facility. It set the standard for how exercises should be conducted and exposed how often our training had been shortchanging us. That's a shame because I served with some incredible men who often had more potential than the system ever gave them the chance to realize.

---

132 Federal Law Enforcement Training Center. "Performance and Accountability Report: Fiscal Year 2004," 2004, Page 16. https://www.fletc.gov/sites/default/files/imported_files/reference/reports/accountability-reports/fletc-accountability-report-2004.pdf.

Back at Fort Stewart, our entire way of doing business was being upended. After returning from Iraq, the 3rd ID began converting its brigades into "Units of Action," the Army's new modular, brigade-centric formations that later evolved into brigade combat teams.[133] This reorganization, which included activating a 4th Unit of Action at Fort Stewart in May and redesignating the remaining brigades over 2004, was part of the Army's largest structural overhaul since World War II.[134][135]

Under the old system, most of us had grown up in battalions that were relatively "pure" in terms of MOS and branch. A typical infantry or armor battalion had roughly 500 to 1,000 soldiers who shared the same basic specialty and relied heavily on separate brigade or division units for artillery, engineers, logistics, and communications. Those supporting units often linked up with the maneuver battalions only when it was time to deploy. That meant battalions and support elements sometimes met each other for the first time in theater and had to figure out how to work together under fire. Maybe not the best idea.

The new Units of Action (UAs), however, were built on a very different concept. Instead of smaller, single-branch battalions plugging into a division, each UA was a self-contained, combined-arms team of roughly 3,500 to 4,000 soldiers.[136] These formations were designed to include their own infantry, armor, artillery, engineers, reconnaissance, signal, and logistics elements, so they could train in garrison as they would fight in combat, and then deploy and sustain themselves with far less day-to-day reliance on higher headquarters. In theory, each UA was a semi-autonomous warfighting package that could be tailored and sent downrange on its own. In other words, it was designed to be much more efficient and sustainable.

But at our level, it felt like someone had taken all the manning documents at Fort Stewart and thrown them into a blender. Practically overnight, what had been pure infantry and pure tank battalions, such as the 3rd Battalion - 7th Infantry and 4th Battalion - 64th Armor, were reorganized into these mixed, combined-arms battalions. Specific armor and infantry companies swapped places between battalions, new units were stood up, engineer companies were attached, and forward support companies moved in to provide organic maintenance and logistics.

Got all that? Confusing, right? Well, we eventually figured it out, but it was a bit hectic for a while. For the old-school grunts and tankers, it was a rude awakening to look around the motor pool and see not just infantry or armor, but a cross section of the division's MOS fields all crammed into the same formation. And for many of us, especially the more prideful grunts, this

---

[133] Global Security. "Brigade Unit of Action," May 7, 2011. https://www.globalsecurity.org/military/agency/army/bua.htm.

[134] Global Security. "4th Brigade Combat Team, 3rd Infantry Division (Mechanized) 'Vanguard,'" May 7, 2011. https://www.globalsecurity.org/military/agency/army/3id-4ua.htm.

[135] AUSA. "Profile of the United States Army: Army Organization." Association of the United States Army, September 1, 2022. https://www.ausa.org/publications/profile-united-states-army-army-organization.

[136] "CRS Report for Congress | U.S. Army's Modular Redesign: Issues for Congress," Every CRS Report, January 6, 2005, https://www.everycrsreport.com/files/20050106_RL32476_bb5da6b51e36f72c1b2bbcf65412cff8ba1eba6e.pdf.

was a tough pill to swallow. We didn't like sharing our time, space, or resources with non-infantry personnel.

As an infantryman in Bravo 3-7, I never interacted with soldiers of a different MOS, except perhaps in the chow hall. Fellow grunts always surrounded me, and we were a tight-knit bunch. As I've mentioned before, we took great pride in the hardships that defined our identity: sleeping in the woods, going weeks without a shower, hazing each other to near-criminal extremes, constantly fighting, and relentlessly trying to one-up one another. We also talked derisively about soldiers of a different MOS, calling them "POGs," short for People Other than Grunts (rhymes with rogues). We loved being infantry and wore our blue cords like a badge of honor.

The devastating shock came when our company received reassignment orders. We were being removed from 3-7 Infantry and were reassigned to the 4th Battalion - 64th Armor Regiment, finding ourselves under the command of LTC Robert Roth and CSM Clarence "Cap" Stanley. We were no longer Cottonbalers; we were now Tuskers. Just like that, Bravo 3-7 became Bravo 4-64. And to be honest, I hated it. I didn't want to be part of a tank battalion. I didn't want to wear a new crest or adopt a new identity. I was a Cottonbaler, by God! What the hell *was* a Tusker, anyway? Some kind of cute mascot? An elephant? I was about to find out.

**20 July 2025 – Reflection of Robert Roth, Commander of 4-64 Armor**

When the reorganization began, I already knew LTC Dave Funk (commander of 3-7 Infantry) pretty well. He pulled me aside and said, "I'm sending you the best company I've got." I remember thinking, *Yeah, right. Why would he send me his best?*

But he wasn't lying. When Bravo Company arrived, they immediately made their presence felt and set the standard for everyone. They led from the front on day one.

When Mike MacKinnon and Mark Barnes came into our armored battalion, I didn't know much about infantry tactics, especially on the mechanized side. Those two became my go-to sources, and I leaned on them heavily. To better understand your jobs, I climbed into the back of one of your Bradleys and rode around with the company during live-fire exercises. I wanted to see it from your perspective. It was uncomfortable, cramped, and miserable. But I learned by doing it.

A few days after the transition, I was walking through the battalion area when I noticed a piece of trash in the middle of the grass. This was our new turf, and we were responsible for keeping it clean at all times. However, I faced a mild dilemma upon seeing it. Anyone who has ever been in the military knows that you DON'T walk on the grass.[137] Sidewalks and designated paths exist for a reason, and violating that sacred rule, even for trash, could earn you an earful from someone with more rank than sense.

---

[137] Milzarski, Eric. "Why Sergeant Major Doesn't Want You Walking on the Grass." We Are the Mighty, October 22, 2020. https://www.wearethemighty.com/military-life/why-troops-keep-off-grass/.

So, I stood there, eyeing the torn magazine cover as if it were booby-trapped. I looked from side to side and saw that the coast was clear. No one was looking. I quickly ran out onto the freshly clipped lawn, snatched the garbage, and darted back onto the sidewalk. My mission had been a success. Or so I thought.

There, taped to the back, was a small note that stopped my heart cold. It read, "Come see me - CSM Stanley."

CSM Stanley wasn't just the highest-ranking NCO in 4-64; he was also a giant bulldog of a man, both in stature and presence. Towering a head above most soldiers, he was built like a tank himself, with a gravelly, thunderous voice that matched his imposing frame. If looks could kill, he'd be fully qualified for the job. He had summoned me to his office with the hidden note, and I knew better than to ignore it. While this man may have been a tanker, he was larger than life and commanded respect from everyone in his presence. Everyone instinctively stood a little straighter, even the grunts.

**CSM Stanley, 4-64 Armor, whose imposing presence matched the authority of his rank**

**Photo courtesy of Clarence "Cap" Stanley**

I was dead. He had set me up, just to make an example out of the brand-new infantryman who'd dared step on the holy grass of his beloved tanker battalion. First, he was going to smoke me in front of everyone. After which, everyone below him in my immediate chain of command would get their turn to inflict punishment. He would probably then assign me a month of CQ or gate guard duty, adding insult to injury. I just knew it.

Standing outside his office, I took a deep breath and loudly knocked. Even though he hadn't been expecting me, his barked "Enter!" came instantly, like he'd been waiting. He stared at me from behind his desk as I tried to spit out what I was doing. Cautiously, I shifted from parade rest to hold the magazine cover up, stammering through every word. How I didn't piss myself, I'll never know.

When I finished, I stood motionless as he continued to stare me down from across the room. My mind raced through all the horrible jobs he was concocting in his brain for me at that very moment. Then, without a word, he stood up, came around the desk, and parked himself directly in front of me, towering over my small frame like a boulder with rank. I kept my eyes straight ahead, resisting the urge to flinch or make a sound.

"Relax, killer," he said in his gravelly voice. He called everyone "killer," though I never knew why. It was just his thing. Instead of a verbal beatdown, CSM Stanley gave me a short but memorable speech on integrity and leadership. It stuck with me far more than any punishment ever could.

Then he grinned and admitted something that floored me: the whole thing had been a set-up. He'd planted the trash with the note attached, just to see if anyone would take the bait. When I knocked on his door, he was pleased that his plan had worked. He hadn't been sure anyone would show up.

He told me he was proud I'd picked up the piece of trash. At that moment, I'd made a decision. Not a big one, but an important one, which was to preserve discipline and order in our area. Even though I'd technically broken a minor rule by stepping on the grass, I'd done so in the service of a greater standard. Clean grounds mattered, and discipline extended beyond polished boots and tight formations, as it was present in the little things, too. Picking up that garbage, he explained, was a reflection of ownership and pride in our battalion. It also showed judgment, initiative, and the moral courage to do what was right, even when no one was watching.

He went further, saying that part of being a soldier, especially one with leadership potential, meant making executive decisions in the heat of the moment. You may be questioned and have to explain yourself, but if you acted with integrity and could stand behind your choice, that was leadership in action.

To a civilian, this simple story might sound funny or trivial. In the military, though, we're constantly tested in small, often invisible ways. The decisions we make under pressure, no matter how minor they seem, shape who we are and how we lead our soldiers. Confidence, integrity, and the willingness to take responsibility are the foundations of becoming not just a disciplined soldier, but a future NCO.

Just before I left, CSM Stanley handed me a challenge coin. It's one of the most treasured keepsakes from my time in uniform, and I still have it. It represents a leadership principle that has stayed with me ever since: take care of your surroundings and do the right thing, even when you think no one's watching.

**3 January 2025 – Reflection of Cap Stanley, Command Sergeant Major of 4-64 Armor**

When the infantry guys came over to 4-64, I had to knock some of that arrogance out of a few of you. We had to function as one team. As for the challenge coin, I picked up that trick from CSM Jack Tilley, the 12th Sergeant Major of the Army.[138] So, in a way, the lesson he taught me got passed on to you. Funny how that works.

Over time, many of us lost our ego trips and grew to respect our tanker brothers. Sure, they did things differently and rode around in tanks, but we started to like a lot of those guys. We began to see the value in what they brought to the table. Additionally, the M1 Abrams tank was equipped with a gas turbine engine, colloquially referred to as a jet engine, and came with several unexpected perks.[139]

More than once during field exercises, when the driving rain was coming down, and we were soaked to the core, they would pull up, idle their engines, and pretend to be hard at work inside. Their feigned delays, especially during the fall and winter months, allowed us to stand behind those super-heated engines, which dried us off and warmed our bones for a few precious minutes.

Small acts of kindness like that went a long way. Camaraderie is built by sharing in the misery. It's also built by those offering unspoken gestures of help, given freely and without expectation.

By this time, I had already been promoted to specialist and joined a tongue-in-cheek network of soldiers affectionately known as the E-4 Mafia.[140] In reality, it's an odd rank in the Army. Specialists are no longer privates, meaning they have been around long enough to know how things *really* work. Additionally, they are known for finding creative, and sometimes lazy, ways to get things done. However, they haven't yet reached a proper leadership role, which typically occurs upon being promoted to sergeant (E-5). So, they are somewhat in a limbo phase, and many soldiers remain there indefinitely, unwilling or unable to assume leadership roles.

One week, we were scheduled for training at the shoot house. This was a large, concrete building in the middle of the woods, featuring metal rails running vertically up and down all the interior walls. Slid onto these rails were large blocks of rubber designed to absorb 5.56mm rifle rounds. The rubber blocks would degrade over time as we shot into them during CQB training, so the rails were designed for quick replacements.

---

[138] AUSA. "12th SMA - Jack L. Tilley." Association of the United States Army. Accessed December 2, 2025. https://www.ausa.org/army/12th-sma-jack-l-tilley.

[139] Honeywell Aerospace Technologies Marketing. "M1 Abrams Still Going Strong With Honeywell Jet Engine." Honeywell. Accessed December 2, 2025. https://aerospace.honeywell.com/us/en/about-us/blogs/m1-abrams-still-going-strong-with-honeywell-jet-engine.

[140] Urban Dictionary. "Urban Dictionary: E4-Mafia." In *Urban Dictionary*. Accessed December 2, 2025. https://www.urbandictionary.com/define.php?term=E4-Mafia.

Along the top of the structure was a catwalk system, where our leadership and range cadre could monitor our progress. They routinely called out any dangerous behaviors, sloppy tactics, or anything that fell short of the rigorous standard. After all, any minor error in CQB could result in someone taking a live rifle round to the face.

The shoot house was a lot of fun, but it was terribly dangerous, with dozens of armed men discharging rifles in tight, confined spaces. Inside the structure, our leadership would set up dummies that were hostages, dummies that held weapons, or sometimes balloons for precision target shooting. Each run was different, as they would move targets, furniture, and other items around to make it more challenging. Many of the iterations promoted competition amongst the troops, with all of us aiming for perfect scores. No one wanted to miss.

I won't go into the specifics of the tactics we used, but they demanded relentless repetition. Day after day, we practiced stacking tightly outside entry points, covering every possible angle, breaching doors, and charging through the "fatal funnel." It's a narrow chokepoint, such as a doorway, where visibility is limited, and any defender inside could easily cut down attackers. It was a lot to remember, but success depended on instinct and reaction. Pure muscle memory was honed through repetition.

During that particular week, SGT Brister was unable to attend training. As the senior member of the fire team, I took charge of everyone and led them through a series of challenging live-fire runs, including in the shoot house. We worked well together, stayed sharp, and ultimately earned praise from leadership for our performance.

A few weeks later, I was summoned to CPT MacKinnon's office. My first thought was that I might be in trouble. Instead, to my surprise, he presented me with an Army Achievement Medal. I was caught off guard.

"Thank you, sir," I told him. "But I'm not sure what I did to receive this."

In the award recommendations to the colonel, the citation read, "SPC Harstad distinguished himself as a team leader during the squad/section live fire. His tactical and technical proficiency enabled him to designate TRPs (target reference points), control rates of fire, and initiate the support by fire position successfully. His actions set the standard for the squad/section live fire, which ultimately received praise from all levels of his command." CPT MacKinnon went on further to add, "SPC Harstad stepped up and took charge as a team leader, a job slotted as a sergeant position. He led the team and squad ultimately to success."

I was proud to receive the award, though part of me felt I was just doing what any soldier should. If my sergeant were to go down, it was my job to take charge and carry out the mission. That's what we trained for. Still, if the captain thought it was worth recognizing, I wasn't about to argue. He was someone I looked up to, and I accepted the medal with gratitude.

The captain's kind words, combined with him presenting me the award, had an unexpected side effect. They nudged me to leave behind my professional wrestling days in the woods and start acting a little more like an adult and become a leader. For reasons I still can't fully explain, he had that effect on me. He made me believe I could be more than just the skinny kid in the platoon and pushed me to go beyond what I thought was possible.

This is really the point in the story where I started crossing that line from inexperienced rookie to someone others could count on. I buckled down and tried to take my job a lot more seriously.

From LTC Roth and CSM Stanley in battalion, to CPT MacKinnon and 1SG Barnes in Bravo Company, all the way down to SFC Hill and SGT Brister in my platoon, it felt like everyone in my chain of command was quietly steering me in the right direction. Looking back, those were the men who helped drag me across that gap from boy to man. They were showing me the way, even if I didn't recognize it at the time.

Our final training deployment arrived in August, and we were sent to Fort Polk's Joint Readiness Training Center (JRTC) for jungle warfare preparation. At the time, it was still called Fort Polk. Since then, the Army briefly rebranded it Fort Johnson and then changed it back again, a neat little case study in how so-called "leaders" will burn time and money on name changes to protect feelings, while treating the one thing that actually matters, winning wars, as an afterthought.[141] But I digress.

The base itself was sprawling, humid, and overgrown, tucked deep in the backwoods of Louisiana. It was known for its dense vegetation and a well-trained OpFor. Units would regularly rotate through the JRTC to receive extra training before heading overseas.

To be blunt, much of the training that we did at JRTC was a waste of time and not enjoyable. The experiences in Korea and at FLETC had a significant impact on our future Army careers, and we could easily see when the exercises were substandard. And substandard they were at Fort Polk. A substantial portion of the gear we were given to use didn't function properly and appeared to be in disrepair. Many of the MILES gear lasers either didn't work at all or would randomly shoot in wild directions for no reason. In fact, on several training exercises, we suffered more casualties from friendly fire incidents than we did from engagement with the OpFor.

It was frustrating, falling far short of the standard we expected. So, I won't waste much more time writing about it. That training deployment doesn't deserve it.

The only truly memorable event during our time at Fort Polk wasn't even part of the training. It involved a Bengal tiger. Yes, you read that right. According to officials, they were unsure where the tiger came from but suspected it was originally someone's "pet."[142] Because, of course, why wouldn't someone in rural Louisiana own an apex predator from India? Maybe it had escaped, or perhaps it had been set loose. Either way, now it was our problem.

We were instructed to report any tiger sightings to local police while conducting our drills. This would've been hilarious if it weren't so deranged. You try maintaining tactical discipline in the middle of the night, pushing through jungle-thick vegetation with night vision goggles and *no live ammo*, all while half-expecting to get pounced on by a massive, striped murder machine.

---

141 Little, Ashlyn. "Two Years After Switch, Fort Polk's Name Returns." *Beauregard News*, July 12, 2025. https://beauregardnews.com/local/two-years-after-switch-fort-polks-name-returns/.

142 Fox News and Associated Press. "Bengal Tiger Seen Roaming La Army Base." *Fox News*, September 1, 2004. https://www.foxnews.com/story/bengal-tiger-seen-roaming-la-army-base.

Every rustling branch, every oddly shaped bush, and every breeze-tickled vine became a potential tiger waiting to reenact the scene from *Apocalypse Now*. Our imaginations were working overtime, with several of us "seeing" the silhouette of the vicious beast in the shadows. Only upon further inspection did we discover it was a swaying plant or other harmless inanimate object.

We never found the tiger, nor did we wish to. For all I know, it's still out there prowling the swamps of Louisiana, stalking the occasional deer hunter. We left Fort Polk with a shared sense of disappointment, not just in the training but in the overall experience. I was more than happy to return to Georgia, where the air was just as humid, but at least the wildlife didn't include escaped jungle cats. More importantly, despite the chaos, I felt ready for Iraq.

**Mata (L) and I running through a frigid and torrential downpour - by morning, I had hypothermia**

We spent the rest of 2004 training… always training. That September, we were stuck out in the field during Hurricane Ivan, slogging through torrential rain and battering winds that made everything absolutely miserable.[143] For once, the commander took pity on us. Soaked to the bone and

---

[143] National Centers for Environmental Information (NCEI). "On This Day: Hurricane Ivan Made Its First U.S. Landfall," September 16, 2017. https://www.ncei.noaa.gov/news/on-this-day-hurricane-ivan-2004.

shivering in the mud, we were transported back to the barracks for hot showers, warm food, and the rare luxury of sleeping indoors until the storm passed. It was the only time I can recall the weather earning us any compassion. Apparently, it only took a full-blown hurricane. Such is the life of a soldier in combat arms.

During one wet, frigid evening, I developed a mild case of hypothermia. I remember being so unbearably cold that it felt as though I could no longer function. My dexterity became nonexistent, I shook uncontrollably, and felt as though I had a terrible case of the flu. Then, out of nowhere, I was no longer cold and almost felt elated. Mata had been keeping an eye on me and wisely had the medics check my vital signs. My body temperature had dropped significantly, so I was put into their APC to warm up.

From that point on, cold and wet became my Achilles' heel. I could handle pain, fatigue, hunger, fear, and even tigers. But the creeping, paralyzing chill of hypothermia was what broke me. That stayed with me forever, and I detested it.

Back in garrison, I tearfully said goodbye to Toby Glass as he headed home to Illinois. His feet had never healed properly, and the Army finally concluded the damage was permanent. He was a great friend, and letting him go was harder than I ever imagined.

PFC Toby Glass outside Bravo Company

That day, I came to a painful realization: the Army would always be a series of goodbyes, and I hated it. I didn't want anything to change. I had already said goodbye to my family when I left for Infantry School, then again to my friends when I broke my foot. Now, I was losing one of my closest brothers. Soon, I knew I'd have to say goodbye to everyone in my platoon. Nothing was permanent; it was all a mirage. People came and went, rotating out every few years. We were all just passing through, like sand slipping through an open hand. And no matter how tightly you held on, the moment always slipped away.

Just before we were scheduled to depart for our second combat tour, a serious incident occurred in the motor pool, where all our vehicles were stored. It was a freak accident that nearly cost the life of one of our young soldiers.

The victim was a forward observer, also known as a FiSTer (Fire Support Team) within the military community, named Donald Urbany. FiSTers work closely with infantry units on the front lines, directing artillery fire and adjusting targeting coordinates in the heat of battle. With a radio and a map, they can rain down thousands of pounds of high explosives onto enemy positions within minutes.

**11 January 2025 – Reflection of Donald Urbany, Bravo 4-64 Armor**

Being a FiSTer, I'd been warned from the start that the grunts hated us. I tried to befriend a bunch of them, but it was pretty clear a lot of guys thought they were better than we were. That rubbed me the wrong way. Eventually, though, I found a few who were actually cool. It just took some time for them to warm up to us.

One afternoon, Urbany and several other soldiers were gathered around the Bradleys in the motor pool. They were performing one of the most grueling maintenance tasks in the Army: replacing the vehicle's treads, with a process known as "breaking track." It involved using a sledgehammer and a specialized metal bar to break the massive metal links apart. Everyone hated doing it because it was exhausting and time-consuming.

As the mechanics hammered away, one of the strikes caused a tiny sliver of metal to shear off from the impact point. That shard became a high-velocity projectile, ricocheting off the Bradley with terrifying force, essentially turning it into a bullet. By sheer bad luck, it was deflected straight toward Urbany. The fragment tore through his uniform and embedded in his abdomen, nicking his aortic artery.

He felt a sharp pain and instinctively lifted his shirt to investigate. Blood immediately began to gush from the small puncture, and he collapsed in shock. To the horror of everyone present, it became clear that he was bleeding out right there in the motor pool.

Fortunately, the soldiers around him acted fast. They applied first aid and called for an ambulance while trying to stabilize him. Urbany lost consciousness, but thanks to their quick response and the efforts of emergency personnel, he was rushed to the hospital just in time for life-saving surgery.

CPT MacKinnon and 1SG Barnes later visited Urbany in the hospital. He was devastated to learn that he wouldn't be deploying to Iraq with the rest of us, but their visit lifted his spirits. He later said it meant the world to him. It showed they genuinely cared about their soldiers, not just as troops but as people.

### 11 January 2025 – Reflection of Donald Urbany, Bravo 4-64 Armor

CPT MacKinnon and 1SG Barnes spent their Friday night sitting with me in the hospital, just shooting the breeze. That meant a lot. They really did care about us, and it showed. It left a huge impression on me and cracked that "us versus them" mentality I'd had about leadership and the infantry. I'll never forget it.

The motor pool, with a variety of tracked vehicles and Humvees

The plan was for Urbany to rejoin us once he had healed. For the rest of us, the incident was a reminder of just how quickly things can go wrong in the military, even when doing something as simple as breaking track. All of us had swung sledgehammers at tracks a thousand times and never seen anything like that happen. It was a freak accident, but we were grateful that he survived.

Little did we know, but Urbany's bloody brush with death was a grim omen of what lay ahead. Sniper attacks, car bombs, IEDs, and even torture awaited us in the sand. The training could only do so much. Now, it was time to face the real thing. It was time to go back to war.

# Reflections

In every man's life, there comes a year, or if fortune smiles, a few, that marks a turning point when the path ahead begins to reveal itself. For me, that year was 2004. It was as if the universe had paused just long enough to let the pieces fall into place. The lessons I learned, the people I met, and the trials I endured began to shape something greater than I understood at the time.

I found myself under the guidance of leaders who didn't just wear the rank; they lived it. They demanded excellence not through brute intimidation but through unwavering example. NCOs like Cap Stanley, Mark Barnes, Harold Hill, Juan Serrano, Brian Torres, and Marcus Brister emerged not merely as mentors but as living blueprints for how the Army should be led.

Then, as if by design, CPT Mike MacKinnon took command of Bravo Company. His presence lit a fire in our ranks and stirred a deep sense of purpose within me. I think it did for all of us. For the first time, I believed that something extraordinary might lie ahead, not just for the unit but for myself. The foundation was being laid, and I could feel that we were ready.

CPT MacKinnon and 1SG Barnes, in particular, were ahead of the curve in preparing us for the upcoming deployment. While I suspect that they received guidance on their plans from superiors, they were the ones who fostered positive company morale and gave us the idea that we were unbeatable when we worked together. They did this through leading by example and deeply caring for their men, and they deserve the lion's share of the credit. Those two were always in the mud with us, and I don't recall ever seeing them take liberties with anyone.

In preparation for this book, I spoke with retired CSM Clarence "Cap" Stanley, the formidable yet inspiring tanker I wrote about with deep respect. Though he was several states away and we would be speaking over the phone, I still hesitated to dial his number before conducting the interview. My palms were sweating as I mentally rehearsed the dozens of questions I'd prepared, determined to show him I'd done my homework. Everything had to be just right. Even over twenty years later, with both of us long out of the military, I didn't want to disappoint him. Leaders like that don't come along often but, for reasons I still can't fully explain, I was fortunate enough to serve under more than one at the same time.

"Hello, Command Sergeant Major," I stammered, finally mustering the nerve to dial his number for the interview. "Thank you for taking the time to speak with me. I'd like to ask you some questions about our time in the Army…"

He cut me off with a familiar tone that immediately brought a smile to my face. "Hold on, killer," he said, his unmistakable voice carrying that same mix of grit and warmth I remembered. Hearing him call me "killer" again, it was as if no time had passed. "Start by telling me about your life: where you've been, what you've been up to, how you're doing. All that Army stuff can wait. And call me Cap; all those formalities are in the past."

We spoke at length, and the conversation reaffirmed everything I had admired about him from the beginning. He reminded me, without even trying, why I had looked up to him so deeply In his TED Talk "Why Good Leaders Make You Feel Safe," management theorist Simon Sinek poses a powerful question: Why are subordinates, especially in the military, willing to give their blood, sweat, and tears for their leader? The answer, repeated time and again, is simple but profound: *"Because they would have done it for me."*[144]

Of course, not all military leaders are cut from the same cloth. Some are lazy, incompetent, unethical, or a combination of all three. However, despite its flaws, the military generally does a commendable job of fostering a culture of strong leadership. This is one area where it gets things right more often than not.

Maybe that's why so many of us look back with such fondness on our time in the military. Was it difficult, dirty, and uncomfortable? Yes, but we shared in the misery and knew that the brothers to our left and right would always be there. They would suffer with us. And, perhaps just as importantly, we learned that our true leaders would do the same.

However, 2004 was also a time of emotional turmoil. I found I had become deeply attached to my unit, my platoon, my friends, and the brotherhood we had forged through hardship. While I should have anticipated the changes, I wasn't prepared for how quickly everything began to unravel. We left 3-7 Infantry and transferred to 4-64 Armor. Soldiers I had trusted and depended on left the military. Toby Glass, one of my closest friends, was medically discharged. Those changes came so fast that it felt like the ground was shifting beneath my feet. If I'm being honest, I didn't yet have the emotional maturity to process it all, and I didn't like any of it one bit.

### 6 January 2026 – Reflection of Toby Glass, Bravo Company, 4-64 Armor

Looking back, the Army was an incredible experience. I wouldn't trade it for anything. If I had the chance to do it all over again, even knowing how it would end, I'd still do it without hesitation. The infantry changed my life and made me a better man.

Years later, I came across an 1826 painting by Joseph-Désiré Court titled *"Scene from the Great Flood,"* and I found myself transfixed.[145] It awakened a feeling I hadn't known how to put into words. It was something primal, unsettling, and profoundly human, and it reminded me of this time in my life.

The oil painting depicts a man standing at the edge of a large rock, with a raging, flooded river churning below him. Within arm's reach is a woman, presumably the man's wife, who is caught in the torrent, desperately holding onto a tree root with her outstretched hand. In her other arm is a baby, unable to fend for himself, and she is trying to lift him above the murky water to the man. However, the man isn't helping his wife or child, even though they are right beside him. In-

---

[144] TED. "Why Good Leaders Make You Feel Safe | Simon Sinek | TED," May 19, 2014. https://www.youtube.com/watch?v=lmyZMtPVodo.

[145] Court, Joseph-Désiré. "Joseph Désiré Court, Scene of Deluge, 1826." Narrative Painting, August 16, 2019. https://narrativepainting.net/joseph-desire-court-scene-of-deluge-1826/.

stead, he is desperately reaching out, nearly falling from the rock into the water. He is reaching for an old man, his father, who is about to be swept away.

What does the painting mean, and why was I so mesmerized by it? Art is always subjective, of course, but I can tell you what it came to represent for me, though I'm certainly not the first to make this observation. It highlights the natural, instinctive human tendency to cling to the past. In the painting, the man's wife represents the present, his father the past, and his infant son the future. Despite having the power to save both his wife and child, who are well within his reach, he risks everything to save his father, the past. He can't let it go. He can't say goodbye, even though his present and future may perish as the flood rises around him.

As I have matured and, hopefully, grown wiser, I have come to realize that I, too, have sometimes leaned on the past at the expense of the present and future. However, it's a mistake to dwell in the shadow of the past. I now consciously focus on my decision-making processes, understanding that true wisdom involves recognizing when a deep, obsessive respect for the past begins to hinder progress. The future requires us to be present, rather than simply nostalgic.

And finally, I think of the similarities between Donald Urbany's story and mine. We were both injured during activities that we deceptively believed were routine: marching and breaking track. Neither situation appeared especially dangerous at first, yet both ultimately knocked us out of action. While his incident turned out to be much more severe than mine, I think it follows a similar parallel.

What struck me most was how Urbany responded. He could have used the injury as an excuse to leave, either to remain at Fort Stewart for the deployment or to depart the military altogether, but he didn't. Despite having endured some tension with the infantry crowd, he still fought hard to return to the unit. He didn't want to be left behind, and he wanted to deploy with the rest of us. An urge drove him to embrace the risks and challenges of the mission ahead. Such unwavering loyalty reveals much about his character.

As fate would have it, Donald Urbany's story didn't conclude there; it was merely the prologue to what lay ahead. In many respects, the same was true for me.

# Part III

## The Human Meat Grinder

---

And I looked, and behold a pale horse: and his name that sat on him was Death, and Hell followed with him.

Revelation 6:8

---

# Chapter XI

## Prosperity

Gathered in the motor pool, the same place where Donald Urbany had nearly lost his life, we anxiously watched CPT MacKinnon climb atop one of the Humvees parked in the lot to deliver his sendoff speech. The time had come to return to the battlegrounds of Iraq, and our leader stood before us to offer words of encouragement, reinforce our mission, and steel our resolve against an enemy more formidable than ever. This was his moment to draw a line in the sand and declare exactly where we stood in the war.

To paraphrase his stirring speech, he reflected on all the hard work we had done over the past year and gave our morale a tremendous boost. Korea, FLETC, JRTC, and the countless trips into the field, the shoot house, and the swamps of Fort Stewart were over. All of it, he reminded us, had been in preparation for Baghdad. His pride in us was palpable as he told us we were among the best soldiers he had ever had the privilege of serving with.

We more than reciprocated that feeling, as he had earned a place in our hearts as the embodiment of Bravo Company. He wasn't just Bandit 6, the company commander's call sign, but carried the spirit of that title in everything he did. Many of us in the company felt that way, and we wanted to deliver nothing but the best results to someone who had given us so much.

As the speech continued, he reminded us that we were not just warriors but also ambassadors. The Iraqi people had been ravaged by endless conflict, and they deserved our genuine help. It was our mission to expose and root out those who sought to destabilize the region. In doing so, we had to be the Americans who could be counted on to do the right thing. We had to act with professionalism, compassion, and integrity.

The tone quickly shifted, however, and he became deadly serious. If anyone attacked us, he warned, we would respond with overwhelming force. Every last insurgent would be slaughtered. He gave us explicit permission to unleash the training we had honed for so long, to dominate the battlefield, and to command the respect of our adversaries. But he also reminded us that our job was to work together to bring everyone in the company home safely. We would leave no man behind.

It was a powerful message that left many of us eagerly anticipating the deployment. It echoed what General Mattis had famously told his Marines, "Be polite, be professional - but have a plan to kill everyone you meet."[146] We knew that we had a difficult task ahead of us and that the bar for success was abnormally high.

I only wish that his address to all of us had been recorded for posterity, as it set the standard for our conduct in the war: fair, helpful, and honorable, but willing to crush anyone who got in the way. It was what we expected from our commander as a motivational message, and he delivered it with sincerity and strength. We knew that he would stand with us every step of the way.

**At Bryant Elementary School (Sioux City, Iowa) just before OIF III with my youngest brother, Brad, talking with many of the same children who had written us letters during our first deployment**

**Photo courtesy of Bill Harstad, Jr.**

As had happened before, we said goodbye to our loved ones at the start of 2005, not knowing if the promise that everyone would come home could actually be kept. My parents drove down to see me off, and they stood there in my living room, both of them shaken, inconsolable, and terrified of what was coming. I tried to reassure them. I talked about how hard we had trained,

---

[146] Mattis and West, *Call Sign Chaos: Learning to Lead.* eBook edition. Chapter 9.

how professional my unit was, and how we knew what we were doing. None of it really helped. Their son was going back to war for an entire year, and we all understood there was a very real chance this would be the last time we ever saw each other. We held on to each other in long, crushing hugs, and traded promises that I would come back. Then they walked out the door and drove away, carrying that fear with them, their stomachs in knots the whole way home.

Buses transported us to Hunter Army Airfield, and we quickly flew overseas, making a brief stop in Europe. Upon our arrival in Kuwait in early 2005, the 3rd ID was on its way to becoming the first U.S. Army division to serve a second combat tour in Iraq.[147]

No one, not even our leadership, knew just how long the conflict would last. We were largely in the dark about the current situation on the ground, having left Iraq in late 2003 when the country was still mired in disorder. We wondered what the conditions would be like. There had been no functioning government, minimal public services, and widespread instability the last time we were there. We would pick up our place again in a leapfrog fashion from those units that had been sent to replace us.

In Kuwait, we quickly established temporary residency and conducted last-minute training on makeshift plywood houses before moving to Iraq. Though flimsy, these improvised buildings served their purpose as we ran through endless iterations of CQB drills to keep our skills sharp. We also had to prepare all the equipment being sent in-country, including Bradleys, Humvees, and other military equipment. There was no shortage of work.

Convoys ran regularly to the ports of entry to retrieve equipment, and I often volunteered to go. As a member of the Heavy Weapons Squad, I manned a machine gun in the turret of a Humvee. In Kuwait, like before, the threat level was low. Our primary job was to deter aggressive local drivers from cutting off or weaving into the military convoy, which we usually did by yelling and waving them off. If they didn't get the hint and persisted in riding our bumpers, we pelted their fancy cars with rocks or water bottles. That tended to quickly back them off.

Some of the transportation soldiers we worked with were squared away and professional. Others, however, were what we in the military called "ate up like a soup sandwich." It was embarrassing how often some of these soldiers seemed completely lost, both figuratively and literally. Directions apparently confused them, basic procedures threw them off, and somehow, getting lost became a regular part of the mission.

### 3 February 2005 – Letter to the Harstad Family

On our second convoy, we had a POG get lost. We had to take our section of the convoy around to look for him. He was on the Saudi Arabian border.

---

147 DVIDS. "Army Unit's Deployment Accelerated." Defense Visual Information Distribution Service, February 15, 2007. https://www.dvidshub.net/news/540139/army-units-deployment-accelerated.

A visibly agitated American transportation lieutenant sat in his Humvee at the checkpoint, arguing with a group of Saudi Arabian guards. They tried to explain, in broken English and with animated gestures, that he had taken a wrong turn, but for some reason, he refused to accept it. Fortunately, they stopped him before he aimlessly wandered too far into their country.

When we arrived and broke the news that he was the one at fault, not the Saudis, he looked genuinely embarrassed, especially as our massive convoy had to perform an awkward, parade-worthy U-turn inside Saudi territory. So, I was technically in Saudi Arabia for maybe thirty seconds, just long enough to fog up the checkpoint with diesel exhaust and confusion. I am absolutely still counting it toward my list of world explorations, because if my boots crossed the border, it goes on the list.

Word passed through 1st Platoon, newly nicknamed "The Gamblers," that they were looking for a couple of volunteers to ride with CPT MacKinnon and a small team on the giant convoy to Baghdad. While most of the company would fly in aboard C-130s to meet at FOB Prosperity, a hand-picked team would make the journey by road. Prosperity was nestled within the International Zone, also commonly referred to as the "Green Zone."[148] It housed much of the fledgling Iraqi government and stood as a fortified stronghold, teeming with American, British, and Iraqi forces.

SGT Vlastelic, SPC Howerton, and I all volunteered to go on the trip. We met up with CPT MacKinnon, and he showed us our hunk of junk Humvee. As I had often experienced on my previous deployment, it wasn't armored and was held together with makeshift equipment that looked dangerously out of place. But, it was to be our home for the next several days as we drove up Route 1 from Kuwait to Baghdad. We had to make the best of it.

Howerton manned the massive M2 Browning .50-caliber machine gun, awkwardly mounted on the rear bed of the Humvee, while I served as a floater armed with my M-240B medium machine gun. Between us, we commanded an impressive amount of firepower. Positioned at the tail end of the convoy, we quickly realized we'd spend the coming days shivering in the open, fully exposed to the biting wind and cold Iraqi air.

A massive sand berm separated Iraq from Kuwait, stretching nearly the entire length of the border and rising roughly a dozen feet high. It was fortified with concertina wire, chain-link fencing, guard towers, observation posts, and parallel rows of trenches.[149] As we drove through a cut in the berm and crossed into Iraq, we were immediately confronted by a bleak, desolate village. Emaciated children, clad in tattered clothes, wandered barefoot along the roadside. We tossed them water bottles and unwanted MRE pouches, particularly the meals we didn't care for. Scenes like that remain etched in my mind and continue to remind me how much I have to be grateful for.

Although the total distance was only slightly more than 400 miles, we made several stops at bases along the way, where we ate, tried to clean up, and slept in a relatively safe area. Whenever a

---

148 Houghton, Richard, and Patrick McDonald. "A Visitor's Guide to Baghdad's International Zone," May 1, 2006. https://bi.gazeta.pl/im/4/4646/m4646484.pdf.

149 Global Security, "The Berm," February 17, 2017, https://www.globalsecurity.org/military/world/gulf/kuwait-the-berm.htm.

vehicle broke down, as they often did, the entire convoy would grind to a halt until a mechanic could get it running again. It was a mind-numbing process, with boredom and the cold as our constant companions.

Another convoy just ahead of us called in an emergency, where an IED detonated beside one of their vehicles. One soldier was killed, and the explosion brought everything, including our convoy, to a standstill until the area could be secured and cleaned up. Hours later, once we could start moving again, we cautiously maneuvered past the site. We saw a blackened crater surrounded by scorched debris and the leftovers of twisted fragments of machinery, a grim reminder of how quickly everything could change.

That night, I rested my machine gun on the Humvee's metal rail above the right rear tire and settled into the familiar green glow of my night-vision, scanning an empty stretch of nothing. Without warning, a thunderclap of a blast slammed into our vehicle. At that very second, my NVGs went instantly black, my pulse spiked, and I started slapping frantically at the goggles, trying to get them to come back to life. While doing this, chunks of debris smacked into my chest and helmet. For a few long seconds, I was certain an IED had hit us with a glancing blow and that the next step was incoming fire. As it turned out, the reality was far less dramatic.

It was the thick Humvee tire directly beneath me that had inexplicably blown out with ridiculous force, launching big chunks of rubber into the air like shrapnel. At that exact moment, though completely unrelated to the exploding tire, my night-vision batteries chose to die, leaving me blind and convinced we were under attack. Later, we laughed about it, especially the fact that both Howerton and I were about a second away from dumping rounds into the darkness over a flat tire. Thankfully, we kept ourselves in check, nobody lit up the night sky, and the crew just shrugged, swapped out the tire, and carried on. So, my first "action" of the deployment wasn't an IED at all, but an over-inflated tire that finally decided it was done with the war.

**19 February 2005 – Letter to the Harstad Family**

I volunteered to be an escort on a Humvee, and we drove from Camp New York, Kuwait, to FOB Prosperity. It was a long, cold trip, and we didn't have much contact (with the enemy). However, one soldier was killed on the convoy ahead of us when an IED blew him up. (On our Humvee), we blew a tire out at night in the middle of nowhere. Right when that happened, my NVGs shut off, so it was a little freaky.

On the last leg of the trip, CPT MacKinnon let us know he appreciated us toughing it out in the back. He tried to pass us warm drinks and food when possible, but it was still a miserable journey. Finally, after what felt like an eternity, I saw Baghdad in the distance and knew that our expedition was nearly at an end. However, we weren't out of the woods yet, and we had another bit of bad luck in store.

The moment we rolled into Baghdad, traffic turned into a full-on nightmare. Cars swarmed around us, cutting in and out, inches from our bumpers. We were hammering the sides of vehicles with water bottles and rocks at a ridiculous pace, trying to keep them from boxing us in, but they just kept coming. It felt like people had lost all fear of the American military, at least in this part of town. That changed quickly once we really started pelting their vehicles. They got the message, but they certainly weren't thrilled about it. Drivers shook their fists, leaned on their horns, and shouted at us as they finally peeled off the highway, clearly impatient and possibly headed for a hot date.

The transportation lieutenant at the front of the convoy, who I'm pretty sure was new and in over her head, suddenly began taking the convoy down the wrong streets, sometimes circling back around for another run, as if the first time wasn't enough. And just to be clear, this wasn't the same one who had gotten us lost in Kuwait. Apparently, the Army kept a whole warehouse of navigationally challenged butter bars and sent them out as needed. She managed to get us progressively lost and led the convoy down some sketchy routes that felt worse with every turn.

While all this chaos was unfolding, we could hear CPT MacKinnon on the radio, trying to figure out where the hell we were and help the lost lieutenant. That was when it dawned on us that we had taken yet another wrong turn and had drifted into a really bad part of town.

The scenery went dark fast. The buildings looked bombed-out and forgotten, with balconies sagging and windows shattered or halfway boarded over. The normal, hectic traffic behind us simply disappeared. It felt like the locals knew something we didn't and wanted no part of whatever was about to happen. The air in the Humvee grew heavy and quiet, the kind of silence where everyone scanned for the first muzzle flash.

CPT MacKinnon, calm as ever, came back on the radio, this time calling for air support. Not long after, the distinctive thump of rotor blades rolled in overhead, and a pair of Apache attack helicopters appeared above us. The mood in the vehicles shifted just a bit. Nobody relaxed, but sphincters unclenched just a notch. With the choppers talking to MacKinnon and guiding from above, the convoy, including the lumbering heavy trucks, slowly worked its way back onto the correct route.

About an hour later, we finally rolled into the Green Zone and through the gates of FOB Prosperity. Looking back, it's a minor miracle no one chose that moment to hit us while we were wandering around completely lost in that part of Baghdad. If an ambush had gone down, it would have been a disaster with all those bulky, slow-moving vehicles. Thankfully, CPT MacKinnon stayed sharp and on top of our position, just as he always had. If we weren't already at our final stop, I'm convinced he would've bumped our Humvee up to the front of the convoy for good.

**20 February 2005 – Letter from Mark Barnes, 1st Sergeant of Bravo 4-64 Armor**

CPT MacKinnon and the wheeled security element made the long trek to Baghdad by road. The transportation lieutenant got lost in Baghdad, but the CO (having been lost in Baghdad enough last time) knew exactly how to get to the link-up point in the Green Zone. So, as usual, he got them all safely to the FOB.

For the next week, we slept on cots inside Al-Salam Palace, a heavily damaged mansion of former President Saddam Hussein in the heart of his old Green Zone compound.[150] The palace contained roughly one million square feet of floor space and was lavishly finished with marble and granite throughout.[151] Parts of the structure had been reduced to rubble by coalition airstrikes in 2003, and those unstable sections were generally avoided.[152] Nearby lay several massive statues of Saddam that had been removed from the palace after the regime fell, their toppled heads and torsos resting in the dust as a sort of accidental monument to his downfall.

**Relics of the old regime gathering dust in our motor pool**

---

[150] Global Security, "Al Salam Palace," September 7, 2011, https://www.globalsecurity.org/military/world/iraq/al-salam-cc.htm.

[151] Evendo. "Al-Salam Palace: A Symbol of Baghdad's History." Accessed December 3, 2025. https://evendo.com/locations/iraq/baghdad/al-jadriya/landmark/al-salam-palace.

[152] Safarway. "Al Salam Palace." Accessed December 3, 2025. https://safarway.com/en/property/al-salam-palace.

Near the compound was a beautiful artificial lake system, reportedly filled with enormous fish. We often heard that the regime disposed of bodies by feeding them to the carp, which would explain their abnormal size. Several people swore it was true, claiming remnants of dead bodies were sometimes found at the bottom.[153] I don't know whether it was an embellished rumor or something genuinely disturbing, but it certainly wouldn't have surprised me. Either way, the whole scene was a striking display of extravagant wealth set against the backdrop of a country mired in poverty and ruin.

**Overlooking the damaged side and dome of Al-Salam Palace**

**Photo courtesy of Derek Hauger**

Unfortunately, opportunists and disturbed individuals immediately started making life difficult for some of our loved ones back home. It wasn't just us who had to stay on guard. Our families did, too.

---

[153] Bourne, Wade. "Fishing Saddam's Waters." Bassmaster, March 4, 2008. https://www.bassmaster.com/news/fishing-saddams-waters/.

**20 February 2005 – Letter to the Harstad Family**

If someone contacts you or anyone in the family about me being killed, injured, or captured, don't believe it at first. Make sure you get authentication of his (the caller's) identity. Some wives at Ft. Stewart have been called, saying their husbands were killed and that they need to call an 800 number for more info. As soon as they do, they're charged with a $100 phone bill.

We rejoined the rest of Bravo Company, many of whom remarked on the dramatic changes they observed upon arriving at BIAP. Gone were the days of no electricity and other amenities. American businesses had begun infiltrating the country, for better or worse, and even fast-food establishments were now popping up.[154] I was skeptical at first, derisively dismissing their stories as exaggerations, but I quickly ate my words when I saw it all firsthand. Some American soldiers could now become blimps in the middle of a war zone, something that I would have considered impossible as we departed in late '03.

FOB Prosperity, too, had running electricity, showers, internet, and food vendors. It was utterly crazy that all of this was available in a battle zone and in what had just been a ruined landscape. Even crazier was that just a short distance away, out beyond the Green Zone, everything was still in a state of decay, although it *was* slowly improving. I hadn't anticipated any of this, and I wasn't quite sure what to make of it.

**7 February 2005 – Letter from Robert Roth, Commander of 4-64 Armor**

Soldiers can take hot showers, and there is a laundry service. In fact, there is a Burger King and a Subway shop on the compound, so we still have the special taste from home.

When Bravo Company finally linked up in full, we were all immediately split back up to meet the Tusker battalion's operational needs. Platoons were shuffled around and reassigned as necessary, often with surprising rapidity. Our own Gambler Platoon fluctuated between Bravo Company and CPT Judge's Delta Company, known as the Death Dealers.

Our duties followed a rotating schedule that included defending the Green Zone and conducting patrols within and just beyond its perimeter, into nearby areas such as Haifa Street, Mansour, and across the river into Karrada, among others. Outside the towering blast walls that defined the Green Zone was what everyone bluntly referred to as the Red Zone. Within the Green Zone, aside from random mortar or rocket attacks, it was relatively safe. Out in the Red Zone, however, the rules changed. It was unpredictable, dangerous, and very much a return to the Wild West.

---

154 Philpot, Robert. "#FlashbackFriday: 20 Years Ago, the Exchange Opened the First Burger King in Iraq." *The Exchange Post*, June 16, 2023. https://www.theexchangepost.com/2023/06/16/flashbackfriday-20-years-ago-the-exchange-opened-the-first-burger-king-in-iraq/.

In addition to patrols and static zone defense, we were tasked with training the newly formed Iraqi Army, manning the access control points at the Green Zone's gates, and executing targeted missions. These included capturing or killing insurgents, destroying stockpiles of old munitions, and attempting to build an intelligence network by earning the trust and cooperation of the local population. That was much easier said than done, though.

**CPT MacKinnon and Bravo Company soldiers move out during one of our countless raids**

**Photo courtesy of Phillip Cornell**

Everyone knew that we couldn't remain in Iraq forever. The Iraqis would eventually need to train, lead, and defend their fragile military and government without relying on constant American support. One way to do this was by temporarily placing some American forces, including us, under the leadership of Iraqi commanders during joint missions. I imagine this decision caused more than a few sleepless nights for our leadership. It probably felt less like sending a child off to college and more like taking the training wheels off a bike while it was already rolling down a steep hill. They had to learn, and this was the only way to do it, but it was nerve-wracking for everyone involved.

**1 March 2005 – Letter from Michael MacKinnon, Commander of Bravo 4-64 Armor**

B Co is taking the lead for the battalion once again by pushing out into sector. The company is operating very independently from the 4-64 battalion. We are currently augmenting the 302nd Iraqi Army Battalion. That's right, we are conducting joint operations with the Iraqi Army. For the combat missions, I am taking orders from the Iraqi Battalion Commander. We have conducted two very successful missions with them so far and are looking forward to continuing our relationship. This is a huge success story for the War in Iraq.

After moving out of Saddam's mansion for good, all of us were assigned newly built trailer rooms to live in for the foreseeable future. The rooms were tiny, with space for only two bunks and lockers, but they were thankfully equipped with air conditioning. Showers and toilets were in a separate building a few hundred feet away. Chow was served on a rotating basis inside the main hall of the lavish palace, though candlelit, romantic dining wasn't on the menu.

Within days of arriving at Prosperity, I had my first encounter with American private military contractors, or, as we would have called them in a previous era, mercenaries. These guys represented a range of American private military companies, hired at staggering rates to provide security for politicians and strategic locations.[155] Most were former military personnel, many with backgrounds in infantry or special operations, making them highly trained and capable. Still, their appearances often clashed with their résumés. Baseball caps, ZZ Top beards, and sometimes long, scraggly hair were familiar sights, an odd aesthetic that somehow became fashionable in their community.[156]

We didn't have any issues with them because they were on our side, but it wasn't long before resentment started to simmer. After all, we were being paid a mere fraction of what they were, and we were being given almost all the lousy assignments, none of which they seemed to be doing. It was hard not to notice who was digging and constructing fighting positions at the checkpoints, pulling long patrols, and going on missions… *and* who wasn't.

That said, I don't want to paint them all with the same brush. Many of the contractors were solid professionals who did their jobs well and were good guys to work with. They just rarely got bogged down with the most miserable work, unlike regular grunt units. Maybe part of my frustration stemmed from envy. I'll admit that. I wanted to be part of the action *and* get paid well for it, too. So, take my criticism with a grain of salt. I know some of it comes from a place of jealousy.

---

155 Cotton, Sarah, Ulrich Petersohn, Molly Dunigan, Q Burkhart, Megan Zander-Cotugno, Edward O'Connell, and Michael Webber. "Hired Guns: Views About Armed Contractors in Operation Iraqi Freedom." The RAND Corporation. Santa Monica, CA, 2010. https://www.rand.org/content/dam/rand/pubs/monographs/2010/RAND_MG987.pdf.

156 McCloud, Gary. "What Do Private Military Contractors Wear?" The Gun Zone, August 10, 2024. https://thegunzone.com/what-do-private-military-contractors-wear/.

**7 May 2005 – Letter from Donald Cyr, Alpha 3-7 Infantry**

Hey, guess who SGT Hall ran into a couple of days ago? None other than (redacted name of a soldier we knew who left the military under questionable circumstances). Yeah, he's working for (redacted company name) as PSD (personal security detail). Can you believe that? He is getting paid over 100 grand a year and doing a fourth of the work we do! Just wanted to let you know…

This experience with private contractors, combined with the influx of corporate money into the country, served as a wake-up call that some questionable and shady dealings were underway. Years later, we would learn of the harsh truth that many of us suspected: a great deal of profit was being made at the expense of many American service members and, likely even more depressingly, on the backs of Iraqi civilians.[157] In a scathing review of how some corporations profit in war while soldiers bear the cost, Major General Smedley Butler once wrote, "It is the only one in which the profits are reckoned in dollars and the losses in lives."[158]

The influx of unnecessary amenities and private contractors was a mixed bag. On the one hand, we all valued sleeping and living in better conditions. We also appreciated any extra help, as there was always a lot to do. On the other hand, there were a lot of opportunities for people to become fat, lazy slobs if their leadership wasn't properly paying attention to curb that behavior. And it was hard to ignore the sight of people reaping oversized paychecks for minimal effort in a war funded by American taxpayers and paid for in soldiers' blood.

Furthermore, it quickly became apparent that much of the equipment we had inherited in-theater was substandard and far less effective than that of our private military counterparts. Our attire was still oddly mismatched: our uniforms were desert tan, while our combat vests featured a dark woodland green camouflage pattern. We stood out like a sore thumb. Many of the Humvees lacked armor, the turrets were either falling apart or nonexistent, and it was difficult for armorers to obtain parts to repair the many failing weapon systems. Our Bradleys and tanks often fared no better, frequently breaking down in the harsh, sandy terrain. It was odd to me that we could obtain a Whopper in the middle of the desert, but not a decent gun turret or camouflage. Wherever those hundreds of billions in the national defense budget were going, it was clear they weren't reaching us.

Around the same time I encountered the private contractors, I came across a curious and elusive creature said to inhabit every FOB across Iraq and Afghanistan. Known to primatologists and grunts alike as the Fobbit, this species was rarely observed outside its natural habitat, preferring the cool confines of its burrow. On occasion, Fobbits could be spotted sneakily "rat-fucking" our

---

157 Beelman, Maud, Kevin Baron, Neil Gordon, Laura Peterson, Aron Pilhofer, Daniel Politi, Andre Verloy, Bob Williams, and Brooke Williams. "U.S. Contractors Reap The Windfalls of Post-war Reconstruction." International Consortium of Investigative Journalists, May 7, 2012. https://www.icij.org/investigations/windfalls-war/us-contractors-reap-windfalls-post-war-reconstruction-0/.

158 Butler, Smedley Darlington. War Is a Racket: The Antiwar Classic by America's Most Decorated Soldier. eBook edition. Skyhorse Publishing, 2013, Chapter 1.

food supplies before being run off, which is military slang for mysteriously making all the good stuff disappear from the chow hall or local vendors. They weren't at all dangerous to humans, but they were incredibly irritating.

Biologists and bored infantrymen have noted that the Fobbit is a temperamental animal, most active during mealtimes, and highly skilled in the art of vanishing whenever danger, or even the prospect of physical activity, approaches. These creatures are notorious hoarders, stockpiling everything in their burrows, from protein bars to, of all things, printer paper, as if the apocalypse were at hand. A true modern marvel of evolution, the Fobbit survives for months or years without ever venturing beyond the walls of the FOB, sustained entirely by those hoarded supplies, intra-species gossip, and a grossly inflated sense of self-importance.

When a Fobbit does emerge into daylight, it does so cautiously, like a vampire wearing sunscreen. Upon realizing its natural predators are nowhere near, it begins to swagger around like a peacock jacked up on Red Bull. The Fobbit puts on a brief show, mainly in an attempt to exert dominance over any other species on the FOB, and then quickly disappears for the rest of the day. It apparently has more pressing matters to attend to deep within the burrow.

Okay, enough zoology. For the civilian reader, let me fill you in on a military secret. A Fobbit, as I'm sure you have guessed, isn't an actual animal at all. It's a portmanteau of "FOB" and "Hobbit." Like Hobbits in Tolkien's *Lord of the Rings*, a Fobbit is a POG who never leaves home, hoards gear, and somehow magically manages to be everywhere except where the danger is.

Fobbits are easily spotted, as they are dressed in freshly laundered uniforms and spotless gear that have never seen a speck of dust. They will regale any audience, willing or not, with tales of valor, such as the time they tripped over a power cable near the chow hall, thus *narrowly* avoiding death by an insurgent booby trap. They exaggerate every minor inconvenience into a near-death experience and often act like their internet outages were frontline trauma. While you may think I'm exaggerating or making this up, I'm not. Fobbits are very real, and there were a *lot* of them de-ployed with us.

One of the easiest ways to spot a Fobbit was by looking at their weapon. Since everyone was required to carry one at all times, it was a dead giveaway. Fobbit rifles were often laughable, decked out with every imaginable accessory: scopes, laser sights, pistol grips, infrared illuminators, flashlights, and other high-tech gadgets they undoubtedly had no idea how to properly use. Believe it or not, sometimes they were even on backwards. They probably didn't even know what half of it was, other than it looked cool.

Remember when I mentioned the misallocation of gear in an earlier chapter? That's right, the Fobbits always had the good stuff. Meanwhile, the grunts were out in the sand with rifles held together by duct tape and 550 cord, often sporting cracked buttstocks, busted iron sights that wouldn't adjust, and at least one part missing entirely.

Versions of POGs and Fobbits have existed in every war, just under different names. Each generation of grunts has come up with colorful terms to describe these men and women who like to hide "in the rear with the gear." Some of the more memorable include. Rear-Echelon Mother Fuckers (REMFs), Chairborne Rangers, Saigon Commandos, Base Rats, Garritroopers (a mash-up of "Garrison" and "Trooper"), Zebras (a nod to their clean, striped uniforms), Tent Queens, PowerPoint Rangers, PX Commandos, and even the Combat Barista.

**The author on patrol in Karrada, with a mismatched uniform, a partially-built turret, and a full load of "please don't break down in this neighborhood" energy**

In *With the Old Breed*, Marine E. B. Sledge describes how the rifle companies in the Umurbrogol ridges had to contend not only with Japanese troops but also with what he calls "rear-echelon souvenir hunters." During quiet moments after firefights, these clean, well-rested POG Marines from the rear would appear, sift through the torn-up terrain, and scoop up Japanese helmets, weapons, and other gear to take home as trophies. Sledge notes that they were instantly recognizable beside the filthy, exhausted infantrymen who had actually done the fighting, and he remarks that the line companies joked about the "hair-raising stories" these men, who had never faced a live enemy or incoming fire, would probably tell after the war.[159]

---

[159] Sledge, With the Old Breed: At Peleliu and Okinawa, Chapter Five.

In *We Few*, Special Forces soldier Nick Brokhausen is even more vicious in his assessment of POG culture. He mocks the way the Army's obsession with order attracts what he sees as the worst personalities in uniform: military police, staff lackeys, ambitious rear-echelon "rising stars," and self-righteous disciplinarians who live to hassle the troops actually doing the fighting. He darkly jokes that only the Viet Cong, with their occasional "122mm doses of reality," keep the rear from being completely overrun by these people.[160] Brokhausen's book is a savage mix of gallows humor, horror, and gritty drama, shot through with a brutally honest view of how war really works, and his sentiment captures *exactly* what every front-line combat grunt thinks about POGs and Fobbits.

**1 August 2025 – Reflection of David Anderson, Commander of Charlie 4-64 Armor**

Word previously came down that the MPs wanted everyone to slow down when driving in the Green Zone. I was a tank commander, and we were rolling into something called Cyclone Circle, the same roundabout where Charlie 4-64 had held position during the initial invasion. I looked at my driver and said, "Let's have a little fun and slide around Cyclone Circle."

He took that as a green light and really gave it some gas. We ended up drifting our Abrams around this huge traffic circle, very much not in compliance with the speed regulations. A moment later, I heard over the radio, "Uh, sir… look behind us." I turned around and saw an Air Force SP (Security Police) car with its lights flashing, trying to pull us over.

I told my driver, "Just go!" and we kept rolling, ignoring them. They followed us all the way to the edge of the Green Zone, but we went right out of the gate and into the Red Zone, where they weren't about to follow. They kept asking the gate guards who that tank belonged to, but the guys at the gate played dumb and never gave us up.

We got away clean.

So, if you ever hear someone say they earned a Purple Heart for paper cuts, I mean, being bayoneted by those ISIS sticky notes, odds are you've just spotted a Fobbit or POG in the wild. Somehow, just somehow, they made it off the FOB and back to the safety of the United States. More than likely, it's the guy sitting next to you in the bar telling war stories or saying that he was a SEAL on missions he can't talk about.

Of all the duties we were assigned, the one I dreaded most was checkpoint security. At several strategic locations around the Green Zone, access control points were established to regulate pedestrian and vehicle traffic entering and exiting the Red Zone. Massive concrete blast walls funneled traffic into narrow corridors, forcing everyone to pass through multiple layers of security.

At the outer perimeter, Iraqi Army soldiers manned fortified bunkers, tasked with questioning drivers and pedestrians. They also searched vehicles and their occupants, and, ideally, inter-

---

[160] Brokhausen, Nick. *We Few: U.S. Special Forces in Vietnam*. eBook edition. Casemate, 2018, Chapter 1.

cepted any threats before they could advance far enough to inflict mass casualties on Americans with a VBIED. It was a nerve-racking situation, made worse by the fact that many Iraqi soldiers were poorly trained and ill-equipped. Terrorists typically preferred targeting Americans but had no qualms about killing Iraqi military or police when the opportunity arose. Given their forward position, Iraqi security forces often absorbed the initial impact of these attacks simply by virtue of proximity.

Deeper within the checkpoint, American troops manned watchtowers and fortified positions that were surrounded by concertina wire. They had to maintain overwatch on the lanes below. This inner zone was where vehicles underwent detailed inspections for explosives, individuals were searched for weapons, and brief interrogations were conducted with the help of interpreters. It always felt like a high-stakes game of Russian Roulette because it was an environment where it wasn't a question of *if* an attack would occur, but *when*.

One evening, as the sun dipped below the horizon, I sat inside a darkened bunker with my machine gun, overlooking a particularly rough neighborhood. Nothing had happened all day, and the buzz of activity was beginning to wind down. Without warning, a loud *crack* rang out, and a geyser of concrete spewed from a nearby block. I instinctively ducked and quickly adjusted my position, hoping to catch a glimpse of a muzzle flash or any sign of movement. But I didn't see a thing.

Staying low, I rapidly scanned every window, alleyway, and rooftop, ready to unleash hell upon the sniper. But the streets were still. Aside from the rumble of approaching Bradleys, there was only silence.

Within minutes, members of my platoon arrived to reinforce our position and began sweeping the adjacent neighborhood. Nothing, not even a spent casing, was ever found. We suspected a random sniper had taken one shot and then quickly fled before we had a chance to return fire. For me, it was terribly frustrating because I knew we couldn't blindly spray rounds into the area. A large number of civilians lived right next to where we were, meaning we were at a significant disadvantage. Insurgents used this to their benefit, knowing we would never mow down innocent people just to get one sniper.

### 4 March 2005 – Letter to the Harstad Family

I was on the gate last night overlooking Haifa Street when someone shot at us. He was a pretty crappy shot, because it hit one of the concrete barrels a few yards away. We think he took one shot and ran. Unfortunately, I didn't see where it came from. He'll be back...

Positioned behind the checkpoints as our final line of defense were several Bradleys. We trusted both the machines and their crews to deliver heavy firepower when needed or to evacuate the wounded quickly. But no amount of armor or weaponry could replace the boots-on-the-ground work we were required to do. We still had to stand out there... exposed, vulnerable, and waiting. This wouldn't be the last time an insurgent took a shot at us.

On the radio at the front of a checkpoint leading into the Green Zone, SGT Brandon Moore returns while I watch his back with my M-240B medium machine gun from inside the reinforced bunker

### 23 April 2005 – Letter to the Harstad Family

We just got assigned a new checkpoint. We have had two sniper attacks out there, and both times they hit our soldiers in the chest. Luckily, our guys always wear the bulletproof plates, and they stopped the rounds.

I quickly grew to hate working at the checkpoints, as most of the time, it meant enduring endless hours of boredom under relentless, suffocating heat. Then, without warning, everything could erupt into chaos, with gunfire crackling through the air or the deafening blast of a car bomb. The worst part was the helplessness. There was almost nothing we could do to prevent it. All we could do was react, but by then, it was already too late.

When not assigned to the checkpoints, we were responsible for maintaining security around FOB Prosperity. A massive concrete wall encircled the base, punctuated by watchtowers at regular intervals. Our schedule included eight- or twelve-hour rotating shifts in those towers, scanning the surrounding neighborhoods for any signs of suspicious activity.

Perched high above the Green Zone, the towers offered a breathtaking view of Baghdad. One of my favorite sights was the Al-Rahman Mosque, visible in the distance through binoculars. Although construction was abandoned in 2003 as the nation was consumed by war, its design was stunning. The mosque had massive domes arranged in a circular pattern around a towering central dome. It looked like something pulled from a *Star Wars* film. Had it been completed, I'm convinced it would have ranked among the most beautiful buildings in the world.

**Al-Rahman Mosque, in the Al-Mansour district of Baghdad, from a tower at FOB Prosperity**

Manning those towers often proved to be monotonous. Being situated in the heart of the Green Zone meant we were relatively safe, too far from most threats to see much action. To pass the time, we shared stories, cracked jokes, and often laughed until our sides hurt. It wasn't the adrenaline-pumping experience some might expect from deployment, but it gave us a chance to bond in ways the checkpoints never could. It also gave us a chance to decompress without the constant threat of exploding into a pink mist. Ironically, it was from one of those quiet, uneventful towers that I first fired my rifle during the early weeks of our tour.

In the middle of the night, I was inside a tower that touched part of an adjoining roadway. The quiet road was lined with strands of concertina wire, some stacked in two or three rows. I never saw anyone walking down there, and it was typically unused by automobiles.

Breaking the lull of the evening, I heard a terrible screaming noise coming from below. Cautiously looking over the wall, I spotted a mangy, feral dog that had somehow ensnared itself in the razor wire. The more he twisted, the more it became apparent that he was embedded in the blades. I sat there in silence and hoped that he could free himself, but he just kept thrashing about and continued to work himself deeper into the strands. It was a pitiful sight and quite disturbing to watch. His howls echoed off the concrete.

After waiting several minutes with the hope that he would be able to extricate himself from the wire, I finally decided to put him out of his misery. I couldn't stomach his cries of pain any longer. I awkwardly leaned out of the tower to get a sight on him, though I had to momentarily pause in place because he would occasionally jerk in another futile attempt to escape. It was a challenging shot due to the angle but, as he relaxed for the final time, I took careful aim and squeezed off a round, hitting him directly in the head. His body momentarily jerked and then went limp. The screaming stopped, replaced by the faint, unsettling sound of blood pooling onto the pavement below. That sound stayed with me for some reason, sometimes showing up in my dreams.

I'd hunted before and killed plenty of animals for food without hesitation, but something about this incident stuck in my mind. Maybe it was the helplessness. Perhaps it was the quiet horror of putting a suffering creature out of its misery, wrapped up in the wire *we* had put down there. Whatever it was, it got to me.

The following day, I returned for another shift in that same tower and looked carefully over the ledge. A large pool of blood had coagulated under the wire, though the dog's body was gone. The soldier I replaced indicated that a pack of wild dogs had pulled its remains free and ripped it to pieces, eating nearly the entire carcass in a matter of minutes. The smell of decomposition lingered in the area for over a week, with flies incessantly buzzing around in search of the stench. After that, I dreaded going into that particular tower and never looked down into the wire again.

While shooting the dog was emotionally difficult, I didn't feel even a fraction of that same sorrow when it came to shooting at hostile threats. There was no sympathy, no hesitation, and certainly no compassion felt. Somewhere along the way, whether it was from basic training or the previous deployment, I stripped them of their humanity in my mind. I didn't see them as fathers, brothers, and sons. No. Instead, they were obstacles to our survival, and they had to be destroyed.

One evening, as the city rushed home to beat curfew, we moved down a dimly lit street a few blocks outside the Green Zone. We walked routes like this all the time, hoping to catch a weapons run or to stumble onto someone planting an IED before they finished the job. We were spread out in a staggered formation on both sides of the road as we approached an Iraqi Police checkpoint just up ahead. They were stopping cars, searching trunks, and working the line with flashlights, hoping to slow down insurgent activity.

Suddenly, without warning, a small, white Nissan hatchback tore through the checkpoint at high speed, heading straight for our position. The Iraqi police officers shouted and waved frantically to no avail. The vehicle just accelerated past them.

Instinct took over for all of us. I frantically dove to the side of the roadway, leveled my M-16, and fired a flurry of precise shots at the speeding car as it screamed toward our column. My world shrank to the front sight post and the target, but I could hear and feel everyone else reacting at the same time, scrambling for cover and squeezing off rounds with me.

In the space of a heartbeat, gunfire erupted up and down the line. The engagement was over almost as soon as it began, but for that brief burst of chaos, it felt like the entire platoon opened up in one simultaneous volley. Tracers lit the air, and sparks flew from ricochets, slicing through the night like wild electricity. I don't know whether the vehicle was disabled or the driver simply stopped, but it came to a smoking halt near the middle of our platoon.

When the shooting stopped, the car looked like something out of a crime scene. The driver's side window had exploded, and other windows were shattered but were still hanging together in their frames. Bullet holes were punched across the windshield and hood, and several long streaks of red blood dripped down the driver's door like something out of a horror movie.

Then, against all odds, the driver emerged from the ruined car. He was bloody and dazed but alive. His hands, shirt, and pants were soaked in blood, yet he appeared somewhat coherent. Despite numerous rounds smashing into the windshield, none had delivered a clear, fatal hit. His injuries seemed to be primarily from shrapnel, glass, and superficial cuts, not bullet holes. I looked down at my weapon in disbelief, heart still racing, and thought, *What the fuck just happened? How is he not dead?*

### 23 April 2005 – Letter to Toby Glass

A couple nights ago, we shot a guy just outside Checkpoint 11. It was dark, and we were on a foot patrol. We were staggered on the sides of the road, and he crashed through an IP barricade and headed straight for us. We shot him up pretty bad. Most of the rounds impacted in the engine block, but three managed to hit him in the chest, hand, and groin area. He survived, but he'll be messed up for a good while. I can't honestly say who shot him, because about 7 of us let go some rounds.

### 26 June 2025 – Reflection of Derek Hauger, Bravo 4-64 Armor

For whatever reason, I was carrying the shotgun on patrol, even though I had zero experience with it. I was near the front when the car came flying through right at us. I went to fire, but because I barely knew that stupid shotgun, I fumbled with the safety and couldn't find it. By the time

I finally figured it out, guys were already in front of me, and I would have been flagging them, so I kept my finger off the trigger and didn't shoot. The driver still got shot up pretty badly.

When we called a cease-fire, I walked up with the interpreter, and the driver claimed he didn't even know we were there. He was clearly smashed drunk, and I could smell the liquor on him.

The Iraqi Police took him away to the hospital, and I never found out what became of him. As far as I could tell, he didn't appear to be an insurgent and didn't have any explosives. In truth, we were all incredibly lucky. Had he been a suicide bomber, he could have taken out half our platoon, and the whole incident was a sobering reminder of how fast everything could go bad. Was he mentally unstable, or just drunk? Or could he have been a wanted criminal trying to evade capture? I don't know. If so, he picked the wrong street to jump the barricade and drive like a maniac toward us.

As I later learned, most of the military, including our unit, was using full-metal-jacket ammunition, which works perfectly for most applications. However, the windshield glass in a car is often laminated and angled, which can surprisingly deflect or fragment these rounds.[161] So, instead of an intact round hitting the driver in the face, he was peppered with tiny pieces of bullet shrapnel, and most skipped harmlessly off into the air.

The solution to this problem is to use bonded ammunition. The process involves chemically or mechanically bonding the core to the jacket, so it won't fragment in situations such as this.[162] Bonded rounds work much better when used on windshields, doors, plywood, and other hard surfaces. In other words, the driver was fortunate we didn't have bonded ammo at our disposal. But it certainly caused a lot of confusion in our platoon as to why he wasn't dead. For the longest time, we couldn't figure out why our rounds had done that. Everything was a learning process.

But why were we so jumpy in the first place? Well, across Iraq, terrorists routinely packed vehicles with explosives and rammed them into convoys, patrols, and checkpoints, detonating in the middle of scrambling troops. This incident took place immediately after a deadly car bomb had rocked Alpha Company, 3-7 Infantry, Donald Cyr's unit, so we were already on edge about anyone speeding toward us, whether on foot or in the checkpoints.[163]

On April 19th, a few days before I wrote that letter home, a platoon from Alpha 3-7 had set up an improvised checkpoint and patrol in the middle of their sector to disrupt insurgent movements. A squad was actively searching trunks and storage compartments while questioning passengers. It was business as usual.

---

[161] Butler, Chris. "Implications of Shooting Through a Windshield." *Blue Line*, June 25, 2022. https://www.blueline.ca/implications-of-shooting-through-a-windshield/.

[162] Winchester. "What Are Bonded Bullets and Why Do You Need Them?" Accessed December 4, 2025. https://winchester.com/Blog/2021/01/What-Are-Bonded-Bullets-And-Why-Do-You-Need-Them.

[163] Vorsino, Mary. "Honolulu Star-Bulletin News." Iraqi Car Bomb Kills Isle Bred Soldier, April 23, 2005. https://archives.starbulletin.com/2005/04/23/news/story2.html.

Without warning, a vehicle accelerated toward their position. As it closed the distance, Corporal (CPL) Jacob Pfister reportedly sensed something was wrong and ran toward the oncoming car, putting himself between the suicide bomber and his fellow soldiers.[164] A moment later, the VBIED detonated in front of the squad. The blast killed Pfister and PFC Kevin Wessel instantly, and wounded four other soldiers, two of them badly enough to require immediate medical evacuation out of the combat zone.

That's what a suicide car bomb was designed to do. A VBIED doesn't just send out a simple, Hollywood "boom." Instead, the supersonic shockwave can violently expand outward from the blast point, knocking people violently to the ground, while rupturing lungs, eardrums, and other organs through sheer overpressure.[165] In the wave, shards of glass and jagged metal tear away from the wreckage at blistering speeds, from massive chunks of engine block to razor-like slivers of door panels. Insurgents often packed the explosives with nails, ball bearings, and other improvised shrapnel to make the weapon even more lethal. Thermal temperatures could reach several thousand degrees, igniting fuel, setting clothing on fire, and burning exposed skin in an instant.[166]

Cyr and his platoon were operating nearby and responded quickly when the emergency call went out over the radio. When they arrived, the scene was chaotic, with troops in a state of stunned disbelief. Even for those not directly injured by flying shrapnel and debris, the concussive blast was still severe enough to cause traumatic brain injury (TBI). Taking control of the area, Cyr's platoon pushed out and established a wide cordon to secure the perimeter and deter any follow-up attack. Thankfully, none came.

### 24 June 2025 – Reflection of Donald Cyr, Alpha 3-7 Infantry

As soon as we got the call, we rushed out there, and the scene was horrific. I remember wondering why it was taking the MEDEVAC (Medical Evacuation, a rescue Black Hawk team) so long to arrive. We held the area and waited while they evacuated the casualties, doing what we could to secure the site.

Afterward, we spread out to pick up body parts. When I say we picked up body parts, I mean tiny fragments, pieces the size of rocks. We were out there for hours, gathering what was left of our friends and trying to find any evidence or intel we could use. After the war, I just tried to push all of that out of my mind. Stuff like that happened all the time.

---

164 Find a Grave. "CPL Jacob Matthew Pfister." Accessed December 4, 2025. https://www.findagrave.com/memorial/15416751/jacob_matthew-pfister.

165 Blast Injury Research Coordinating Office. "What Is Blast Injury?" US Department of Defense, September 23, 2024. https://blastinjuryresearch.health.mil/index.cfm/blast_injury_101/what_is_blast_injury.

166 Brevard, Sidney, Howard Champion, and Dan Katz. "Combat Casualty Care: Lessons Learned From OEF and OIF: Weapons Effects," 2012. https://medcoeckapwstorprd01.blob.core.usgovcloudapi.net/pfw-images/borden/ccc/UCLAchp2.pdf.

So, for some, firing at a car speeding through our patrol late at night might seem like an outrageous overreaction. But for us, the threat was all too real, and our friends were dying. We were fully aware of the deadly stakes surrounding us, and this was no drill. Thousands of insurgents were actively trying to kill us, and maintaining constant vigilance was the only way to stay alive.

That reality seeped into the small rituals we adopted. Instead of wearing dog tags only around our necks, the way you normally do, many of us started threading one tag through our boot laces. The logic was simple and brutal. If a car bomb or IED ever tore us apart, maybe they would at least find that tag in a boot and be able to figure out who the pieces had once belonged to.

### 24 June 2025 – Reflection of Donald Cyr, Alpha 3-7 Infantry

Pfister was a good soldier. He was always buried in tactics and infantry manuals, and always seemed to know what to do. Wessel was a good kid, too, full of energy and bouncing off the walls, always ready to go and eager to be involved in everything. It's a real shame, because both of them would have grown into good men.

Between manning checkpoints and conducting patrols, the workload was relentless. On top of that, we were training Iraqi forces and carrying out targeted missions, leaving little time to rest. It was only then that I truly understood why our drill sergeants had been so tough on us during basic training. I found myself silently thanking them every single day I was there.

### 25 April 2005 – Letter to the Harstad Family

Today was a rough one. We had a long foot patrol, about 12 miles, and then we came back and did the checkpoint for eight hours. I'm beat!

Our command did a great job of keeping us motivated. They understood that the checkpoints were tedious and dangerous, but they consistently emphasized the need to protect the transitional government. By April, the newly elected Transitional National Assembly had selected Jalal Talabani, a Kurd, as President, and Ibrahim al-Jaafari, a Shiite, as Prime Minister.[167] Each political milestone, however, was met with a new wave of violence, as insurgents sought to derail the country's path toward stability.

---

[167] RadioFree Europe / RadioLiberty. "New Iraqi Leaders Take Office, Name Prime Minister." *RadioFreeEurope / RadioLiberty*, April 7, 2004. https://www.rferl.org/a/1058309.html.

**1 March 2005 – Letter from Michael MacKinnon, Commander of Bravo 4-64 Armor**

(Writing to the Family Readiness Group) Be proud of all your husbands. They are very brave and honorable. I'm truly humbled and privileged to have the opportunity to command a company of American heroes.


**14 June 2005 – Letter from Cap Stanley, Command Sergeant Major of 4-64 Armor**

Your soldiers stand on the frontiers of freedom, and do so with a sternness and personal pride that is so hard to describe. You can't photograph it, video it, paint it, copy it, or make it available in any other modern medium... you can simply see it in their eyes and hear it in their voices. Thank you for letting us have the privilege of leading your loved ones!


As strange as it may seem, children from nearby neighborhoods often acted as runners for American troops at the checkpoints, whether sent by their parents or acting on their own initiative. Since we weren't allowed to leave our posts while on duty and the area around us had no shortage of delicious food and drink, these kids quickly found an opportunity. They knew we carried American currency, which was worth far more than Iraqi dinars, and offered to fetch whatever we wanted in exchange for a little bit of cash.

It didn't take long for this informal system to catch on. In return for a few dollars, the kids would bring back cold drinks, ice, and piles of fresh Samoon, a locally baked yeast bread that was absolutely delicious. Samoon is traditionally baked in stone or brick ovens, sometimes on a bed of small pebbles, and has a distinctive diamond shape. Its soft, airy texture is somewhat reminiscent of pita bread. Like pita, it can be filled with honey, hummus, kebab meat, butter, or cheese.[168] Its outer crust is chewy and slightly tough, while the inside remains moist and fluffy, making it perfect for eating on the go, especially at checkpoints.

These young runners were far more economical than buying directly from street vendors, who often charged us grossly inflated prices. Knowing we had American money, the vendors would demand several dollars for a single sandwich-sized piece of Samoon. The trick was to hand a kid a $1 bill far from the vendor's prying eyes and ask for as much bread as he could get. Minutes later, he'd return with a large plastic bag filled with a dozen or more steaming loaves. We'd usually give him a couple of extra dollars or American candy as a thank-you. I never once felt cheated by the children, nor did they ever tamper with the food. They likely relied on that money to help support their families and didn't want to jeopardize a steady source of income.

Still, looking back now as a parent myself, I can't fathom the idea of sending my own children into an active military checkpoint. It was a wildly dangerous place where gunfire and car

---

[168] Agence France-Presse. "Iraqi Cuisine's Crown Jewel: The Diamond Shaped 'Samoon' Bread." NDTV World, January 24, 2023. https://www.ndtv.com/world-news/iraqi-cuisines-crown-jewel-the-diamond-shaped-samoon-bread-3721119.

bombs were a constant threat to men, women, *and* children. The violence didn't discriminate. It speaks volumes about the desperation those families faced and the strange, unspoken trust that sometimes developed amid chaos. We genuinely grew fond of some of those kids and did our best to look out for them.

**Checkpoint diplomacy: Iraqi girls bring me flowers and, in return, I share my stash of candy and MREs**

Whenever possible, I gave the kids clothing we had on hand, most of which was donated by American businesses and churches. Before I deployed, my father sent me off with several baseball caps bearing his company's logo, *Harstad Carpet Cleaning*. It's entirely possible that, to this day, a few folks in and around Baghdad still have some Harstad swag in their regular wardrobe rotation. As of this writing, he has yet to receive a long-distance call from the Middle East requesting carpet or rug cleaning, but we're still holding out hope. Joking aside, we genuinely did our best to help, routinely handing out candy, food, medicine, and clothing to those in need. And that seemed to be everyone.

Many Iraqis turned out to be warm, generous people who truly admired aspects of American culture and hoped for a better future. At first, they were understandably cautious, especially as we patrolled their neighborhoods with heavy weapons and armored vehicles. However, over time, they began to realize that we were there to help, and small acts of kindness gradually built trust on both sides. One of the most effective tools we had for breaking the ice was handing out soccer balls.[169] The local kids went absolutely wild for them. We even tried to join in a few games, though chasing after a ball with 60 to 100 pounds of gear strapped to our backs was no easy task. Even the little kids usually kicked our asses.

SSG Brian Torres, modeling a smashed *Harstad Carpet Cleaning* hat destined for some really cool Iraqi kids

**23 May 2005 – Letter from Cap Stanley, Command Sergeant Major of 4-64 Armor**

My interpreter, "Hawk," has been pestering me about a poster that is seen around the building. It is a reenlistment poster that has a picture of Uncle Sam on it in the "I Want You" pose that is famous from the World War days. He stated that he has always dreamed of being in the States. I got him one for his room.

---

[169] DVIDS. "1st Sgt. Barnes." Defense Visual Information Distribution Service, October 17, 2005. https://www.dvidshub.net/video/6839/1st-sgt-barnes.

The people of Iraq have dreams, too. Most dream of a time when they can be safe and live a life that is suitable for a human being, and they desire not much more. They don't want much, even though they live in the shadows of great palaces that are constant reminders of the man that squandered this country's vast riches.


Building relationships with the local Iraqis was very much a double-edged sword. Insurgents could easily blend in with the civilian population, observing our movements without raising suspicion. For Iraqis who were seen as "too friendly" with the Americans, the consequences were brutal. Kidnappings, sadistic torture, and even murder were tragically common. For the average civilian, it was a no-win situation. They were caught between the brutality of insurgents and the imposing presence of foreign soldiers. Everyone was truly trapped between a rock and a hard place, and far too often, they paid for it with their lives.

### 15 May 2005 – Letter from Robert Roth, Commander of 4-64 Armor

And we need to remember we are not suffering or sacrificing alone in our efforts. I ran across an old friend of mine from when I was here almost 2 years ago. I was hoping he would hear the Tuskers had returned - and he did. So, he sought me out, and we finally linked up. It was a wonderful reunion.

Last time I was here, he was our electrician, plumber, and all-around fix-it man. Unfortunately, several gunmen drove up in front of this friend's house and killed him, his father, and his two sons. All because he worked with the Americans.

His name was William. He was a former Iraqi Army LTC and fought during the Iran-Iraq War. He was proud to be an Iraqi, and his battle scars were numerous. He became an English professor after that war but quit his job the day the Americans took Baghdad in April 2003. He wanted to help the Americans any way he could. Even when he and his family were faced with death, he never relented. Now, his wife and daughters are left to fend for themselves.


Long before American forces and insurgents turned Iraq's streets into a battleground, life under Saddam Hussein's regime was far from better. For some on the wrong side of power, it was brutal and inhumane. Beatings, torture, and murder were woven into their daily life. Stepping out of line or simply being in the wrong place at the wrong time could land someone in a horrific prison, subjected to savage beatings and interrogations over fabricated charges.

At FOB Prosperity, we employed a diverse range of Iraqi contractors, including construction workers, food vendors, and internet technicians. An alarming number bore the scars of that era, including missing fingers, mutilated tongues, and burned-out eyes. The damage was often visible and always sobering.

**17 May 2005 – Letter to his church from James Barnett, Bravo 4-64 Armor**

I work with a guy who Saddam put in prison and he has no fingernails or toenails. They were ripped off and cauterized. I've also seen men who have no tongues.

One night, our patrol rolled through our area of operations in Humvees and stopped at a secret, abandoned prison once used by Saddam Hussein's regime. We had visited the site several times before, but never after dark. In the pale moonlight, the place felt downright creepy. The plan was to remain hidden and observe the surrounding neighborhoods, hoping to catch signs of insurgent or criminal activity in progress.

Tucked away inside a walled compound, the dusty tan-stone structure stood oddly untouched, avoided even by the locals. As SFC Hill put it during an interview, "This is one place in Iraq that stays empty. Nobody tries to move in and make a home out of it." According to nearby neighbors, the prison had been the scene of horrific and unspeakable atrocities, which they were hesitant to discuss. They believed it was cursed, forever haunted by the ghosts of those who had suffered and died in there. No one wanted to go anywhere near its blood-soaked walls.

During an interview and video tour, which is still available online, our platoon sergeant offered a sobering reflection: "Stuff like this is what people back home, I think, really need to see. There's some pretty rough stuff written on the walls in there. Some of it says, 'God, help me,' or 'I'm wrongly accused,' or 'I think they're going to kill me today.'"[170]

I walked carefully through the prison, scanning my surroundings through the dead, green glow of my NVGs. Prisoners had carved countless tally marks into the cell walls, seemingly marking the passage of time inside their agonizing torture chambers. Interspersed among the scratches were haunting messages, translated for us by our interpreters and Iraqi Army counterparts.[171] Some were pleas, and others were warnings. Each one was a fragment of human suffering, forever frozen in stone.

Inside the cramped cells, as many as ten prisoners were packed together within the stifling walls. According to the unit we had replaced, the prison guards abandoned their posts and fled as soon as American forces approached the city during the invasion, leaving everyone trapped inside. Allegedly, it was twenty or thirty days before the prisoners were discovered and rescued. By then, many had already died from starvation or dehydration.

Though I cannot claim to believe in ghosts, I crept around with the genuine hope of encountering a spectral remnant. If the supernatural existed anywhere, I reasoned, surely it would linger here. There were too many devices of torture and too many ingrained memories in this place

---

[170] DVIDS, "Saddam's Hidden Prison B-roll," Defense Visual Information Distribution Service, April 16, 2005, https://www.dvidshub.net/video/1813/saddams-hidden-prison-b-roll.

[171] Defense Visual Information Distribution Service. "Saddam's Hidden Prison Package," April 16, 2005. https://www.dvidshub.net/video/1823/saddams-hidden-prison-package.

for it not to possess… something. Something ethereal. Was I right to expect it? Or was my imagination simply craving a distorted reality?

**Prison cell interior, its walls covered in tally marks and scratched pleas for help and mercy**

**Photo courtesy of Derek Hauger**

Much to my genuine disappointment, I didn't see anything that offered a glimpse into the realm of the unexplainable. There were no flickers in the shadows or quiet whispers in the dark. Instead of being otherworldly, the prison was a stark reminder that true and real evil exists within the hearts of men. There was no need to invoke supernatural entities, as human beings were responsible for all these events. Wickedness was already here, and it was on full display within the confines of that wretched structure.

Taking shifts, I slept for a few hours inside one of those empty cells with my back against the scratched wall. My weapon rested reassuringly against my lap, always ready should it be needed, as we waited for daybreak. In the morning, the first soft glows of light peeked out from behind the horizon. The sun's rays soon swept through the compound, revealing a distressed landscape that was left dirty and barren. Dust clung to every surface and wouldn't relinquish its tenuous grip.

When I spoke with my old battalion commander, Robert Roth, about the prison twenty years later, his voice broke briefly, carrying a sense of long-buried pain. As he reflected on that site, he had to pause to collect himself before continuing. He remembered the horror.

**27 July 2025 – Reflection of Robert Roth, Commander of 4-64 Armor**

I can't remember the name of that prison, but it was a notorious torture site. My interpreter and I spoke with an old man from the neighborhood who told us he had once been a prisoner there. As we walked through the grounds, he described what had happened inside.

In the basement, there were no windows. They would flood the room nearly to the ceiling with prisoners still inside, forcing them to stand upright for days. If they grew tired or fell asleep, they would drown. High up on the walls, there were scratch marks.

Before we departed for our bunks at the FOB, I took one final walk through the prison's courtyard, hemmed in by those oppressive walls. What I saw there chilled me more than any ghost story ever could. In the center of the courtyard, bolted to the sunbaked brick, stood a large metal cage, placed there deliberately so that the sun could torture its captives without pause. There was no shade, no mercy, and no escape from the heat. And when night fell, who knew what fresh cruelty awaited?

# Reflections

For any veteran, I suspect the most amusing part of this chapter was my description of the Fobbit. Naturally, I had to approach it as a primatologist might describe a monkey, simply by observing them in their natural habitat. The funniest part is that much of the description could just as easily apply to an actual animal as to a Fobbit.

Iraq was a particularly unusual combat zone in that, logistically, it differed significantly from many previous wars. There was no defined "front line." Indeed, danger was always a possibility no matter where you were, but life on the FOB was relatively secure, only occasionally interrupted by the rare, poorly aimed mortar strike or sniper attack.

As a result of the perceived danger, real or not, many Fobbit troops gravitated toward the safest place they could find, often the sandbag-lined interiors of their rooms, where they would remain for an indefinite period. This tendency drove the rest of us crazy, especially when certain groups always seemed to vanish whenever real work needed to be done.

Because Fobbits had little to no exposure to life beyond the wire, where it was *actually* dangerous, even something as routine as walking to the chow hall felt like a perilous odyssey, as if they were gearing up for a life-or-death bout of hand-to-hand combat. They often struggled to discern what was truly hazardous and what wasn't. I realize I keep returning to this theme throughout the book, but it bears repeating: the vast majority of troops were not combatants. Some people fight, but most are there to support those who do. Sometimes support troops do come under attack or find themselves in harm's way, but that's not the same as going out and actively looking to pick a fight.

There's a guy I met years after my time in the military who deployed to a FOB somewhere in the Middle East back in the day. He was a nice enough guy, but had a story for everything: ambushes, near-death experiences, and daily high-stakes drama straight out of a war movie. After sitting through yet another action-packed tale, I finally asked what his MOS was. I figured it *had to be* Delta Force or maybe even something so classified he couldn't technically tell me. But no, not even close. Turns out he wasn't in combat arms at all. He was a full-blown POG. Ever since, a few of us have taken to calling him "Jack Reacher," and he remains blissfully unaware that his legendary exploits are the stuff of barracks folklore.

Enter Nick Brokhausen, the hilarious Special Forces soldier I quoted earlier in the chapter. He has a gift for saying out loud the things the rest of us only mutter under our breath. In his memoir *We Few*, he launches into another rant, this time against MPs, calling them "the same the world over, little self-important assholes," and then takes it a step further by half-seriously suggesting he would rather throw in with the Viet Cong than with the guys in whistles and armbands.[172] Hey, he said it, not me. I was just the guy who burst out laughing, nodding along while I read it because it's true. Ask any grunt.

---

[172] Brokhausen, *We Few: U.S. Special Forces in Vietnam*, Chapter 1.

Of course, I'm painting with a very broad brush here. For every self-proclaimed Jack Reacher hiding on the FOB, there were support troops quietly doing unglamorous and thankless work that never made it into the war movies. They were the supply guys who always "found" one more set of plates, the intel weenies who actually knew what they were doing, and the MPs who were competent professionals instead of the whistle-and-armband dickheads Nick was talking about. Combat-arms types love to sneer at Fobbits and POGs, but the truth is that a lot of these men and women were good soldiers, too. The teasing is just part of the ecosystem. So, if I'm not making fun of you, it probably means you weren't important enough to notice in the first place.

I also took a few jabs at the swarm of contractors who flooded into the country at the behest of the United States government. It may have come across as bitter, but that wasn't my intent. The war exposed a great deal of questionable behavior, including what many observers would call blatant war profiteering (more on that in the next chapter). Still, I want to be fair. Not everything I witnessed was negative. My intention is simply to reiterate that certain individuals, corporations, and even entire industries were making a substantial amount of money, often without everyday citizens knowing about it.

As I mentioned earlier, we regularly encountered a revolving door of American companies and private military contractors. The exact cost of their presence is anyone's guess, but it is safe to say the price tags were significant, all ultimately paid by the American taxpayer.

I don't harbor any resentment toward the contractors themselves. Most were prior service members, and I commend them for capitalizing on an opportunity to provide for their families. My concern lies in a broader observation: the system created two distinct camps. A growing bifurcation of financial, logistical, and operational resources was emerging between the U.S. military and the privatized forces operating alongside it. At the very least, we should be asking whether that is a direction we want to keep going.

Even my best friend, Donald Cyr, went to work for a private military company after his honorable discharge from the Army. As he told me, it wasn't just the military contractors making serious money. Regular, everyday civilian contractors were raking it in, too. It was widespread, and everyone seemed to be cashing in.

**30 March 2025 – Reflection of Donald Cyr**

There was a guy I knew who was working for (redacted company name), and his entire job on the FOB was pumping gas into vehicles. Nothing special, nothing high-risk. He just fueled whoever drove up, like an old-fashioned gas station attendant. For that, he was making over $10,000 a month tax-free. Some days, he would even pay a local a few bucks to do it for him while he sat back and watched.

Contracting companies regularly hired third-country nationals as civilian workers to perform manual labor, laundry services, janitorial duties, construction, and other service-related roles. So, when I walked into the chow hall on the FOB, it typically wasn't American soldiers serving the food. It was foreign workers. Why? Because flying them in from overseas was far cheaper than paying high American wages.

**1 August 2025 – Reflection of Noor "Nora" Aljassim, Interpreter for the U.S. Military**

As Iraqi interpreters, most of us didn't really understand the American tax or payment system. I worked for (redacted), an American contracting company, and they were the ones issuing my pay. Eventually, we discovered that our supervisor had been stealing half of our checks. He was later sent to prison in the United States, but the whole ordeal showed just how badly some interpreters and contractors were being taken advantage of.

Most of these workers came from Southeast Asia, though some hailed from Eastern Europe and other poor regions. Countries like India, Thailand, Pakistan, the Philippines, Bangladesh, Sri Lanka, Nepal, Egypt, Uganda, and Turkey were common points of origin. All of them were employed by U.S. defense contractors operating under the multi-billion-dollar Logistics Civil Augmentation Program (LOGCAP).[173]

Now, keep in mind, I'm writing all this from the perspective of an enlisted soldier in the desert doing the grunt work, not from the viewpoint of a business owner, a commanding officer, a migrant worker, or a politician. I'm sure they each have valid arguments of their own. As such, I am not trying to take a political stance in this book. Much as with my reflections on Korea, my goal is simply to document what I witnessed and to allow the reader to draw their own conclusions. At the very least, these are issues we should be willing to have honest conversations about as a country.

And now, on to the more somber parts of this chapter. Looking back as a man in my 40s, I'm still not entirely sure why shooting that trapped dog settled so deeply into my memory, but it did. Of all the things that happened on that deployment, that is one of the scenes that repeatedly returns on the nights when sleep won't come.

I think it was the helplessness. He was tangled in a barrier we had put there, starving and half feral in a city that had been battered by years of sanctions, war, and neglect. His suffering felt like a small reflection of the place itself. Iraq seemed caught in its own kind of razor wire. Every attempt to break free only cut it deeper. You could see it in the kids picking through trash, the stray dogs roaming in packs, and the burned-out neighborhoods that never quite recovered. It all added up to a slow, grinding misery that we were never going to fix with rifles and tanks alone.

Every now and then, something would slip past the emotional armor we wore. For me, it was usually the children and the animals. I could compartmentalize the adults, tell myself they were the sum of their choices. If they picked up a rifle or planted an IED, I didn't have to picture them as real people. Instead, they were threats, and I never lost sleep when they were removed. But kids and dogs never fit into that neat little box. They were just caught in the blast radius of other peo-

---

[173] Phinney, David. "Blood, Sweat & Tears: Asia's Poor Build U.S. Bases in Iraq." CorpWatch, October 3, 2005. https://www.corpwatch.org/article/blood-sweat-tears-asias-poor-build-us-bases-iraq.

ple's decisions. That dog in the wire felt like one more innocent that had wandered into the machinery and been crushed.

Minutes after I pulled the trigger, the sergeant in charge of the towers, who wasn't even from our company, came storming over to investigate the shot. When he found out I'd killed a dog, he threw a tantrum and said I should've just left it to die in the wire, apparently unconcerned that it was perishing in a most gruesome way. And while I did feel guilty about putting that dog down, I also knew it was the right thing to do. If taking an ass-chewing was the price, then so be it.

Though he was more than 6,000 miles away, I was able to talk to my dad about the incident just days after it happened. As always, he listened patiently, letting me vent without judgment. I think he understood that I needed to get my head back on straight. After all, there was still a long road ahead. Carrying that weight indefinitely wouldn't do me any good. Thanks to that honest conversation, I was able to process it, let it go, and move forward.

Another memory that continues to haunt me is the night we spent in that abandoned Iraqi prison. That place sits at the center of how I think about the war. These days, it has become fashionable for people on *both* the left and the right to loudly condemn the decision to invade Iraq and topple Saddam Hussein. I understand why. The justifications were flimsy, the planning was a mess, and the human cost on all sides was staggering. On my more cynical days, I look at the wreckage that followed and think we lit a match in a room full of gasoline, and then acted surprised when it exploded.

Other days, my mind goes back to those cells. I can still picture the scars on the walls, the scratch marks, the embedded blood stains, and the pits in the concrete where people had clawed or beaten at it in desperation. I think about the stories our interpreter relayed from former prisoners, the casual cruelty of having tongues cut out, eyes burned, and fingernails ripped off as a matter of routine. In those moments, it's hard not to feel that removing Saddam was, at least in a narrow moral sense, the right thing to do. Leaving him in power would have meant that particular engine of torture and terror kept running, day after day, year after year. When I stand mentally in that prison, it is difficult to argue that the world was better off with him left in charge.

So, I tend to bounce back and forth. Some mornings, I read about the chaos that followed, the sectarian bloodletting, the rise of new monsters in the vacuum we helped create, and I think we shouldn't have gone. Other days, I remember the faces of Iraqis who quietly thanked us, or I imagine that old man outside the prison describing his torture to my colonel, and I think that standing aside would have been its own kind of crime.

What I'm left with is tension that never fully resolves. Yes, the war was devastating. Yes, our government made unforgivable mistakes. But the alternative wasn't some peaceful, stable Iraq where everyone lived happily under a strongman who only bullied people on weekends. The alternative was more prisons like that one, more rooms full of screaming, and more lives methodically destroyed in the dark. I still don't know for sure if we chose the lesser evil. But I do know that once you have walked through a place like that for yourself, it becomes very hard to pretend that doing nothing would have been the better choice.

In 2023, I took my family on a vacation to Europe to visit relatives. While we were in Munich, we visited the Dachau concentration camp established by the Nazis. Walking through the gate bearing the infamous inscription "Arbeit Macht Frei" ("Work sets you free"), I immediately broke out in goosebumps.[174] I was hit with a vivid flashback to that prison in Iraq. The feeling was eerily familiar.

In a journal I later published, I reflected: "Several days ago, I wrote about why it is that we spend our hard-earned money to travel across the globe. For many of us, it is because we are seeking rest and relaxation in a tropical location, soaking up the sunlight on a beautiful beach. Today wasn't one of those days, though. Instead, we voluntarily subjected ourselves to images and locations of sadness, pain, and death."

Before the visit, my wife and I sat down with our daughters, then ages 16 and 11, and prepared them for what they might see. We told them it would be difficult, but sometimes, you have to look evil directly in the face to understand that it truly exists. We assured them that if they ever felt overwhelmed, they could close their eyes or we could step into a different room. To their credit, they handled it with maturity and dignity. Still, my eldest daughter, Laylah, paused in the courtyard and quietly said, "Something feels weird in this area." She didn't know why, but I think she felt it, too. It was the lingering weight of what had happened there. It was the same thing I had experienced in Baghdad.

I concluded the day's entry with a paragraph that I intended to leave a lasting impression. I wrote, "In the end, we most often travel to get away from life and relax for a little while. It is good to recharge your batteries. Just sometimes, though, we need to travel so we can slowly creep up and look into that dark hole in the corner of the room. The one you really don't want to look into. For inside, you know that you won't like what you see."

Finally, I need to discuss the first losses of the tour that had an impact on my life. At the time, I knew that a couple of soldiers from our old battalion, 3-7 Infantry, had been killed, but I didn't know their names. I figured that if it had been Donald Cyr, someone like Brian Torres or a family member would have reached out to let me know. It wasn't until August that I found out it had been Pfister and Wessel.

I had met both of them through Cyr, though I wouldn't claim we were close. I only knew them in passing. They were friendly guys, as I recall, and Cyr spoke fondly of them. And for me, that was enough. If Cyr liked someone, it meant they were solid people, so I liked them, too.

When I finally learned who it was and realized I had known them, I wasn't sure how to process it. It felt surreal, and I'm certain I went through a bit of shock. The only outlet that made sense at the time was to write. In hindsight, though, I wish I had spared my family from the full weight of what I was experiencing and some of the graphic details, at least until I was safely back home. Too often, they ended up carrying the burden through my letters home.

[174] Blakemore, Erin. "Stolen 'Arbeit Macht Frei' Gate Returns to Dachau." *Smithsonian Magazine*, February 22, 2017. https://www.smithsonianmag.com/smart-news/stolen-arbeit-macht-frei-gate-returns-dachau-180962252/.

**19 August 2005 – Letter to the Harstad Family**

On a sadder note, two of my buddies were killed in a car bomb. One of them totally blew up, from what I was told, so they couldn't even send his body home to his family. The other was cut in half and had all his clothes burned off. I guess I don't know how I'm feeling right now. A little bit of shock, and a lot of rage.

In *Surprise, Kill, Vanish*, Annie Jacobsen recounts an interview with former Green Beret Lew Merletti about the lasting grief of losing friends in combat, especially his close friend Michael Kuropas. Merletti describes standing in Vietnam, suddenly overwhelmed by the blunt reality that his friends had been alive one moment and dead the next, with no comfort, no anesthetic, and only pain in between.

What stayed with him most was the sense of obligation that followed. As he put it, he wanted to "live up to certain expectations of myself, for him. For Mike."[175] From that point on, as Jacobsen explains, Merletti treated every difficult moment in his own life as a kind of test. Whenever he faced something that felt hard or unfair, he would stop and think of Mike. The very fact that he had problems at all meant he was still alive to feel them. Mike Kuropas no longer had that option. For Merletti, pushing forward, living fully, and not wasting the time he had left became a way of honoring his friend's sacrifice.

And so it goes. Those of us left behind are often faced with two paths. We can let grief consume us, leading us into destructive habits: drugs, alcohol, and apathy. And if I'm being honest, that dark path is easy to fall into, and few would blame us for taking it. Or, we can choose to honor their sacrifice by becoming better versions of ourselves, holding ourselves accountable, and striving to live with purpose. We can live not just for ourselves but for those who gave everything.

I still think about Pfister and Wessel from time to time. I know Cyr does, too. We often wonder who they might have become, where life might have taken them. Perhaps most tragic of all, Pfister's wife was expecting their first child, a baby girl due in June.[176] He was killed before he ever had the chance to hold her.

Sadly, theirs were not the last names added to the list. Many more lives would be lost or forever altered before our time in the Middle East came to an end.

---

[175] Jacobsen, Annie. *Surprise, Kill, Vanish: The Secret History of CIA Paramilitary Armies, Operators, and Assassins*. eBook edition. Little, Brown and Company, 2019, Chapter 14.

[176] Daytona Beach News-Journal. "Jacob Pfister Obituary." Legacy, April 27, 2005. https://www.legacy.com/us/obituaries/news-journalonline/name/jacob-pfister-obituary?id=26763357.

# In Memoriam

---

CPL Jacob Matthew Pfister[177]
26 September 1977 (Buffalo, New York) – 19 April 2005 (Baghdad, Iraq)


PFC Kevin Scott Kenani Wessel[178]
5 January 1985 (Honolulu, Hawaii) – 19 April 2005 (Baghdad, Iraq)

---

[177] Military Times, "Army Spc. Jacob M. Pfister," Honor the Fallen, https://thefallen.militarytimes.com/army-spc-jacob-m-pfister/803668.

[178] Military Times. "Army Pfc. Kevin S.K. Wessel." Honor the Fallen, https://thefallen.militarytimes.com/army-pfc-kevin-s-k-wessel/803665.

# Chapter XII

## Summer of Explosions

I peered into the murky, stagnant water that pooled over the lower steps descending into the abandoned hotel basement, and my stomach twisted into a knot. Leaning down near the water's edge, I aimed my flashlight beneath the partially submerged door, its open metal frame just barely visible above the flood. The water had filled the basement almost to the top of the doorway, but I had it in my head to investigate what was on the other side. *This is a bad idea*, I thought to myself.

About an hour before I found myself contemplating this swim through the stinking, black water, I was resting comfortably near the radio at headquarters. We were watching a pirated movie on a cheap DVD player while scanning the traffic for any emergency broadcasts. Our task was QRF, so, in essence, we served as a first-responder detachment, ready to mobilize if something serious went down, like a car bombing or an ambush on a friendly patrol. Usually, we just relaxed with our weapons and gear close at hand, playing cards or watching movies. No one was allowed to leave, not even for chow. One guy always sat by the radio, listening intently, just waiting for that call: enemy contact.

While sometimes boring, QRF was a crucial component of maintaining American morale. Every soldier knew that the moment a unit came under fire, friendly faces and war machines would be on their way, rolling out to provide backup with overwhelming force. We simply had to stay alive long enough to allow for a flood of grunts, tracked vehicles, and air power to saturate the area. But the enemy understood this, too. That's why hit-and-run ambushes became their tactic of choice. They'd strike hard and fast by killing as many as possible in thirty seconds or less, and then vanish into the civilian population. If they lingered too long and gave us time to react, or worse, stuck around long enough for QRF to arrive, they were cut to ribbons. No one could survive a direct fight with the American juggernaut.

In a matter of seconds, we went from laughing at a '90s comedy to piling into our vehicles, roaring out the gates of Prosperity. A sister unit had requested immediate support following a suspected sniper attack, reportedly coming from the roof of an abandoned hotel. Within minutes, we were on-site, sealing off every possible escape route. The hotel loomed above us, standing at least a dozen stories tall, its dark, empty windows staring out over the dusty Baghdad skyline like hollow eyes. Heavily armed Humvees and Bradleys swarmed the surrounding streets and sidewalks, forming a tight perimeter.

**At the Victory Arch with two members of my Heavy Weapons team, Izonia Chism (L) and Brandon East (R)**

By this time in the tour, I had already been designated a team leader in the Heavy Weapons Squad, a role usually reserved for a sergeant. Because of the tight, close-quarters environment, we ditched our heavier machine guns and went in lighter, armed with M-16s, SAWs, and shotguns. As our platoon closed in on the building, I directed my team into position, and we began clearing rooms with stack-and-breach precision. All those muscle-memory drills we'd run a thousand times before became silky smooth.

However, the deeper we went, the more it hit us that this place was a monster. The hotel was far larger than we'd expected, and slowly our tactics became sloppier. Floor after floor, hallway after hallway, with doors everywhere, we were tiring out. There were simply too many rooms to search, and locking down individual sectors was an impossibility. We needed more people to conduct a thorough search, but this was the best we had available.

After sweeping countless rooms, we were ordered back outside. Other squads had leap-frogged ahead and were now navigating the dark corridors, moving carefully with their weapons up, but still encountering nothing. There was no gunfire, no voices, and no movement encountered by anyone. It was just endless silence, and an eerie, oppressive stillness pressed down on everything.

After a short break, I led our team back into the building and discovered a staircase tucked beside the lobby that led down to the basement. As we peered down the steps, we saw that the lower level was completely flooded, with water near the tops of the doorways. The power was out, leaving everything below in total darkness.

At the bottom of the stairs, just before stepping into the water, I spotted something through my NVGs. One end of a synthetic rope was tied to the brass railing, and the other trailed through an open doorway and disappeared into the murky water. I couldn't make out much detail through the darkness and green hue of the goggles, but the rope appeared to lead into what looked like a boiler room, or perhaps a large maintenance closet. Whatever was on the other end, it seemed to be connected to something, possibly a duffel bag.

We went back upstairs and regrouped, carefully discussing what we had seen. The rope felt out of place. It was suspicious, but none of us was eager to plunge into pitch-black, stagnant water to find out where it led. Was it filled with sewage, mold, or even worse?

As I deliberated our next move, word came in from the other squads that every room up-stairs had been cleared, and we were preparing to leave. No sniper had been found, and there were no signs of gunfire. It made us wonder if there had even been a sniper at all. Were we chasing an imaginary gunman? Or, had he already slipped away before we arrived? We didn't know.

One young member of our platoon, who will remain anonymous, casually walked down the steps and tugged at the rope, much to my momentary horror. We'd all just completed that FLETC training on booby traps, and I couldn't help but imagine the humiliation of blowing our-selves up in a flooded basement by doing something as stupid as that. Thankfully, nothing hap-pened, as the rope was still firmly snagged on whatever it was beyond the doorway.

I didn't want to go into the water myself, for obvious reasons, but I wasn't about to order one of my guys to do it either. Someone had to quickly get wet, or we'd have to call it off and leave the basement a mystery. So, I decided to take matters into my own hands and do it myself. Was it a ridiculous idea? Absolutely.

"Doc," our trusty combat medic, loaned me his pistol as I dropped all my gear at the top of the stairs. My plan was simple. I'd wade into the shallow end of the water and try to unhook the rope, but nothing more. I wasn't about to go deeper.

My primary concern was the possibility of live electrical wires submerged somewhere in the basement. Drowning by electrocution didn't sound like a good time. My secondary concern? That someone was waiting in the dark on the other side of that doorway, ready to put a bullet in my face the moment I popped up through the water.

I stepped cautiously into the black water and immediately felt the cold seep into my boots. I wasn't being electrocuted, so that was good. Slowly creeping down the steps, I swept my flashlight in a wide arc, trying to pierce the murk. The water rose to my beltline, and from that angle, I could just barely see through the top of the flooded doorway. The rope was definitely attached to a canvas bag, but from where I stood, I couldn't reach it, let alone free it.

I maneuvered as close to the doorframe as I could and gripped the rope, my feet now dangling in the open water. Slowly, I drifted into the blackness. Taking a breath, I closed my eyes and dunked my head beneath the surface, emerging on the other side of the doorway. Once inside the adjoining room, I braced myself, half-expecting that the moment I clicked on my flashlight, I'd see a muzzle flash and then nothingness.

Fumbling with the light, my heart skipped a beat as I nearly dropped it. When I finally got it on, I was relieved to find I was alone. The room was exactly what we'd guessed, a maintenance area. The water was deeper than six feet because I couldn't touch the floor. A foul layer of air lingered two or three feet below the ceiling, apparently the only breathable pocket. The concrete walls were coated in green slime, with mold creeping over every exposed surface. The rope was snagged on a piece of rusted metal, so I rocked it back and forth until it came loose.

From the stairwell, I heard SGT Wiseman yelling, "Get him back over here!" Dunking into dark floodwater and entering an unknown room solo hadn't gone over well with the higher-ups.

"Hold on," I called back. "It's free. Just pull it in."

The bag floated loose, but it kept catching on the inner frame of the doorway. Though partially submerged, it had enough buoyancy to snag on every obstruction. I braced myself against a piece of metal and gave it a hard shove from behind. That did the trick. The guys on the other side pulled it through, and I followed close behind, popping back into the stairwell, soaked but in one piece.

Outside, we opened the sack and found several firearms wrapped in greasy plastic bags, along with a large chunk of buoyant foam wedged inside. The discovery reignited our suspicion that someone *had* been on the roof after all, but if they were ever there, they were long gone by the time we arrived. The weapons were confiscated and stashed behind the seats of one of our Bradleys, where they remained for months.

Although the incident ended without a single shot being fired, it taught me a valuable lesson: leadership didn't come naturally to me. I tended to take on everything myself, such as swimming through the flooded basement, instead of delegating responsibilities to my subordinates, especially when it came to doing anything dangerous. With our platoon understrength from the start of the deployment and the constant troop reorganizations, I was thrust into yet another leadership role and found myself learning how to take charge on the fly. Despite the challenges, it was an invaluable experience and one that helped prepare me for the upcoming sergeant's exam.

I found that being a leader in a position typically reserved for a sergeant was very much a double-edged sword. On the one hand, it brought a sense of pride and accomplishment, knowing that I was responsible for overseeing a small team. It felt good to be trusted and to know I was helping carry the platoon's weight. On the other hand, because I was still only a specialist, I often felt I was held to a harsher standard than some sergeants in similar roles. Every minor misstep my guys made seemed to land squarely on me, and certain NCOs came down hard, living by the old saying: "Shit rolls downhill."

Captured hardware: the author with a confiscated Grease Gun

**20 June 2005 – Letter to the Harstad Family**

Yesterday was a headache. I got my ass chewed because one of my soldiers left his machine gun at the base and took his M-16 instead.

But I didn't take it personally. I understood that most NCOs in my platoon were trying to shape me into a better leader. By early May, I had earned a solid reputation and was hand-picked to serve as CSM Stanley's driver and de facto bodyguard. That is, if I were to want the job. As much as I admired and respected the man, I had no desire to leave the infantry line company. That was where I belonged.

**1 May 2005 – Letter to the Harstad Family**

I was offered a new job as the sergeant major's driver. I had no choice but to go and talk to him about it, even though I turned it down (as a permanent job) immediately.

It was pretty easy, and we just walked around and bullshitted with the guys at the checkpoints. There are a lot of good things about the job, but I couldn't stand not being infantry even for a few months. Infantry is what I signed up to do, and that is what I will do.

Still, I'd be lying if I said I didn't have a few regrets about turning down the job, especially when I was baking under the sun at the checkpoints. Even worse, explosions were going off across Iraq, numbering in the tens of thousands, with Baghdad being hit especially hard.[179] If I had to sum up the spring and summer of 2005 in a single acronym, it would be this: IEDs. Car bombs, road-side explosives, and terrorists wearing suicide vests seemed to be everywhere. It felt like things were slowly spiraling out of control.

**3 May 2005 – Letter to the Harstad Family**

In the morning, we conducted a traffic control point. Basically, we set up roadblocks (in the Red Zone) to slow down traffic. After the roadblock, we went on a patrol. As soon as we did that, a giant car bomb exploded just down the street. We went flying towards it and were the first ones to respond. There were people lying everywhere and screaming.

It was pretty bloody, and there was a lot of damage. Anyhow, we formed a perimeter around the blast and kept all the people out, as the firefighters tried to save them and control the fire to the buildings. They were pulling people out that were completely black, and every time they did that, the crowd went crazy. People started screaming and crying and tried to get to their loved ones, but we had to hold them back for security reasons. Not a good situation.

While some of my writings were graphic, they still fell short of capturing the full drama and intensity of those days. During this particular incident, a terrorist detonated a vehicle packed with explosives in the densely populated Karrada district of southern Baghdad. The neighborhood was lined with apartment buildings and storefronts, all busy with heavy foot traffic. We were just down the street when I heard the massive explosion, and I turned to see smoke and debris erupting into the air. Immediately, we piled into our vehicles and were on the scene almost instantly.

What was left of what appeared to be a pickup truck lay twisted and obliterated outside an apartment complex. Smoke poured from gaping holes in the building's façade, now partially collapsed into a heap of rubble. Balconies hung at odd angles, threatening to fall onto the civilians below who were desperately trying to reach the wounded inside. The road was littered with chunks of metal, shattered concrete, and gruesome pieces of human remains. Near the blast site, a white shirt lay soaked in black bile and bright red blood, with entrails smeared everywhere across the pavement.

---

[179] Reuters, "US Army: Iraq Attacks Jumped in 2005," *Al Jazeera*, January 23, 2006, https://www.aljazeera.com/news/2006/1/23/us-army-iraq-attacks-jumped-in-2005.

**The immediate aftermath of a VBIED attack in Karrada**

Several of us pushed a few feet into the shattered apartment complex, staying low as thick, noxious black smoke poured out, making it nearly impossible to breathe. In what I assumed had once been a living room, a teenage boy lay face down, partially pinned beneath a collapsed section of wall. His body was nearly pitch black from smoke and fire damage.

Another soldier tugged at his shirt while I grabbed his arm and pulled, eyes watering and lungs burning as I coughed uncontrollably. In a grisly instant, the skin on his arm, apparently seared by the VBIED blast, sloughed off in my hand, peeling away like a snake shedding its skin. Beneath it, a raw layer of yellow fat and pink muscle was exposed.

Squinting through the stinging smoke, I forced my eyes open and reached for his upper body. Just then, a civilian paramedic in blue appeared beside us, and together we managed to free him, but he did most of the work. The boy was loaded onto a stretcher and rushed away. To this day, I don't know if he survived. I honestly can't imagine that he did. But we were in no condition to keep pushing deeper into the building. We didn't have masks or air tanks, just our bare faces in that choking black cloud, so we backed out into the street and hoped the Iraqi rescuers could get inside and reach whoever was still alive.

As firefighters and paramedics began arriving in force, we pulled civilians back to give them space to work. I coughed and gagged a bit more, spitting out the last of the smoke, but quickly got my bearings. We shifted from rescue mode to securing the site, making sure no follow-on attacks occurred. According to reports, nine people were killed in the blast, and ten others were wounded.[180]

**A fire rages inside an apartment complex from the VBIED attack - the smoke was overwhelming**

**Photo courtesy of Derek Hauger**

Across the street, reporters began to arrive, filming and snapping photos of the aftermath.[181] Word had it we even made the news back in the States, though the story quickly faded

---

[180] News Agencies. "Many Killed in Deadly Baghdad Blasts." *Al Jazeera*, May 3, 2005. https://www.aljazeera.com/news/2005/5/3/many-killed-in-deadly-baghdad-blasts.

[181] AP Archive. "Several Feared Killed in Shopping District Blast," July 21, 2015. https://www.youtube.com/watch?v=JFsxA87QmPo.

from the headlines. Car bombs were going off so frequently that it had become almost routine. Three exploded in Baghdad that day alone.[182]

**A minivan damaged by shrapnel near the blast site - the author is standing on the corner near the Humvee**

**Photo courtesy of Derek Hauger**

You'd think that would have been the most intense part of the day, but Iraq always had a way of throwing in one more surprise. After wrapping up the patrol, we headed straight to the checkpoints. Always the damn checkpoints.

### 3 May 2005 – Letter to the Harstad Family

Right after the car bomb, we went to the checkpoint for 8 hours. I had to go to priority search. Basically, people with high-ranking badges come through our gate, and we search their vehicles and

---

[182] RadioFreeEurope / RadioLiberty. "Militants Target Baghdad With Bomb Attacks." *RadioFreeEurope / RadioLiberty*, May 2, 2005. https://www.rferl.org/a/1058701.html.

their bodies. I was standing there when a car rolled up. I was doing a visual inspection on the car, and the dog handler was using the dog to sniff out bombs. Anyhow, the dog sat right where the rear seat was. When the dog sits, that means he smells explosives.

As soon as I saw the dog sit, I knew we might have a serious problem. The two Iraqi men, whom we had previously ordered to stand by the wall, were now glancing around nervously. They were whispering to each other. Not wanting the situation to escalate, I calmly ordered them to sit on the pavement, raise their hands so I could see them, and cross their legs. They complied immediately, apparently speaking English.

It wasn't until that moment that I realized the civilian dog handler and his K9 partner had vanished without saying a word, leaving me alone with a potential VBIED and two military-age males. *Shit*, I thought. *This is really bad.*

I tried to get SSG Rider's attention because he was overseeing our side of the checkpoint, but he was too far away to hear me. The minutes dragged by, tense and silent, while I nervously edged away from the car, but I had to keep the men in my sight. SSG Rider finally approached, apparently having been briefed by the skittish dog handler who had run away. As I kept a close eye on the two men, Rider moved in, zip-tied them, and conducted a thorough body search. Nothing suspicious was found.

Carefully, I began inspecting the car for any signs of explosives. Were there exposed wires? Unusual odors? Electronic triggers or pressure-sensitive pads? Every inch had to be scrutinized. Anything out of place could mean disaster.

Finding nothing out of the ordinary, I moved on to using sample pads to test for explosive residue. Warily, I crawled under the vehicle near the rear quarter panel, the exact spot where the dog had alerted. My heart pounded in my chest as flashes of those car bomb videos from FLETC replayed in my mind, paired with the piles of red goo I had just stepped over earlier in the day. I swabbed the areas most likely to conceal explosives, each movement slow and deliberate. Sweat poured down my face as I inched my way back out, every second stretching longer than the last. Like a field scientist, I meticulously tested each sample. One by one, they came back negative. Nothing.

Undeterred, I cautiously swabbed the door handles, seatbelt clips, and other high-contact areas in the back seat. This time, the result was very different, and I'm sure my eyes went wide as saucers.

**3 May 2005 – Letter to the Harstad Family**

The dog handler got the hell out of the area right away, but I had to stick around to make sure there wasn't a car bomb. I pulled out these sticky pads and took samples in the car. It immediately turned dark brown, which means TNT. This meant it was either a car bomb ready to go off, or it recently had explosives in it. Nothing (explosives) was ever found. They took the two guys in for questioning, and I'm not sure the results.

I'm convinced those two men were up to no good, particularly because of the way they nervously talked to each other when the dog sat next to their car. While we didn't locate any direct evidence of explosives, we couldn't just let them go on their way, especially after the double-positive tests. An MP unit arrived and transported them to an interrogation team. As usual, we never found out what happened. Were they released? Did they confess? We had no idea. Had we done a good job or not? The incident just disappeared into another endless black hole of unanswered questions.

It was during these relentless bombings that acts of heroism and quiet gestures of honor took place. More often than not, they went unrecognized, simply because everyone was performing admirably, day in and day out. Still, it was always appreciated when recognition made its way down the chain of command.

**31 May 2005 – Letter from Robert Roth, Commander of 4-64 Armor**

This has been a tough past 10 days for the brigade and the battalion. One of our sister battalions was attacked in our area of operations this past week, and the attack killed three soldiers.[183] They were from 1-76 Field Artillery Battalion and were engaged by a roadside bomb called an IED.[184]

Once word of the attack reached the Tusker Battalion Operations Center, we immediately dispatched reinforcements with medical support to the area. In heavy traffic, we made it to the scene and found three soldiers had passed away, but found a fourth critically wounded soldier who needed to get to a hospital fast. In no time at all, and through heavy traffic, the battalion command sergeant major, CSM Stanley, led the medical evacuation convoy to the American Hospital and was able to save the life of the fourth wounded soldier. The soldier was from the Georgian Army, who was accompanying the 1-76 FA convoy.[185] Several days later, the unit presented a certificate to the command sergeant major, making him an honorary member of their battalion.

After additional medical assets arrived to pick up the fallen soldiers who were killed in action, they convoyed back to the Green Zone slowly. As the convoy entered one of our Entry Control Points leading into the Green Zone, the Tusker soldiers on duty had been alerted of their approach and were standing ready. The soldiers came out and lined the route, and saluted our fallen comrades who had just passed away. I have never been more proud than I was at that moment!

While American troops were frequent targets, Iraqi Army and police forces were being hit just as hard, if not harder. Even worse, civilians were maimed and killed in shocking numbers, of-

---

[183] AP Archive. "Humvee Hit in IED Attack," July 21, 2015. https://www.youtube.com/watch?v=_-18yDeXNSg.

[184] L.A. Times Archives. "Military Deaths." *Los Angeles Times*, June 5, 2005. https://www.latimes.com/archives/la-xpm-2005-jun-05-me-wardead5-story.html.

[185] 4th Brigade, 3rd Infantry Division. "Vanguard Point: Volume II, Issue 16." DVIDS Hub, December 2005. https://media-cdn.dvidshub.net/pubs/pdf_0640.pdf.

ten just for being in the wrong place at the wrong time. The terrorists showed no hesitation in slaughtering innocent bystanders, and we often had to secure and pick up those sites.

### 20 June 2005 – Letter from Robert Roth, Commander of 4-64 Armor

Just outside checkpoint 2, a suicide bomber exploded the bomb he carried in a restaurant where many Iraqi Police go for lunch.[186] The terrorist simply waited outside for a group of policemen to enter the building, then followed them inside - detonating the bomb. Five Iraqi policemen were killed, 15 Iraqi civilians were killed, and about 15 wounded. The blast was severe and destroyed the restaurant and adjacent buildings. There were many sad and angry policemen today, and our thoughts and prayers are with them.[187]

For Iraqis who had lost everything, including their homes, vehicles, and even loved ones, many were left destitute, trapped in grief and crushing poverty, clinging to hope for lasting peace. We did what we could to help, but it was nearly impossible to kill insurgents while simultaneously delivering food, medicine, clothing, and other essentials.[188] The sheer volume and variety of tasks were overwhelming, especially for the grunts not trained for that sort of thing.

### 8 June 2005 – Letter to the Harstad Family

Today we went on a MEDCAP (Medical Civic Action Program).[189] Basically, we set up a medical station in the middle of Baghdad, and let the people come to us with problems they are having. Our job is to make sure the doctors and medics are secure and aren't blown up or anything. It went smoothly.

My writings from that time reveal a growing frustration with the situation, especially when many Iraqis responded with blatant disrespect and even overt hostility, despite receiving essential supplies from American troops. Goods were sometimes handed over to scowling faces, many of whom made obscene gestures to us as they walked away. More than once, I felt the strong desire to slap one of the men in front of everyone, but thankfully resisted the urge. As Colonel (COL) Car-

---

186 BBC News. "Many Dead in Attacks Across Iraq," June 20, 2005. http://news.bbc.co.uk/2/hi/middle_east/4109908.stm.

187 DVIDS, "5 Iraqi Police, 13 Civilians Killed by Suicide Bomber," Defense Visual Information Distribution Service, June 20, 2005, https://www.dvidshub.net/news/2205/5-iraqi-police-13-civilians-killed-suicide-bomber.

188 DVIDS, "Generator and Clothes Help Baghdad Orphanage," Defense Visual Information Distribution Service, September 18, 2005, https://www.dvidshub.net/video/6171/generator-and-clothes-help-baghdad-orphanage.

189 DVIDS, "Medical Civil Action Program B-roll," Defense Visual Information Distribution Service, March 9, 2005, https://www.dvidshub.net/video/1914/medical-civil-action-program-b-roll.

don, the brigade commander, had previously instructed us, we were to make no new enemies and treat everyone with dignity and respect.

### 23 June 2005 – Letter to the Harstad Family

This morning, we got an emergency call for some VBIED attacks. We had to rush out and secure the area. Three went off, and we were able to disarm the fourth.[190] Then we had to chase around some other cars that we thought were bombs (info from civilians), but they turned out to be clear.

This evening, we brought a shitload of food and water to the Iraqis that were left homeless. They, of course, swarmed our convoys, but we did our best to keep them at bay. You get really irritated when you try to help them, yet they still seem to be at odds with Americans in general. They have been oppressed for so long that they don't understand what it is to be free, and I don't know if they ever will.

**Distributing truckloads of food to Iraqi civilians in Baghdad — a goodwill mission that nearly turned into a riot when the crowd swarmed the trucks**

**Photo courtesy of Derek Hauger**

---

[190] CBS News. "Four Car Bombings Rock Baghdad," June 23, 2005. https://www.cbsnews.com/news/four-car-bombings-rock-baghdad/.

My writings also reflect an overly simplistic worldview on my part, likely shaped by my youth and inexperience. Looking back, I can understand why many Iraqis held the United States at least partially responsible for the hardships they faced. At the time, though, I was too focused on staying alive and trying to do the right thing to view the situation with any real objectivity. So, I couldn't understand why they were so angry with us, especially when we were the ones assisting, and it was the terrorists, not us, who were still destroying everything.

As frustrating as it was, there was still an abundance of good Iraqi civilians who were also just trying to do the right thing. While some may have blamed us for their circumstances, most didn't support the terrorists destroying their neighborhoods. They wanted the bombings to stop, too.

**27 June 2005 – Letter from Robert Roth, Commander of 4-64 Armor**

Last Sunday, a man strapped explosives to his chest and walked into a restaurant where Iraqi Policemen eat. He detonated the device and killed about 20 Iraqi Police and citizens combined. Many more were wounded.[191]

Several days later, four car bombs exploded.[192] Two exploded in front of a Shia mosque, one exploded next to a police car that was responding to the scene, and the fourth exploded in front of several shops and office buildings. A fifth car bomb was found and defused before it detonated.

When the first two bombs detonated, several children in the area hid behind a few cars and vans. One child looked inside the van he was hiding behind and saw a huge bomb in it. He grabbed another friend, and together they ran about 2 miles to checkpoint 11. The children told the soldiers on duty about the bomb. They even hopped in a HWMMV (Humvee) and went with the soldiers to show them the location. Because of their bravery, the bomb never exploded.[193] It was placed there because the insurgents know that when Americans arrive on scene, we set a cordon to keep people out. If that bomb detonated, there is a very high likelihood that it could have killed several US soldiers. The actions of these children potentially saved the lives of US Army soldiers. We are attempting to find these brave children and thank them and their parents for their bravery.

Some of my frustration also stemmed from persistent internet issues on the FOB, and we often went days or even a couple of weeks without regular access. That connection home was one

---

[191] Reuters. "Baghdad Diner Bombed." *Times of Malta*, June 19, 2005. https://timesofmalta.com/article/baghdad-diner-bombed.86590.

[192] Associated Press, "Four Car Bombs Explode in Western Baghdad, Killing 23," *Deseret News*, June 22, 2005, https://www.deseret.com/2005/6/22/19898859/four-car-bombs-explode-in-western-baghdad-killing-23/.

[193] CNN, "American Morning Transcripts," June 23, 2005, https://transcripts.cnn.com/show/ltm/date/2005-06-23/segment/02.

of the few ways to decompress and stay sane. The worst part was that our families would panic whenever they didn't hear from us, immediately assuming something terrible had happened.

The wives in the Family Readiness Group (FRG) did a fantastic job of keeping everyone informed. Still, communication blackouts usually signaled one of two things: a dangerous mission was underway, or someone had been seriously wounded or killed. Rumors constantly swirled, eventually drifting back to our FOB. So, in addition to the death and destruction we faced daily in Iraq, we sometimes had to contend with the emotional fallout and drama unfolding back in the States. In that sense, the internet was a mixed blessing.

**3 July 2005 – Letter to Toby Glass**

My Dad has been bugging the hell out of me lately. I let him know I would be home soon (for leave), and he gets all excited about it. Then, of course, he tells me to be safe the next few weeks, as if I'm not safe all the time. But I understand where he is coming from. Just a parent wishing the best for his kid.

The situation was made more chaotic when one of the Iraqi internet contractors, a guy I had always found a bit shady, was permanently banned from all American bases. His removal created a temporary void and stirred up a minor uproar. However, the story of what happened was a bit comical. One of our interpreters, a fiercely loyal Iraqi, didn't take kindly to the remarks being made by this contractor.

**8 June 2005 – Letter to the Harstad Family**

We have been having serious internet problems here on the FOB. The guy that runs the internet was beaten to a pulp by one of our interpreters for badmouthing the US.

On top of everything else, I was dealing with some minor but nagging injuries that made daily life a bit uncomfortable. During downtime, I often attended a Combatives class on base to keep my skills sharp. At one sparring session, I suffered a boxer's fracture in my right hand, or, more specifically, a break of the fifth metacarpal. The knuckle looked sunken, pushed back into my hand. It wasn't unbearable, but it certainly affected my dexterity.

Our medic advised me to see a doctor and get a cast, but I declined. That would have taken me out of action for a few weeks, and I wasn't about to let that happen. Instead, I just wrapped it up and carried on as usual.

**28 May 2005 – Letter to the Harstad Family**

I'm just sitting here trying to type a letter the best I can with one finger on my right hand. It doesn't hurt too badly, but it has been a little difficult typing.

Around the same time, I had a terrifyingly close call that left me pretty shaken. A foot patrol of American National Guard soldiers was exiting a checkpoint and heading into the Red Zone, while I was coming back from that direction on foot. I'm still not entirely sure what happened, but the SAW gunner suddenly fired a short burst that chewed into the concrete barrier right next to me. It was an accidental discharge, a trigger pulled unintentionally, but I don't know if it was nerves or simple incompetence.

One of the rounds struck extremely close to where I was standing, sending a spray of concrete shards into my face. I was left with several superficial cuts on my cheek and a few pieces of embedded debris, but otherwise escaped with only a ringing ear. I was extremely lucky.

Realizing the severity of what had just happened, the SAW gunner and his team leader rushed over to check on me. Aside from being irritated that I'd nearly been shot by one of our own, I was fine. After rinsing the grit out of my eye, I went back to work, and we never spoke of the incident again.

Later, back at the FOB, 1LT Turcotte, our platoon leader, asked me what had happened after seeing my face. I told him the story but asked him to keep it between us. I didn't want the SAW gunner punished over the incident. He laughed and said it looked like a cat had scratched me, which, to be fair, it did. Nothing ever came of it until I later talked to my Dad on video chat.

Seeing my face, he naturally asked what had happened to me. Embarrassed and not wanting to admit I'd nearly been seriously injured in a friendly fire incident, I lied and said an Iraqi had done it. Obviously, that didn't go over well. He didn't care who blasted me in the face, he was freaked out that it had happened at all. I shouldn't have lied to him, but I did. The only reason I'm admitting it now is that I suspect that soldier, had he stayed in the Army, would be retired by now. It was an accident, and I didn't want to see him dragged through the wringer, especially when we had no shortage of actual enemies to deal with. Sometimes shit happens in a combat zone.

Injuries and death during deployments don't always come from enemy fire. More often than people realize, casualties stem from stress, fatigue, hunger, and sleepless nights. A warzone is hazardous in more ways than most can imagine.

Any time we rolled out in our Humvees, I served as the turret gunner. One of my duties was to alert the driver and vehicle commander to any visible hazards. Hundreds of spliced power lines and loose cables dangled precariously across the roads. They posed a serious threat not just to people on foot but also to vehicles like ours. Mounted on the back of each Humvee were two tall radio antennas. If they made contact with a live wire, it could be deadly.

So, whenever I spotted overhead wires, I had to climb onto the rear of the moving Humvee and pull both antennas down by hand, then duck back into the turret and hold them firmly in place. Only once we cleared the hazard could I let the spring-loaded antennas snap back into their upright positions. Clearly, it was a dangerous job, especially doing this atop a moving vehicle rolling across Baghdad's terrible roadways.

One day, while entering a narrow neighborhood, I reached for the thick metal antennas, each taller than I was, and pulled them down. I settled back into the turret, scanning left and right for threats, gripping the antennas tightly in both hands. Everything appeared quiet, and nothing seemed out of the ordinary. Then, without warning, it felt like I'd been hit by lightning.

My entire body seized. My head jerked back, and I couldn't move or let go. I was frozen, arms locked in a death grip around the antennas. An electric current surged through my body, and I tried to scream, but nothing came out. I heard SFC Hill talking over the radio and struggled to get his attention, but I couldn't speak or move. I was being electrocuted, completely helpless.

The shock probably lasted just a few seconds, but it felt like an eternity. The moment SFC Hill released the handset, the current stopped, the antennas snapped from my hands, and I collapsed into the Humvee. The guys inside panicked, temporarily thinking I'd been shot. I could barely gasp that I was okay and that I'd been shocked through the antennas.

As I discovered, touching both antennas wasn't a problem, *unless* someone keyed the radio. In that moment, the transmitted signal surged through both poles and directly into my hands. The standard procedure was for me to call out before grabbing the antennas so that the crew wouldn't key up the radio. But we were deep in a sketchy neighborhood, and I had done it so many times already that I figured our platoon sergeant knew what I was doing. He didn't. That was my mistake, and I paid for it.

Our medic checked me out and found nothing serious aside from mild burns on both palms and a healthy dose of embarrassment. I felt wiped out, as though I'd just finished a long run. But my strength slowly returned, and we continued the mission. From that point on, I made sure to loudly announce whenever I grabbed the antennas, probably louder than necessary, but I wasn't about to repeat that experience.

Not all turret injuries involved electricity. On a night patrol I wasn't part of, my friend Noel Mata had a close call of his own. His crew was running "lights out" through a residential neighborhood to reduce visibility, which made things especially tricky for the driver. That's because night vision goggles eliminate depth perception. With enough experience, you adapt, but mistakes still happen, especially at higher speeds.

In the darkness, the Humvee plowed straight into a concrete barrier. For someone unbelted and half-exposed in the turret, that's a nightmare scenario. Mata slammed headfirst into the turret shield but, somehow, stayed in the vehicle. His face was bloodied and swollen, but by some miracle, he didn't suffer any lasting injuries.

As if all that weren't bad enough, I was also battling a nearly debilitating illness, though I stubbornly refused to miss work. Call it pride, or maybe just sheer obstinacy, but I kept going because I didn't want to let anyone down. What was the cause of the illness? Let's just say I have my suspicions, and I'll let you guess what it was.

One of Bravo Company's primary responsibilities was training Iraqi National Guardsmen in basic combat skills.[194] Thankfully, the Gambler Platoon was mostly spared from this duty, falling

---

[194] Wilsoncroft, Emily. "Iraqi Army Patrolling Streets With Task Force Baghdad." Defense Visual Information Distribution Service, April 13, 2005. https://www.dvidshub.net/news/1576/iraqi-army-patrolling-streets-with-task-force-baghdad.

squarely on the shoulders of Bravo Company's other platoons. On the occasions when we were tasked with it, the experience for us often felt like a futile exercise. The quality of recruits we received was far too frequently alarmingly poor. Many couldn't read or write, and some struggled with even the most fundamental tasks, such as tying their boots, driving, loading weapons, firing accurately, or marching in formation. We had to dig even deeper into their development, covering not only military essentials but also basic hygiene and simplistic life skills. It sometimes felt like teaching a group of overgrown toddlers.

**Noel Mata and the author before a night operation**

Some of these challenges stemmed from stark cultural differences, but certainly not all. A significant part of the problem was that many of the candidates simply lacked the intelligence or basic aptitude for military service. Iraq was clearly not sending its best and brightest to join the ranks at that time.

For example, a row of portable toilets was installed near their barracks after the area became heavily contaminated with human waste. Many of the trainee guardsmen had taken to defecating outside and casually kicking dirt over it, which made it nearly impossible to walk anywhere without stepping in feces. The hope was that the toilets would solve the issue. They didn't.

Eventually, we had to start physically forcing them to use the portable toilets instead of relieving themselves on the road or in the dirt. While that might sound like a win, it wasn't. Most were accustomed to squatting over a hole or using squat toilets, and many didn't understand how to use a Western-style lavatory. Even after demonstrations, yes, actual demonstrations, where American soldiers sat bare-assed on the seat to show them the proper method, the problem persisted.

Time and again, we'd open a stall door and find feces smeared on the seat, spackled on the walls, or piled on the floor. It became clear that they weren't sitting on the seats at all. Instead, they were climbing up and attempting to hover over the opening, like paranoid germaphobes in a gas station bathroom, but stuck within the tight confines of the portable toilets. Their aim was horrendous, and they never cleaned up after themselves, leaving the stinking mess for the next guy.

So, the next logical step was to remove the toilet doors. The idea was that they wouldn't engage in that behavior if someone were watching. Would they? If an American soldier caught an Iraqi trying to hover, we were expected to yell at him like drill sergeants and correct him immediately. This was easier said than done, though. They got sneaky, slipping in when no one was looking and treating the stall walls like they were conducting artillery fire with flying turds.

As absurd, and perhaps even harsh, as some of this may sound, these were the real conditions we had to work under while training so-called soldiers. We couldn't allow such unsanitary and disease-spreading practices to continue. That's saying something, especially coming from infantrymen who were used to living in absolute filth. It was just a new level of disgusting, and, unfortunately, it was a constant battle, both for their hygiene and for their ability to defend a fragile government teetering on collapse.

**22 September 2005 – Letter from Mark Barnes, 1st Sergeant of Bravo 4-64 Armor**

(Bravo Company) is training Iraqi soldiers. Sounds simple, huh? Well… it's not. We have been literally potty-training them lately because they have trouble understanding what a port-o-potty is used for. We taught them to pick up trash, too. They haven't grasped the concept fully, but we are making it in their best interest to pick it up.

The good news story in all this is that our teams are doing a phenomenal job of bringing them up to standard in tactics so that they can eventually run their own operations. It's still a ways off, but every day we make progress.

So, you might be wondering, how exactly did I end up with *dysentery* in the middle of a war zone? I can't say for sure, but trudging through those filthy, unsanitary grounds probably had something to do with it. Call me crazy, but I bet I'm right.

For the uninitiated, dysentery is an exceptionally unpleasant gastrointestinal infection caused by ingesting food or water contaminated with fecal matter, typically the result of poor sanitation. It's common in developing nations and combat zones, and the symptoms are exactly as fun as they sound: intense abdominal cramps, severe dehydration, uncontrollable bloody diarrhea, and the sudden, humiliating inability to trust a single muscle below your belt. In other words, disagreeable liquid projectiles are launched at random intervals. You know, just what you want to be dealing with during a firefight.

I was sick for about a week, and it made my life absolutely miserable. The showers and toilets were in a separate building from my room, so when the urge hit, I had to awkwardly sprint down a gravel path with clenched butt cheeks and desperately hope that I made it in time. Twice, I didn't. I soiled myself on the way, a humiliating mess of foamy, bloody diarrhea.

To say it was embarrassing would be an understatement. I just stepped straight into the showers, clothes and all, and did my best to clean up while shivering through fever and cold sweats. It was a wretched existence and one of the lowest points of my deployment.

Being sick is bad enough. Being sick with explosive diarrhea, while feverish and dehydrated, in the middle of a mission? That's a special kind of hell that I will never forget.

Another reason I was determined to stay in the fight was that, inexplicably, I'd been promoted again, at least temporarily and in spirit, to squad leader. It was a role typically reserved for a staff sergeant (E-6), two pay grades above me. Now I was attending daily briefings, responsible for two teams of men, and juggling even more duties on top of everything else.

On one particularly uncomfortable mission, we were tasked with capturing or killing a suspected car bomb manufacturer at his residence. Due to our superior firepower, the Heavy Weapons Squad was assigned to set up the outer cordon and provide overwatch while the rest of the platoon kicked in the doors. There was a long lull as they sifted through intel, ultimately finding no sign of the target. He had fled before our arrival.

Unfortunately for me, that was when I felt a sharp, twisting pain in my gut, like a dagger ripping through my intestines. Cold sweat broke out across my forehead, and I had to go… *right now!*

Luckily, our interpreter caught on to my distress. He began frantically pounding on the door of a small, tan home near where we'd established our overwatch position. I don't think he wanted me exploding on the sidewalk any more than I did.

A middle-aged man cautiously opened the door, his eyes wide at the sight of heavily armed soldiers outside his home. He clearly assumed the worst. But our interpreter quickly explained the situation, and, bless him, the man welcomed us inside without hesitation, kindly leading me to the restroom as if I were royalty.

Isn't it strange how, when you *really* have to go, your brain seems to signal your bowels the moment your hand touches your belt? It doesn't wait for you to sit down, grab a newspaper, and get comfortable. No, your body says, *go right now.* Well, that's precisely what happened.

In a desperate struggle for my dignity and, possibly my life, I tore off my overloaded combat vest, frantically flung my Kevlar helmet aside, and fought to get my pants down in time. Thankfully, I made it, but just barely.

There was just one problem: this was my first encounter with a squat toilet, a simple, small porcelain hole in the ground. There was no seat and no stall. There was just gravity and an insecure sense of hope. What was that old British saying? *Proper planning and preparation prevents piss-poor performance.* Yeah, that.

On the verge of disaster, I braced myself against the wall and let nature take its course. I came dangerously close to total humiliation, but somehow, I managed to avoid soiling myself in front of everyone. That day, I counted my blessings… every last one of them.

Fighting through the sweats, I opened my eyes and turned around to assess the damage. I knew I couldn't just leave their squat toilet in a horrid state, especially after that explosion, and I'd have to clean it up. But then panic set in. *Where did it go?*

To my utter confusion, there wasn't anything there, and the porcelain was still pristine white. I started a frantic search for… well, *it.* Where had it gone? I checked my pants. Nothing. *Where was it?* Somehow, like Larry Bird sinking a three-pointer from half-court, I must've scored a direct hit into this tiny hole with nothing but net. How I managed that amazing feat, I'll never know.

Grabbing the napkin from my MRE pack, I cleaned myself up as best I could and tossed it into the bowl, only to instantly remember: they *don't* flush paper products here. There wasn't even a flusher! It was just a tiny hole in the ground, so anything like toilet paper or napkins would clog the pipes.

Panicking again, I scrambled to flood the bowl with water. Miraculously, it worked. Crisis averted, for now, at least. Still, I couldn't shake the creeping fear that I might've left behind a ticking time bomb of a plumbing disaster. *What was I thinking?* Clearly, I wasn't. The delirium of sickness was getting to me.

Emerging a few minutes later, still woozy but alive, I was taken completely off guard. In the short time I'd spent in the lavatory, the man, his wife, and their many children had arranged a beautiful rug in the center of their modest home. On it sat a delicate spread: small glasses of steaming black tea, freshly baked Samoon, and other humble delicacies. I stood there, stunned. Their hospitality, offered so freely and so quickly, nearly brought me to tears.

My interpreter, wearing a proud smile, told me they wanted me to stay for tea and dinner. As graciously as I could, I explained that we couldn't stay long because of our activities in the neighborhood, but that we'd be honored to join them for a cup of tea. In Iraqi culture, as in many homes across the Middle East, guests are treated with great respect, often likened to royalty. For those few minutes, I had become their guest of honor.

While monitoring the radio, I nibbled on some of the plain bread, hoping it might calm my stomach, and carefully sipped the strong, hot tea. Through the interpreter, I did my best to express my deep gratitude for their kindness. Though our visit lasted only five minutes, we probably spent half of that time thanking them profusely. The man, his face weathered by war, but still kind, simply nodded and smiled.

The truth is, I'll never be able to repay them for their generosity. I don't know their names, and I couldn't point out their home on a map if I tried. That part of the memory is lost to time. But somewhere in Baghdad, there was a family who welcomed a sick and desperate foreign soldier into their home and took care of him without hesitation. I'll never forget them.

(L–R) The author, SSG Gasman, and SGT Dvorak enjoying Iraqi tea and bread in a family's home in southern Baghdad — and for the record, this *wasn't* the dysentery house

**24 June 2005 – Letter to the Harstad Family**

I just got back from the medics, and they are saying that I need to come in again tonight for more tests. They are talking about sending me to the Baghdad hospital or Germany. They think I have dysentery.

As it turned out, I did have dysentery. A round of antibiotics cleared it up, and thankfully, I stopped shitting myself. So, that was a win. I never went to the hospital, and I didn't take a single day off from work. Not that I really had much of an option.

So, to circle back to why I was so frustrated with those disrespectful Iraqis while we were trying to help them, I hope I've done a decent job explaining the context and where I was coming from. I don't say all this to throw a pity party, as everyone was dealing with their own injuries and illnesses. Out there, everyone had something wrong, and no one was unscathed. But we all had a job to do, and we got through it the only way we knew how: together, like brothers.

**20 June 2005 – Letter from Robert Roth, Commander of 4-64 Armor**

D Company (Dealer) and B Company (Bandit) continue to conduct a series of searches and patrols in search of insurgent activity. Their schedule is virtually non-stop, and they have had some success detaining insurgents in their area of operations. They both work hand-in-hand with HHC to incorporate the newly training Iraqi Army Battalion into the sectors.[195] The cooperation has gone very well, and the soldiers are performing superbly. While this sounds easy, it isn't, so job well done.

Unfortunately, around this time, the Tusker family suffered a terrible loss. SSG Victor Cortes III, a dedicated mechanic who worked at the 703rd Forward Support Battalion, tragically took his own life on the FOB. I didn't know him personally, but those who did spoke highly of his commitment to his work.

From physical injuries to illness and invisible wounds, none of us was immune to the relentless pressures of life in a combat zone. The strain was constant and sometimes, it became too much to bear.

**13 June 2005 – Letter to the Harstad Family**

We did some patrols last night and set up a checkpoint on a couple bridges, where we stopped random cars. There were quite a few people out past curfew, so we stopped a lot of them. Of course, everyone has an excuse, but we try to filter out the bad guys from the good ones.

I stopped one vehicle that was flying down the road at gunpoint. When I got to the vehicle, the man stepped out and had tears in his eyes. He was pointing to his wife, and I looked into the car. His wife was holding a baby girl, maybe half a year old, and she looked dead. I just told them to get to the hospital as fast as they could, and gave them my bottle of water. I don't know why, maybe just an act of desperation on my part.

Man, I hate seeing shit like that.

Although we were hitting multiple targets night after night, scouring neighborhoods for the masterminds behind IED production, explosions continued to erupt across the city. Our efforts appeared to have little impact on the number of detonations. On July 15, 2nd Platoon was hit directly by a VBIED, one of twelve that day alone.[196]

---

[195] DVIDS, "Task Force Baghdad Unit Thanks Local Iraqis for Support," Defense Visual Information Distribution Service, August 4, 2005, https://www.dvidshub.net/news/2631/task-force-baghdad-unit-thanks-local-iraqis-support.

[196] Frederiksen, Paul, Emily Fall, and Patrick Baetjer. "Iraqi Security and Military Force Developments: A Chronology." *Center for Strategic and International Studies,* May 26, 2006. https://csis-website-prod.s3.amazonaws.com/s3fs-public/legacy_files/files/media/csis/pubs/060526_isf_chron.pdf.

According to a newsletter from the FRG, two NCOs, Gonzalez and Hanson, were wounded in the blast on the Karrada Peninsula during a mosque observation, along with two specialists, Tumlinson and Stevens. Stevens was flown to Germany for further treatment, but thankfully, no Americans were killed. It was a close call and could have easily ended in tragedy for us.

Years later, Mark Barnes told me that when the car bomb detonated, the blast launched the suicide bomber's foot into the air, and it came down in the bustle rack on the back of the Bradley. They only found it days later, once it started to decompose and fill the air with a terrible stench, a grisly reminder of how close they had come to disaster. Unfortunately, at least one Iraqi civilian died in the blast, and eight were wounded.[197]

### 7 June 2005 – Letter from Mark Barnes, 1st Sergeant of Bravo 4-64 Armor

We've done a few raids. So far, we haven't been able to tell how valuable the people we grabbed are. A couple nights ago, we did raids on suspected VBIED garage sites. Found some people who may have known something, but no car bombs. It was an eerie night because, as the mission was wrapping up, a dust storm engulfed us.[198]

2nd and 3rd Platoons did a good job taking down their target buildings, and the tankers kept all the riff-raff from bothering them while they did it. 1st Platoon was nearby with Dealer (Delta Company), making it happen, too. They hit a house that a sheikh supposedly lived in. He wasn't home. I doubt he'll be hanging out there much anymore.

### 6 June 2005 – Letter to the Harstad Family

Last night we had a mission, and of course, there had to be a giant sandstorm.[199] I was still coughing up dirt and shit this morning. That really can't be good for you.

The air was almost always thick with dust and sand, like breathing through a dirty rag. On particularly bad days, though, the sandstorms did more than just blow through. They swallowed everything. The sky disappeared, the horizon vanished, and the world shrank to a few feet of swirling reddish-brown. Sometimes you could barely see your own hand at the end of a fully outstretched arm. We would yank balaclavas over our faces and squint through the grit, trying to keep moving and working, feeling the storm pressing in on all sides.

---

[197] Mroue, Bassem. "Car Bombs Target U.S., Iraqi Forces, Killing 29." *Lawrence Journal-World*, July 16, 2005. https://www2.ljworld.com/news/2005/jul/16/car_bombs_target_us_iraqi_forces_killing_29.

[198] NASA, and Jeff Schmaltz. "Dust Storm in Saudi Arabia." NASA Science, June 6, 2005. https://science.nasa.gov/earth/earth observatory/dust-storm-in-saudi-arabia-15004/.

[199] Davidson, Brian. "Desert Dust in the Wind." US Air Force, June 6, 2005. https://www.af.mil/News/Article-Display/Article/134257/desert-dust-in-the-wind/.

Gunk built up in the corners of everyone's eyes until it felt like sandpaper every time you blinked. Guys hacked and coughed for hours, lungs burning, while the fine dust worked its way into every crevice of our gear. Weapons started to feel like they were full of gravel, every moving part grinding just a little more with each shot or dry fire. But we still had to work. Insurgents building IEDs didn't take a day off because the air wanted to choke us.

**18 July 2005 – Letter to the Harstad Family**

There was an explosion outside the Green Zone on one of the bridges we patrol. It was strange because they detonated a car bomb right between two Iraqi Army bunkers (instead of hitting them directly). Nothing we could really do, just secure the area and let the Iraqis start cleaning it up.

As I mentioned in my previous entry, we responded to the massive explosion on a bridge spanning the Tigris River. President Jalal Talabani's residence was nearby, and though he was home at the time, he was unharmed.[200] He wasn't believed to be the target.

The scene was chaotic and initially confusing because we couldn't find any bodies. There was plenty of destruction and gore, but surprisingly, no one was there that we could help. It wasn't until later that we learned the blast had killed three security guards and wounded nine others. Strangely, by the time we arrived, they had all vanished, gone without a trace. The fog of war was always present.

While it may have seemed on the surface that we were losing ground to the insurgents, each incident became a hard-earned lesson in how to fight smarter. We remembered, and we adapted. Tactics were constantly evolving. At the same time, the Department of Defense was working overtime to equip us with new technologies designed to counter the ever-growing threat of IEDs.

**25 July 2005 – Letter from Robert Roth, Commander of 4-64 Armor**

The battalion recently received a new piece of equipment called the "Buffalo."[201] It looks like a huge dump truck and has been designed to find roadside bombs and disarm them before they detonate.

The "Buffalo" was a massive Mine-Resistant Ambush-Protected (MRAP) vehicle, built with a V-shaped hull to deflect blasts from below. It was outfitted with thick armor, ballistic glass,

---

[200] Associated Press, "Attacks Kill 29 Across Iraq," Fox News, July 15, 2005, https://www.foxnews.com/story/attacks-kill-29-across-iraq.

[201] Military.com. "Buffalo Mine Protected Route Clearance Vehicle." Accessed December 7, 2025. https://www.military.com/equipment/buffalo-mine-protected-route-clearance-vehicle.

defensive machine guns, and a distinctive articulating robotic arm used to uncover and disable explosives from a safe distance. Although we grunts didn't operate them ourselves, our support personnel had great success locating and disarming countless bombs intended to kill those of us on the ground.

As effective as it was, Buffalos were in short supply. Even worse, the IED threat was everywhere. In many ways, it felt like applying a Band-Aid to a gunshot wound, but it was a start.

**A Buffalo clearing a dangerous route of IEDs for the infantry**

**Photo courtesy of Derek Hauger**

**10 September 2005 – Letter from Robert Roth, Commander of 4-64 Armor**

I had not been on patrol with this new equipment yet, so I hopped on board one day and went with 1st Platoon (BEAST Engineer Company). I was given instructions by SSG Taylor, SGT Baker, and SGT Walsh on the procedures. I had a window seat and an area to scan as we drove about 10 mph up and down a stretch of road known for many IED attacks.

At first, you listen to the soldiers perform their duties as a synchronized team inside the vehicle. "Large box, right side of the road near the curb."

"Got it!"

"Box is clear, continue to move."

When we would stop to investigate anything suspicious, you notice the security force quickly establishes a 360-degree perimeter around the Buffalo.

Though we were being hit especially hard, we also had plenty of successes. Too many to count, in fact. One incident stands out clearly in my memory, when a combination of American technology and training, paired with Iraqi manpower, prevented what could have been a devastating attack, right at a checkpoint we were guarding.

The day started uneventfully until a car bomb detonated just outside the checkpoint perimeter near an Iraqi Police station. Moments later, Iraqi soldiers and police officers spotted two military-aged males rushing toward the gate, clearly intent on breaching it. They opened fire, striking both men. As it turned out, the men were wearing suicide vests.

The first guy detonated upon being shot, spraying the area with a mist of blood and gore. As is common with suicide vests, the blast decapitated him instantly, leaving his body blown apart in a grotesque display of violence. Thankfully, he was far enough away that the blast didn't kill anyone. Well, other than himself, of course. The second bomber apparently decided at the last second that martyrdom wasn't for him, spun around to run, and caught a couple rounds in the legs for his trouble, dropping him flat and very much stuck with his new career as a failed suicide bomber.[202]

While we provided overwatch, the Iraqis took swift, disciplined action. They called in one of their own explosives teams, secured the area, and isolated the wounded bomber. For reasons we never fully understood, he was unable to detonate himself. The explosives team safely disarmed his vest and transported him to a hospital.[203] He reportedly survived and was later interrogated.

The Iraqis deserve full credit for the heroics that day. We simply supported from the sidelines, making sure the situation didn't spiral out of control. [204] Their training had been paying off, after all.

---

[202] American Forces Press Service. "DefenseLINK News: Suicide Bomber Stopped Near Checkpoint Before Attack." Global Security, July 14, 2005. https://www.globalsecurity.org/military/library/news/2005/07/mil-050714-afps01.htm.

[203] DVIDS, "Suicide Bomber Captured, Attack Thwarted," Defense Visual Information Distribution Service, July 19, 2005, https://www.dvidshub.net/news/2446/suicide-bomber-captured-attack-thwarted.

[204] DVIDS, "Interview With LTC Roth/MSNBC," Defense Visual Information Distribution Service, July 15, 2005, https://www.dvidshub.net/video/4219/interview-with-ltc-roth-msnbc.

**15 July 2005 – Letter to the Harstad Family**

I was working the checkpoint last night. We had a lot of time to sit there, since someone came in with a bomb strapped to himself. The Warlock system, which is a radio-frequency device that jams bombs and other explosives, set him off before he could get in. So, we shut the checkpoint down until morning.

The Warlock system was an electronic countermeasure device we used at checkpoints, on select vehicles, and around high-value targets. It was built to jam the radio signals commonly used to detonate IEDs remotely, everything from cell phones and handheld radios to pagers and garage door openers.[205] In theory, if someone needed a wireless signal to set off a bomb, Warlock stood a good chance of smothering it.

While effective in many cases, Warlock was far from perfect. It sometimes interfered with our own radios and other equipment, and it couldn't stop a bomber using a hard-wired or locally triggered vest.[206] Someone who was close enough to press a button or close a switch could still kill everyone in range. Whether or not Warlock played a role in setting off one of the bombers that day, I cannot say for certain. That was our initial suspicion, but it is just as likely that a stray rifle round hit his vest, he triggered it himself, or something else entirely happened. Whatever the cause, we were simply grateful they never made it inside the checkpoint.

As time went on, many Iraqis in our area of operations began to recognize our genuine commitment to making a positive impact in their communities. Whether it was removing insurgents from their neighborhoods, delivering supplies to those in need, or simply building trust through daily interactions, we were gradually winning many hearts and minds.[207] Time and again, locals brought us gifts and tokens of gratitude, many of which we couldn't accept due to regulations, but we always made a sincere effort to show our appreciation.

One gift came from a man who wanted to thank Bravo Company for everything we'd been doing. He proudly presented a local delicacy to CPT MacKinnon: Pacha, a sheep head stew.[208] This was perfect because MacKinnon was a notorious germophobe and was disgusted by, well, pretty much everything, especially food that still had a face. He did his best to be gracious, smiling and

---

[205] Wilson, Clay. "CRS Report for Congress - Improvised Explosive Devices (IEDs) in Iraq: Effects and Countermeasures," February 10, 2006.
https://www2.law.umaryland.edu/marshall/crsreports/crsdocuments/RS22330_02102006.pdf.

[206] Barry, John. "Iraq's Real WMD," *Newsweek*, March 26, 2006, https://www.newsweek.com/iraqs-real-wmd-106531.

[207] DVIDS, and John Queen. "An Iraqi School Girl Looks Through the School Supplies." Defense Visual Information Distribution Service, February 23, 2005. https://www.dvidshub.net/image/4084/iraqi-school-girl-looks-through-school-supplies.

[208] Choudhury, Indrajit Roy. "Pacha: A Taste of History, Culture, & Iraqi Hospitality." Indrosphere, May 6, 2017. https://indroyc.com/2017/05/06/pacha/.

thanking the man, but there was no way he was going to eat it. He held that bowl like it was radio-active while the rest of us tried not to laugh at the situation.

One Iraqi who worked with us wrote a long speech, hoping LTC Roth would deliver it to all the 4-64 Armor soldiers. He did. While the translation was rough, we understood exactly the sentiment he wished to convey.

**31 July 2005 – Letter from Robert Roth, Commander of 4-64 Armor**

"Dear heroes, soldiers... I know you're very long away from home and overloaded, and your departure of your dead people is unbearable and a heartbreaker. But all that is for all Americans. Today, we are blessed and lucky, since we all are led by a wise American leadership, ensuring us a decent, generous life for both American and Iraqi people, wherever we are. There is nothing but sacrificing, for our lives and for our families and children. To me, I will not forget and keep remembering the favor of those American soldiers who had protected my wife when she was pregnant, laying in the hospital. Many thanks to CPT Judge and all his soldiers who take care of me. They are really honest and loyal, well-bred soldiers. Thank you very much."

One bright, early morning before most of the city had stirred, a group of Tuskers, including soldiers from Bravo Company, climbed into three vehicles led by an M113 armored personnel carrier. Packed carefully inside the convoy were toys and a celebratory cake, destined for a local kindergarten class in Baghdad that was holding its graduation ceremony.

Riding in the front vehicle was none other than Donald Urbany, the now-recovered FiSTer who had nearly died back at Fort Stewart. After months of recovery, he had rejoined us and wasted no time in volunteering to help, eager to contribute in any way he could.

Though exhausted from lack of sleep, Urbany joined the public relations mission without complaint. Up to that point, he'd mainly been tasked with radio guard shifts and crater analysis. This was a chance to do something different, something that felt meaningful. As the convoy rolled out, he climbed into the open sky hatch of the M113 with an Iraqi interpreter and kept a watchful eye on the terrain. The roof hatch provided excellent visibility but came with obvious risks: standing exposed above the armor made soldiers vulnerable to enemy fire or explosions. Still, it was the best vantage point for spotting danger.

As they rolled out of the relative safety of the Green Zone and into the Red Zone, Urbany scanned the unfamiliar stretch of Route Wild, his senses sharpening. The morning sun flashed off the pavement when a lone car parked tight against the curb caught his attention. It sat buried in a dense neighborhood, pedestrians and buildings pressing in from both sides, but something about it felt wrong. As the convoy crept closer, entering Al-Nasr Square, he narrowed his eyes.

"Hey, I think that's a..." Urbany began, the interpreter turning toward him.

Before he could finish, the vehicle violently exploded, showering the area in a storm of red-hot shrapnel. The Iraqi interpreter screamed as the blast tore into her, nearly destroying one of her arms. It was left hanging by ruined tissue. Several fragments of metal punched straight through her hat, leaving large, ragged holes that somehow missed her face. She collapsed backward into the M113, wounded but still alive.

Urbany, too, was hurled back into the vehicle, disoriented and soaked in blood. Jagged fragments of metal had shredded the right side of his face and torso. His right eye was now a bloodied mess, a large piece of shrapnel buried deep inside it. He couldn't see and was heavily concussed, unsure of what had just happened.

The mission was over before it began. For Urbany, though, the battle was just beginning.

**11 January 2025 – Reflection of Donald Urbany, Bravo 4-64 Armor**

I was standing outside the hatch with the interpreter when we turned down a street we hadn't been on before. I spotted a vehicle parked along the side of the road. It was unoccupied, and as we got closer, I saw that it was filled with pieces of metal. Just random chunks of metal. Unsure, I started to say, "Hey, I think that's a..."

Then I remember a loud pop. The next thing I knew, I was back inside the 113. The interpreter was screaming, but I had no idea what had happened. I couldn't see, so I instinctively tried to reach for my face. Another soldier had to hold me back and punch me in the arm because I kept grabbing at a jagged piece of metal sticking out of my eye. It didn't hurt at all. I just felt really tired and wanted to go to sleep, but they wouldn't let me.

Immediately after the blast, a column of thick black smoke boiled up over Baghdad and hung in the air like a storm cloud.[209] Windows shattered more than a hundred yards away. Dozens of shops were ripped open, glass and debris were thrown into the street, and at least fifteen cars were either torn apart or were burning.[210] Arriving Iraqi Police fired their rifles into the air to drive back the crowd, trying to push the shouting, panicked onlookers away from the wreckage.[211]

The parked VBIED had detonated roughly between the lead M113 APC and a trailing Humvee. The Humvee suffered catastrophic damage, came to a stop, and began rapidly leaking fluids onto the pavement. Leaders on scene started pushing out dismounted infantry to secure the area. They encountered the carnage that filled the street, with twisted pieces of vehicles, shattered chunks of buildings, and human remains from nearby pedestrians scattered across the blast site.

---

[209] CBS News, "Two Car Bombs Kill 7 in Baghdad," May 10, 2005, https://www.cbsnews.com/news/two-car-bombs-kill-7-in-baghdad/.

[210] News Agencies, "Twin Blasts Rock Baghdad," *Al Jazeera*, May 10, 2005, https://www.aljazeera.com/news/2005/5/10/twin-blasts-rock-baghdad.

[211] Associated Press, "At Least Seven Dead in Two Blasts in Baghdad," AP Newsroom, November 26, 2007, https://newsroom.ap.org/editorial-photos-videos/detail?itemid=ecc42ce4f540b5c0c8337b57c9463b1f.

**8 December 2025 – Reflection of Phillip Cornell, Platoon Sergeant, Delta 4-64 Armor**

The blast hit and knocked out one of the Humvees in our formation immediately. The license plate from the VBIED actually flew up into the air and landed in one of our tracks. We had to stop because of the Humvee, so we pushed the infantry out on foot to secure the area while we assessed the damage. The interpreter was found in bad shape, and so was Urbany, but I have to give him credit, he tried to stay in the fight. He kept apologizing for getting hurt and wanted to keep going.

We quickly hooked one of the tracks to the front of the Humvee and dragged it out of the immediate kill zone as fast as we could, but we couldn't get very far with it. While this was happening, we were consolidating all the casualties into one track. A few others had been wounded in the blast, including SGT Robinson and 1LT Clausen, though none as badly as Urbany and the interpreter.

Clausen had taken a piece of shrapnel to the back of his arm, right through the tricep. It looked like someone had swung an axe into it. When he tried to help carry the interpreter on a stretcher, the added strain must have torn his muscle even further because he lost his grip, dropping her. Someone else jumped in, and we quickly loaded everyone up.

Before we scooped up the infantry and pulled out of the kill zone, just as the Iraqi Police were arriving, I saw people going about their morning business, stepping over the body parts of people who had just been blown apart. I could not believe what I was seeing. Then I just told the track with the casualties in it, "Go, get to the hospital!" The rest of us remained on scene for a short time until the police could take over.

As our soldiers broke contact, they made a direct push to the nearest hospital, urgency overriding everything else. In an interview nearly two years later, Urbany recalled the chaos and determination of that ride: "They were driving over cars and firing their weapons into the air to clear traffic. It was like Moses parting the sea."[212]

However, the severely damaged Humvee, which was hooked behind the APC, could go no farther. It was basically destroyed, with the front end shattered and the wheels refusing to turn, sparks shooting in every direction. Since it was just being dragged, the casualty track dropped it at a bridge near the entrance to the Green Zone. Shortly thereafter, follow-on forces secured it and brought it the rest of the way through the checkpoint.

Now only semi-conscious, Urbany repeatedly asked for water, but they wouldn't allow it. It was standard procedure in case of internal injuries. As the M113 arrived at the hospital, CPT MacKinnon, having heard what happened, immediately jumped into the back of the vehicle. He knelt beside Urbany, offering words of reassurance and telling him he would be okay. Then several medics helped rush him into the emergency room.

---

[212] Womack, Patrick. "Injured Iraq Veteran Starting New Life in Mebane." The News of Orange County, April 24, 2007. https://www.newsoforange.com/mebane_enterprise/news/article_84a19242-d420-5be8-86ff-d9b12c700ae3.html.

**The rear of the M113 APC carrying Donald Urbany, showing shrapnel scarring from the VBIED**

**Photo courtesy of Phillip Cornell**

From a nearby hallway, the agonized screams of the wounded interpreter echoed through the facility as she was treated for her injuries. Somehow, the doctors managed to save her arm, though it was left severely damaged.

**8 December 2025 – Reflection of Phillip Cornell, Platoon Sergeant, Delta 4-64 Armor**

When I arrived at the hospital, the interpreter was on a gurney and about to go into surgery. They had stabilized her, dressed her wounds, and given her painkillers. She called me over and asked me to find her purse.

I said, "Why do you need your purse? I'll make sure everything is secured."

And there she was, lying there with blood in her hair and looking like a mess, saying, "I want to fix my hair and make-up. I don't want anyone to see me like this."

Doctors also worked swiftly to treat Urbany in a scene that echoed his previous brush with death back at Fort Stewart, but this time, his injuries were no accident. Someone had tried to kill him. As the trauma team assessed the damage, they discovered, almost unbelievably, that the shrapnel had missed all major arteries and vital organs. Still, the damage was severe, with metal embedded in his face, right arm, and right midsection. He lost consciousness as surgeons began delicately removing the fragments embedded near his eye.

**11 January 2025 – Reflection of Donald Urbany, Bravo 4-64 Armor**

They brought me into the emergency room, and I remember that I couldn't see. I had an eight-inch piece of metal sticking out of my right eye. The chaplain was there, trying to offer encouragement as the doctors immediately went to work. They were poking and working around my eye, but I don't remember feeling any pain.

Then I felt something touch my hand. It was probably just a piece of gauze or something, but I freaked out. I was convinced my eyeball had just fallen out and landed in my hand. I passed out, and I don't remember anything after that.

Miraculously, no other Americans were killed or seriously wounded in the blast, though many suffered minor injuries, including a ruptured eardrum. After the incident, word spread quickly throughout the unit, and we anxiously awaited updates on Urbany's condition. Of all the people to be wounded, we thought, it was the guy who hustled to get over here after already being hurt.

**8 December 2025 – Reflection of Phillip Cornell, Platoon Sergeant, Delta 4-64 Armor**

Once they were in surgery, another soldier and I took the track around and found a wash rack. I didn't want it to become a tourist attraction. It was covered in blood and tissue, both around the hatches and inside the vehicle. We hosed everything out and threw away the items that had become contaminated.

I have to give a lot of credit to PFC Pendergrass, "Teddy P." He was the driver of that 113. He navigated that entire route quickly and got them to safety with no guidance, since the commo was trashed and the lieutenant's helmet was busted.

**10 May 2005 – Letter from Robert Roth, Commander of 4-64 Armor**

This morning, a B/4-64 AR patrol was enroute to a kindergarten graduation ceremony with toys and cake to celebrate. On the way to the ceremony, a car bomb exploded, and three soldiers were injured. Everyone will recover! However, one soldier was evacuated for higher-level medical care. The Rear Detachment has been notified, and they are contacting the families of those injured.

Also in this attack, several Iraqi citizens were killed, and a Bandit Company Iraqi interpreter was severely injured. Our prayers are with all of them and their families.

Tragically, at least seven civilians were killed in the blast, and dozens more were wounded.[213] The dead included two children, described as shoe-shine boys.[214] One news agency suggested the toll may have been as high as twenty-five.[215] Whatever the final number, the human cost that day was devastating along Al-Saadoun Road.

### 1 August 2025 – Reflection of Noor "Nora" Aljassim, Interpreter for the U.S. Military

My good friend was the interpreter injured in the attack. Her arm was shattered, and she needed emergency surgery. When they brought her to the American hospital, the staff didn't want to treat her there because she wasn't a soldier. They wanted to send her to an Iraqi hospital instead.

CPT MacKinnon yelled at them. He told the staff that she was going to be treated there like one of his soldiers. They did what he said. She had a metal bar and screws put into her arm to hold it together.

After the surgery, they wanted to transfer her to the other hospital again, but CPT MacKinnon refused. He made them keep her there for about a month, until she was finally able to be released.

Urbany survived the initial surgery in Baghdad and was soon airlifted to northern Iraq. From there, medics worked to keep him stable as he was transferred to Germany for advanced care. He spent a week in a military hospital there while doctors monitored his condition and prepared him for the next leg of the journey. He was also personally awarded the Purple Heart by General Peter Schoomaker, then Chief of Staff of the Army.[216] A transatlantic flight was risky, and they needed to be sure he wouldn't medically crash midair over the ocean. Once deemed stable, Urbany was flown to Walter Reed Army Medical Center in Washington, D.C., the Army's premier treatment facility at the time.

---

[213] IrishExaminer.com. "Seven Reported Killed in Baghdad Blast." *Irish Examiner*, May 10, 2005. https://www.irishexaminer.com/world/arid-30201825.html.

[214] Iraq Body Count. "6-7 by Suicide Car Bomb, Saadoun Street, Baghdad (K1278)." Accessed December 9, 2025. https://www.iraqbodycount.org/database/incidents/k1278.

[215] Murphy, Caryle, and Jonathan Finer. "Shiites Dominate Committee Chosen to Draft Constitution." The Washington Post, May 10, 2005. https://www.washingtonpost.com/archive/politics/2005/05/11/shiites-dominate-committee-chosen-to-draft-constitution/91d40773-1ad3-4e72-ad29-5e3c76472475/.

[216] DVIDS, "Army Chief of Staff Pins Purple Heart on Vanguard Soldier," Defense Visual Information Distribution Service, May 19, 2005, https://www.dvidshub.net/news/1877/army-chief-staff-pins-purple-heart-vanguard-soldier.

Though he had survived the ambush, the war was over for him. His military career had ended on a roadside in Baghdad. Now, he faced an entirely new battle: learning to live as a wounded combat veteran with the loss of one eye and the heavy weight of what he had endured.

**8 December 2025 – Reflection of Phillip Cornell, Platoon Sergeant, Delta 4-64 Armor**

For days after the VBIED attack, SGT Robinson kept sneezing out tiny shards of glass and plastic. In a blast like that, everything gets pulverized into fine particles you end up breathing in. That can't be good for you.

Then, not long after that attack, we got a tip about where the car bomb had come from. It led us to an IED factory that was cranking out vehicle bombs. We hit the place and found one car already halfway rigged. We grabbed the guy running it and turned him over to the Iraqi authorities. They were in no mood to be gentle with him. With kids and other civilians dead, they were already "working him over" in the back of the truck as they drove away.

**11 January 2025 – Reflection of Donald Urbany, Bravo 4-64 Armor**

I was at Walter Reed when CPT MacKinnon called me. He told me to hang in there and that I was a good soldier. He told me he was proud of me. I'll never forget that. Ever.

# Reflections

Donald Urbany's story echoes through countless generations of warriors, transcending both time and nationality. While his personal tale is obviously rooted in the American experience of fighting an insurgency in the Middle East, it reflects timeless themes of military service worldwide. At its core, it follows the familiar arc of a young soldier answering his country's call, doing his best to serve with honor, despite the omens that suggest he may be headed for tragedy.

Erich Maria Remarque's semi-autobiographical masterpiece, *All Quiet on the Western Front*, remains, in my opinion, the pinnacle of war literature. Set amid the maddening brutality of World War I trench warfare, the novel follows Paul Bäumer, a young German who enlists at his trusted schoolteacher's urging. From the very beginning, subtle signs hint that the romantic vision of war he's been sold will soon unravel. A creeping sense of dread eventually gives way to overwhelming and demoralizing destruction. By the novel's end, the haunting phrase "All quiet on the Western Front" evokes not peace, but a final, exhausted silence.[217] It's a merciful end to the horror.

Unlike Bäumer, however, Urbany survived the war. But looking back, the warning signs were all there before we even deployed. The accident in the motor pool might have been a fluke, or maybe it was something more. I can't help but wonder.

Still, rather than take the easy road and remain at Fort Stewart, Urbany pushed forward. As soon as he was physically cleared, he rejoined our unit in Iraq. It was a choice he didn't have to make, and no one would have blamed him for staying behind. That he didn't speaks volumes. It was an act of courage, commitment, and something else, too. It was an act of brotherhood.

For his troubles, Urbany spent the next 19 months as an outpatient, slowly healing and grappling with an uncertain future. With one eye, visible scars, lingering physical pain, and invisible wounds that would never fully fade, what was he supposed to do next? In an interview for this book, he admitted to relying too heavily on painkillers during that time, drifting through the days in a groggy haze, unsure of where life would take him. He fell deeper into depression.

Then came a heartbreaking yet unforgettable moment in Urbany's life. It is one, he says, that will stay with him forever.

**11 January 2025 – Reflection of Donald Urbany, Bravo 4-64 Armor**

I was feeling low and depressed about my situation. I missed the guys, and I missed being overseas with them. It felt empty here.

Anyhow, I ended up being paired with a Marine who had lost both of his arms in an explosion. One day, as we were sitting together, I heard, "Daddy!" I looked over and saw this little kid running up to the Marine. All he could do was awkwardly lean over as the little kid gave him a big hug.

---

[217] Remarque, Erich Maria. *All Quiet on the Western Front*. Ballantine Books, 1987.

I realized right then that this man would never be able to hug his baby again, yet he was still so happy just to see his kid. Instantly, it put everything into perspective and hit me hard. *What did I have to complain about?*

**Donald Urbany with General Peter Schoomaker, receiving his Purple Heart**

**Photo courtesy of Donald Urbany**

Although I didn't know Urbany all that well during our time together in Bravo Company, I remember him as a young, friendly guy who was always smiling. After the car bomb, I mostly lost track of him until years later. His wounds had healed, but from his scars, anyone can instantly see that he has been through war. Yet instead of letting those injuries define or limit him, as so many tragically do, he found a new purpose in life. As unbelievable as it might seem, he joined a local hockey team. He then took on one of the most demanding roles on the ice: goalie. All this he did, and continues to do, with one eye. His wife later published an article detailing his journey from FiSTer to hockey player.[218] It's an incredible story.

The takeaway for me is simple: if Urbany can lose an eye and still suit up as a goalie, what excuse do I have? It's the same hard truth he once learned from the Marine. Every time I revisit

---

[218] Urbany, Jennifer. "Interview With Donald Urbany." Cross Creek Cats, April 27, 2021. http://crosscreekcats.org/interview-with-donald-urbany.

that article or call him, it reminds me to quit complaining and keep going. Life hands all of us set-backs. We can either let them become crutches, or we can dig in, push through, and discover what we are still capable of becoming. Donald Urbany taught me that.

When I asked him how he coped with Post-Traumatic Stress Disorder (PTSD), his answer surprised me. "I have four children," he said. "I can't fail." That was it, simple and resolute.

We can let the past consume us, or we can move forward and lift others as we go. While being a victim is strangely applauded today, we must understand that we can still be the heroes in our own stories. Although we may stumble along the way, I choose the latter, and so does Donald Urbany.

A month after the incident, Urbany wrote to CPT MacKinnon through the FRG, expressing heartfelt thanks for his words of encouragement. His message reflected the deep bonds we had formed and how much we had all grown together.

**16 June 2005 – Letter from Donald Urbany, Bravo 4-64 Armor**

Sir, thank you for your kind words. Not only am I a Tusker, I'm a Bandit. I learned that no matter what, whether we are Field Artillery, Medics, or Infantry, we are a family. I will never, ever forget the times I endured with you all. I thank you for everything you have done! God bless you all!

BANDITS FOR LIFE!

We both vividly remember the first half of 2005. It was marked by the constant sound of explosions, car bombs, and IEDs thundering across Baghdad. Sustaining the fragile Iraqi government was a crucial and extremely challenging task. We were fighting to prevent the country from descending into complete chaos, and had we failed, the consequences would have been catastrophic. Looking back, I'm still astonished that we managed to keep things functioning at all, especially considering the unreliable equipment and sometimes the subpar Iraqi soldiers we had to work alongside.

Several years ago, a wave of corporate media reports exposed the widespread use of fake bomb detectors overseas. One of the most infamous was the ADE 651, a so-called bomb detection device manufactured and distributed by a British company, Advanced Tactical Security & Communications. These devices were widely deployed by Iraqi military and police forces at checkpoints and on patrols in the early years of the war.

The ADE 651 featured a rudimentary design, including a black plastic handgrip with a swiveling silver antenna that resembled the kind found on old cars. It had no power source and supposedly "charged" through the user's movement. The manufacturer claimed it could detect explosive materials by tuning into their frequencies, as if by some kind of mystical resonance. However, there was no scientific basis for any of it.

Despite its apparent flaws, tens of millions of British pounds were spent on these worthless devices, and they became a common sight on the battlefield. Iraqi soldiers routinely waved them over vehicles and pedestrians, falsely believing they were being screened for explosives. While I haven't been able to find any personal photographs of it, I'm nearly positive we had one of these devices, or something similar, at one of our checkpoints, though I cannot say that with certainty. It was, in reality, no more effective than waving a magic wand. Tragically, lives depended on this sham.

The device's inventor, James McCormick, was eventually prosecuted by the British government for fraud, and the UK imposed an export ban.[219] Unfortunately, the damage had already been done. Even after the ban, Iraqi forces continued using the ADE 651, unaware or unwilling to accept that it was a complete scam. Some continued to swear by their validity.

The ADE 651 is just one example of the many scams born from war. How many people were killed or maimed because of this device and others like it? We'll never know. What is certain is that war attracts charlatans, those who exploit chaos and desperation for profit. It's a sickening thought, especially to those of us who stood at those very checkpoints, watching lives depend on a lie. As I mentioned before, we were, quite literally, playing Russian Roulette every single day.

Similarly, when I think back on those explosive test kits we were issued, I still wonder, did they actually work? Or were they the equivalent of the ADE 651? Honestly, I don't know. We had to trust that the equipment we were given was reliable and that someone, somewhere, had properly vetted it.

When we did receive legitimate gear, like the Buffalo MRAP, we made full use of it. The problem was that there was never enough to go around. And even then, no piece of equipment could replace good old-fashioned grunt work on the ground. Technology helped, but in the end, it still came down to boots on the pavement and eyes scanning for danger.

In retrospect, I remember that time with a sense of astonishment. As I mentioned earlier in the chapter about my first deployment, I never truly felt fear. Sure, my heart pounded, the adrenaline surged, and I ran faster than ever during tense moments, but *genuine* fear? It never really registered. Strangely, I'm more afraid now of what I did back then than I ever was in the moment. I know that might sound nonsensical, but at the time, fear simply didn't occur to me. It was as if my mind didn't allow room for it.

There is a quote I have seen time and again, though no one seems to know who authored it. Some say Guy de Maupassant wrote it, and others suggest it comes from a Special Forces soldier or a Marine. Perhaps it doesn't really matter. It rings true, regardless of the source. "You have never lived until you have almost died. For those who fight for it, life has a flavor the protected will never know."

---

[219] Hawley, Caroline. "The Story of the Fake Bomb Detectors." BBC News, October 3, 2014. https://www.bbc.co.uk/news/uk-29459896.

# In Memoriam

---

SSG Victor Manuel Cortes III[220]
2 May 1976 (Erie, Pennsylvania) – 29 May 2005 (Baghdad, Iraq)

---

[220] Military Times, "Army Staff Sgt. Victor M. Cortes III," Honor the Fallen, https://thefallen.militarytimes.com/army-staff-sgt-victor-m-cortes-iii/888302.

# Chapter XIII

## Samurai Swords, Extortion, and Dysfunction

Approaching the midpoint of our anticipated year-long deployment, everything seemed to be falling into place. We had found our rhythm. We had also grown confident in our roles and built a strong track record of success. The Gambler Platoon was constantly executing countless raids, capturing insurgents, and seizing cash, weapons, and bomb-making materials. But just as we hit our stride, word came down that our mission was about to change.

Before that transition, though, I had finally earned enough trust from my superiors and accumulated enough promotion points to go before the board for advancement to the rank of sergeant (E-5). Advancing from E-1 to E-4 had been relatively straightforward. Simply stay out of trouble, demonstrate competence, and serve the required time in each rank. But the jump to sergeant, a noncommissioned officer role, was different. It required passing a formal promotion board composed of several first sergeants and the battalion command sergeant major.

Bravo Company's 1SG Barnes would be the only "friendly" face in the room, though even he wouldn't go easy on me. He had no interest in promoting weak NCOs into his ranks. Meaning, I had better do well, or else…

In my limited free time, I had buried myself in the appropriate textbooks and reference guides, determined to master every aspect of my profession. Several NCOs in my platoon also took time to prep me. This was done partly to help and partly to avoid the embarrassment of sending someone in unprepared. The board was specifically designed to apply pressure, to see how well a candidate could think and respond under stress.

Outside the door, I stood in my sharpest, most squared-away uniform. I knocked loudly and was granted permission to enter by CSM Stanley, the bulldog tanker who'd given me his coin the year before. Marching crisply into the room, I took my position at the front of the table, where the four first sergeants and command sergeant major watched me like hawks, scrutinizing my bearing, uniform, and composure.

One by one, they drilled me with questions. They wanted to know specifics about my duties as an infantryman, my combat experience in Iraq, and the kinds of leadership knowledge expected from an NCO. Those days spent as a team leader and squad leader began to pay off. I was also quizzed on current events in the news, required to awkwardly wail the Army Song, and had to demonstrate my familiarity with multiple weapons systems. It was all business, no wasted time.

To my relief, I answered nearly every question confidently. If I didn't know the answer, I didn't try bullshitting them as they would have seen right through the deception. I admitted my flaw, told them I would remedy the situation, and moved on to the next question.

It became clear early on that I was well-prepared and that they were comfortable advancing me. The entire assembly probably lasted no more than ten minutes, much to my surprise. I was prepared for a grueling session, but I suspect my preparation and confidence won the day early on.

The only part that made me squirm was singing the Army Song. My beautiful singing voice, or lack thereof, left much to be desired. You may wonder why aspiring sergeants are asked to sing a song. It's partly a test of memorization, just one more thing expected to be committed to memory. More importantly, it's about discomfort. Singing is nerve-wracking for most people, and doing it in front of a row of scowling faces only makes it worse. Pressure, pressure, pressure. I know they found it amusing, but not one of them broke character. Instead, they remained stone-faced and professional throughout.

To reinforce my desire to grow, I made a point of recalling the questions I had stumbled over. Digging through the textbooks later in the day, I then found the information and committed those elusive answers to memory. Next, I returned to CSM Stanley and shared the corrected responses with him. I believe that small act left a lasting impression.

Shortly after passing the board, I was officially promoted to sergeant in the United States Army, finally earning those coveted stripes I had worked so hard to achieve. It was one of the proudest moments of my military career, and wearing that rank brought a profound sense of accomplishment. However, it also came with a weighty sense of responsibility, not just to the men in my platoon, but to the standards and expectations that came with being an NCO.

I asked SSG Brister, a leader I deeply admired, to be the one to affix my stripes. He put the rank on my collar with a smile. I had finally joined the ranks of sergeant, making me think back to that early lesson at Fort Stewart, "In a year or two, you will probably be in my position and will likely have your stripes if you keep working hard." He was right.

The promotion ceremony took place in a secluded building, away from any prying eyes or easily offended onlookers, and definitely away from the lower enlisted soldiers. In that makeshift setting, those of us entering or rising within the NCO Corps endured the time-honored ritual of "blood-pinning."[221]

By the late 1990s, the services had formally banned blood-pinning and similar hazing rituals in policy.[222] So, of course, I will never reveal who was there or who participated. In fact, *no one* participated. Let's just say I did it to myself, which was purely symbolic, of course.

---

221 Madrid, Michael. "Military Tradition Banned: The Shocking Reality of Blood Wings." Veteran Life, July 29, 2022. https://veteranlife.com/military-history/blood-wings.

222 Matthews, Miriam, Kimberly Curry Hall, and Nelson Lim. "A Commander's Guide to Hazing Prevention." RAND Corporation. Accessed December 8, 2025. https://www.rand.org/content/dam/rand/pubs/tools/TL100/TL168/RAND_TL168.pdf.

**The aftermath of our pinning ceremony - one of the greatest moments of my life**

The tradition called for the sharp metal prongs of our newly earned rank to be pressed and then punched into our chests. Anyone already holding the rank of sergeant or higher was invited to take part. One by one, they stepped forward, shaking our hands and then striking the pins with enough force to leave behind a pair of bruised, bloodied welts. Some took it easy on us, though others took great delight in throwing haymakers. The most painful part wasn't the punching itself, but removing the prongs afterward, now deeply embedded in the skin just under the collarbone.

While this may sound barbaric, especially to an outsider not associated with the military, especially the infantry, it was an essential rite of passage. It forged camaraderie in war through shared pain. Meaning, it wasn't so much about hazing. Like being taped to that tree in the woods, it was about earning your place. To me, the ritual symbolized our willingness to bleed for each other and for our men. We all went through it, and it was a declaration of loyalty, honor, and brotherhood. I don't expect everyone to understand it, but I do hope they can appreciate what it meant to us. And yes, given the chance, I'd do it all over again.

Now, some VIP with a lot of stars on their collar will probably read this and declare that what we did was completely unacceptable. That's fine. He can keep his opinion and his policy binder. I will keep the tiny scars on my chest and the memory of the men standing next to me, and I still don't give a shit what anyone else thinks about it.

Months after I had been promoted, I was walking through the FOB on my way to chow when a man on a bike came screeching to a halt, stopping next to me. "Holy shit!" I heard him exclaim. I looked over and saw a familiar, friendly face. It was SFC Hassel, my old platoon sergeant, whom I hadn't spoken with since our days on Rear D.

**8 October 2005 – Letter to the Harstad Family**

I just ran into my old platoon sergeant from Rear D. He hadn't seen me since I was a private in his platoon. I was walking past him, and he saw that I was a sergeant, and he about shit his pants. He screeched to a halt on his bike and ran over to me, shaking my hand.

It made me feel really good. And I'm sure he was proud one of his soldiers went on to have success, and that he had made a difference (in helping me get there).

Back at FOB Prosperity, we were assigned an additional rotating duty of guarding the Al-Rasheed Hotel and the numerous VIPs residing there. Formerly a luxurious hotel under Saddam's regime, it became the go-to place for foreign diplomats, high-ranking military officials, wealthy contractors, and intelligence agencies. At the entrance, Saddam had a mosaic of President George H.W. Bush installed so that visitors would trample over his face upon checking in, but Americans removed this before I had a chance to see it for myself.[223]

**1 August 2025 – Reflection of Noor "Nora" Aljassim, Interpreter for the U.S. Military**

My mom's friend was Layla al-Attar, an artist and painter. One day, men from Saddam's regime came to her house and told her she had been "hired" to create a mosaic of President Bush for the Al-Rasheed Hotel. She couldn't say no. She designed the mosaic for the front entryway, knowing that in Iraq it is deeply disrespectful to walk on someone's image, which is exactly why Saddam wanted it there.

In 1993, President Clinton ordered missile strikes in Iraq, and one of the missiles hit her house. Layla was killed, along with her husband. One of her daughters survived but needed extensive plastic surgery.[224] Many Iraqis were furious and accused Clinton of killing her because of the Bush mosaic.[225]

---

223 Morrison, Kenneth. "The Al Rasheed Hotel: 40 Years of an Iconic Baghdad Landmark." *Al Jazeera*, November 10, 2022. https://www.aljazeera.com/features/2022/11/10/the-al-rasheed-hotel-40-years-of-an-iconic-baghdad-landmark.

224 Ibrahimi Collection Cultural Enterprise. "Layla Al-Attar (1944-1993)." Accessed December 8, 2025. http://www.ibrahimicollection.com/node/114.

225 Kandel, Didhiti. "The Empowering Legacy of Iraqi Artist Layla Al-Attar." Arts Help. Accessed December 8, 2025. https://www.artshelp.com/layla-al-attar/.

**Overlooking Baghdad from the roof of the hotel**

At the hotel, I had my first encounter working alongside soldiers from Georgia, the country, not the state with onions and Waffle Houses.[226] Hearing them talk, I initially thought they were speaking Russian. So, in my infinite cultural wisdom, I asked one of them, "Hey, are you guys Russians?"

Big mistake. One of them, who spoke broken English, gave me a death glare and snapped, "No, we *aren't* fucking Russian dogs," before spitting on the ground with a sneer.

As I discovered, Russia and Georgia didn't exactly get along very well. In fact, they were simmering with tension that would boil over into a full-on war just a few years later in 2008.[227] So, I

---

[226] Carney, Stephen and Center of Military History. "Allied Participation in Operation Iraqi Freedom." Library of Congress, 2011. https://tile.loc.gov/storage-services/master/gdc/gdcebookspublic/20/23/69/31/56/2023693156/2023693156.pdf. Page 63.

[227] Pruitt, Sarah. "How a Five-Day War With Georgia Allowed Russia to Reassert Its Military Migh." History, August 8, 2018. https://www.history.com/articles/russia-georgia-war-military-nato.

smiled awkwardly, nodded like I totally knew that, and slowly backed away from my international incident. Good talk.

In all seriousness, once we skipped over that little speed bump, they turned out to be great soldiers. We got along surprisingly well, communicating through a combination of broken English, exaggerated hand gestures, and the universal man language of describing beautiful women and girlfriends back home. Nothing bridges a cultural divide like air-drawing curves and saying, "Yes, very pretty."

**16 October 2005 – Letter from Robert Roth, Commander of 4-64 Armor**

The Georgians were switched out by another Georgian Army battalion, so we started over, training this new battalion on how things work in the IZ. And do you think there would be more translators to help with the language barrier? Of course not. In fact, it is worse than before.

The BEAST company commander went down to check on the ECP this past week, and no one was on duty who spoke English. So, using sign language, he was able to find out that the Georgia lieutenant on duty spent many years in Germany. Well, so did CPT Emenheiser. So, they spoke German to communicate.

Soldiers then began taking designated leave, or R&R. Some went to luxurious spots in the Middle East, such as Dubai, but most returned home to the United States. This schedule created manpower issues and was part of the reason why I had been made a temporary squad leader. Nonetheless, it was a brief reprieve from the madness of Iraq.

**17 July 2005 – Letter from Robert Roth, Commander of 4-64 Armor**

If I can share one thought from a Tusker soldier whom I had the pleasure to talk to the other day, it is an awesome story. The soldier had just returned from R&R leave. I asked him if he missed us, and the response I received is what I hear every time I ask a returning soldier.

"Sir, I didn't miss Iraq, but I missed my squad. I needed to get back so I could be with them and complete the mission, so we can all come home together."

To mark the midpoint of our deployment, the Tusker battalion was granted a few days off for some much-needed rest and what the brass affectionately called "mandatory fun days." Surprisingly, those days lived up to the name. We spent our time swimming, downing absurd amounts of candy, and feasting on steak and lobster like kings in desert camouflage.

The battalion also hosted a friendly competition between the companies, featuring soccer matches, foot races, and other physical challenges. Much to our satisfaction, and maybe a little to everyone else's annoyance, Bravo Company took home the gold. CPT MacKinnon, as usual, was beaming with joy and had a perpetual smile.

Standing in a massive battalion formation near the palace, LTC Roth presented the trophy to CPT MacKinnon. What happened next would go down in 4-64 Armor lore. CPT MacKinnon grabbed the trophy, hoisted it high above his head like a gladiator, and took off running around the entire formation, grinning like a kid who had just unwrapped the best present on Christmas morning.

With his wide, mischievous smile, he slowed just enough to parade the trophy in front of each company commander, shaking it triumphantly, rubbing it in their faces, and savoring every second of the moment. Bravo Company erupted with laughter and cheers. That was *our* captain, and he was loving it.

**LTC Roth and CPT MacKinnon – around the mandatory fun day celebration at 4-64 Armor**

**Photo courtesy of Phillip Cornell**

**1 July 2025 – Reflection of Mark Barnes, 1st Sergeant of Bravo 4-64 Armor**

I used to run into soldiers from the engineer company years after that (the battalion competition), and they always tried to talk shit. I'd say, "Remember that day CPT MacKinnon rubbed the trophy in y'all's faces?"

**1 August 2025 – Reflection of David Anderson, Commander of Charlie 4-64 Armor**

You guys (Bravo Company) dominated over in 3-7 Infantry, so when you came to 4-64 Armor, we were always competing to see who was best. It was a great rivalry between Bravo and Charlie, and Mike MacKinnon and I were always fighting for bragging rights.

Unfortunately, the celebration couldn't last forever. We eventually had to return to the grind in the desert. Still, those few days were a rare and much-needed morale boost. Looking back, they were among the most enjoyable moments of the entire deployment and my entire career.

Two significant changes loomed on the horizon. First, CPT MacKinnon was set to relinquish command of Bravo Company on July 8th, with CPT Dugan slated to take his place. We truly hated to see MacKinnon go, especially in the middle of the deployment. We had grown so close under his active leadership, but we were proud to send him off with one final victory, the battalion trophy, as a parting gift before he moved on to his new staff role. CPT Dugan was certainly a capable infantry officer and a good guy, but we all knew things wouldn't feel quite the same.

The second change hit even closer to home. Our steady, no-nonsense platoon sergeant, SFC Hill, was also rotating out to a new position within the battalion. In his place, SSG Serrano, affectionately known as "Papi," was about to pin on E-7 and take charge of the platoon. As with MacKinnon, we hated to see Hill go, but there was comfort in knowing we'd be led by someone we respected. Papi had taken on a fatherly role for many of us, myself included, so we were excited to see the Gambler Platoon's new trajectory.

**20 July 2005 – Letter from Harold Hill, Platoon Sergeant, Bravo 4-64 Armor**

I have had the pleasure of not only leading these fine young men you call sons and husbands, but also of calling them my friends, brothers, and Soldiers. They are our family, and we will forever be linked to one another.

**21 July 2025 – Reflection of Juan Serrano, Platoon Sergeant, Bravo 4-64 Armor**

On that deployment, you were 21, but I was 45. My oldest kid was the same age as you. I think the next oldest guy in the platoon was 30 or 31, so I saw all of you as my kids. You were young and inexperienced, and I felt responsible for protecting all of you.

On the third anniversary of my enlistment, I wrote home, doing my best to express my appreciation for my family's support. Along with that, I tried to explain the reasons behind my continued resolve to stay in the fight.

**31 July 2005 – Letter to the Harstad Family**

Just wanted to say thanks for all the support you have given me these few years I have been away. I know it's been especially difficult on you, and I wish that I could come home every week-end. But that's not why I'm doing this (serving in the military). You have to understand that there is a cause much more important than just me.

That is not to say you are not important to me, but in order for my future kids and grandkids to be safe, I have to do something to secure that. Generations of men and women have fought, bled, and died to make sure we can enjoy the small things in life. It is my turn to fight.

Just after reaching my third year of service, I was awarded the Army Good Conduct Medal. To earn it, I had to complete at least three years of honorable service, free from disciplinary actions such as Article 15s, and receive a recommendation from my company commander. CPT MacKinnon gave me that recommendation. While the award isn't exactly prestigious, it meant a great deal to be recognized once again by a leader I deeply admired and wanted to emulate.

While I look back on this time with great fondness, not everything was going smoothly for everyone. Just south of us at FOB Falcon, units were engaged in what we would soon learn was a fight for their lives. This included Donald Cyr's unit, and I hadn't seen or heard from him in months.

Although I wouldn't hear the following story until after the war, it perfectly encapsulates the kind of chaos we were about to face. During one mission, Cyr's platoon was tasked with establishing the outer cordon during a raid. This included blocking escape routes for any "squirters," people attempting to flee, and providing cover for the assault teams moving in.

**24 June 2025 – Reflection of Donald Cyr, Alpha Company, 3-7 Infantry**

We were running raid after raid, and everyone was exhausted. Our platoon was on cordon security while another platoon conducted the hit in Baghdad. A sergeant and I hunkered down near a house to watch the distant activity, with the plan that the assault platoon would pick us up on their way out. Somehow, the plan got messed up. Everyone left, unaware that we were still there alone, and our radio wouldn't work.

As it started to get dark, we hopped a fence and quietly climbed onto a roof, trying to stay silent and out of sight. Hours went by, and all I could think about was someone cutting my head off on TV while my parents watched. We sat there without moving as people walked around below us, completely unaware we were overhead.

Just before sunrise, we finally heard our company sweeping back through the area with bullhorns, calling our names and trying to find us. After they found us and brought us back, I was extremely paranoid, holding a pistol between my legs. Following the war, I buried that memory so deep I had almost forgotten it, until someone asked me about it years later, and it all came rushing back.

Even within the relative safety of the Green Zone, some soldiers were struggling. Some more than others. While I don't know every detail of the following incident, I've confirmed its general accuracy through multiple senior sources. Still, certain specifics may be off, as I am writing it based on what I was told at the time. When I re-read my own letters years later, I almost thought I had fabricated the story, because I have virtually no memory of it.

**8 August 2005 – Letter to the Harstad Family**

There was some soldier (not in the battalion) that was on tower guard a few weeks ago. He got drunk and stole a Hummer, and managed to drive into the "Red Zone." He drove fast as fuck and crashed into a wall, killing himself. Anyhow, the Hadjis stole his weapon, his gear, and all the equipment in the Hummer, then took off. They obviously didn't stop to help him.

I want to address two important topics related to the previous entry. First, I used the term "Hadji," a quasi-racial slur used in a derogatory context. But I will examine that more thoughtfully in the Reflections section of this chapter. Second, it's essential to review the General Orders that were in effect during that period, as they help frame the environment and expectations under which we operated.

General Orders issued by U.S. Central Command and its subordinate headquarters laid out a long list of prohibited actions and items in the war zone. The rules ranged from genuinely serious to what felt nitpicky and absurd. As an example, under the version in effect during our deployment, it was technically a violation to adopt a stray dog or cat, make a pet out of any animal, or even feed or water it.[228] Break General Order Number 1, and you could be charged under Article 92 of the Uniform Code of Military Justice (UCMJ) for disobeying a lawful general order.

Thankfully, our chain of command showed a little decency and chose to look the other way on that part. For a while, our platoon had a couple dogs named Rocket and Scully, who eventually went to live with one of our interpreters. The company also had a lazy pet donkey, though I can't remember his name, and he lived in relative luxury on the FOB, eating more fresh food than he had probably seen in his entire life.

The same order banned alcohol, illegal drugs and the misuse of prescription meds, pornography, gambling, having relations with the locals, privately owned firearms and explosives, entering mosques without proper authorization, and taking personal war trophies, among other things.[229] On top of that, commanders enforced strict movement control policies that limited when and how anyone could travel off-base without authorization, which was basically never. A lot of this was

---

228 Clark, Alicia. "You're Gonna Get Caught …General Order 1B Is in Full Effect!" Defense Visual Information Distribution Service, August 20, 2008. https://www.dvidshub.net/news/22678/youre-gonna-get-caught-general-order-1b-full-effect.

229 Department of the Army. "General Order Number 1 (GO-1)," November 1, 2005. https://img.slate.com/media/1/123125/123073/2133676/2150683/GO%201%201%20NOV%2005.pdf.

common sense if you wanted to stay alive and keep the host nation reasonably happy, even if some didn't always follow the rules to the letter.

**With Scully at FOB Prosperity: future insurgent killer and, per general orders, "not" one of our mascots**

News articles and personal memoirs from the special operations world have begun to describe a quiet culture of self-medication from that era, with some troops turning to alcohol and prescription sleep aids to cope with chronic pain and the grind of repeated deployments. Delta Force veteran Tom Satterly, for example, has talked about washing down Ambien with beer while he was downrange.[230] I don't mention this to single them out for criticism. I mention it only because, from the perspective of regular Army units, it often felt as if General Order Number 1 was enforced ruthlessly on us. While, at the same time, similar behavior in more elite circles was often handled behind closed doors or simply ignored like our pet policy.

In a way, I'm glad it was enforced so harshly on us. I know for a fact most of us would have been slamming beers and popping pills if we'd been given anything close to official permis-

---

[230] Satterly, Tom, and Steve Jackson. *All Secure: A Special Operations Soldier's Fight to Survive on the Battlefield and the Homefront.* eBook edition. Center Street, 2019, Chapter 16.

sion. Instead, we lived on legal stimulants. We pounded Red Bulls, chewed caffeine pills like candy, and some guys smoked and dipped so constantly it was hard to picture them without a cigarette or a wad of tobacco in their mouth. I'm convinced the U.S. military helped pay for more than a few yachts and vacation homes with all the money we poured into that stuff.

So, if my foggy memory and the letter I wrote at the time are accurate, that single incident involving the stolen Humvee included multiple violations of General Orders. Fortunately, events like that weren't common, but they did happen from time to time. When they did, the military typically acted quickly and quietly, aiming to resolve the matter without drawing public attention or risking embarrassment.

Amid all this, the rumors we'd been hearing for the past few weeks finally came true. We were officially being relocated from FOB Prosperity in the Green Zone to FOB Falcon in southern Baghdad. Once again, it was the Gambler Platoon that was being shuffled around.

### 22 August 2025 – Reflection of Matt Turcotte, Platoon Leader, Bravo 4-64 Armor

CPT MacKinnon pulled me aside and told me he was attaching Gambler Platoon to the California National Guard battalion down south at FOB Falcon. They had been having a lot of problems, and I didn't like the sound of it at all. He said, "I need my ace platoon to go down there and help clean that place up. It's going to be rough, but I know you guys can handle it."

But before diving into the chaos that followed, a bit of backstory is necessary to understand the drama that quickly unfolded. In August 2004, the Army mobilized the 1st Battalion - 184th Infantry Regiment, a National Guard unit based in Modesto, California. A bit of an experiment was underway to determine whether Guard and Reserve units could help relieve pressure on active-duty forces, who were being stretched thin by the relentless deployment cycles between Iraq and Afghanistan.[231] For many of us on active duty, the future seemed to hold a predictable, potentially unsustainable pattern. That being one year deployed, one year home training. Naturally, this raised serious concerns for soldiers with families, and many contemplated leaving the military altogether.[232]

So, after several months of pre-deployment training, 1-184 was sent to Iraq in early 2005 and attached to the 3rd ID. They took up residence at FOB Falcon and almost immediately found themselves in serious trouble on several fronts. On one side, they were handed a notoriously violent slice of Baghdad, a buzz saw where IEDs, ambushes, and sectarian killings were a regular part

---

[231] Foss, Adriane. "National Guard CSM to Urge More Overseas Training, Talks About the Reserves' Changing Role." National Guard, May 8, 2008. https://www.nationalguard.mil/News/Article-View/Article/573235/national-guard-csm-to-urge-more-overseas-training-talks-about-the-reserves-chan/.

[232] RAND Corporation. "Military Actions in Iraq and Afghanistan Have Stretched U.S. Army Thin." *RAND Corporation*, July 13, 2005. https://www.rand.org/news/press/2005/07/13.html.

of the landscape. On the other, the battalion began racking up disciplinary and conduct problems, the kind of issues that hinted this wasn't just a case of bad luck or a tough neighborhood.

Commanding the battalion was LTC Patrick Frey, a special-education teacher by trade and, back in California, a decorated "Teacher of the Year" who was widely regarded as both charismatic and intensely patriotic. A Vietnam veteran who later fought as a mercenary in the Rhodesian civil war, he eventually built a second career in the classroom before rising through the ranks to take command of the California Army National Guard's 1-184.[233] By 2004, he was leading around 700 of these so-called "weekend warriors" into what was supposed to be a year-long combat tour in Baghdad, the centerpiece of that broader experiment to push National Guard units from the rear into the front lines.

Even before they joined us overseas, serious cracks were showing in his command. Some soldiers swore they would follow Frey anywhere and described him as a gifted, old-school commander. Others saw him as erratic, theatrical, and dangerously self-absorbed, and they went so far as to complain to reporters about poor training and inadequate preparation for IED threats during the mobilization phase in New Mexico.[234] Army records from that period even quote one general, after a clash at a combat training course, warning that Frey would "get soldiers killed," a judgment that captured the deep unease some senior leaders felt about his style.[235]

Years later, Frey himself would say that the Army singled out his battalion as a "case study in dysfunction," a label that matched the rumors we had already heard before we made the jump to Falcon.[236] I never met Frey in person, so I can only rely on what was being said in our circles at the time and what has since come out in news reports. Opinions about him still span the full spectrum, from admiration to outright contempt, and even now, most people who were there seem reluctant to talk about just how bad things got.

As their casualty numbers began to climb, stories started circulating to surrounding FOBs, including our own, about serious misconduct inside the battalion. Some of it sounded so outrageous that I initially assumed it was just bored private rumor-mill material. One story in particular, about a 1-184 soldier walking the streets of Baghdad with a Samurai sword, seemed too ridiculous to be true. Later, I discovered it had been reported in the press almost exactly that way. According to a Los Angeles Times report, one of Frey's soldiers really had been caught patrolling with a Samurai sword swinging from his belt.[237] From the perspective of a trained grunt, the sword was pure

---

[233] Tempest, Rone, and Scott Gold. "Controversy Surrounds California Guard Officer." *Los Angeles Times*, August 1, 2005. https://www.latimes.com/archives/la-xpm-2005-aug-01-me-guard1-story.html.

[234] Tempest and Gold, "Controversy Surrounds California Guard Officer."

[235] Gold, Scott. "A Look Inside a National Guard Unit Known for Valor, Dysfunction." *Los Angeles Times*, September 14, 2011. https://www.latimes.com/archives/la-xpm-2011-sep-14-la-me-national-guard-iraq-20110914-story.html.

[236] Gold, "A Look Inside a National Guard Unit Known for Valor, Dysfunction."

[237] Gold, "A Look Inside a National Guard Unit Known for Valor, Dysfunction."

theater, likely to intimidate Iraqis, as it served no real purpose. It added weight, snag hazards, and almost no combat utility. We were fighting in Baghdad, not reenacting a scene from feudal Japan.

The stories only got worse. In addition to some of the soldiers engaging in turf wars with other units, one guy was found in possession of a so-called "death-list" that included some of his own people, not just enemy targets.[238] We also began hearing that elements of the battalion were shaking down Iraqi-run businesses in the Green Zone. Again, reporting by the Los Angeles Times confirmed that soldiers from 1-184 had approached several Iraqi-owned shops and demanded off-the-books "rent" to raise money for a "soldiers fund." Army spokesmen put the total at around $4,000, while soldiers inside the battalion told reporters the scheme had brought in more than $30,000. Several of those same soldiers described the arrangement in plain terms as "extortion," even though official statements shied away from that word.[239] Whatever label you put on it, the effect was the same. We were supposed to be protecting Iraqis, not running a two-bit "protection" racket that made us look like a uniformed mafia.

The unit seemed to stagger from one crisis to the next. At one point, the brigade learned that SPC Jorge Estrada, a 1-184 Guardsman on emergency leave for the birth of his wife's daughter, had been shot to death in a California parking lot. News stories later filled in the details: the baby's biological father, her former partner, confronted Estrada outside their apartment and opened fire, killing him in front of his family.[240] It was the kind of senseless, stateside killing that never shows up in battle maps or statistics. But it hit morale all the same and fed the perception that this battalion was being hammered from every direction.

Then there was the detainee-abuse case that briefly pushed 1-184 into the national headlines. One night during security operations near a power plant, soldiers from the battalion detained four Iraqi men. According to Army investigative records and subsequent press reports, at least two of the detainees were handcuffed and blindfolded, and one was shot with a stun gun, kicked, and zapped in the genitals, with the entire incident captured on video.[241] Whether it was abuse, torture, or something else entirely, one soldier remarked about it, "They did a pretty good job on them."[242]

The four Iraqi men from the power plant were found with no weapons and were later assessed as noncombatants or petty thieves, rather than insurgents. Most were eventually released

---

[238] Gold, "A Look Inside a National Guard Unit Known for Valor, Dysfunction."

[239] Gold, Scott. "Guardsmen Took 'Rent' From Iraqi Businesses," *Los Angeles Times*, August 6, 2005, https://www.latimes.com/archives/la-xpm-2005-aug-06-me-guard6-story.html.

[240] San Diego Union-Tribune. "Murder of Soldier in Murrieta Parking Lot Gets National Attention." The San Diego Union-Tribune, January 1, 2006. https://www.sandiegouniontribune.com/2006/01/01/murder-of-soldier-in-murrieta-parking-lot-gets-national-attention/.

[241] Gold, "A Look Inside a National Guard Unit Known for Valor, Dysfunction."

[242] Tempest, Rone, and Scott Gold. "Army Probes Guard Unit." *Los Angeles Times*, August 1, 2005. https://www.latimes.com/archives/la-xpm-2005-jul-27-me-guard27-story.html.

without charge.[243] Then, a soldier who hadn't taken part in the incident later stumbled across the footage on a computer. He was apparently uncomfortable with what he was seeing and turned it over to his higher chain of command. This triggered a full-blown criminal investigation, and all these incidents began to unravel.[244]

In a separate incident in another company, a first sergeant was relieved of duty after being accused of conducting a mock execution on an Iraqi detainee. According to soldiers who spoke on condition of anonymity, the NCO first shot a nearby water heater with his pistol. Then, he allegedly turned to the detainee, told him "You're next," held his weapon to the man's head, shifted it a few inches to the side, and fired another round.[245]

In an interview for this book, I spoke with a U.S. service member who had knowledge of 1-184's tour in Baghdad, though I won't reveal his identity. He told me that the cases which reached the newspapers were only part of the story. According to his account, the final straw for the troubled unit actually occurred during an incident near the University of Baghdad, when students were crossing the road between classes. It doesn't take a genius to recognize that many of those students probably had relatives in influential positions within the newly formed Iraqi government. Yet, that reality apparently didn't deter some members of 1-184 from making a reckless decision. As he remembers it, because the students were not moving out of the way quickly enough, patrolling soldiers allegedly fired rounds from their vehicles into the air and into nearby buildings. He recalled a scene of mass panic on campus and what he described as a near-international incident.

I have not been able to find any publicly available reports or official documents confirming this university incident, and it appears to be absent from the public record. The story rests entirely on this source's recollection and should be read with that caveat in mind. I include it here only because his account is consistent with the broader pattern of problems already documented and because I have no reason to doubt his credibility.

Other controversies involving the battalion also surfaced in news coverage still available online. Later reports described additional allegations of misconduct, concerns about its partnership with a notoriously heavy-handed Iraqi Police unit, and further examples of questionable judgment by both soldiers and leaders. Taken together, those accounts reinforced the picture of a battalion that had become the poster child of dysfunction.

Roughly a dozen soldiers were charged with misconduct related to detainee abuse, with several NCOs sentenced to confinement, loss of rank and pay, and bad-conduct discharges, while others received shorter terms of hard labor or administrative punishment.[246] The episode never

---

[243] Tempest and Gold, "Controversy Surrounds California Guard Officer."

[244] Tempest and Gold, "Army Probes Guard Unit."

[245] Tempest and Gold, "Army Probes Guard Unit."

[246] Amnesty International. "Iraq: Beyond Abu Ghraib," March 2006. https://www.amnesty.org/en/wp-content/uploads/2021/08/mde140012006en.pdf.

approached the worldwide notoriety of Abu Ghraib, which itself was mostly a "weekend warrior" scandal involving Army Reserve MPs under poor leadership.[247] But inside our brigade, it was seen as yet another self-inflicted wound that damaged America's reputation and undercut whatever trust the local population still had in us.

But the 3rd ID, overseeing the 1-184 Task Force, didn't stop their house-cleaning there. The most prominent figure relieved of command would be the 1-184 battalion commander, LTC Frey, who was removed in the wake of the detainee-abuse case and replaced by LTC William Wood, a regular Army officer brought in to take over the California Guard battalion.[248] Rather than promoting from within, higher headquarters installed an outside active-duty commander, a clear sign of how seriously they viewed the unit's problems.

Frey wasn't the only leader who was removed, though. Alpha Company also saw a change of command, and CPT Michael MacKinnon, our highly respected former Bravo Company commander from 4-64 Armor, was reassigned at the last minute to take charge of the struggling Guard company. The hope was that new active-duty leadership with fresh eyes could steady the battalion, reduce casualties, and restore some sense of order and discipline.

In fairness, not all the battalion's problems were of their own making. They deployed into a highly volatile area, with insurgent activity spiking throughout the Karrada and Al-Dora districts of Baghdad. Then casualties began mounting at an alarming rate, ramping up the pressure. By the time their tour ended, the 1-184 had suffered the most combat casualties of any California unit since the Korean War.[249] Even seasoned active-duty units, including our own, would later struggle to tamp down the violence in that sector.

That said, many of their difficulties cannot be chalked up solely to bad luck or a challenging mission set. A significant part of the blame can be ascribed to insufficient pre-deployment training, uneven leadership, and a command climate that sometimes seemed out of step with the counterinsurgency reality on the ground. In later years, the battalion was reportedly cited in Army training materials as a kind of cautionary example, a case study in "what not to do" when a unit starts to come apart under pressure. It was a harsh environment, and they were thrown into the fire, ready or not.

Much about this troubling period can still be found on the internet, although strong opinions and hindsight clearly color many accounts. Some sources describe the unit and many of its soldiers as out of control and undisciplined, emphasizing the scandals and leadership failures. Oth-

---

[247] Weissman, Maddie. "Abu Ghraib Torture Scandal." EBSCO, 2022. https://www.ebsco.com/research-starters/military-history-and-science/abu-ghraib-torture-scandal.

[248] McCool, John. "Operational Leadership Experiences in the Global War on Terrorism: Interview With COL (P) Edward Cardon." Combat Studies Institute, October 24, 2006. https://cgsc.contentdm.oclc.org/digital/collection/p4013coll13/id/279/rec/1.

[249] Rone Tempest and H.G. Reza, "Arms Now Used Only for Hugs," *Los Angeles Times*, January 17, 2006, https://www.latimes.com/archives/la-xpm-2006-jan-17-me-guard17-story.html.

ers, however, highlight the battalion's numerous awards and commendations, as well as the heavy casualties it sustained while operating in what many considered one of the most dangerous sectors in Iraq at the time. I suspect there is truth on both sides of that divide.

In a Los Angeles Times essay by 1LT Robert C. J. Parry, a returning soldier from 1-184, he recounted powerful stories of heroism from within his unit that the media largely overlooked. He described SGT Thomas Kruger, who, despite having his legs shredded by an explosion, dragged himself through debris in a desperate attempt to save a mortally wounded comrade. He also wrote of SFC Tom Stone, who used his own body to shield a wounded soldier from further harm.[250] Countless stories like these exist within the ranks of 1-184, though most never reached the public eye.

Amidst this upheaval, our Gambler Platoon was sent to FOB Falcon alongside Charlie Company, the hard-charging active-duty tank unit at 4-64, led by CPT David Anderson and 1SG Alan Gifford. As much as all of us dreaded it, we were temporarily placed under the direct command of that National Guard battalion, about whom we had heard all the terrible rumors.

It became the Gambler mission to operate alongside our armored counterparts on the ground and root out insurgents in the area. As active-duty soldiers, we also had to work shoulder to shoulder with the remaining National Guard personnel still stationed at the FOB, under orders to help fill the gaps, steady their ship, and stabilize the region. But that was easier said than done. As future Secretary of Defense Pete Hegseth, who served alongside us at Falcon as an Army officer, once observed, it was a *"meat grinder"* there. He wrote of us, "These guys were facing a buzz saw every day – and you could see it on their faces every time they rolled through one of our gates."[251]

Ultimately, the 1-184 continued to suffer significant losses. The battalion left Iraq with more than 100 wounded, including 80 Purple Hearts awarded for combat wounds, and seventeen soldiers under its command were killed in action over the course of a single tour.[252] Whether the Army's unofficial experiment of integrating National Guard units into prolonged front-line roles was ultimately effective remains debatable. Personally, I tend to believe it wasn't. Others may see it differently, but, as always, I leave it to the reader to make their own decision.

---

[250] Parry, Robert C.J. "The War You Didn't See." *Los Angeles Times*, February 12, 2006. https://www.latimes.com/archives/la-xpm-2006-feb-12-op-parry12-story.html.

[251] Hegseth, *The War on Warriors: Behind the Betrayal of the Men Who Keep Us Free*, 2024, Chapter 10.

[252] Gold, "A Look Inside a National Guard Unit Known for Valor, Dysfunction."

# Reflections

My grandfather, William Harstad Sr., served in the U.S. Navy from 1943 to 1946 as a Hospital Corpsman, essentially a medic in sailor clothes. He was stationed aboard the USS *Yavapai* (APB-42) and saw action in the Pacific at places like Iwo Jima and Okinawa.[253] For anyone who knows even a little about those battles, they are the kind of names that live in history books and nightmares. His ship had been built to land tanks on hostile beaches, then converted into a self-propelled barracks ship and floating clinic for the wounded. He worked his way up to the equivalent of an E-5, the same rank I would eventually hold as a sergeant, and he spent his days assisting doctors and treating wounded sailors, soldiers, and Marines as the U.S. closed in on Japan.

For all of World War II, he received a *single* personal decoration: the Good Conduct Medal. That was it. It certainly wasn't because he or his fellow corpsmen lacked courage. They dealt with kamikaze attacks, vicious sea battles, and a constant flow of shattered bodies. The reason was far simpler. Back then, doing your job didn't come with a round of applause. You did what needed to be done. Medals were reserved for the truly extraordinary or for those who went far beyond what anyone could reasonably ask.

Jump ahead about sixty years, and there I was standing in formation at Fort Stewart, receiving an Army Achievement Medal for my leadership in the shoot house and on the live-fire ranges. Later, I was awarded the Good Conduct Medal while deployed in Iraq. I still remember thinking to myself, even in those moments, whether I had really done anything to deserve them.

After I left the Army and had some time to grow up a bit and look back, I realized many of the awards I collected didn't mean all that much to me. They looked good enough on a uniform, but I didn't feel as if I had done anything especially heroic to earn them. Meanwhile, some of the hardest things I ever did, the things I *was* genuinely proud of, slipped by with little or no notice, which is oddly funny when you think about it.

When I wore my Class A uniform, all those colorful ribbons and badges looked impressive to people who had never worn one. They would stop and ask what this or that one was for, pointing at my chest, and I would usually answer with something short and simple. The truth was that some of the stories behind those decorations were kind of underwhelming. I felt awkward explaining them because they didn't match the pride I felt in the things for which nobody had pinned anything on my chest. I noticed the same disconnect with other soldiers, too.

Over time, I came to believe that most awards in life say very little. They are symbols, nice to look at, but they rarely tell the whole story. Years later, when I finished my Master of Science degree, people around me treated it like a significant life milestone. Maybe they were right. Maybe I should have been celebrating more. To me, though, it felt like an oversized, very expensive piece of paper with my name printed on it. The real achievements were the knowledge I gained along the

---

[253] Naval Historical Center. "USS Yavapai (APB-42)." Department of the Navy - Naval Historical Center, July 27, 1999. https://www.ibiblio.org/hyperwar/OnlineLibrary/photos/sh-usn/usnsh-xz/apb42.

way and the people I met and worked with. That was what mattered to me. Not the ribbon, not the certificate, and certainly not the ceremony. The work itself, the growth, and the kind of person you become in the process are what count.

I bring all of this up because sometimes I look at my medals and feel a little uneasy. It seems strange, and more than a little unfair, that my grandfather, whom I loved very much, received only one of those small strips of ribbon for serving in the largest war in human history. His experience was brutal and harrowing, and far too large to be summed up by a single little rectangle of fabric and metal. If I could transfer some of my recognition to him, I would do it without hesitation. But I already know how he would react. He would smile, shake his head, and quietly refuse. I think there's a good lesson there.

Next, I need to address my use of the slur "Hadji," along with other derogatory terms that countless service members have used throughout the long history of warfare. The military never handed us a list of approved insults, of course, but our training absolutely pushed us toward being comfortable with the idea of killing. Part of that desensitization came on its own. We started talking about the enemy in ways that made them feel less human, and the language naturally followed.

The word *Haji* (or *Hajji*) is, in its original context, an honorific title given to a Muslim who has completed the Hajj, the pilgrimage to Mecca in Saudi Arabia.[254] In many Muslim communities, it is a respected form of address, a mark of religious devotion and accomplishment. At the same time, dictionaries now explicitly record that in modern American usage, the word has also become an offensive slang term, used as a disparaging label for Arabs or Muslims, and it was widely used that way by U.S. troops in Iraq and Afghanistan. I was part of that. I used it as a slur, not as an honor.

Somewhere along the way, people sometimes swapped the first "j" for a "d" and spelled it like Jonny Quest's cartoon sidekick. However it was spelled, the meaning in our mouths wasn't respectful. It was said with venom, as it was in my own letter home.

We used slurs like "towel heads," "rag heads," "Hadjis," and worse. I certainly used them, and I think most of us did. I'm not saying this to excuse it. I'm saying it because it's true. Veterans from other units have openly described how infantry culture leaned on these words to lump every Iraqi or Arab person into a single, faceless mass. It's something that has happened in virtually every war, partly because it's easier to kill someone once you have convinced yourself that they are nothing like you.

In World War II, Germans were "Krauts," and the Japanese were "Nips." In Korea, the enemy became "zipperheads." In Vietnam, it was "gooks." None of those words were neutral. They were racial slurs, weapons of language that helped turn entire populations into caricatures. And if you go back even further, the whole "Yankee Doodle Dandy" thing started as a British way of calling American colonists ridiculous.[255] It sounds almost cute now, but at the time, it was prob-

---

[254] Publishers, HarperCollins. "Hajji." In *The American Heritage Dictionary of the English Language*. Accessed December 11, 2025. https://www.ahdictionary.com/word/search.html?q=hajji.

[255] Volle, Adam. "Yankee Doodle." Encyclopaedia Britannica. Accessed December 11, 2025. https://www.britannica.com/topic/Yankee-Doodle.

ably the eighteenth-century equivalent of saying "your momma." Historians and linguists now talk explicitly about how these kinds of labels were used to mock, demean, and dehumanize the people on the receiving end, especially in wartime.

But these names were not just insults. They were also coping mechanisms, shields we held up between ourselves and the moral weight of what we were doing. They helped us tell ourselves that "they" were a category, not individual human beings with families and inner lives. As long as there is war, I suspect this pattern will repeat in one form or another, because it's always easier to pull a trigger, drop a bomb, or plant a charge when you have first convinced yourself that the person on the other end is not fully human.

In his book *On Killing*, LTC Dave Grossman argues that the language of war is often re-shaped to put emotional distance between soldiers and the act of killing. He notes that "most soldiers don't 'kill'"; instead, they talk about the enemy as being "knocked over," "wasted," "greased," and so on.[256] Those kinds of euphemisms are not random. They serve a very real psychological purpose by sanitizing violence, so it is easier to live with.

Grossman also emphasizes that this kind of linguistic detachment is not unique to American forces. He describes a former Viet Cong guerrilla who had defected to the South. This man, like his new American allies, referred to his former comrades as "dinks" and "gooks." When asked why he did this, the man shot back, "Do you think the Americans are the only ones who do that?" He went on to explain that his unit used to call Americans "Big Hairy Monkeys."[257]

Without even realizing it, your mind starts to create distance. *Don't let them get too close. Keep them at arm's length.* But it wasn't that simple with everyone. We didn't feel that way about all Iraqis. Our interpreters, for example, often became like family.

That's where things got really complicated. Many of our interpreters were, in the ways that mattered most, more "American" than a lot of people actually born here. They worked hard, told the truth, and cared deeply about the mission and the people standing next to them. So, we had to perform a kind of subconscious mental gymnastics, constantly sorting who we saw as "good" and who we feared might be "bad." It was messy and inconsistent. Nothing about it was black-and-white. Then again, what *is* in war?

**3 January 2025 – Reflection of Cap Stanley, Command Sergeant Major of 4-64 Armor**

We had a lot of excellent interpreters. One of them was Nora. More than once, when a soldier went down and was out of the fight, she threw herself over him and used her own body as a shield until the shooting stopped. She was a warrior.

---

[256] Grossman, *On Killing: The Psychological Cost of Learning to Kill in War and Society*, Chapter 7.

[257] Grossman, *On Killing: The Psychological Cost of Learning to Kill in War and Society*, Chapter 7.

**1 August 2025 – Reflection of Noor "Nora" Aljassim, Interpreter for the U.S. Military**

I didn't want the jobs on the bases. I liked being out in the streets where the action was, especially on Haifa Street, where I lived and spent most of my time. Once, a grenade got tossed at us, and I caught a few fragments in my foot. Mostly, I was just mad that it ruined my boots.

Nora sitting on the back of an armored vehicle - she earned the nickname "Warrior Princess" for a reason

Photo courtesy of Noor Aljassim

Now, let's address the elephant in the room: the 1-184 and our interactions with them. While I've tried to present a fair view from both sides, I feel obligated to be honest about what I saw firsthand and what I came to understand through my research.

Being an infantryman in an active-duty unit like the 3rd ID was a full-time job. Not only were we constantly training, but the variety and repetition of that training far exceeded what our Reserve or National Guard counterparts could realistically achieve. The logic is simple: they were civilians most of the year, holding down regular jobs. The old motto, "one weekend a month, two weeks a year," might be sufficient for domestic support roles, but it didn't translate well to the demands of overseas combat. I saw that personally. In my view, they simply needed more time and training to function effectively in a combat environment.

I've read statements from leaders who were in charge during that time, many of whom expressed confidence in the 1-184's training and capabilities. The tone is often optimistic, sometimes even glowing. I can't help but suspect that some of those remarks were aimed more at maintaining morale and preventing further breakdowns than offering an unfiltered assessment of the situation. But, I'll admit, that's speculation on my part. All I can speak to is what I personally observed, and from that vantage point, I often felt unease during our interactions with the unit.

Donald Cyr, my buddy from 3-7 Infantry, also had occasional dealings with them at FOB Falcon. When we spoke about it nearly twenty years later, he described some of the soldiers he met as "goofballs" and "undisciplined." I agreed. Even seemingly minor things raised eyebrows, such as their use of juvenile mission names like "Operation Camel Toe."[258] Sure, on its own, that kind of humor might be harmless, but when viewed alongside other warning signs, it started to feel like a symptom of a deeper issue. Piece by piece, the strange behavior and questionable professionalism made us both wonder what was really going on over there. We recalled the rumors and drama swirling around Falcon during the summer of 2005, and none of it left a good impression.

That said, many of the individual soldiers were genuinely good people who wanted to serve their country with honor. In many ways, I sympathize with them. They were caught in a no-win situation, just as I had been during my time at 30th AG and with Rear D at Fort Stewart. When leadership fails and structure breaks down, even the most motivated and disciplined troops struggle to succeed. That's why I don't aim my criticism at the individual Guardsmen. They deserved better from the Army. My frustration is directed at the unit as a whole and the systemic issues that plagued it. It was, frankly, a mess.

Speaking strictly as an enlisted grunt, I've always believed that the National Guard didn't belong on the front lines in Iraq or Afghanistan. I still don't. They should have been supporting from the rear or been kept stateside altogether. But I also recognize that I wasn't the one making decisions at the division or corps level. With two wars running simultaneously, manpower was stretched dangerously thin, and the Army had to adapt. In that sense, deploying National Guard and Reserve units into combat may have been a necessary experiment. If it was, I'd argue the results were, at best, questionable.

---

[258] UPI, and Ken James. "Anti-Insurgent Operations in Iraq." UPI, June 17, 2005. https://www.upi.com/News_Photos/view/upi/d0f54c8b018eb4bfbe173f05723459c3/ANTI-INSURGENT-OPERATIONS-IN-IRAQ.

Many readers, particularly those accustomed to Western sensibilities, may find some of the reports cited in this book shocking or even outrageous. *How could American soldiers do that?* It's a fair question, but one that needs to be approached with perspective. To those who react this way, I would urge you to read more widely and educate yourself on the history of warfare. War, by its very nature, is brutal, and terrible things happen. Even in wars we consider morally justified, where we clearly define heroes and villains, atrocities have occurred on all sides. This isn't a defense of wrongdoing, but a recognition of reality.

In *With the Old Breed*, E.B. Sledge recounted a horrifying scene in which a fellow Marine tries to pry gold teeth from the mouth of a wounded, partially paralyzed Japanese soldier who had been shot in the back. The man was still alive, kicking and thrashing, while the Marine went to work on him with his knife. The first attempt slips, driving the blade deep into the man's mouth instead of the tooth. Cursing, the Marine hacks at the sides of his face, then braces a boot on the man's jaw and keeps going. Blood pours out as the soldier gurgles and struggles, trapped and helpless.

Eventually, another Marine ends the Japanese soldier's suffering with a bullet to the head, and the tooth collector moves on to look for more trophies. Even for those who view World War II as a necessary and righteous war, the scene forces an uncomfortable question: what could drive a man to do this to another human being?

Sledge offers a haunting answer: "Such was the incredible cruelty that decent men could commit when reduced to a brutish existence in their fight for survival amid the violent death, terror, tension, fatigue, and filth that was the infantryman's war."[259] In other words, the battlefield didn't just kill bodies. It could warp the moral code of ordinary men, dragging their conduct far from anything that would have been tolerated back at division headquarters.

And this wasn't an isolated incident. In *The Rifle*, GWOT Marine veteran Andrew Biggio interviewed elderly World War II veterans. In one conversation, he sat down with CPL Joe Drago, an old neighbor he had grown up next to. Drago tells him, "On Okinawa, there was no such thing as a war crime!" Sitting back in his recliner, he then directed Biggio to open a nearby drawer, where Biggio found a small bag of gold teeth taken from a long-dead Japanese soldier. Drago went on: "Something like 40 percent of the Japanese bodies in the South Pacific were missing skulls and teeth."[260]

It's a reminder that even in the so-called "good wars," the darkness of battle doesn't discriminate. It consumes anyone who lingers too long in its shadow. The point isn't that they were uniquely evil, but that they were frighteningly normal. They were ordinary Americans, the kind who went home and became somebody's father or grandfather… maybe even yours.

In *Ordinary Men*, historian Christopher Browning presents a reality that is even more disturbing. Through meticulous research, he shows how ordinary German police officers, many of

---

[259] Sledge, *With the Old Breed: At Peleliu and Okinawa*, Chapter 5.

[260] Biggio, Andrew. *The Rifle: Combat Stories From America's Last WWII Veterans, Told Through an M1 Garand*. eBook edition. Regnery History, 2021, Chapter 1.

whom had *not* been steeped in Nazi ideology from childhood, nevertheless became active participants in mass murder. Some did so eagerly, most did so reluctantly, but the overwhelming majority took part. Only a tiny fraction refused.[261]

What does this tell us? It tells us that, under the right conditions, almost anyone can commit unspeakable acts. War strips away the layers of civility and conscience, exposing the darker instincts beneath. That is why strong, principled leadership is essential, not just to issue orders, but to restrain the chaos that war unleashes.

In the end, that is why I think so many of their failures come down to leadership. Combat leadership is absolutely critical. You need a strong, disciplined commander at the top to set the tone and shape the unit's direction. Just as important is leadership at the junior officer and NCO levels. That is the backbone. Without it, a unit is lost before the first shot is even fired.

Good leaders do more than tell you where to go and what to do. Their real job is to harness that chaotic storm that inevitably arises and keep it from spinning out of control. I often think back to CPT MacKinnon's speech in the motor pool, the one that echoed General Mattis. Beneath that cold pragmatism lay a more profound message: maintain control. Don't let the madness of war consume you, because once it does, there may be no coming back. So, while we were sad to see MacKinnon leave to take command of Alpha 1-184, we also knew he was the one person who might turn things around. He was the guy who could save the sinking ship.

After taking all of this into account, did 1-184 truly do anything egregious? I'll leave that for you to decide. Again, this isn't about excusing bad behavior, but about forcing a harder question: *What would I have done in that situation, with my friends dying all around me?*

If, after everything you have read, you still believe you're incapable of doing something truly horrifying, I don't know what more could convince you. I hope you never find out. But I suspect it is one of two things: either you're lying to yourself, or you haven't allowed your imagination to go far enough. The truth is, under the right pressure and without proper guidance, any one of us can become a monster. *Even you.*

---

[261] Browning, Christopher R. *Ordinary Men: Reserve Police Battalion 101 and the Final Solution in Poland.* Harper Perennial, 2017.

# Chapter XIV

## Falcon

At our new FOB, I stood at parade rest, my eyes glued forward. Dust swirled in aimless circles, and the flies attacked from every direction. Then I heard it: "Does your mom know you're here?"

I clenched my jaw and cut my eyes sideways without turning my head. *Not again*, I thought. Striding through our formation was 1SG Gifford, the senior NCO of our newly assigned foster-parent company, Charlie Company, the Cyclones. A faint smirk tugged at the corner of his mouth.

"Yes, First Sergeant," I answered, dry as dust. By then, I was used to the digs about my boyish face. Unlike with the woman at the checkpoint, I wasn't about to make a smartass remark.

It didn't take long to realize that behind the wisecracks, 1SG Gifford was a solid leader and a good man. He and the affable company commander, CPT Anderson, pulled us into the 4-64 Cyclone family with humor and a little light razzing. The welcome was warm, but it didn't last long. Once the meet-and-greets were done, the jokes stopped, and the tone shifted.

We were given our next assignment, and it was serious. Our platoon would be embedded in a specially formed task force charged with pacifying the Al-Dora district, working alongside elements of special operations, infantry, and armored units. As I mentioned earlier, and much to our shared concern, the whole effort would fall under the direct command of 1-184, the National Guard battalion already neck-deep in problems. The good news, though, was that our days of manning checkpoints were over. The bad news was that we were stepping into something much more dangerous.

We quickly saw why 1-184 had been getting chewed up. Al-Dora was a patchwork of incredibly poor villages and sketchy neighborhoods, with crumbling mud huts, overgrown vegetation, tangled groves of trees, and more than a few hostile faces. It was a shocking sight, depressing and horrifying at the same time.

**21 July 2025 – Reflection of Juan Serrano, Platoon Sergeant, Bravo 4-64 Armor**

We took over the area from a unit that had been doing some crooked shit, and we could feel a lot of people there hated us. Not just the Army, but Americans.

Every time we left the gate, we knew we were going to get hit. It wasn't a question of *if*, just *where* and *how bad*. Because we had Bradleys and tanks, we automatically became the QRF for everyone, on top of our regular operations. We were the battalion's guardian angels. Guys knew that if a patrol got blown up or pinned down, we were the ones who would come to rescue them.

That kind of trust felt good, but it was also a heavy weight to carry and a lot of stress.

SSG Thaler with a poor but friendly Iraqi family at the front door of their mud hut

### 15 September 2025 – Reflection of Jason Sturm, Bravo 4-64 Armor

On one mission, we were searching for weapons and intelligence. We blew a brand-new padlock off a door with a shotgun, and the stench hit us immediately, ammonia and human waste. Inside, we found a young Iraqi girl chained to the floor, surrounded by feces, urine, and rotting food. Her hair and clothes were matted with filth, and she couldn't communicate with us at all.

Later, we learned she was mentally ill, and her family had no way to care for her. Chaining her up was the only way they knew to manage her condition. It wasn't done out of cruelty, but out of desperation. We called for medical personnel, and higher command took over, but I never found out what became of her.

On top of the poverty and overtly hostile people, much of the area was open farmland, sliced by massive dirt roads that doubled as dikes. Those raised roads were the only real paths for vehicles to take, meaning they were also perfect for burying IEDs. The fields on either side were soaked and muddy, often impassable even for tracked vehicles. Everything funneled onto those dikes. So, it was a natural choke point, and the insurgents knew it better than anyone.

SPC Belton (top left) pulls security as the tank crew fights to free their M1 Abrams from the mud – our area of operations was a maze of flooded fields, elevated dikes, and poor farmland

**22 August 2025 – Reflection of Matt Turcotte, Platoon Leader, Bravo 4-64 Armor**

When we arrived, we sat down and went over every route and corner of our sector. Many of them were flagged as exceptionally dangerous, marked as red or black. When we asked when they'd last sent patrols down those roads, the answer was often, *"Never."* There were whole stretches that hadn't seen an American soldier in months, if ever.

That was what we were walking into. We had to push into places we knew nothing about, where the insurgents had been given plenty of time to build their defenses and ambushes. In some spots, they had layered IEDs so thick that they were literally stacked on top of each other.

To put this in context, 1-184 was being ordered to operate in that volatile part of southern Baghdad in Humvees, not behind the mass and protection of our tanks or Bradleys. At the time, thousands of Humvees in Iraq were still not armored for true combat conditions, and the stopgap steel kits were often inadequate against the kinds of bombs insurgents were fielding.[262] The Army was already scrambling to buy and field up-armored models specifically designed to counter the IED threat, which tells you how quickly the battlefield was outpacing the vehicles. But even then, top-of-the-line models were still being blown apart, showing that while they were agile, they couldn't withstand a heavy punch.

That reality helps explain why the National Guard had to shrink their footprint. Every time they pushed out, they were gambling against an enemy toolkit that included conventional roadside IEDs with a variety of triggers, incendiary attacks, and shaped charges designed to punch through armor and spray molten metal into a vehicle's cabin. And then there were the new, especially nasty threats we started seeing more of: explosively formed penetrators, or EFPs. This is a shaped-charge variant that uses an explosive to collapse a concave copper or steel liner into a high-velocity "slug" built specifically to defeat armor.[263] They could even disable and, in some cases, penetrate vulnerable areas of heavily armored vehicles, including the Abrams tank.[264] Meaning, these devices would certainly demolish a Humvee.

But they didn't just use command-wire, manually detonated explosives. We also encountered timed devices, passive triggers like tripwires and other sensor-based initiators, pressure plates and contact triggers, and radio- or remote-controlled setups, along with countless variations. If there was a way to set off an explosive, they were using it.

So, I'll give 1-184 a lot of credit here. They were enduring genuinely hard conditions with limited resources. Later, in writing about the experience, LTC Roth summed it up perfectly.

**29 August 2005 – Letter from Robert Roth, Commander of 4-64 Armor**

Cyclone has moved to FOB Falcon and works with the National Guard battalion from California. A platoon from Bandit is attached to them.

---

[262] Wilson, "CRS Report for Congress - Improvised Explosive Devices (IEDs) in Iraq: Effects and Countermeasures."

[263] Lecappelain, Josh. "Birth of an Explosively Formed Penetrators; Death of Innocent Civilians." Defense Visual Information Distribution Service, September 25, 2008. https://www.dvidshub.net/news/24119/birth-explosively-formed-penetrators-death-innocent-civilians.

[264] Guest, Tim. "Raining Down From Above." *European Security & Defence* (blog), January 24, 2025. https://euro-sd.com/2025/01/articles/42229/raining-down-from-above/.

Their area is mostly farmland and very poor, yet it tends to get more than its fair share of road-side bombs (IEDs).

**The author in Al-Dora, Baghdad - one less captured insurgent AK-47 pointed at us**

As it turned out, that was an understatement. This place was IED central. It was virtually impossible to drive anywhere without getting hit by one or sometimes several. They were everywhere and nearly impossible to detect. That's when we discovered the insurgents had developed a grimly effective tactic. Under the cover of darkness, they tunneled sideways into the dikes, buried the explosives, and ran thin detonation wires hundreds, sometimes thousands, of feet through the waterlogged fields. Meaning, you couldn't see any signs of tampering in the roadway, nor could we see the insurgents themselves as they were often hidden too far away to be seen. Then they waited for the next unlucky vehicle to roll through. As soon as they clacked off the IED, they vanished, melting into a nearby village or slipping away through a drainage ditch.

As much as we wanted to put boots on the ground, that was nearly impossible with much of the terrain. The moment we stepped into the fields, which seemed to be everywhere, we began sinking into ankle-deep muck, sometimes even up to our knees. Our heavy armor plates, Kevlar, reserves of ammunition, and other gear weighed us down far more than a light Iraqi farmer. Crossing a single football-field-sized plot could take an absurd amount of time, with soldiers either getting stuck and needing to be pulled out by their buddies or crawling just to move forward. So, we tended to avoid the fields whenever possible, especially when they were wet. To make matters worse, the thick crops and overgrown vegetation cut visibility down to only a few feet, making it nearly impossible to see what, or who, was ahead.

### 21 August 2005 – Letter to the Harstad Family

We have been operating in a new sector that is more dangerous than the Karrada Peninsula that we were originally assigned to.

So far, we have conducted several missions, mainly reconnaissance, and have had a few successes. On one particular mission, we disarmed a roadside bomb that was underneath a pedestrian bridge. I also found a fired but undetonated rocket that we then blew up. It was located in a small village, lined with kids.

The enemy's tactics posed a dire, relentless threat to us, but they were undeniably ingenious. David Anderson, our tanker commander, captured the feeling of inevitability in his book *Ride the Cyclone* when he wrote, "We would soon be striking IEDs almost every day." He describes one blast that "sprayed shrapnel all over the turret," shredding the gear strapped to the outside, and leaving "a large chunk of metal" lodged in the tempered glass near his hatch. Then, as if it were routine, he adds, "Not long after that, we were hit again. We rode on each time."[265] That blunt rhythm is what made the roads feel cursed. The bombs were not rare events; they were ever-present and extremely deadly.

### 3 January 2025 – Reflection of Cap Stanley, Command Sergeant Major of 4-64 Armor

You had to go out and hold the terrain, or the enemy would maneuver with no resistance. We couldn't allow that to happen.

It was a tough decision for everyone involved, but it had to be done.

Even back at the FOB, there was no safety.[266] Each day brought mortar fire, the attacks growing more frequent. Sometimes one or more insurgents would drop two or three mortar rounds

---

[265] Anderson, David. *Ride the Cyclone*. Biblioscholar, 2012. Pages 7-8.

[266] Global Security, "Camp Falcon / Camp Al-Saqr," September 1, 2012, https://www.globalsecurity.org/military/world/iraq/camp-falcon.htm.

into a tube in quick succession. Before anyone could react, they had vanished, the tactic rarely changing.

The sound was unmistakable: that deep *whoomp whoomp whoomp,* eerily similar to distant fireworks being launched on the Fourth of July. The moment we heard it, we'd dive for cover or scramble inside. At best, you had a few seconds before impact. All you could do was hope it didn't land near you.

### 20 August 2005 – Letter to the Harstad Family

Well, as I'm typing this, a mortar just landed outside. It broke my window, but I don't think it did any real damage. The sandbags are in the windows for a reason.

FOB Falcon sat directly beside a densely populated residential area. In fact, it was so close you could look across the street and see into people's homes. This proximity meant we couldn't return fire at the insurgents who operated with near impunity just outside the wire. Whether we were working in the fields out in the sector or back at the base, it was a terrible situation, and it often felt like everything was stacked against us.

### 3 January 2025 – Reflection of Cap Stanley, Command Sergeant Major of 4-64 Armor

The insurgents had motorcycles and vans that they were using. They welded a mortar tube to the front of a motorcycle, and inside a van, they concealed another, revealing it only when the sliding door was opened. They'd drive into the neighborhoods surrounding Falcon, right up next to civilian homes, and fire a few rounds before slipping away.

We kept requesting friendly artillery or mortar fire to stop them, but command would always call "check fire" because of the insurgents' positions. So, we had to find a way to lure them into the open.

### 1 August 2025 – Reflection of David Anderson, Commander of Charlie 4-64 Armor

We regularly called for the Paladins (artillery) or a mortar team to return fire. Command had a drone up in the sky watching, and we were all frustrated because we could almost never get approval to shoot back. We'd hear it come back down from battalion: "Uh, there's a cow in the field, and we don't want collateral damage." A cow. After a while, it started to feel absurd. We understood the concern about civilian casualties, but at the moment, it was frustrating. What were we supposed to do then, just stand there and take it? Yep, that's what we did.

**Nora on patrol in a dangerous neighborhood with Bravo Company, standing next to an M113**

**Photo courtesy of Noor Aljassim**

Some of our existing interpreters accompanied us during the move from Prosperity to Falcon, but not all. Many of them used Americanized nicknames or call signs.[267] In part, it was for security. Using a real name could put them, and sometimes their families, at greater risk. It was also practical. Arabic includes consonants that don't exist in English, and some names were genuinely hard for English speakers to pronounce quickly over the radio. So short, memorable nicknames made day-to-day communication smoother and helped everyone stay on the same page.

### 1 August 2025 – Reflection of Noor "Nora" Aljassim, Interpreter for the U.S. Military

Many of the interpreters didn't want to use their real names. The American military even recommended we cover our faces when out on patrol, but I didn't. I wasn't worried about it.

---

[267] Peter, Tom A. "An Iraqi Interpreter as Chronicler of the War." *The Christian Science Monitor*, August 13, 2008. https://www.csmonitor.com/World/Middle-East/2008/0813/p06s01-wome.html.

Most of the time, interpreters could choose their own names. With mine, my real name is Noor, but "Nora" is already a nickname. So, I just stayed with that.

These nicknames often reflected something unique about the interpreter, such as an American movie character, a personal interest, or some other familiar reference. For example, our interpreter Forrest got his name from Tom Hanks' role in *Forrest Gump*. Another, named Chevy, took his nickname from the Chevrolet car brand. Then there was Jersey, who was an overweight interpreter who looked like an Italian mobster, talked too much, and spoke with a strange accent. The name fit him perfectly.

**11 December 2005 – Letter to the Harstad Family**

We've got this interpreter that's pretty cool... his name is (redacted). We just call him Forrest, because he runs everywhere.

Anyhow, he's hooked on nude magazines, so I bought him a subscription of Playboy for Christmas (if it actually makes it through the mail). He's helped us out quite a bit, so that's a small price to pay.

Unfortunately, many of the interpreters who had already sacrificed so much began to be hunted.[268] One day, a fellow platoon member handed me a gruesome photograph. Jersey, our loudmouthed but well-meaning interpreter, had reportedly met a horrific end after being captured by terrorists. But from the photograph alone, I had a hard time verifying that it was truly him.

The full details of his abduction were never known, at least not by me, but his body was found bloated and discarded along the roadside, not far from our base. Our area had become a dumping ground for sadistically tortured bodies.[269] His arms were handcuffed above his head, as if he had been suspended, a blindfold wrapped tightly around his face. It was clear he had been tortured, likely with power tools, as had so many others.[270] His corpse had been abandoned in the sand, now blackened and soaked with thick, congealed blood. Flies swarmed the body, feeding on the fluids that had drained and pooled into the surrounding dirt.

It was a gruesome, deliberate message, likely sent because he had helped us. We suspected that his loud, braggadocious nature may have drawn the attention of al-Qaeda operatives, possibly

---

[268] Kane, Jim. "Iraq: Translators Dying by the Dozens." CorpWatch, May 19, 2005. https://www.corpwatch.org/article/iraq-translators-dying-dozens.

[269] Hegseth, *The War on Warriors: Behind the Betrayal of the Men Who Keep Us Free*, 2024, Chapter 10.

[270] Knickmeyer, Ellen. "In Balad, Age-Old Ties Were 'Destroyed in a Second.'" The Washington Post, October 22, 2006. https://www.washingtonpost.com/archive/politics/2006/10/23/in-balad-age-old-ties-were-destroyed-in-a-second-span-classbankheadsectarian-battles-drive-out-sunnis-create-state-of-siegespan/b816d0c4-64c5-4109-85d5-454c15dcd7b0/?

linked to Abu Musab al-Zarqawi's group, which was reportedly still hiding in the Haifa Street area and remained an active threat.[271] Unlike most interpreters who kept a low profile for their own safety, Jersey stood out. That was just who he was: talkative, animated, impossible to ignore. He didn't deserve to die like that, though. No one did. Following the interpreter incidents, our leadership issued a new order: all personnel would be required to live full-time on the FOB for their own protection.

**1 August 2025 – Reflection of Noor "Nora" Aljassim, Interpreter for the U.S. Military**

Many linguists and their families were threatened for working with the Americans. Someone took my photograph while I was on patrol, wearing a military vest, and started posting it around Haifa Street. My father, coming home from his job as a dentist, saw one of the posters on a mosque and called me immediately to say there was a price on my head.

I asked, "How much are they offering?" He didn't find it funny. He said, "Do you want us to get killed?"

Eventually, the threats from al-Qaeda forced me to flee the country. Not long after, my family began receiving envelopes in the mail filled with bullets, and they had to flee, too. None of us wanted to leave our country, but there wasn't a choice. I felt like it was my fault, because of me and my job.


While speaking with a high-ranking official for this book, he explained that, like many bases across Iraq and Afghanistan, there was a persistent fear that someone inside the FOB was feeding information to the enemy. Recalling a previous breach, he told me, "They had bumper numbers, detailed layouts of the base, and someone on the inside." As it turned out, the mole was an Iraqi plumber. He was later made, in the official's words, "very dead" during a targeted operation.

While we dealt with those head-choppers, power-drill enthusiasts, and the nightly rounds of mortar Frogger, tensions began to surface within our task force. Even though the Gambler Platoon stayed on good terms with Charlie Company, thanks to the steady guidance of CPT Anderson and 1SG Gifford, we often clashed with certain individuals from 1-184. Some higher-ups within the task force made baffling decisions, perhaps driven by incompetence or indifference, but this is by no means unique to the military. Maybe you have encountered similar forehead-slapping situations at your own workplace, even if yours probably didn't come with the added perk of getting your head sawed off next to the water cooler.

One day, during a company formation, not long after we returned from a mission where we'd captured several insurgents, our platoon was reprimanded for how we had our gear set up. A

---

[271] Georgy, Michael. "Dangerous Haifa Street in Iraq Defies US Military." Arab News, February 23, 2005. https://www.arabnews.com/node/262736.

higher-ranking National Guard soldier, who will remain unnamed, insisted that every soldier configure their combat vest exactly the same. Pouches had to line up, ammo had to be in the same spots, and everything had to match. Because… well, the enemy needed to know that we had some fashion sense, right? And if we were going to get blown up, we were apparently expected to do it while taking home the "Best Dressed" trophy.

Of course, this makes no sense. Soldiers have different roles and different needs, such as left-handed versus right-handed, or riflemen versus machine gunners. Medics, grenadiers, drivers, turret gunners, and pretty much every other role have unique gear. Thus, gear placement must fit the individual and the role if you want them to be fast and effective in a fight. Uniformity for the sake of appearance might look great at a stateside inspection, but in a combat zone, it was absurd. So, we didn't comply because it was nonsense, but our leadership took a lot of heat right away.

### 1 August 2025 – Reflection of David Anderson, Commander of Charlie 4-64 Armor

1SG Gifford was furious. He told me, "Sir, this doesn't make any sense. We need to do something about it." He and I sat down with the battalion leadership and tried to make them understand. Eventually, they backed off and let us do what we needed to do, but it was a terrible fight at first. Gifford really led that charge.

### 7 October 2005 – Letter to the Harstad Family

Our platoon has been getting really frustrated with some of our leadership.

In the Army, you run into many kinds of people, including those that genuinely care about other soldiers and the mission. You also have those that are only concerned about not looking bad in front of the populace or earning that next promotion. Unfortunately, we are dealing with the latter right now.

It is very frustrating not being able to do your job and instead being put out on parade. Let me do my job while you sit behind your desk in the AC.

Years later, I spoke with retired CSM Stanley about the incident. Unfortunately, he and LTC Roth were no longer in our chain of command, at least temporarily, and they had no direct authority over what was happening to us. He told me he had witnessed similar behavior from many high-ranking officials throughout his career, and he shared my disbelief. *Why would anyone think this is a good idea?*

### 3 January 2025 – Reflection of Cap Stanley, Command Sergeant Major of 4-64 Armor

I watched people in charge try to mandate a parade-ready uniform in a combat zone, and I wasn't having it on my watch. It made no sense. Tankers need a different loadout than infantry, and infantry need something different than medics. Why would we prioritize looking good if it comes at the expense of actually fighting?

As this guy walked through the formation, he stopped in front of me and flippantly asked why I had a knife strapped to my vest, as if we hadn't just gone hands-on with insurgents. I replied that it was a backup in case I was separated from my firearm. He ordered me to remove it. I reluctantly did so, though only temporarily. After the formation ended, I put it right back on.

I assumed this guy was still bitter over his unit getting reprimanded for the Samurai sword incident. And how did he handle it? By taking it out on the active-duty guys. But let's be clear, this wasn't some flashy Samurai sword strapped to my vest. It was a small, lightweight blade that served a real purpose. Even soldiers carrying basic Swiss Army knives and Leatherman multi-tools were told to remove them. It was some personal regulation he had concocted, taken to an illogical extreme.

This same individual also insisted that we keep a round out of the chamber, meaning our weapons would technically be unloaded, unless we were directly attacked. He eventually walked this back, but I don't believe anyone actually complied with this insanity in the first place. The rules of engagement were already, at best, confusing. Now, however, they appeared suicidal.

Pete Hegseth, now the Secretary of Defense but then a lieutenant with us at FOB Falcon, described the same kind of vibe in his book *The War on Warriors*, and it felt uncomfortably familiar. He recounts a JAG-led rules-of-engagement brief where the class was shown a man holding an RPG. The JAG officer then asked if they were authorized to shoot him. When the soldiers said they'd fire, they were immediately corrected.

The JAG's warning was, "You are not authorized to fire at that man, until that RPG becomes a threat. It must be pointed at you with the intent to fire."[272] *What?!*

All around our happy little task force and FOB, it seemed like these leaders were making similarly strange calls. Maybe it was in the water. Or maybe they just had their heads so far up their own asses they were getting high on the fumes.

### 5 October 2005 – Letter to the Harstad Family

We got in trouble today. We moved our Bradleys over to an abandoned meat packing plant and sat there for a little while, trying to cool off (during a patrol). (Redacted) found us and yelled at all the leaders in the platoon, including me, for "not doing our jobs."

On the day I wrote that letter, our platoon had just gotten thoroughly chewed out for not staying in constant motion during a 12-hour patrol. Apparently, stopping for bathroom breaks or water, even after we'd spent the day literally roasting inside our Easy-Bake ovens, was considered a luxury we hadn't earned. So, we adapted.

---

[272] Hegseth, *The War on Warriors: Behind the Betrayal of the Men Who Keep Us Free*, 2024, Chapter 10.

At least once, with one guy about to have a full-blown emergency and no permission to stop this mindless rolling patrol, we popped open the little swinging troop door embedded in the ramp while we were still moving down the road. We grabbed onto his gear and held him in place, just so he didn't tumble out, and he backed up to the opening and let 'er rip onto the idyllic Iraqi countryside. Yes, while driving, and the whole time, all we could do was hope he wouldn't take a piece of shrapnel or a sniper bullet to the bright white moving target that was his ass. That is not an exaggeration. That is exactly what we were being ordered to do.

It just seemed to be a case of excessive oversight gone awry, with a healthy dose of unexplained animosity thrown in for good measure. My best assessment is that much of what we were witnessing was likely the result of overcorrection, driven by the desire to prevent further embarrassment to the unit. I still don't fully understand what was happening behind the scenes, and I probably never will, but for many of us, daily work had become the equivalent of walking on eggshells.

Papi, our platoon sergeant, took the brunt of much of the heat. He respected the rank but didn't shy away from calling out poor tactics and questionable decisions by certain individuals, and his simmering frustration was obvious. We all knew he was right, as we were the ones out there doing the actual work. However, despite his efforts, the decisions were being made above our heads and outside our control. My letter from later in the deployment summarizes much of the frustration.

### 15 October 2005 – Letter to the Harstad Family

We were tasked with making sure that the Iraqis had the chance to vote for the upcoming Constitution referendum.[273] We had to provide security for the polling sites and escort all personnel in and out of those compounds.

The first day was pretty hectic. We were attached to Delta 1-184, a National Guard unit from California. They were so FUBAR (Fucked Up Beyond All Recognition) that no one had any idea what the fuck was going on. So, basically, we just sat there the entire day pulling security in the middle of the sun while the lieutenants tried to figure out what was up.

Finally, Lieutenant Turcotte (our platoon leader) made a command decision and told us to take up quarters in a school. We slept on the floors and, let me tell you, it wasn't nice.

As you can imagine, our morale began to decline rapidly, bleeding over from their battalion. It was a constant pendulum swing, from strict micromanagement one moment to utter disorganization the next. It felt like we were fighting a losing battle. Strangely, it wasn't going out into the sector that was bothering me, or any of us that I could tell. It was the feeling of being hand-

---

[273] Wong, Edward, and Dexter Filkins. "Iraqis Vote on New Constitution." The New York Times, October 15, 2005. https://www.nytimes.com/2005/10/15/world/middleeast/iraqis-vote-on-new-constitution.html.

cuffed while trying to do our jobs. Fortunately, I ran into a sorely missed friend just in time. I think he felt the same way, too.

**14 August 2005 – Letter to his family from Donald Cyr, Alpha Company, 3-7 Infantry**

As I went to chow today, being tired and burned out, a soldier from 4-64 said hi to me. He said to go to a certain barracks, and I will find a friend. So, I went after I ate to look for this friend.

As I make it to the barracks, I noticed he was working outside. I took my Gatorade bottle, as he hadn't noticed me, and threw it at him as hard as I could, scaring the bejesus out of him! I came around to attack him, and he caught me.

It was my best friend, Jeff Harstad, from Sioux City, Iowa. We ran and gave each other a hug like two brothers that have not seen one another, literally, for 8 months. We were both happy that we were still alive and shared some of the events of the last 8 months.

Just when I felt down, my friend gave me all the motivation that I needed.

**Donald Cyr in the FOB at night - it was great catching up with him (2005)**

**Photo courtesy of Donald Cyr**

**14 August 2005 – Letter to the Harstad Family from Donald Cyr, Alpha Company, 3-7 Infantry**

I was very burned out and tired of this place. Too much has happened that people were not meant to see or ever talk about. It's like *Groundhog Day* out here, and every month someone is getting hurt. Less people – more work.

All that pain, agitation, and irritation inside went away after I saw Jeff and hugged him. I find it amazing that sometimes all you need is to see your friend. It's something the stress people (psychiatrists) and chaplains can't help you with.

He's out in the wild now (Al-Dora district), so he will see more action, but nothing a 240 (machine gun) can't fix.

Cyr had also been promoted to sergeant. We quickly caught each other up on the unbelievable experiences we'd had since arriving in Iraq and discussed the challenges we were facing in our respective missions. He had his problems, and I had mine. Sometimes, advice from an outside source proved invaluable, while other times, there just wasn't much that could be done. Still, it was the perfect chance to vent freely, with no need to worry about prying ears or someone passing judgment.

Although we were spending more time outside the FOB than ever before, I made a point of attending Combatives classes on base whenever I could. Along with trips to the chow hall, it was one of the few ways to escape the drama.

**21 August 2005 – Letter to the Harstad Family**

I have been doing weight training and a martial arts class. I had to wrestle some military police chick a few nights ago. I choked her unconscious in about ten seconds. Everyone (the instructors) was like, "I can't believe you just did that!" I was confused. Didn't you just pair me up with her? Was I supposed to go easy?

I wrestled in four other bouts, winning the first three and coming to a draw in the fourth. The fourth was called because the guy busted my nose and I was bleeding everywhere.

As NCOs, I think most of us tried to keep the lower enlisted personnel motivated and somewhat out of the loop. I'm sure they were quite well aware of what was happening, though. Still, weight training, Combatives classes, and regular PT were good ways to keep busy and avoid stressing over things outside our control. Plus, we were constantly wrestling around, coming up with elaborate pranks, and doing our best to keep the mood light.

**6 October 2005 – Letter to the Harstad Family**

We did some PT this morning. All we did was run around the FOB a couple times and try to get our privates to throw up. They did pretty good, though, and I was probably closer to puking than they were.

Lately, I have been doing a lot of crap to get stuff organized within our squad. I took East (one of my soldiers) down to get a notarized power of attorney for his house. He just got married and now has a mortgage and everything. Damn kid.

Hallway Fight Club, Iraq Edition: Wiseman vs. Harstad, with a Gasman surprise attack from the top rope

**17 October 2005 – Letter to the Harstad Family**

This morning, I did PT with the rest of the guys in my squad. SSG Gasman told everyone in the squad that they would get the next two days off if they put me upside down in a trash can. There was no way in hell that this was gonna happen.

He and SSG Wiseman attacked me and tried to stick me in there. I wiggled my way free and choked them both at the same time. The rest of my team watched in absolute entertainment.

**6 January 2026 – Reflection of Derek Hauger, Bravo 4-64 Armor**

I remember you and Wiseman got into a fight in the hallway. I was off to the side doing what I do best, riling everybody up like, "Damn, Wiseman! Harstad just whupped your fuckin' ass!" Then Gasman decided to jump in, and you handled both of them at the same time.

Afterward, Wiseman was pissed because I wouldn't stop giving him shit. He turns to me, all heated, and goes, "You want to fight?" I said, "Sure, let's fight. You'll beat me, so I'll just tap out immediately. It really doesn't matter to me." With that comment, I just took his ammunition away. He was really mad about getting worked over in front of everyone and from my comments, and he wanted revenge.

The constant scrapping, nonstop pranks, and general chaos in the barracks weren't personal. We were just a tight group of guys who fought like brothers. We argued and went at each other all the time, but the moment someone from outside the platoon was dumb enough to insert himself into our business, we closed ranks. Suddenly, he wasn't dealing with one of us; he was dealing with *all* of us. In that way, even with our differences, we defended each other fiercely against anyone from the outside.

**21 July 2025 – Reflection of Juan Serrano, Platoon Sergeant, Bravo 4-64 Armor**

I tried to get everyone to relax the brain and made things as fun as possible in our platoon when we were back at base. I let you guys play cards, and we pulled a lot of pranks, as long as everyone stayed out of trouble.

All the crazy, funny shit I did had a reason behind it. You didn't know what I was doing at the time, but I did. I'd seen it before. It's like those kids who go through a school shooting. That one event can give them PTSD for the rest of their lives, but we were in shootings, explosions, and crazy stuff every single day.

I needed guys who wouldn't end up at sick call from emotional stress. I needed you all to show up and do your jobs so we could go home. We had to keep people relaxed. It worked, but I took some serious heat over it.

Still, receiving orders for the day was often baffling. Conversations with my squad leader regularly went something like this:

"So, where are we headed today?" I'd ask. "Snatching anyone? Anything fun?"

He'd just shake his head. "Nope, not today. Route Irish and Route Mercedes."

"What? We got blown up on Mercedes yesterday. Why the hell are we going back? You know there's going to be another IED waiting for us there."

He sighed. "I don't know, man. I don't like it either, but that's what came down. We're running it again and going through all the houses out there." But the look on his face said it all.

"Are they *trying* to kill us?" I asked. "If the insurgents know we're gonna run the same routes every day, they're gonna hit us again."

"Shit, I don't know… probably," he said. "But I don't make the calls, I just follow what was given to me. Maybe we'll get lucky this time. Put a positive spin on it for your guys."

So, we'd roll out, hoping the inevitable IED would miss us or be too small to do any real damage. Like clockwork, we'd get hit again. Back at the base, we would assess the damage, try to repair what could be fixed, and curse the situation.

What's on the schedule later today? Running Mercedes. And tomorrow? Running Mercedes. And the day after that? *Running Mercedes.*

Sitting in the back of a Bradley feels like being entombed alive, especially when you know what's coming. The air is thick with heat and humidity, heavy enough to feel like it's pressing down on your chest. It's grown stale and sour, seven men packed inside the back, rebreathing the same dusty, rancid air with every breath. The engine's deafening scream drowns out every other sound, leaving you sealed in a thick metal box with nothing but the roar and your thoughts. There are no windows, only narrow periscopes to glimpse the outside world, as if through a keyhole in a coffin lid. Everything shakes and shudders as you tear down the narrow and cratered roads, blindly tossed about with each quick turn.

We sat shoulder to shoulder on uncomfortable benches on either side, packed in like undead corpses in a mass grave, sweat mixing with diesel fumes, our weapons pressed against our chests, waiting. Always waiting.

Waiting for the ramp to drop. Waiting for the earth to erupt beneath us. Waiting for the explosion and fire to come searing through the hull like a blowtorch through tinfoil. Waiting for the vehicle to flip, metal shrieking as it crumples inward, turning the troop compartment into a burning steel coffin, where your body is crushed and flayed open, your lungs fill with smoke and fire, and your screams are lost beneath the roar of a dying machine. And through it all, you sit there, silent, sweating, praying the tomb doesn't permanently seal shut today.


**20 July 2025 – Reflection of Robert Roth, Commander of 4-64 Armor**

What happened in 1-184 IN Battalion's Area of Operations involved serious tactical mistakes. They simply were not prepared for counter-insurgency operations. During an insurgency, you have to remain unpredictable. If that means driving through someone's front yard, so be it. Nothing should resemble "normal" operations. This was a problem for the leadership of 1-184 IN to understand.

**1 August 2025 – Reflection of David Anderson, Commander of Charlie 4-64 Armor**

We were being micromanaged and told we had to keep a patrol moving in our sector 24 hours a day. But the battalion insisted it couldn't be just a "presence patrol." I kept telling my platoon leaders, "Know your task, and make sure you've got a purpose." The problem was that I understood the task but couldn't see the purpose. Why are we doing this? What is the goal? It turned into a constant fight.

Just when the fun had started, it was time for my designated leave. I decided to head back home to Sioux City, where my family was eagerly awaiting my arrival. I was just as eager to see them, and it was going to be a much-needed break after everything we had been through.

But, of course, there was a catch, as if you haven't already caught on to the pattern of this book. To go on R&R, I had to leave all my ammunition, gear, and weapons behind at the FOB. I would then have to walk unarmed to the helicopter platform, fly across Iraq, and take a commercial flight home while still in uniform. It felt like 2003 all over again.

To say I was unnerved by this would be an understatement. Actually, I'm officially changing my previous statements in the book about never being afraid. I *was* afraid. I thought it was a ridiculously stupid idea, especially if we were shot down in this part of the country.

I had one of my soldiers, PFC East, accompany me to the landing site while I carried my weapon and one magazine of ammunition. When the chopper landed, I reluctantly handed him both items and bid him farewell. As I climbed aboard, I was completely unarmed, feeling more vulnerable than I had in 2003. At least then I had something to hit terrorists with! Here, I had absolutely nothing at my disposal, but this is how all of us did it.

Making it safely back to the States, I made a beeline for the Cinnabon stand in the airport. I hadn't had any American food since 2004, despite the presence of fast food around the Green Zone. The kind lady at the counter gave me two of the super-sweet cinnamon rolls for free, all because I was still in uniform. I scarfed them down in record time, but this turned out to be a terrible decision. As delicious as they were, my body wasn't used to American food anymore, and I almost immediately vomited into the trash can.

The first day or two at home, I was fine, hugging a lot of family members and sharing sanitized stories. But it didn't take long before I began feeling this overwhelming sense of impending doom. I needed to go back, and I needed to be with my brothers. They needed me, but more importantly, I needed them. As crazy as it sounds, I felt utterly naked without my weapon, my uniform, and a buddy by my side.

**23 April 2025 – Reflection of my father, Bill Harstad, Jr.**

I remember you were a little distant. You were home, but it didn't feel like your mind was really here with us. Mom and I figured you probably just needed some time to decompress. Then you started telling me what was going on in your new area of operations, and that's when I got genuinely worried. When you said you were afraid guys in your platoon were going to die, it hit hard. We didn't know how to help you from where we were, and it left us feeling kind of powerless.

**Two good dogs, one tired soldier on leave in Sioux City: Rox (left) and Stormin' Norman (right)**

So, as too many soldiers do, I started drinking heavily. I drank so much that I don't remember most of my R&R, with the memory just a blur. I wish I could recall what I did or who I talked to, but the truth is, I don't. It's a shame because I wasted that week, or however long I was in the United States.

To make matters worse, while I was home, I got word that my platoon had been hit hard by an IED. There were injuries, and somehow, we'd narrowly avoided a slew of KIAs. My heart sank. *Fuck*, I thought, *I have to get back*. Derek Hauger, who would also be promoted to sergeant just shortly after the incident, relayed the news to me.

**18 July 2025 – Reflection of Derek Hauger, Bravo 4-64 Armor**

We were out in a field late at night, walking around with night vision. At one point, SPC Balicki said, "Hauger, my boot is stuck on a wire. Should we cut it?" I told him not to move and went to get SSG Brister.

We started tracing the wire carefully. One end ran toward the road. The other ran straight into a deep drainage ditch, and we couldn't see the bottom. It was steep. SGT Swears and I decided to go down into it together, back-to-back, covering opposite directions. We counted down quietly and jumped in. I was sure we were about to land on top of somebody, and I remember thinking I'd have to fire my SAW the moment we hit the bottom. But the ditch was empty.

We followed the wire as it zigzagged through the ditch for roughly 200 meters. It felt like trench warfare, because every turn felt like a kill zone. No one was there, though. The wire simply ended, so we collected it and took it with us. Balicki finding it may have kept a future patrol from getting hit.

After that, we searched the area and hung around in a patch of trees until dawn. We'd been out all night, and everyone was exhausted. Once it got light, we started moving back.

After wrapping up the reconnaissance mission, everyone piled into the back of the Bradleys and started heading for home. While they hadn't caught anyone, they were still excited to learn the location of the wire and would continue to actively monitor the area. With a bit of luck, they could kill or capture those involved.

Now, to understand what happened next, it is essential to grasp the layout of a Bradley to fully appreciate the severity of the situation they then found themselves in. A Bradley is divided into three main sections: the driver's compartment, the turret, and the troop compartment.

At the front left of the hull sits the driver, positioned in a semi-reclined seat, with a hatch overhead and a periscope for forward visibility. He can't really see left or right and must rely on the vehicle commander for directions. To the driver's right is the engine compartment, which takes up the front right side of the vehicle.

The center section houses the turret, which can rotate 360 degrees and is manned by two crew members: the gunner, seated on the lower left, and the commander, seated higher up on the right. Each has dedicated sights and firing controls for the vehicle's primary weapons. They can and sometimes do stick their heads out of hatches above them, which increases visibility, but during combat or in areas with IEDs, this is extremely dangerous.

Behind the turret is the troop compartment, where six or seven dismounted infantrymen sit on two narrow benches running along the left and right sides of the interior, facing inward. There are no seatbelts, and as I previously described, the ride is often rough and disorienting, especially during combat or high-speed maneuvers. Entry and exit are through a powered rear ramp. If the ramp loses power, a small personnel door is built into it that can be manually opened using a heavy metal lever, although doing so can be quite challenging. Technically, there is a small passage connecting the troop compartment to the driver, often informally called a "tunnel." It is incredibly cramped and not realistically usable while wearing body armor or carrying full gear. In practice, the rear infantry squad, the turret crew, and the driver remain separated and compartmentalized during operations.

Just before 0800 on August 26, the Gambler Bradleys, stretched out in a line, turned onto Route Mercedes, moving back to the FOB. On the right, the road ran alongside a deep concrete irrigation canal, filled to the brim with fast-moving water. They were still within earshot of the spot where they'd found that suspicious wire.

### 22 August 2025 – Reflection of Matt Turcotte, Platoon Leader, Bravo 4-64 Armor

We'd taken our Bradleys into the southern sector of our area of operations, where we hadn't conducted many patrols. After a few hours, we decided to head back to the FOB and tried numerous egress routes, but each one led to a dead end. The only way back was the same way we'd come in. We were trapped.

Army tactics teach you to ingress and egress on different routes. Worse, the route we'd used was a chokepoint: a bend in the road surrounded by trenches, the kind of terrain that could give insurgents cover and concealment at a distance. But we'd exhausted our options. There was no other way out. After talking it over with the other Bradley commanders, we decided the best course was to punch through it. My track went first, and we quickly learned it was the wrong decision.

*Boom*! Without warning, an IED detonated, slamming directly into the front left side of the lead vehicle, right where SPC Little sat in the driver's seat. The blast ripped through the armor, and shrapnel tore into the Bradley, injuring him. Smoke, dust, and engine oil sprayed into the driver's compartment as the 30-ton vehicle rumbled forward, heading toward the canal.

### 22 August 2025 – Reflection of Matt Turcotte, Platoon Leader, Bravo 4-64 Armor

I heard and felt a tremendous explosion. We got crushed. Smoke filled the compartment, and it was impossible to see. When I finally got my bearings, I heard SPC Little screaming in pain from the driver's seat. I looked over to my gunner, SGT Johnson, and saw a gash on his head from where he'd smashed into the turret. I scanned for a secondary threat and caught sight of someone in the distance, but the figure bolted out of view before I could react.

### 18 July 2025 – Reflection of Derek Hauger, Bravo 4-64 Armor

I was sitting on the left side of the Bradley, on the bench as far in as it would go, with my ass right next to that tunnel. A few feet off to my left was where Little sat, driving us back home. I'd dozed off, trying to grab a quick nap on the ride back to base.

Then, all of a sudden, everything exploded, and I got flung into the middle. Dust was everywhere, and my bell was rung. I was in a daze, trying to figure out what had just happened.

The vehicle immediately lost all power, yet it continued rolling, its dead weight drifting toward the canal. Though stunned and injured, Little managed to steer just enough to avoid plunging into the water. He accomplished this despite having no powered steering, and the Bradley came to a halt, its front right end hanging over the canal's edge. The right track was partially submerged. The rest teetered on the brink.

"Get out!" the men in the turret shouted to the grunts in the back, throwing open their overhead hatches. "We've got no power!"

### 18 July 2025 – Reflection of Derek Hauger, Bravo 4-64 Armor

The vehicle lost power, but I guess the ramp still worked. Little hit the button, and it started to drop. I remember hearing SGT Swears yelling, "Hauger, let's go!" Everyone was piling out onto the roadway, but the female interpreter was right in front of me, and I couldn't get past her. I shoved her as hard as I could from behind, and we both stumbled out onto the road.

That's when I looked down, and I remember thinking, *That's weird. What is that?* From the concussion and us being in a daze, it didn't even register at first that we were all standing right beside a massive pile of unexploded ordnance, rigged with a switch to go off the moment we spilled out of the Bradley. It was a big pile of artillery rounds sitting in a hole, right next to us, with wires and a detonator hooked to it.

### 22 August 2025 – Reflection of Matt Turcotte, Platoon Leader, Bravo 4-64 Armor

I quickly looked over the track and saw our front end hanging over a trench, but luckily it wasn't still creeping forward. Then I checked the rear and saw the blast hole with exposed, unexploded ordnance, so we couldn't back up either. By then, SPC Little came over the radio and said the track was dead and couldn't move anyway. We were sitting ducks. My mind went straight to the seven infantrymen standing in the back.

### 1 August 2025 – Reflection of David Anderson, Commander of Charlie 4-64 Armor

The insurgents had set a second IED for the dismounts as they came out of the Bradley. They'd stacked three 155mm rounds together and wired them to another detonator. When the first IED went off, it ended up disabling that one. It was pure chance that everyone spilling out of the back of that Bradley wasn't killed.

"There he is!" SGT Johnson yelled from inside the turret.

Everyone turned to see him pointing at an Iraqi man off in the near distance, who had just jumped up from his hiding place. The crew couldn't spin the turret as it had been immobilized in the blast, and instead pulled out their M-16s and M4s in an attempt to shoot the fleeing insurgent. Johnson squeezed off a round or two, but his rifle jammed. In a split second, the man disappeared.

SGT Swears, armed with an M-16 and an attached M203 grenade launcher, sprinted toward the spot where the insurgent had last been seen. SPC Hauger, carrying his heavy SAW, and the disoriented female interpreter ran parallel to him. They ran so fast and so far that they quickly lost contact with the rest of the platoon, who had already begun frantically maneuvering away from the pile of unexploded ordnance and into a perimeter, its detonator having failed to go off.

Both Swears and Hauger were strong runners and managed to catch up to the fleeing man, despite his significant head start and the 80-plus pounds of gear they were carrying. Rather than shooting him, as many likely would have in the same situation, they took him down by force, beating him into submission and taking him into custody. He was zip-tied and blindfolded on the spot.

Upon returning to the scene, they realized just how narrowly they had escaped death. Not only had they miraculously survived the IED and avoided the pile of artillery rounds, but they now saw that the edge of the Bradley was hanging over the canal. Had the vehicle gone just a few inches farther, it likely would have flipped into the deep, fast-moving water.

The driver and the two men in the turret would have had virtually no chance of escape, trapped as the vehicle rapidly filled with water. The soldiers in the back might have had a slim chance, but only if they could pry open the heavy rear hatch while upside down, with the full force of the canal pressing against it. But it's highly doubtful anyone would have survived. Ten lives could have been lost in an instant.

**1 August 2025 – Reflection of David Anderson, Commander of Charlie 4-64 Armor**

We took over for a unit that had been running tanks along those same canals. One night, they were driving with night vision and missed a turn. The tank went into the canal, and the main gun jammed into the dirt. This meant they could no longer turn the turret. The driver's hatch was directly beneath the main gun, and because of that, the driver couldn't open his hatch or get out. He drowned.

After hearing that, we strapped garden hoses along the sides of our tanks, just in case it ever happened again. The idea was that we could feed a hose down to the driver so he could breathe until we could get him out. It would be awful to breathe through a hose, but it might give him a fighting chance.

I don't think you guys had anything like that on your Bradleys, though. They would have all died.

**The front edge of the disabled track over the canal, which was filled to the brim with rushing water**

**Photo courtesy of David Anderson**

**21 July 2025 – Reflection of Juan Serrano, Platoon Sergeant, Bravo 4-64 Armor**

I was the turret commander in the second Bradley, right behind Little's vehicle. There was a sharp turn to the left, and a water ditch ran along the right side of the road. I saw the explosion and watched them start rolling toward the water. If they'd gone in, they would've gone right into the canal, or maybe even flipped. They would've all drowned.

Somehow, Little wrestled it back away from the water. It all happened so fast. I sent SPC Foutch (an infantryman in our platoon) to check on Little, since he'd been a nurse before joining the Army. At the same time, I sent people after the guy who took off running.

Upon a closer inspection of the site, the Gamblers discovered a wire leading back to a hidden detonation switch. Embedded in the ground just ahead of where the insurgent had been lying was a wooden stick with a crude notch carved into it, a makeshift aiming device. Additional evidence from the attack was carefully photographed, collected, and turned over to the intelligence personnel at FOB Falcon for analysis.

**The insurgent's sight picture: a simple stake with a notch carved into it that he used to line up our Bradley**

**Photo courtesy of Derek Hauger**

To their surprise and genuine dismay, the entire group of Gamblers and Cyclones were not commended for their actions; instead, they found themselves facing accusations of abuse by the end of the day's operations. The National Guard task force launched an official investigation, and personnel were questioned about their involvement in the incident. It seemed the military had suddenly turned into a circus.

### 1 August 2025 – Reflection of David Anderson, Commander of Charlie 4-64 Armor

Lieutenant Turcotte actually saw two guys detonate the first IED. Then two of our infantrymen tackled one of them, and we eventually caught the other. When a 200-pound, jacked-up infantryman hits you while you're trying to sprint down a road, especially when wearing full kit, you're going to get hurt.

But this was right on the heels of Abu Ghraib, and the military was hypersensitive to anything that could turn into another controversy. It didn't help that the two men we captured turned out to be the sons of a powerful local sheikh. They started making accusations that the interpreters burned them with cigarettes and threw in a bunch of other lies. They knew exactly how to work the system.

It disgusted me, but an investigation was opened anyway, and the Army made me read my guys their rights. In the end, the battalion ordered us to release both men, even though we had witnesses. They didn't want the political headache with the sheikh or the risk of an abuse scandal, even if the allegations weren't true. They said, "We want to calm the problems in the area, so we need to let them go."

Throughout all the drama, the steadfast leaders of 4-64 stood by us, especially those from Bravo and Charlie companies. Even though certain higher-ups in the Army seemed to always be on a witch hunt, our leaders took care of their men. We could only shake our heads in mutual frustration. What exactly did they expect us to do? Then, quite comically, the allegations of abuse were suddenly dropped for some reason, and the Gamblers were now regarded as heroes.

**22 September 2005 – Letter from Mark Barnes, 1st Sergeant of Bravo 4-64 Armor**

SGT Swears and SPC Hauger demonstrated some physical dominance the other day to someone who blew up the Bradley that SPC Little got hurt on. They ran his ass down through the fields with all their equipment on and took him for a ride back to the FOB.

For their actions, SGT Shawn Swears and SPC Derek Hauger were each awarded the Bronze Star. If they hadn't moved as fast as they did, the insurgent and his brother, who was also taken into custody, would have escaped and likely carried out follow-on attacks. Their speed, strength, and quick thinking made all the difference.

But hold on. Let's stop and think about this for a second and see if it makes any sense. Just playing Devil's Advocate, let's pretend some higher-ups genuinely didn't believe those two men were insurgents, and that everyone had made the whole thing up. Sure. We apparently blew up our own Bradley for fun, Little faked his injuries by using a prosthetic leg and ketchup packets, and those 155mm artillery rounds were actually avant-garde art pieces we "found" at a Yoko Ono convention. And the two detainees we chased down? Just wholesome members of the Mickey Mouse Club on a pilgrimage to Disneyland. So, what do we do with them? We were ordered to release them.

Fine. Let's go along with this bizarre but apparently possible scenario. Here's the problem: not long after we were told to let them go, our soldiers were awarded *Bronze Stars* for capturing them in the first place, which aren't exactly everyday medals. So, which is it? We had to release them because they supposedly weren't a threat… yet we're decorated for taking them off the battle-field? Let the head-scratching begin.

### 18 July 2025 – Reflection of Derek Hauger, Bravo 4-64 Armor

One day I'm facing a court-martial, and the next they're pinning a Bronze Star on my chest. When they called us out to formation, nobody told us what it was for. I was scared. Then, out of no-where, they award me a Bronze Star. *What*? Weren't you trying to get me into trouble yesterday?

**LTC Roth (left) and CSM Stanley (right) present a challenge coin to SPC Little, and stop in to check on us**

SPC Jason Little was awarded the Purple Heart for the injuries he sustained in the blast, primarily to his leg. And yes, I saw that leg. It wasn't a prosthetic, and his wounds were very real, even if they weren't severe. Despite those injuries, his quick actions likely saved everyone in the Bradley. Though he eventually made a full recovery, the truth is, it could have ended very different-ly. We were lucky… this time.

**30 August 2005 – Letter from Cap Stanley, Command Sergeant Major of 4-64 Armor**

Speaking of heroes… and we have MANY… SPC Little of B Company was injured in an IED the other day. His Bradley was hit with such force that it cracked the hull, and the vehicle teetered on the verge of falling into a canal. SPC Little, although injured, maintained control of the vehicle and ensured it remained on dry land, thus saving the lives of his vehicle-mates.

**10 September 2005 – Letter from Brian Dugan, Commander of Bravo 4-64 Armor**

1st Platoon left FOB Prosperity with Charlie Company to join the 1-184th IN from the California National Guard down south at FOB Falcon.

The men of 1st Platoon may not live here with us on FOB Prosperity, but they are with us each and every day in our thoughts and prayers. As you may know, 1st Platoon has been on some rough patrols, as they have hit several IEDs, with one causing minor injuries to SPC Jason Little.

1SG Barnes and I recently made a trip to FOB Falcon to visit SPC Little and the men of 1st Platoon. SPC Little seems to be doing fine. He had some cuts to his leg and a very bruised ankle from the explosion. When we saw him, he was up, limping around, which is a good sign. It is tough to keep a Bandit down.

Just after this incident, I emerged from my drunken stupor that was R&R and returned to the desert. While it may seem strange, I finally felt at home with my brothers. I had my weapon, my friends, and my purpose back. Now, I needed to get back to work.

But before I move on to our dark days in September, a couple things need to be said about this event. Although the two brothers were ultimately ordered released, there's a small silver lining to the story. A rare moment of cosmic justice, one may call it.

**1 August 2025 – Reflection of David Anderson, Commander of Charlie 4-64 Armor**

I don't remember what happened to the one brother, but I had orders to bring the other one back, which we reluctantly did. A few weeks later, we pulled that same guy out of one of the canals. He had been tortured and killed by someone, but we never learned who did it.

In an effort to kill or capture more insurgents planting IEDs along that route, CPT Anderson and 1SG Gifford established a concealed observation post near the site of Little's incident. A small team of grunts manned the position around the clock, hiding and watching for any signs of enemy activity, while the tankers waited on standby farther down the road. It was a bold plan, and it would go on to become a hilarious Cyclone legend.

**1 August 2025 – Reflection of David Anderson, Commander of Charlie 4-64 Armor**

We had a rotating squad of infantrymen set up in a hide overlooking that intersection, and we sat down the road in our tanks just waiting for the word. You guys had to stay in there and not move around for days at a time. At first there wasn't any activity, but eventually we had some success.

SSG Santiago was out there on the radio, and he quietly said over the company channel, "There are two men unrolling a wire in the roadway." I immediately called battalion on their frequency and let them know about it, just to keep them updated. The battalion commander, on the other end, kept calling me back and asking questions as I'm trying to manage the scene.

Eventually, he was bugging me so much that I couldn't focus on what I was doing. Frustrated, I said over the company channel that I didn't think he was monitoring, "Battalion can eat a bowl of dicks." Well, unfortunately, my statement, made in the heat of the moment, went across the airwaves and was very much heard by the commander.

Then I hear, "Cyclone 6, this is Stalker 6. I can eat a bowl of what?" I thought, *Oh shit.* I immediately radioed back and said, "I'm sorry, sir. I'll have to eat them myself."

Fortunately, the commander had a pretty good sense of humor about it and didn't yell at me too much. He told me he understood it was said during a stressful situation, but that I needed to control myself on the radio. A few days later, a general was visiting our unit. Standing there with him, the commander said to me, "Hey, tell the general about the bowl of dicks!" So, I had to embarrassingly explain to the general how I had told the battalion commander to eat a bowl of dicks. After that, our working relationship greatly improved, and I think he started to trust me. It was a bit of a turning point.

Although we remained locked in a constant struggle with both the local insurgents *and* our own task force, this time, we had come out on top.

## Reflections

**Back from R&R, standing with Derek Hauger - I missed this guy!**

In retrospect, I now realize that what I was experiencing during leave was the early onset of what psychiatrists now call PTSD. In the past, it went by other names, such as shell shock, combat fatigue, war neurosis, soldier's heart, or operational exhaustion.[274] But they all describe the same thing, that being the body's response to intense mental or emotional trauma. For some, it can stem from a single life-altering event, but in our case, it came from hundreds of incidents, all stacked back-to-back with no time to process, reflect, or recover. There was no time, simply because we had to keep fighting.

I should have known better than to drink myself into oblivion, but that's exactly what I did. It made me feel better. But, as much as I hated the situation we were dealing with at FOB Fal-

---

[274] Friedman, Matthew. "History of PTSD in Veterans: Civil War to DSM-5." U.S. Department of Veterans Affairs, March 26, 2025. https://www.ptsd.va.gov/understand/what/history_ptsd.asp.

con, I knew one thing for sure. I knew I'd rather die out there with my brothers than sit safely back in the States. That might be hard for some people to understand, and I get that. But I think most of us felt the same way. We were in a dark and dangerous place, and there was no room for slacking off or letting our guard down. We couldn't afford it.

When interviewing people for this book, especially those familiar with our mission at FOB Falcon, we all shared a sense of exasperation and growing frustration with the situation. Across the board, officers and enlisted alike, no one actually believed that anyone was intentionally trying to put us in harm's way. Instead, the general consensus was this: 1-184 had been through hell. They'd had multiple leaders relieved of command, several soldiers facing disciplinary and even criminal charges, and they'd suffered heavy losses. Their men had been killed and wounded while trying to hold the line in Al-Dora's buzz saw.

Now imagine being part of that unit, carrying all that baggage, and knowing you might have to go home with your tail between your legs. They were under immense pressure to turn things around quickly, to reclaim their reputation, to prove themselves, and to find some kind of redemption, but the damage had already been done. And while many of the tactics being employed weren't malicious, even if at the time we thought they were, they were instead just desperate and sometimes downright foolish. That was Hanlon's Razor in real time. We mistook desperate, clumsy decisions for deliberate malice because the consequences felt the same either way.

In *It Doesn't Take a Hero*, General H. Norman Schwarzkopf described a leadership lesson that echoed something I once learned from Marcus Brister. As he put it, "Nobody at West Point had told me you can learn as much from a lousy leader as from a good one...."[275] I took that idea to heart because my own education often came the hard way, by watching what *not* to do and then choosing a different path. Those hard examples stayed with me. They shaped how I carried myself around the lower enlisted, and they followed me long after the uniform came off, into my life as a husband and a father.

One of the peculiarities of the American military is that rank is carried across all components, including Active Duty, Reserves, and National Guard.[276] In other words, a sergeant in the National Guard, who may train less than 40 days in a typical year, still wears the same stripes as a full-time active-duty sergeant. That meant we were expected to show the same respect for the rank, even if we were far more experienced or strongly disagreed with the tactics or decisions of those who wore it.

To better illustrate the unusual mix of components and the imbalance in experience that we encountered, a brief example may help. In active-duty service, a soldier typically begins at the rank of E-1 (private) and can reach the rank of E-4 (specialist) within two to four years, assuming

---

[275] Schwarzkopf, Norman. *It Doesn't Take a Hero: The Autobiography of General H. Norman Schwarzkopf.* eBook edition. Bantam, 1993, Chapter 6.

[276] Department of the Army. "Army Command Policy: Army Regulation 600-20," July 24, 2020. https://www.ncolcoe.army.mil/Portals/71/administration/eeo/ref/AR-600-20.pdf.

steady progression.[277] Keep in mind, this rank is not a leadership position, but it usually precedes assignments that involve leading junior soldiers and is the steppingstone to becoming an NCO.

While active-duty military personnel are technically on duty 24/7, 365 days a year, let's set that aside for a moment and consider what a "normal" work schedule looks like in practice. Assuming a standard five-day work week and no deployments or extended field time, an active-duty soldier works roughly 260 days per year. Doing the calculation, that equates to 520 to 1,040 full-time workdays of hands-on experience by the time they reach E-4.

Now compare that to an E-7 (sergeant first class) in the National Guard or Army Reserve. These part-time soldiers often serve one weekend per month and two weeks per year, totaling about 38 or 39 training days, assuming no deployments, mobilizations, or extended training exercises.[278] If it takes 15 years to reach E-7 under that standard schedule, they could have as few as about 585 cumulative training days, near the lower range of days an active-duty E-4 may accumulate in just a few years.

Yes, this is a simplified calculation that doesn't account for individual variation, basic training, prior active duty, or deployments. But the point stands: a National Guard E-7 platoon sergeant, responsible for 30 to 40 soldiers, may very well have fewer hands-on training days than a young, active-duty E-4 specialist with no formal leadership role. In other words, many junior active-duty soldiers in their first enlistment can accumulate twice as many on-the-job days as a mid-career Guardsman. Let that sink in.

To further illustrate, imagine working at a full-time job for several years, only to have a co-worker who works just two days a month come in and tell you what to do. How do you think that would play out? If your job requires any skill at all, you'd likely realize that those skills start to fade when they're not regularly used. Instructions from the previous month are forgotten, simple tasks take longer to recall, and each task needs to be redone to "knock the rust off."

Of course, none of this is meant to suggest that the National Guard or Reserves don't bring value to the military. It's quite the opposite. They provide essential capabilities, especially in areas where civilian experience aligns with military roles. For example, civilian police officers often transition effectively into Military Police units, and financial professionals can apply their expertise in military finance positions.

Another advantage is that National Guard and Reserve troops are generally older than their active-duty counterparts. While this doesn't always equate to enhanced maturity, it does reflect more years of accumulated life experience. This can bring valuable life skills, civilian job experience, and other benefits from their diverse backgrounds.

The challenge, in my opinion, tends to arise in combat arms roles, where there is little to no civilian equivalent. Outside of something as extreme and clearly unrealistic as a mob hitman,

---

[277] U.S. Army, "U.S. Army Ranks," accessed December 14, 2025, https://www.army.mil/ranks/.

[278] U.S. Department of Defense, "The Eleventh Quadrennial Review of Military Compensation - Chapter 17: Overview of Reserve Component Compensation and Benefits," accessed December 14, 2025, https://militarypay.defense.gov/Portals/3/Documents/Reports/SR20_Chapter_17.pdf.

there's simply no civilian job that mirrors the responsibilities of an infantryman, whose sole mission is killing people and blowing things up. The same goes for tankers and artillery crews. These are uniquely military occupations, requiring constant, hands-on training that civilian life just can't replicate.

So, it wasn't unusual to hear one of our infantry or tanker privates glance over at the National Guard and ask, "What are they doing?" Even the junior enlisted could see that something wasn't right, and at times, it wasn't good for anyone involved.

I know some people may not agree with or like what I'm writing, but my intent is not to offend. Whenever I bring this up, I often hear, "But, but, but…," followed by a confusing series of non-sequiturs, deflections, and false equivalencies. Some might even counter my points by mentioning that units like the 1-184 underwent training just before deployment. This is true. However, it doesn't replace the experience of active duty, no matter how it's framed. I saw it firsthand, we all did, and no amount of wishful thinking can change that.

To put it into perspective, being on active duty meant we were incredibly close, not just in physical proximity, but also in the amount of time we spent together. We literally spent all day, every day, with each other. We ate together, worked together, went to the clubs together, and sometimes even went on leave together. I knew everyone in my company so well that I could recognize them from a distance, just by the way they walked, even in the middle of the night.

Most of the time, especially with the people on my team, I didn't even need to have a conversation to know exactly what they were going to do. For example, I could tell that one guy didn't like to buttonhook in CQB, so I knew exactly which way he would move. We didn't need to talk; instead, I could read the small body language cues he'd give me. I mention this because time spent together is invaluable, especially in combat. It can't be replaced.

# Chapter XV

## Dark Days of September

While we were dodging IEDs and mortar attacks around FOB Falcon, the Tusker battalion kept our families updated with regular emails about what was happening in and around the Green Zone. The updates were well written and appreciated, but they caused some confusion. I remember people asking, *"Wait, are they still at the Green Zone or a different FOB? Are they still working the checkpoints?"* The logistics of being temporarily attached to another unit weren't always easy to explain, especially when that unit was *also* attached to someone else.

As I tried to clarify to my family, our platoon, along with Charlie Company, had been temporarily swapped with a light infantry company from 1-184 Infantry. We were placed under the command of the National Guard, while the 1-184 troops we had replaced operated under the Tuskers. It was essentially a one-for-one exchange.

In one of the emails from CSM Stanley, he mentioned the new senior NCO who had just taken over at the brigade level, who expressed his impression of the assignment. It was encouraging to see some recognition for how brutal and thankless those checkpoint missions really were.

**Undated, Summer of 2005 – Letter from Cap Stanley, Command Sergeant Major of 4-64 Armor**

His (CSM Torres) first day as Brigade CSM, having no clue as to our mission, he visited the checkpoints and the Tusker soldiers.[279] His response, simply: "They are good, very professional. I have a new respect for what your battalion does. I never knew just how tough this mission is."

Well, I do know, and I'm constantly in awe of your Soldiers and Marines. With temperatures soaring in the 120s, standing watch for as much as 12 hours at a time, they maintain their professionalism, tenacity, and constant vigil. Same as you, I expect nothing less of them.

---

[279] McCool, "Operational Leadership Experiences in the Global War on Terrorism: Interview With COL (P) Edward Cardon."

Our brigade also circulated an email reporting that $188 million had recently been allocated to 109 local projects, with most of the funding directed toward infrastructure. New roads, water lines, sewage systems, electrical wiring, and other essential services were being installed daily. Slowly but steadily, Iraqi ministries, backed by American funding and oversight, began rebuilding Baghdad. While the media largely overlooked these efforts, it was undoubtedly a step in the right direction toward improving the lives of the Iraqi people. LTC Roth followed up on the email, highlighting several positive developments that the United States military had directly contributed to.

**24 October 2005 – Letter from Robert Roth, Commander of 4-64 Armor**

I want to share some interesting facts that the media never seem to report, just to set the record straight:

1. Since the new Iraqi government assumed control in April 2005, more than 3,700 Iraqi civilians and security forces have been killed by terrorists.

2. There are many protests in and around the Iraqi capital of Baghdad – yet, only one protest over the past 10 months was directed against the US military remaining in Iraq. It lasted a whopping 30 minutes, and approximately 50 people attended.

3. There is more electricity produced in Iraq today than ever before under Saddam Hussein... but the demand for electricity has grown at an even faster rate. People are buying air conditioners, heaters, computers, stoves, and refrigerators by the bundle.

4. Hundreds of meters of new sewer and water pipelines have been replaced in the city of Baghdad. The city has not seen a reconstruction effort of this magnitude in almost 20 years – and there is still more work to be done.

5. Hundreds of mounted and dismounted patrols – US and Iraqi – are on the street each week, with less than 1% of them actually being attacked. If they are attacked, it is by roadside bomb or suicide bomber, because the terrorists know they cannot win a direct confrontation with US or Iraqi forces. Iraqi security forces control most areas of all the major cities in Iraq.

6. The Iraqi Army is the success story of the year. They are committed, dedicated, and can fight. They protect the people and are proud of it.

Funny, if you listened to the national news media – you wouldn't know that peace is breaking out. (Still), this country has a long, long way to go.... We are attempting to teach them how to live amongst each other- in peace, and respect each other's differences and values.

In August, Iraqi leaders finalized a draft constitution, which marked a significant step toward establishing a democratic government. The following month, the Iraqi Transitional National Assembly approved the draft and submitted it for a national referendum. In October, millions of Iraqis participated in that referendum, casting their votes on the proposed constitution.[280] It was ultimately approved by a majority of voters, laying the foundation for Iraq's future political system.[281]

**11 September 2005 – Letter from Cap Stanley, Command Sergeant Major of 4-64 Armor**

An Iraqi woman was coming through the checkpoint, carrying her small child; the first thing I noticed was his big old chubby cheeks. He was quite a cutie. I mentioned this to the soldiers working there; the woman understood English. Her face lit up with the biggest smile!

The young PFC standing there with me said the most profound statement, and it caught us all (including the Iraqi woman) by surprise: "Yeah, Sergeant Major, and he was born free."

Out of the mouths of babes… wow! I didn't even think about this simple, beautiful baby, and the possibilities of being the first generation in Iraq made free by the sweat and blood of America's finest. The Iraqi woman walked away with her beautiful child in tow, and a smile that could light up New York City. I also carried around a smile for the remainder of the day.

But at FOB Falcon, the progress being made in other parts of Iraq felt as distant as if we were in another country entirely. We were locked in a relentless struggle for control of our sector, facing an insurgency that was overwhelmingly Iraqi but amplified by outside networks.[282] Sunni extremists tied to al-Qaeda in Iraq (AQI) drew in foreign volunteers and facilitators, many of whom entered Iraq through Syria. Captured AQI personnel records showed that Saudis made up the largest share of these foreign recruits, followed by Libyans, with additional fighters from Syria, Yemen, and other countries.[283] At the same time, U.S. executive-branch and congressional reporting described Iran's IRGC-Qods Force providing weapons, funding, training, and advice to Iraqi Shi'a militant groups, with Lebanese Hezbollah instructors reportedly providing training and other sup-

---

[280] DVIDS, "Iraqi Referendum," Defense Visual Information Distribution Service, October 15, 2005, https://www.dvidshub.net/image/11101/iraqi-referendum.

[281] ABC News. "Iraq Constitution Vote Accurate, UN Says," October 25, 2005. https://www.abc.net.au/news/2005-10-26/iraq-constitution-vote-accurate-un-says/2132156.

[282] U.S. Government Publishing Office. "Policy Options for Iraq (Senate Hearing 109-312)," July 2005. https://www.govinfo.gov/content/pkg/CHRG-109shrg26137/html/CHRG-109shrg26137.htm.

[283] Bergen, Peter, Joseph Felter, Vahid Brown, and Jacob Shapiro. *Bombers, Bank Accounts, & Bleedout: Al-Qa'ida's Road in and Out of Iraq.* Combating Terrorism Center at West Point, 2008. https://ctc.westpoint.edu/wp-content/uploads/2011/12/Sinjar_2_FINAL.pdf.

port to some of those militias.[284] So, the foreign element wasn't the bulk of the fight, but it could be disproportionately lethal.

Our interpreters, many of whom spoke multiple languages, would often say about an insurgent we captured, "He's not Iraqi. He's…" and then fill in the blank with the country of origin, picking up on subtle clues from an accent, choice of words, or even small mannerisms. These hints were far outside our level of awareness, so we grew to trust their judgment.

We were killing fighters not only during our targeted missions but also during routine patrols. Yet, it felt like battling a hydra. Every time we cut off one head, more seemed to grow back, often stronger and angrier than before. No matter how many we took down, replacements continued to arrive, fueled by that hidden network of foreign fighters, local recruits, and fanatics eager to fill the void. There was a relentless, almost endless supply of bad guys waiting for their chance to take a shot at us. And when they finally did, we gladly took them up on their offer.

Every day, the Cyclones and Gamblers went out into sector, looking for a fight. Our platoon typically rolled with four Bradleys, while the tankers often deployed with three or four Abrams tanks and occasionally a wheeled vehicle sandwiched in the middle for protection. These configurations provided us with the optimal balance of firepower and maneuverability while minimizing exposure. Too many vehicles in one spot was risky, as an IED could halt a convoy and create deadly kill zones.

There was at least one CIA targeting officer on base who supplied us with intelligence, working alongside our Army sources. We had photographs of high-value targets we wanted to kill or capture, locations to monitor, vehicles to track, and a wide range of other intel. Whenever we captured a fighter alive, we typically brought him straight to the small interrogation compound on base, where the CIA kept its office. What happened to them after that, I have no idea.

Shortly after Little's near-death experience on Route Mercedes, another explosion rocked the company. This time, one of our Humvees was moving through the field right next to where that incident had happened. They were trying to stay inside the older tank tread marks that had already been cut into the vegetation. The Humvee hit what was believed to be an anti-tank mine, which effectively functioned like an IED. The blast tore the front end of the vehicle completely apart, and the driver suffered a broken femur. Everyone else somehow escaped with only minor injuries, except for the gunner, who later developed a blood clot in his leg from the blast and had to be evacuated.[285] Once again, we were fortunate. After that, we went back over the area with detection assets and treated the entire stretch as if it were seeded with booby traps.

Following that incident, the task force had another close call. While on patrol in tanks, CPT Anderson spotted a vehicle containing several military-aged males acting suspiciously. They were sitting in an intersection, watching intently, almost as if challenging us to pursue them, but something felt off. As our patrol approached, the vehicle sped away abruptly, only to stop a short

---

[284] Congressional Research Service, and Kenneth Katzman. "Iran's Foreign and Defense Policies," April 29, 2020. https://nsarchive.gwu.edu/sites/default/files/documents/r1xbvd-nzgpd/20200429%20R44017.pdf.

[285] Anderson, *Ride the Cyclone.* Pages 7-8.

distance ahead. Then they took off again before we could close the distance, crossing over into Delta 1-184's area of operations.

Instead of pursuing them deeper into the National Guard-controlled sector, which would have been futile with a tank, we turned back into our own area and continued the patrol. Unless absolutely necessary, it was always better to operate within familiar ground for several reasons. You know the houses, bridges, and terrain, and you're far less likely to run into a friendly fire situation.

But the Cyclones didn't get far. A massive explosion ripped through the air, coming from the direction we had just left. CPT Anderson, immediately understanding what had happened, turned the patrol around and sped back to the scene. One of the National Guard Humvees, about thirty meters into their own territory, had hit an IED, shredding its body and immobilizing it.[286]

The gunner was dancing around, seemingly elated. Apparently, he had just taken a piece of shrapnel to his groin, though I don't believe any of it actually penetrated his body. Instead, it had embedded itself in the thin Kevlar covering his most sensitive region. He was incredibly lucky to be wearing it, especially since many infantrymen avoided that gear because it restricted movement and was too thin to stop a bullet. Had that been me, I probably would have been dancing around, too.

On the ground, however, was a bloodied NCO, missing part of his foot, but he was alive. His agony was apparent, thrashing about as medical aid was rendered. If I remember correctly, everyone survived this ordeal, but the insurgents responsible were never found.

That stretch of roadway was then flagged as particularly dangerous. But still, patrols had to run through there. The following day, a Delta Company soldier was killed near that same intersection.[287] They had chased another suspicious vehicle when an IED detonated near their Humvee.

Over the next few days, more soldiers would perish in attacks. In one incident, an IED destroyed a National Guard APC and flipped it, pinning a soldier underneath. Under blistering enemy fire, his comrades hooked a chain around the APC to pull the massive vehicle off him. They were successful, but unfortunately, he succumbed to his injuries.[288]

Shortly afterward, another patrol was hit when an IED detonated near an APC, and the troops again came under small arms fire. One soldier was killed immediately, and another died the following day.[289] People were being killed or seriously wounded at an alarming rate in our area of operations.

While all this was happening, we were working closely with Navy SEAL and U.S. Army Special Forces teams, including a dog team. On raids, they focused on the highest-value targets, while we were tasked with capturing or killing lower-value targets and providing outer security. To

---

[286] Anderson, *Ride the Cyclone*. Page 8.

[287] Bailey, Eric. "Army National Guard Staff Sgt. Alfredo B. Silva, 35, Calexico; Killed in an Explosion in Iraq." *Los Angeles Times*, September 25, 2005. https://www.latimes.com/archives/la-xpm-2005-sep-25-me-silva25-story.html.

[288] Parry, "The War You Didn't See."

[289] Union-Tribune, San Diego. "Two Local Men Killed in Iraq." San Diego Union-Tribune, September 26, 2005. https://www.sandiegouniontribune.com/2005/09/26/two-local-men-killed-in-iraq/.

accomplish this, we positioned Bradleys, tanks, and other vehicles in a wide, circular perimeter, often surrounding an entire village. As their special operations teams went to work, we hit additional targets at the same time, creating mass confusion for the insurgents. It was demanding work, but we were all highly effective.

On a hot Iraqi day, September 16th, CPT Anderson and his tanker patrol returned to FOB Falcon after capturing two blacklisted insurgents in our area of operations.[290] Their return had been delayed by several hours due to the capture and processing of the prisoners. In the meantime, 1SG Gifford and battle crews rolled out past the wire in three Abrams tanks. Using this leapfrog approach, we maintained a constant vigil, though it was challenging given the sector's large size and our limited manpower.

Meanwhile, I was fast asleep in my bunk, even though it was the middle of the afternoon. The Gambler Platoon had been active with fighting over the past several nights, and most of us were trying to catch up on some much-needed rest. The Cyclones had given us recuperation time, meaning we wouldn't be disturbed for any operations until they felt we were properly rested.

With the jet-like roar of their engines and the distinctive clacking of the treads, the Cyclone patrol turned west onto Route Corvette, a road we all knew and dreaded because of the countless IEDs buried along its length. Like so many routes in our sector, it ran high above the fields, its steep slopes offering little room to maneuver. Once you were on it, there was almost no escape. It was only forward or back.

**29 July 2025 – Reflection of Travis Nelson, Charlie 4-64 Armor**

I was in the gunner's hole of the lead tank, Red 2, scanning forward.

Route Corvette was a concern for us, and it had been pointed out by the unit we'd taken over for. It was dangerous because there was a canal on one side of the road and a steep drop-off on the other, where you could easily roll a tank.

The (National Guard) battalion had us doing area saturation patrols, which meant we couldn't stop, and had to keep patrolling for 12 hours at a time. They wanted the Iraqi people to constantly see us, and we weren't allowed to stop or do overwatch.

We were all disgruntled about it because we felt like we were just waiting to get hit, and it seemed like they didn't really care about us.

**4 August 2025 – Reflection of Heath Hutchison, Charlie 4-64 Armor**

I was in the middle tank, Red 1, right after Nelson and that crew, and it was uneventful as we turned down Corvette.

---

[290] Anderson, *Ride the Cyclone.* Page 9.

In the middle of the dirt road, there was a boggy, soft spot. We drove over it, and I saw two males lying prone, side by side, on the ground out to our left, almost 400 meters away. They were eyeing us suspiciously, like they were ready to attack. Then they jumped up and sprinted away.

I switched the machine gun to fire, but I was told not to shoot them.


Without warning, a massive explosion larger than anything anyone in the patrol had ever witnessed erupted beneath Red 3, the call sign of the rear tank in the formation. They, too, had been forced to drive over that soft spot in the road, with nowhere else to go. Sand and a giant fireball tore outward in every direction. The turret, weighing well over 20 tons and secured to the hull by a heavy turret ring and massive bolts, was unbelievably ripped free and hurled dozens of feet into the air. Two men were inside the turret as it was launched skyward, while one was in the gunner's hole and another in the driver's compartment.

The turret reached its zenith before plummeting back to earth, screaming down toward the mangled hull that lay crumpled within a smoking, gigantic crater. The tank's barrel struck first, jamming violently into the dirt like a lawn dart, followed by the crushing impact of the turret itself onto the hull. Unexploded 120mm tank rounds lay scattered across the area, mixed with equipment and debris flung hundreds of meters in every direction.

The two remaining tank crews were slammed by the concussion, the deafening blast reverberating through the steel of their vehicles. As they turned to look, the sight they witnessed defied belief: Red 3 lay utterly destroyed. The following sequence of events became chaotic, with actions overlapping, confused, and purged from memory.


**4 August 2025 – Reflection of Heath Hutchison, Charlie 4-64 Armor**

I thought *our* tank had just been hit by an IED because the explosion was so powerful. It violently shook everything. I looked through the sights and saw nothing but dirt and dust flying. When I switched to thermals, there were heat signatures everywhere from the blast, but you couldn't see anything.

Then, as the cloud of dust cleared, my brain couldn't process what I was looking at. I just said, "Holy God. The turret is off."


1LT Kroells, SSG Noto, and SPC Hutchison quickly dismounted from their nearby tank, middle in the column, and sprinted back toward the shattered wreck of Red 3. Other than the tank's engine somehow still running, there was no indication of movement, let alone life. They hesitated for a brief moment, trying to grasp what to do next. Over the radio, someone's strained voice broke through the static, calling back to the command post. It was urgent. They needed help, *now*.

**29 July 2025 – Reflection of Travis Nelson, Charlie 4-64 Armor**

The explosion was so huge that I thought it was us that had been hit. Then I heard someone say, "Oh my God, the turret is off."

We stopped where we were, at the front, and I began scanning for the enemy through my sights while other people dismounted and ran to help them. I didn't see any insurgents at first.

**A destroyed Red 3 on Route Corvette with the turret blown off - a canal on one side and an open field with thick vegetation on the other**

**Photo courtesy of Travis Nelson**

1SG Gifford and SSG Marek, aboard the now-demolished Red 3, had been standing in their seats, trying to gain better visibility, their upper bodies above the open turret hatches. The explosion struck while they were fully upright, scanning the terrain, leaving them dangerously exposed at the worst possible moment. Incredibly, they remained inside the turret as it was launched skyward and then came crashing back down onto the hull. When the dust finally began to settle, their men rushed to the wreckage and found them still in their hatches, slumped over and unresponsive.

The lieutenant forced himself onto the turret, which was precariously lodged on the rear of the tank's hull, directly above the exhaust vents. With the engine still running, super-heated fumes of around 900 degrees Fahrenheit discharged into the turret compartment like a blast furnace.[291] Fighting the heat and smoke, the men hauled a bloodied 1SG Gifford from the loader's hatch and dragged him clear, but there was no pulse. Shouting over the roar, 1LT Kroells ordered someone to kill the engine, though reaching the shutoff was easier said than done.

### 4 August 2025 – Reflection of Heath Hutchison, Charlie 4-64 Armor

1SG Gifford had his loader's hatch locked to the back, but the blast was so massive that it snapped the lock. There was speculation that the lid hit him so hard in the back of the neck that it killed him instantly.

Back at the FOB, CPT Anderson was wrapping up paperwork when an NCO burst into his office, wide-eyed and breathless. "Sir, Red 3's been hit… they said the tank is fucking destroyed!"[292]

The words barely registered before Anderson was on his feet, barking orders to mobilize the rest of the company and locate 1SG Gifford, unaware he had been on that very patrol. Within moments, he and SFC Marrero, along with their crew, jumped into their Abrams and tore across the FOB toward the gate. They roared at the guards to clear the path, the urgency in their voices leaving no room for hesitation. The guards jumped out of the way, but what they were doing was dangerous to the extreme. A lone tank barreling into the most hostile sector of Baghdad was terribly risky.

The tank blasted across Route Jackson, a heavily traveled roadway by military and civilians alike, and crashed over the center median, driving the wrong way down traffic. Arriving at their turn, they pointed in the direction of Red 3. Help was coming.

"Wake up!" someone screamed from the barracks doorway. "We need infantry on the ground! A tank's been destroyed! Grab your shit!"

My roommate, SGT Howerton, and I sat upright, groggy and confused, but we started throwing on our gear. Everything had been laid out in advance, ready for a moment like this, but our minds were still trying to emerge from deep sleep. We didn't fully understand what was happening, only that we had to run and get to the Bradleys immediately.

---

[291] Forecast International. "Honeywell AGT1500," March 2009. https://www.forecastinternational.com/archive/disp_pdf.cfm?DACH_RECNO=180&.

[292] Anderson, *Ride the Cyclone*. Page 9.

Some of the Gamblers had been eating down the road at the chow hall when the call came in. Without hesitation, they abandoned their dinner trays and bolted from the building, sprinting back to the barracks to grab their gear and prepare for combat.

Back on Route Corvette, the men had already pulled 1SG Gifford from the wreckage, but it was clear he was gone. As the crews fought to shut down the engine, SPC Bowen, the gunner, was found on his hands and knees beside the tank. He was dazed and disoriented, but somehow alive.

**4 August 2025 – Reflection of Heath Hutchison, Charlie 4-64 Armor**

"Bo Bo" (SPC Bowen) told me he remembered scanning inside the turret, holding onto the handle, and then feeling the blast. Then he found himself flying up in the air. The explosion somehow blew him just to the south of the tank, and he landed on the dirt.

That's when I found him, on his hands and knees, looking down at the ground. He was out of it.

SPC Hutchison helped his friend steady his knees and partially sit up. Bowen, a muscular bodybuilder and one of the largest men in the platoon, was difficult to move. But with effort and teamwork, they managed to get him on his feet.

Seconds later, a burst of rifle fire split the air. A round tore into Bowen's thigh, narrowly missing bone and major arteries. The fully exposed tankers atop Corvette immediately dove for cover, instincts kicking in as chaos erupted around them.

**4 August 2025 – Reflection of Heath Hutchison, Charlie 4-64 Armor**

After they shot at us, Bo Bo just started screaming, "My leg is on fire!" I knew it wasn't because I was looking right at it, and I told him he was going to be okay.

I pulled him around to the other side of the tank to protect him and do first aid. He was hurt, but I think that because he was so muscular, and it didn't hit his bone, the rifle round didn't do more damage.

Not far away, insurgents had quickly repositioned among a cluster of trees and opened fire on the men struggling to save those still trapped in the tank. The Cyclones roared back with their .50 caliber and 7.62mm coaxial machine guns, trying to hold off the ambush while simultaneously scanning their sectors, dragging out the wounded, rendering first aid, and calling for a MEDEVAC. Everything blurred together in a storm of madness.

**29 July 2025 – Reflection of Travis Nelson, Charlie 4-64 Armor**

The National Guard battalion wouldn't allow us to use the main rounds of our tanks, so we were restricted to using only machine guns. Hutchison then came over the radio and said that they were taking small-arms fire.

When the shooting started, I couldn't see where it was coming from. I just started shooting everything, using suppressive fire to try and keep their heads down. Then I saw a guy running away down by the chicken coop. I shot and saw him fall, and then just sprayed that whole area.

Instead of doing a few seconds of bursts, like we were supposed to, I panicked and just held the trigger down. The barrel got so hot it was glowing, and I could actually see the rounds shooting through it. Then, since it was so hot, the rounds just started cooking off without me even pulling the trigger, and I had a runaway gun. It wouldn't stop, and it just kept shooting.

I had to break a link on the machine gun belt to cease fire, then quickly switched out the barrels to resume firing.


Meanwhile, SSG Marek, commander of Red 3 that day, was still partially trapped in the turret, unconscious and burning alive. Ignoring the searing heat, 1LT Kroells clawed him free and dragged him to safety. Marek was somehow still breathing, but one look at his injuries told them the truth. He was slipping away, and they didn't expect him to survive. More than 20 percent of his body was burned from the explosion and exhaust, his right arm and leg were partially gone, one of his ears was missing, and it was clear he had a traumatic brain injury.[293]


**4 August 2025 – Reflection of Heath Hutchison, Charlie 4-64 Armor**

I'm not entirely sure how SSG Marek was burned so badly. The fireball that came off that bomb was astronomical, and I remember how much heat I felt, even though I was probably over a hundred yards away. I think the majority of his burns came from that, but the exhaust might've done it, too.

I put a tourniquet on both his arm and his leg. I think, in the explosion, there was stuff flying everywhere, which is what caused these injuries. His arm was completely shattered, and his leg was detached below the knee. Only a small piece of skin was still connected, so his bone was sticking out.

He had clear liquid coming out of his ears, and I remember in basic training they told us that a clear, sweet fluid would come out if there was a severe head injury.[294] I tasted it, and it was

---

[293] Daily, Space Coast. "RUSS MAREK: A Space Coast Hero." Space Coast Daily, May 25, 2015. https://spacecoastdaily.com/2015/05/russ-marek-a-space-coast-hero/.

[294] More, Daniel. "Difference Between a Runny Nose and a CSF Leak." Verywell Health, October 23, 2025. https://www.verywellhealth.com/cerebrospinal-fluid-rhinorrhea-83123.

sweet. I kept telling him, "You can't die here. You promised you would teach me how to surf in Florida." We just laid him on the ground and tried talking to him.

The men forced themselves back into the wreckage yet again, this time hauling SPC Ford from the driver's seat. Unfortunately, the young tanker was found unresponsive, like 1SG Gifford, and would die on scene. Being closest to the ground, his body had absorbed much of the blast, which caused catastrophic internal injuries. The grim tally was now two dead and two gravely wounded, with SSG Marek barely clinging to life, his survival measured in minutes rather than hours. There was no time to dwell on the horror, though, and they needed to evacuate everyone from the area immediately.

Bowen and Marek were carefully positioned beneath the shattered turret, shielded as best as possible from further harm while they clung to life. Just a few feet away, the bodies of Gifford and Ford lay still, their final moments had already passed. Overhead, the distant thump of rotor blades grew louder. At any moment now, the Black Hawk would descend to take the casualties away.

Still thundering down the road in his tank, CPT Anderson and his crew reached a checkpoint at the same time as another tank platoon that had been out in sector. Both units seamlessly merged into one, fully aware of the unfolding disaster. No longer alone, they swung onto Route Mercedes, the same treacherous stretch where Little's track had been damaged only weeks before. As CPT Anderson later wrote, "Route Mercedes was another hot IED route with numerous holes and small craters caused by past explosions. I could see the holes at the upcoming second turn."[295]

An elderly man and a woman were walking along the roadside near the turn. A narrow canal paralleled the road, and across the water stood a small group of men tending a herd of goats. Then, without warning, an IED detonated beneath CPT Anderson's tank, the blast lifting the 70-ton vehicle off the ground. Inside, the crew was dazed but, miraculously, uninjured. The battle tank was damaged but seemingly unfazed, never slowing.

Another tank crew, following behind, later recounted the horror: the old man walking the road had been right next to the device when it was triggered by insurgents. He was blown apart instantly, his shredded body violently flung into the canal. The woman beside him suffered the same fate, with the exception that she was reduced to a grotesque pile of bloodied flesh and torn clothing that lay heaped along the roadside.[296]

Almost immediately after the explosion, muzzle flashes erupted from a line of trees, but Anderson's crew didn't slow. They pushed forward through the smoke and gunfire, answering with the roar of their own weapons. SFC Marrero dropped one insurgent with precise bursts, while another's head exploded in a red mist under the impact of CPT Anderson's rifle fire. The ambush

---

[295] Anderson, *Ride the Cyclone*. Page 12.

[296] Anderson, *Ride the Cyclone*. Page 13.

collapsed as quickly as it began. It had utterly failed, other than taking the lives of two innocent pedestrians.

As this unfolded, we screamed out of Falcon in our Bradleys, heading straight for Route Corvette. Moments later, CPT Anderson reached the scene, confronted by the still smoldering wreckage of Red 3. Just as he had arrived, the MEDEVAC was leaving with the two dead and two severely wounded tankers.

Anderson quickly directed air assets to block any insurgent escape along the Tigris River, the most logical path, as it snaked nearby. Then, over the radio, came the report: the insurgents had broken contact, either fleeing into the shadows or lying dead where they'd fallen.

The company commander surveyed the devastation with growing horror. Spent machine gun casings and the twisted fragments of what had once been Red 3 lay scattered about the battlefield. Then came the order that would forever haunt CPT Anderson.

With fresh tank crews arriving and infantry forces closing in, and with a lull in the action, Anderson looked into the faces of the men who had just endured hell. They were bloodied, shaken, and marked by the loss of friends killed and wounded before their eyes. He turned to 1LT Kroells and told him to take the two surviving tanks and another vehicle back to the FOB. They had endured enough for one day. Kroells protested, unwilling to abandon the fight, but eventually he nodded and led the battered survivors away to regroup and recover.

Though still dangerous, the scene had transitioned into a tense recovery operation. The area needed to be secured until a wrecker crew could retrieve the destroyed tank. With their mission over, 1LT Kroells and the remnants of his platoon pulled away, their tank crews methodically scanning the terrain, still hoping to catch sight of any insurgents slipping away into the distance.

**29 July 2025 – Reflection of Travis Nelson, Charlie 4-64 Armor**

We were still pointed in the same direction, going west on Corvette, and I saw a jerry can next to the road up ahead. We had been hit by one IED in the past, but never two at the same site. Still, I told the driver what I saw and was concerned that it might be a marker for another explosive.

He just gunned it, and we launched by the area. Nothing happened. Red 1 was right behind us, and then they drove over the same area.

*Bam*! Another massive explosion tore through the air, this one erupting roughly 400 meters from the original blast site, still along Route Corvette. The tank struck was now Red 1, 1LT Kroells', part of the very crew that had just been ordered back to the FOB. The IED detonated directly beneath them, crippling the vehicle and leaving them momentarily stunned.

Although the second explosion was smaller than the first, it had completely torn the left track off the tank. The blast and immediate aftermath carved a deep groove into the earth beneath it, leaving the tank tilted at an alarming diagonal, as though it might roll off the elevated roadway at any moment. Like before, dirt, debris, and parts of the machine lay scattered in all directions.

### 4 August 2025 – Reflection of Heath Hutchison, Charlie 4-64 Armor

The blast slammed me into the roof, and I fell onto the ground inside the tank. I could hear the engine was opened all the way up for some reason, like we were trying to go as fast as possible, but we weren't moving anywhere. We were stuck.

I yelled to SGT Deckard, who was driving, "Turn it off! We're going to catch on fire!"

Somehow, everything had gone from bad to worse. There were now two crippled tanks stranded at separate locations, divided by a long stretch of exposed, elevated roadway that offered little cover. CPT Anderson struggled to gauge the full extent of the second blast from where he stood, but the fog of war refused to lift its suffocating veil.

In the wake of the second IED, a fresh barrage of small-arms fire erupted, forcing men at both sites to again dive for cover. *Where was it coming from?* Inside the second crippled tank, 1LT Kroells called out to his crew. Everyone responded except SGT Deckard, the driver. From him, there was only worrying silence.

### 4 August 2025 – Reflection of Heath Hutchison, Charlie 4-64 Armor

I kept calling him, but Deckard wasn't responding to me. Since the turret was at the wrong angle and wouldn't move, I couldn't get to him, but I could just barely reach my arm through the passage where he was.

I could feel him moving around in there, but everything was really hot. I pulled my arm out, and it was covered in blood.

Once again, Kroells dismounted, bullets snapping around him as he moved. He quickly maneuvered to the front of the tank, where the driver's hatch sat sealed shut with SGT Deckard inside. There was no response from within, and the hatch wouldn't budge. It had been combat-locked from the inside, a safety feature designed to keep enemy soldiers from opening it to kill the driver. Kroells frantically signaled to CPT Anderson, standing at the other site, for help.

From his position, Anderson acknowledged the calls for assistance, his mind racing to figure out what to do, but he knew he couldn't bring his own vehicles forward due to the mangled tank blocking the way. The only option was to go on foot. He ordered SSG Spangaro to follow him, then instructed his tank gunner to provide covering fire as they prepared to sprint across nearly four football fields of open ground atop Route Corvette. Before leaving, Anderson gave clear instructions: if he fired a flare into the air, they were to call in a MEDEVAC to that second location.

Anderson and Spangaro sprinted across the open route, weighed down by their gear as insurgent gunfire cracked around them. The tank crews from the first site answered with a barrage of heavy machine-gun fire and grenades, blasting the suspected enemy positions in the thick foliage. In the confusion, with dust and smoke obscuring everything, it was nearly impossible to pinpoint exactly where the fire was coming from. Tactically, it was a disastrous position. The insurgents remained concealed in the undergrowth below, while American forces were caught out in the open, perched atop an exposed stretch of raised roadway with nowhere to hide.

**29 July 2025 – Reflection of Travis Nelson, Charlie 4-64 Armor**

Even "Speedy," our interpreter, had picked up an M-16 and was firing into the woodline, trying to drive the insurgents back.

He's an American citizen now. He deserves it.

The commander arrived at the site of the second blast as we, the infantry, closed in from an alternate route. A flare almost immediately arced high into the sky, signaling the urgent need for a second MEDEVAC. Papi, our platoon sergeant, unwilling to risk another run down Route Corvette for obvious reasons, had taken us a different way around. Unfortunately, though, we arrived on scene only to find ourselves on the wrong side of that irrigation ditch, swollen with filthy, rushing water nearly up to the edge.

**21 July 2025 – Reflection of Juan Serrano, Platoon Sergeant, Bravo 4-64 Armor**

There were already two explosions on Corvette, and I knew a different way to get there. I didn't want to risk being hit again. When we pulled up, I looked at what the IED did to that tank (Red 3). What would it have done to us in our Bradleys? We would have been cut in half, exploded, and melted to the ground. There wouldn't have been anything left.

In the meantime, SPC Hutchison, the gunner in Red 1, had stumbled out of the turret after the second blast, limping and disoriented from a concussion and a wrenched back. Desperation drove him as he scrambled for the set of tools mounted on the tank's exterior, his only hope of prying open and unlocking the driver's hatch from the outside, which was still combat-locked.

Reaching SGT Deckard from inside the tank was nearly impossible with the turret jammed in place from the explosion. The turret was surrounded by thick metal shields designed to prevent anyone from being crushed as it rotated around inside the hull during combat. When properly aligned, however, these shields left a small gap, creating a narrow passage to the driver's position in an emergency. Now, because of the turret's position, that route was sealed tight by the layers of metal, preventing rescue.

**4 August 2025 – Reflection of Heath Hutchison, Charlie 4-64 Armor**

I initially tried getting to him through the turret by clawing at the screen with my Gerber, but it didn't work.

After I got out of the tank, I grabbed the tanker's bar, hoping to pry the driver's lid off. But I think we could have tried that all day, there was just no way to get inside. Then they started shooting at us again. I screamed at them, "Just give me a fucking break!"

Inside the tank, other crew members clawed and pried at the metal barriers, desperately trying to force a way through. If they could just rotate the turret, the passage would open, but it was frozen in place, and every system was immobilized. Accessing him from the inside was proving futile, no matter how much they struggled. They figured their only real chance was to pry open the hatch from above, but time was slipping away, and no one knew how long Deckard could hold on.

The next problem was that the turret had been facing forward when the explosion occurred, partially obstructing the driver's hatch from the outside. To fix this, they needed to slightly rotate the turret to allow the hatch to open, but once again, it was immovable. As a result, they were unable to reach Deckard, either from the inside *or* the outside.

Despite the hatch still being combat-locked from within, they managed to lift it open about two inches, just enough to peer inside. Through the narrow gap, CPT Anderson spotted SGT Deckard, eyes open, blood on his face, and saw him move. They kept shouting to him, promising they'd get him out. *Just hold on.*

Anderson later wrote of frantically trying to save Deckard. Looking through the partially opened hatch, he described the horror of what he saw, "My focus returned and I saw his left eye. There was no white showing. It was as if someone had replaced his eye with a reddish black marble."[297]

Half of the Gambler Platoon, including Derek Hauger and James Barnett, dismounted from their Bradleys on the other side of the deep canal and rushed to the water's edge. They had to cross to reach the tank, no matter the risk. Several grunts, weighed down by gear, dropped into the torrent and, in their attempt to reach the other side, were nearly swept away by the current. On the opposite bank, CPT Anderson spotted their potentially deadly struggle as other tankers continued to work on the lid. Without hesitation, he lowered himself partway down the slippery concrete embankment, bracing against the mud and debris, using his own body as a makeshift ladder to pull them across.

**24 July 2025 – Reflection of Derek Hauger, Bravo 4-64 Armor**

We were on the wrong side of this ditch filled with water, but it was like farm water. It was filled with mud, sticks, shit… Just full of nasty farm animal shit. We tried getting across, but it was steep

---

[297] Anderson, *Ride the Cyclone.* Page 19.

and slippery, and people were going to drown from being swept away. Someone, I don't remember who, used himself as a ladder, and we started linking together to climb up.

**27 July 2025 – Reflection of James Barnett, Bravo 4-64 Armor**

I think we beat the actual QRF because we hauled ass to get out there. When I got in, I went completely under, swallowing a bunch of "tiger's water." On the other side, I volunteered to lay down on the walls and let the rest of the squad climb up me like a human ladder.

Miraculously, all climbed to safety, thanks to the captain's quick thinking and the infantry working together. Now at the site of the second explosion, the Gamblers spread out into defensive positions, forming a wide perimeter as they scanned the vicinity for any sign of gunfire, but the shooting had gone silent as the infantry saturated the area. Several grunts were called over to assist with the rescue effort. Everyone had safely made it out of the tank except SGT Deckard, who was still trapped in the driver's seat.

**24 July 2025 – Reflection of Derek Hauger, Bravo 4-64 Armor**

SGT Vlastelic went into the turret to try to help pull off those shields and move cables so they could get to him from inside.

I climbed on the tank, and since I was skinny, I could stick my arm inside the hatch just a little bit. I thought I could hear him mumbling, and I just kept telling him, "We're gonna get you out. Just hang on buddy."

All I could do was barely touch him and try to reassure him. It was horrible because he kept coughing up blood, and we had to watch him slowly die in front of us. We couldn't get it open.

The other half of the infantry platoon, including myself, was pushed into the tree line where we believed the insurgents were hiding. We were told, quite bluntly, to kill them and viciously prevent anyone from attacking the rescue crews on the tanks. As soon as our squads hit the ground, we moved into bounding overwatch, just like we'd practiced on those lanes back at Fort Stewart. Suppressive fire cracked through the trees as we leapfrogged forward and hosed down the area, each man covering the next. I thought there was movement through the woodline, as if the remaining insurgents were fleeing, but everything was happening so fast.

The fields were littered with wires, with some half-buried and others lying exposed like traps waiting to be sprung. Jumping over the wires and dodging holes dug in the dirt, I half-expected that the insurgents might have rigged the ground with IEDs, counting on us to venture straight into their kill zone. But somehow, we crossed without incident. All the wires seemed to be headed toward the road. Within minutes, we had closed in on the edge of a farmhouse and its surrounding property.

Around the area we had just blasted, spent shell casings littered the ground, glistening in the receding sunlight, while patches of blood and scattered remnants of the battle painted an intense picture. The air was heavy with the acrid scent of burning trash, mingling with the unmistakable stench of scorched hair and the pungent, fetid odor of animal dung. Each breath was a struggle, the odors seeping into my lungs and overwhelming my senses with their foul intensity.

Breathing heavily in the thick, humid air, I instinctively stacked up along a farmhouse wall with other members of my team. We still had no idea where the insurgents were, and every corner felt like the mouth of an ambush waiting to snap shut. The sun was sinking fast, shadows twisting and jerking like living things, each flicker forcing my brain to burn precious energy just to decide what was real. Farm animals scattered in every direction, adding to the already chaotic scene.

Someone kicked in the first door, and it splintered off its flimsy frame. We surged inside as one and slammed into an unarmed man standing in the entryway. His eyes went wide just as our momentum hit him, sending him tumbling backward before he hit the floor with a heavy thud. Taking no chances, we swarmed him, controlled his arms, and snapped zip ties tight around his wrists. Once he was restrained, we dragged him outside and placed him against the wall under guard.

Door after door, we pushed through the rooms, sweat pouring off us and stinging our eyes. Each time we braced for a fight, hands went up, and voices rose in surrender. More zip ties. More detainees to search and question. Not a single one of them tried to face us head-on.

Near a barn, a sudden shot cracked through the air, and several of us unleashed a storm of rifle and machine-gun fire, shredding the walls with bullets, thinking that's where the threat originated. Moving cautiously around the perimeter, another team leader lobbed a grenade from his M203. It detonated inside with a deafening blast, spraying the interior with shrapnel. When we stormed in, the place was empty. Had there even been anyone in there? Or had they slipped out the back while we were circling? Everything was happening at a breakneck pace, blurring into a state of confusion.

Above us, Apache attack helicopters roared in, circling overhead, clearly chasing the same targets we were after. They began gun runs, strafing the area with their deadly cannons. From our position, we couldn't see the river, but we assumed our prey had to be hugging the Tigris closely. Not wanting to get caught in the storm of firepower above, the order came down to hold position at the farm. So, we began setting up hasty fighting positions while combing the property for anything of evidentiary value. What could we find to link them to the explosions?

Back at the second tank, Anderson and Hauger kept talking to Deckard, murmuring reassurances as a desperate, improvised plan took shape. Another Abrams was brought up close, its main gun tube lined up and pressed against the disabled tank's barrel at a perpendicular angle. With a surge of power, the engine roared, and then a sharp, metallic pop echoed across the field as the jammed turret finally broke free. Using the hand crank, they managed to rotate it just enough for someone to reach an arm inside the driver's compartment, unlock the hatch, and power down the tank.

Groups of hands hauled him through the now-unlocked hatch, and the sight froze every-one for a moment. The top of his head was severely wounded, with blood streaming from his ears. At first, they thought they detected a pulse, but if it was there, it was faint. But, as it turned out, they were wrong. He was already gone, with Anderson left holding him as the medic shook his head, and he was soon placed into a body bag. Still, everyone scrambled to get him onto a stretcher, with the captain, Hauger, and the front of the tank now smeared with Deckard's blood.

### 4 August 2025 – Reflection of Heath Hutchison, Charlie 4-64 Armor

After the explosion happened, the adrenaline had me going, but I soon crashed. Even before they got the hatch open, my body had given out. I collapsed on the side of the tank, and the medic was checking on me, thinking I had been hit because of Deckard's blood on my arm.

He was scared to death, and he was trying to give me an IV, but he was shaking so badly he couldn't do it. I had to do it to myself.

Then I must have blacked out, because I don't remember anything until later.

As they worked, the second MEDEVAC began its descent, guided by the plume of a smoke grenade tossed nearby. Then CPT Anderson spotted what was about to happen, and his heart jumped into his throat. The drifting smoke cloud was partially obscuring a tangle of overhead electrical wires. If the Black Hawk clipped those lines, the entire area would be engulfed in fire and death.

### 27 July 2025 – Reflection of James Barnett, Bravo 4-64 Armor

We realized the helicopter was too close to the power lines, so I ran out into the field to pick up the IR strobe and throw it farther, but the bird was still coming down toward my location. So, I shined my Surefire (flashlight) up into the cockpit to get his attention.

SFC Davis also reacted, simultaneously sprinting into the open and waving his flashlight like a beacon, signaling the pilot to shift course.[298] Miraculously, the helicopter veered away just in time, touching down in a rough, uneven field nearby. The tankers and infantry rushed to carry Deckard aboard, Hauger at the front of the stretcher.

Hauger would later recount what happened next in a raw, pained voice, the weight of it still burdening him even 20 years later. It's a memory that we both will never forget.

---

[298] Anderson, *Ride the Cyclone*. Page 21.

**24 July 2025 – Reflection of Derek Hauger, Bravo 4-64 Armor**

I was carrying the stretcher right where his head was at. He was still bleeding everywhere, and my entire side was covered in blood.

Walking as quickly as possible, I was struggling to get through the field because of the weight I was carrying. I had all my gear, my SAW... and the stretcher in my hands. The field wasn't smooth, and there were all these small ditches. It was really uneven. I was sweating profusely and was still soaked from crossing the canal, so my hands were wet.

Just as we went down into one of those depressions to get to the MEDEVAC, all the weight shifted to my end and... the stretcher slipped from my hands. I dropped him. He rolled out of the stretcher and onto the ground.

I heard, "Oh my God!" I looked up and saw a female soldier from the flight crew at the Black Hawk's door. She had seen what happened and saw us struggling to get him to the helicopter through that field. SSG Brister came up from behind and helped me, and he carried him the rest of the way.

I struggle with that to this day. I dropped him.


Sadly, despite the heroic efforts of everyone that day, SGT Deckard didn't survive. With the total number of KIAs at three, others feared the worst for SPC Bowen and SSG Marek. Given their critical conditions, we assumed they would soon follow.


**4 August 2025 – Reflection of Heath Hutchison, Charlie 4-64 Armor**

Waking up in the helicopter, I saw a female crew chief working nearby and felt something dripping onto my face.

I looked up, and above me, there was SGT Deckard in the body bag. I was in shock, but I just had to move my face a little bit to keep his blood from landing on me. That still bothers me to this day.


Around the farm, we launched into quick interrogations of the few prisoners we had taken. There was no way they were uninvolved, as every wire we'd traced led straight back to their area. It was also clear this was where the small-arms fire targeting the tankers had originated. Adding to the evidence, we discovered undischarged explosives, coils of wire, tools for building IEDs, and other incriminating materials scattered about. In short, we wrecked the place, leaving it incapable of producing more IEDs or doing much of anything ever again.

**SSG Santiago holding the detonation wire with Red 3 behind him**

**Photo courtesy of Travis Nelson**

By the time total darkness settled in, the order came to fall back to the wreckage of the destroyed tanks. Our mission was now to secure the scene until morning, ensuring no insurgents could snatch up unexploded ordnance or plant more IEDs under the cover of night. It was now a mission of playing defense and waiting to see if anyone would come back.

A large, shadowy drone hovered above, tracing a steady, repetitive pattern through the night sky. Its peculiar hum became increasingly familiar as darkness deepened. While it had been present for much of the day, the drone's distinctive sound had largely gone unnoticed amidst the chaos unfolding around us. Now, it felt as though an invisible observer was quietly monitoring our every move, its surveillance camera diligently scanning for any sign of suspicious activity. The atmosphere was thick with tension, an unspoken game of cat-and-mouse unfolding in the shadows.

**The eerie remnants of Red 1, destroyed and dangerously tilted onto its side as night fell**

**Photo courtesy of Travis Nelson**

The steady hum of the Bradleys' diesel engines provided a low, comforting backdrop as they stood sentinel, offering overwatch. Gunners carefully scanned the area through their thermal visions, hoping to spot a human-sized heat signature sneaking into our kill zone. On the ground, we spread out into defensive positions, scanning through our NVGs for the slightest flicker of movement. Every so often, we would silently shift positions, creeping into the fields to catch any sign of enemy activity.

High above, illumination rounds fired from a nearby FOB drifted across the night sky, glowing eerily as they cast the terrain in a surreal, alien green. When they flared overhead, we froze in place, our bodies pressed to the ground, eyes unblinking, watching. The mangled wreckage of Red 3 sat just behind me, silent, a terrible sight to behold. As the light of the illumination round faded and the world plunged back into darkness, we moved again. Slow, deliberate, hunting.

By pure chance, one of the burning illumination rounds landed on the roof of a house we hadn't destroyed, just before it fizzled out. The structure caught fire, perhaps a touch of poetic justice. Flames licked upward, sending the shadows into an even wilder frenzy as we relished the sight.

**24 July 2025 – Reflection of Derek Hauger, Bravo 4-64 Armor**

That was the longest night of my life. I was covered in blood and animal shit, and the mosquitoes were eating me alive. But I couldn't do anything about it. We just had to wait, and we all had to suffer out there. I remember taking a break inside the back of a Bradley. I was really upset and talked to SSG Brister, telling him I had let SGT Deckard down by dropping him.

He said, "No, you stepped up and helped him. You were there for him. It's not your fault." I really needed to hear that.

Then afterwards, I don't know why, we started talking about aliens and stuff. It was so dark and creepy out there.

I, too, remember having strange, disjointed conversations, especially after particularly painful and traumatic events like this. Odd exchanges, which any outsider would deem to be lunacy, seemed to be our way of compartmentalizing the grief, sealing it off and locking it away, buried beneath fragments of random talk that made no sense to anyone but us. Aliens just happened to be part of the whispered discussion.

I also vividly remember the giant mosquitoes, which feasted on all of us that night, but there wasn't anything we could do about it. We simply had to lie still in the mud, watching for any signs of the enemy. It was miserable.

As the sun broke over the horizon, we realized that, aside from the marauding swarms of mosquitoes, we'd had no contact with the enemy. No one had dared to breach our lines during the night. In the distance, the low rumble of a wrecker crew signaled their arrival to deal with the wreckage. Soon after, another company moved in to take our place, giving us the chance to head back home and recover. But before any of that, we still had one task left. We had to make it out of there alive. IEDs were everywhere around us, and, as we eventually discovered, the entire area had been rigged with traps and ambushes.

**24 July 2025 – Reflection of Derek Hauger, Bravo 4-64 Armor**

I don't remember why, but we had to ride back to the FOB on top of one of their tanks. I'm not sure if there wasn't room in the Bradleys, or... I just don't remember.

But I had to ride on top of this thing. They told me, "If we hit any IEDs, we aren't stopping. We are just going to go as fast as possible and get away, so hold on." Do you know how dangerous that was? But we had to do it.

Like Hauger, Barnett expressed his confusion twenty years later as to why we rode on the top of Abrams tanks as we left the scene. It's just one of those strange memories hidden in the confusion. It may be due to the detainees being brought back to the FOB by our usual Bradley transports, but that may be incorrect. Regardless, as we rolled away, we turned off Route Corvette and took an alternate route, hoping to avoid any more hidden bombs. The captain had split us into multiple groups, each taking a different path back to the FOB to reduce the risk of being jammed up.

### 27 July 2025 – Reflection of James Barnett, Bravo 4-64 Armor

On the way back, we saw a squirter at the far end of a field. I remember asking for permission to engage, but it was denied. Given the situation, trying to get out of there, it was probably for the best.

On the way back, CPT Anderson discovered his tank, still damaged from the IED strike less than 24 hours earlier, could barely crawl away at 10 mph. Before long, his group was forced to halt less than a kilometer down the road when an EFP was spotted along their path. The device, potentially capable of punching through his Abrams, had to be detonated in place by an explosives team before they could move again.

Later, as they turned onto a road lined with civilian traffic, CPT Anderson stood exposed in the hatch, guiding his driver to avoid crushing local vehicles. Without warning, another IED detonated beside them, the second to hit the tank in a matter of hours. Shrapnel sliced into his elbow, and the force of the blast slammed his face against the butt of his machine gun.[299] Yet, miraculously, they were still alive, mostly uninjured, and still moving.

### 24 July 2025 – Reflection of Derek Hauger, Bravo 4-64 Armor

When we left Corvette, there were explosions going off in every direction. We had to hope that we wouldn't pick the wrong way to go. Then the mortars started coming in. It was crazy.

At last, after what felt like an eternity, all surviving soldiers and vehicles returned to FOB Falcon. For us, the ordeal was finally over. The cleanup on Route Corvette would fall to other units who were fresh and rested. Our only job now was to regroup, collect ourselves, and get ready to step back into the fight. When would that be? No one knew, except that it would likely be very soon.

### 24 July 2025 – Reflection of Derek Hauger, Bravo 4-64 Armor

I remember making it back to the FOB. When we walked into the barracks, the chaplain and the mental health team were already there. When the chaplain saw me, because I had so much blood on my uniform, his eyes became huge. He was like, "Oh, my goodness." They all wanted us to stand around and talk about what had just happened. I asked, "Can I go change first?"

---

[299] Anderson, *Ride the Cyclone.* Page 23.

**17 September 2005 – Letter to the Harstad Family**

Today, the chaplain and mental health people came to talk to us. They each sat down with us individually and tried to get us to talk about our "feelings." I have a hard enough time talking about my feelings to someone I feel comfortable with, so I surely didn't say anything to these fucks. They tried to get us to show our emotions, and when some people did, it REALLY came out. A bunch of sick and twisted stuff, and I think it even shocked the mental health staff. They told us that most of us will have PTSD, but I think they are a bunch of crackheads.

As per standard procedure, a communication blackout was enforced. No one was permitted to contact the outside world until the Army had completed all official death and injury notifications. In the meantime, all we could do was continue our resupply and prepare to head back into the killing fields of Dora.

**17 September 2005 – Letter from Robert Roth, Commander of 4-64 Armor**

Two days ago, at approximately 1800 hours on the 15th of September (incorrect date), insurgents attacked a CYCLONE patrol.

Insurgents attacked this patrol with several large bombs that had been dug into the road, resulting in seven casualties. It brings me great sadness to report to you that three soldiers were killed in the attack. Those soldiers are: 1SG Gifford, SGT Deckard, and SPC Ford.

The three wounded include: SSG Marek, SPC Bowen, and SPC Hutchison, who is expected to return to CYCLONE soon. SSG Marek remains in critical condition and needs your prayers. Their translator was the seventh person injured.

I know this will be tough on everyone, but that is why we are a family. We are here to help one another during these tragic moments. There are many heroes in this attack as well, and we will honor their achievements for doing what they could to try and save their friends' lives.

When the blackout was finally lifted, I wrote home about what had happened. It may be hard for some to understand, but information didn't flow freely, not because anyone was hiding anything, but simply because we were all overwhelmed. We were too busy surviving, reacting, and doing our jobs to stop and piece together a clear picture of the entire event. Remember the fog of war? It was in full effect.

As a result, I mistakenly believed that the second tank destroyed was part of the QRF. I didn't realize it had been part of the original patrol. I only learned the truth many years later, during a conversation with someone else who had been there. So, keep that in mind when reading the following passage. Things were unfolding so fast that accurate details were nearly impossible to track, let alone communicate.

**17 September 2005 – Letter to the Harstad Family**

Friday, at about 1800 hours (6 PM Iraq time), the first sergeant and a small patrol were headed down a dirt road in our sector. There are a lot of dirt roads in our area that are built up on farmland, so they are elevated off the ground. On either side of the road, it drops off about 10-15 feet, with crops on both sides. Anyhow, his Abrams tank was moving along when insurgents detonated an anti-tank mine, along with a great deal of explosives, underneath his vehicle. 1SG Gifford and SPC Ford were instantly killed.

Upon the blast, the bottom half of the tank ripped open, and the turret flew into the air. The tank sunk halfway into the hole created, and the turret landed on the back of the main body, digging its barrel into the ground. The crater was so large that the entire tank could fit inside it. It was about 20-25 feet across and about 20 feet deep.

Upon this detonation, the QRF was called, and they responded. The insurgents had another bomb waiting for this crew, as well. They detonated it under the driver's hatch, killing SGT Deckard. As soon as all this happened, they unleashed with AK-47 fire. The tankers acted bravely and fought their way out of the ambush, firing .50 caliber and 7.62mm ammunition into the woodline.

We were on recuperation time when it happened, so they woke us up to get the infantry on the ground. We rolled onto scene and broke the platoon into two parts. One part was responsible for pulling out the wounded and killed, while my part was to provide security and clear the villages.

By the time the choppers got everyone out, it was probably 11 PM. Half of our platoon, including myself, stayed on scene until 8 AM to provide security for the destroyed vehicles until a wrecker crew could arrive. We found plenty of undetonated bombs, wire, and houses that were suspicious.

Those tanks had pieces thrown several hundred yards away. If one of our Bradleys had rolled over that (IED), we would have exploded and landed in pieces miles away. We all swore that we would not let these three individuals die in vain.

Subsequent analysis of the site by an explosives team revealed that the device wasn't an anti-tank mine, but rather a 500-pound bomb typically dropped from an airplane. In addition to that, an estimated 500 pounds of extra explosives had been packed around it.[300] This massive detonation caused the turret to be blown off the tank's hull.

The final battle damage assessment reported three American soldiers killed, three others wounded, and a knee injury sustained by "Speedy," our interpreter. Two Abrams tanks had been completely destroyed. We recovered the bodies of two insurgents, both armed with AK-47s, and much of the surrounding village complex lay in ruins. It's likely that more enemy fighters were killed or injured during the counterattack, but an accurate count will probably never be known.

---

[300] Anderson, *Ride the Cyclone*. Page 16.

**September 2005 – Letter to his family from Derek Hauger, Bravo 4-64 Armor**

I have some bad news. Our company got hit pretty bad last night. We had 3 KIA, including our first sergeant. We had to stay there all night, even though we were up at 6 AM already that morning, so I was up way past 24 hours.

I'm fine, but if that would have hit our Bradley, which has less armor, everybody would have been turned into soup. I have to go to bed. I have to get blood work tomorrow because we had to go through some nasty sewage water to get to the guy that we helped.

**4 August 2025 – Reflection of Heath Hutchison, Charlie 4-64 Armor**

I spent four days in the hospital for the concussion, the internal bleeding caused by the blast, and my back. When I hit the roof, all the soft tissue in my neck and spine was smashed or hyperextended.

CSM Stanley and another NCO took time to visit me. They gave me challenge coins, I think to try and boost my spirits. But I dumped them on the ground, I was just so upset about losing my friends.

Originally, the medical staff said I was going to be shipped to Germany or back to the United States. After they had me under observation for those four days, they changed their minds and decided I was OK to return to work.

CPT Anderson showed up with a fresh set of uniforms and said to me, "Hey, let's get back to work." So, I did. I'd follow that guy anywhere. He led by example, and we couldn't have asked for a better commander. We all still had a job to do.

Unsurprisingly, the incident only heightened an already tense atmosphere for those back home. In hindsight, while I'm grateful to have documented these memories, part of me wishes I hadn't shared any of it in real time. It was an incredibly difficult experience for everyone involved, including loved ones in the United States.

**22 September 2005 – Letter from Mark Barnes, 1ˢᵗ Sergeant of Bravo 4-64 Armor**

As you know, 1st Platoon is about 15 minutes south of us at FOB Falcon in a sector of Doura with the 184th Infantry. Their performance the other night when C Co lost their 3 soldiers was absolutely outstanding. They secured the bomb sites, helped evacuate the casualties, and made sure that nothing came in and hurt anyone after the bombs went off.

Many of you will be, and should be, rightly concerned for their welfare, as am I and the commanders. Doura is more dangerous than Haifa Street, for those of you who remember when we were up there. But 1st Platoon is sticking together. They are motivated and eager to send the terrorists a message for what they did to Charlie Company.

I feel with certainty there will be some dead terrorists in the near future, compliments of 1st Platoon and others. I would go into detail about what they are doing regularly, but I will just summarize by saying that they are conducting A LOT of combat operations. They are down in the scary places every day and every night, and they are conducting themselves in a way that would make you all very proud. I don't want you to be afraid for them. I want you to know that they are capable of dealing with terrorists.

1st Platoon really needs your love and understanding. They cannot afford to be distracted by petty things. If you could spend one day with them out on patrol, you would take a lot more time saying, "I love you, and I'm praying for you." I tell you this because some of you have been desensitized to their predicament.

**Undated – Letter from my grandmother, Toni Harstad**

About the cross (a small, silver cross taped to her letter): When your G'Pa and I were married a long time ago, our minister gave Bill and me a cross to carry anywhere we pleased. I carry mine in my purse, but now I would like you to carry it. God watch over you and your company.

Love you dearly, Grandma Toni

On September 21st, the battalion held a memorial service for our fallen comrades.[301] It was a gut-wrenching experience, one of the hardest days we faced, and I doubt there was a single dry eye among those in attendance.

At the memorial, three M-16 rifles stood inverted on a makeshift monument, their barrels driven downward in solemn tribute to our fallen brothers.[302] Atop each weapon rested a combat helmet, familiar to all, now forever still. Their names and ranks were displayed with quiet dignity, but it was the dog tags, swinging faintly in the breeze, that seemed to speak the loudest. Each metallic clink was a whisper of a life cut short. Framed images of our comrades sat below, their boots placed neatly beside them, silent reminders they would never again stand by our side. That patch of ground, once just dirt and dust, had become sacred. It was etched with grief, silence, and the unbearable weight of goodbye.

LTC Roth delivered a eulogy for the men. Everyone was heartbroken, but I think his message landed. It still echoes in the way I, too, try to live my life.

---

[301] DVIDS, "Spc. Nathan Parks Speaks at a Memorial Service," Defense Visual Information Distribution Service, September 21, 2005, https://www.dvidshub.net/image/10895/spc-nathan-parks-speaks-memorial-service.

[302] DVIDS, "Col. Edward Cardon Pays His Last Respects," Defense Visual Information Distribution Service, September 21, 2005, https://www.dvidshub.net/image/10894/col-edward-cardon-pays-his-last-respects.

**21 September 2005 – Excerpted remarks by Robert Roth, Commander of 4-64 Armor[303]**

I have a philosophy that I live my life by… I believe that for every horrible thing that happens on this earth, something very good comes from it.

I remember the night of the attack, at the hospital, seeing one of the injured soldiers, SPC Bowen. The first thing he said to me that night was, "Give it to me straight, sir, how is everyone?" When I replied they were all being attended to by doctors, and he needed only worry about himself, the next thing he said to me was, "I'm sorry, sir!" Now, here is a young soldier, bleeding internally, with a nasty wound, who experienced probably the most traumatic event in his life. And he is worried about his crew, his buddies, his leadership, more so than himself. Then he apologizes to me, as if he let me down.

The single most important thing we have on this earth is each other. And the actions I saw that night, and the following days, by those who were injured, those who were not injured, families and friends back home, and all the rest who had an opportunity to share in the lives of these great men… these three heroes made us remember that all we have in this sometimes cruel world is each other.

Step back and see the incredible love that came from everyone, everywhere! The quick reaction from many soldiers across the Brigade to secure the site and evacuate the destroyed tanks; the air MEDEVAC crews who never gave up and did whatever it took to save these soldiers' lives; the Apaches hovering above, scanning the perimeter to look for those responsible; or to the Artillerymen who fired illumination in support of those securing the site.

How about the families and friends back home surging to take care of the spouses and children of those who perished? What about the people who don't even live in Georgia, wanting to go to the hospital in Washington, D.C., and visit the injured as they return? Or how about the soldiers in 4-64 AR who would do anything for any of the casualties or their families at the drop of a dime? Ladies and gentlemen, that is love for one another, and it is the most powerful force on earth.

Walk away today and honor 1SG Gifford, SGT Deckard, and SPC Ford by talking about them. Thank these heroes for their sacrifice, and for reminding us that all we have is each other.

All of us carried the weight of grief, but for the tankers, the pain ran deeper. It was etched into their very souls. They had lived as brothers, sharing long nights in the field, joking over hot dogs and hamburgers after training, and laughing in the kind of way only those who've truly bonded can. Their loss wasn't just of comrades. It was of family.

For many of us, myself included, the most painful part of the memorial came near the end. We have a tradition in the military called the "Final Roll Call." In it, CSM Stanley stood before the

---

[303] Balda, Dan. "4th Brigade Soldiers Honor Three Fallen Comrades." Defense Visual Information Distribution Service, October 11, 2005. https://www.dvidshub.net/news/3262/4th-brigade-soldiers-honor-three-fallen-comrades.

company with a roster in his hands and began calling names, one by one. Each soldier who was there answered with a loud, automatic "Here!" The responses snapped through the air with the crisp rhythm of any formation, like muscle memory. Then he reached the bottom of the list.

He barked, "First Sergeant Gifford." No answer came back, and the silence felt heavier than anything that had been said all day. He tried again, louder. "First Sergeant Gifford!" Still nothing. Then, a final time, he called it in full, as if the name itself could pull him back into the world: "First Sergeant Alan Nye Gifford!" After that came a long pause, the kind that makes you painfully aware of every breath, every shifting boot, and every swallow. The message was unmistakable. He was no longer with us.

CSM Stanley did the same for each of the fallen. Each name was called, and each one was met by that same empty space where a familiar voice should have been. Then came the Firing of the Volleys, sharp and final, followed by "Taps," echoing across the area as we raised our hands in a final salute to our fallen brothers.

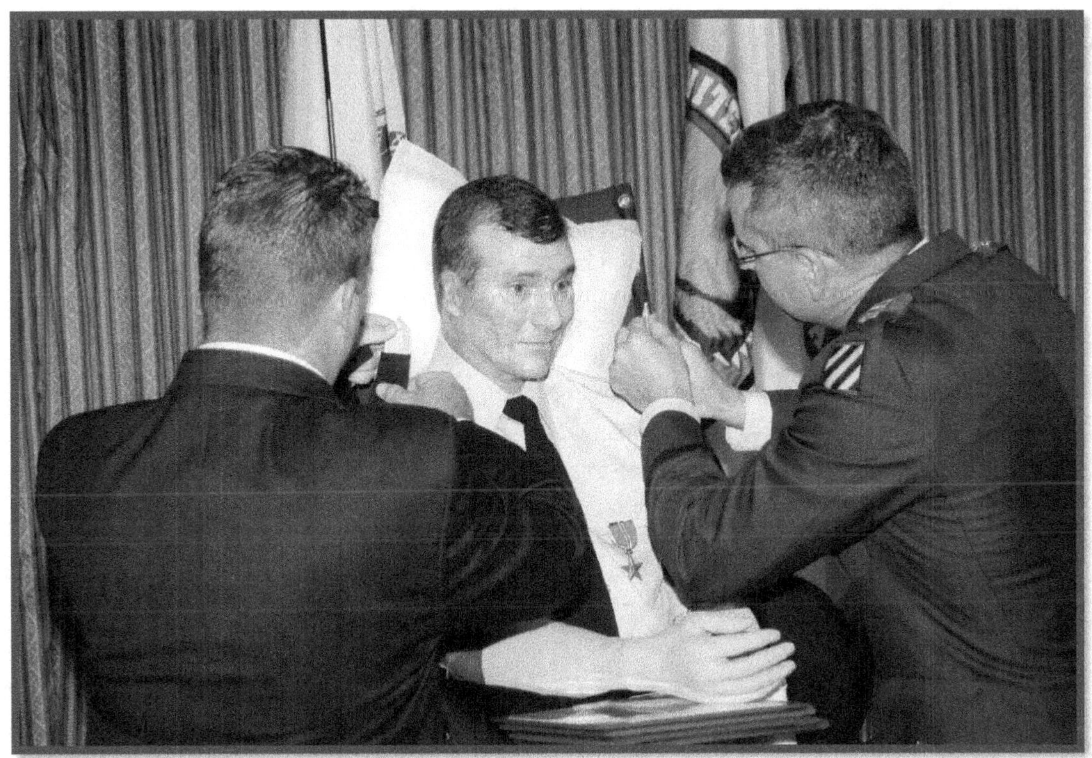

**LTC Roth (right) pins a severely wounded SSG Marek with the Bronze Star and Purple Heart during a later awards ceremony**

**Photo courtesy of Robert Roth**

Then there is one image I can never erase: CPT Anderson standing before the three memorials, his hand raised in a trembling salute, his face contorted by a grief too heavy for words. A photograph was taken of that terrible moment, but I have no need to ever see it again. The scene is permanently seared into my memory, having personally lived through it. It's unshakable, eternal, and re-emerges time and again. And for some, beyond the grief of loss, the weight of second-guessing decisions that had been made and the burden of survivor's guilt would linger for a lifetime.

**21 September 2005 – Letter to the Harstad Family**

I just got back from the memorial service for those three soldiers. It was really rough for everyone, especially the tankers. It was quite long, about two hours, and it really sucked. I hope I never have to go to another one.

My hopes were unrealistic and short-lived. More death and destruction waited for us just around the corner.

**25 September 2005 – Letter from Robert Roth, Commander of 4-64 Armor**

BEAST was hit by a roadside bomb a few days later (after the C Co incident). This was a powerful explosion that resulted in SSG Bradley and PFC Amaya being evacuated to the U.S. (three others were slightly wounded, but all have been treated and released).

All of the injuries this week took a heavy toll on the battalion, as you would expect. Yet the soldiers understand that what we do here is important and that life is precious. Times like these make us realize how vulnerable and human we are. Now, we say goodbye to our friends, remembering all the good things they brought to this earth. While they are no longer among us each day, they will always be with us in our hearts. Celebrate their lives, and remember to take care of those still here.

As for 1SG Gifford, SGT Deckard, and SPC Ford? See you on the high ground, gentlemen.

# Reflections

When I spoke with Derek Hauger on the phone during an interview for this book, it was clear that the trauma of that night still weighed heavily on him. Witnessing a fellow soldier die slowly before his eyes had been horrific enough, but the memory of dropping him on the battlefield was a burden he could barely speak about. Not wanting to add to his pain, I quietly asked if he wanted me to include that part of the story or leave it out entirely.

"What do you think?" he asked, deferring the decision to me. "It happened. I just don't want to hurt anyone else." It was clear he was thinking about Deckard's family.

I paused for a moment, then told him, "Let me write about it, but only if you approve it before anything goes to print. I know how to tell it with the care it deserves. I think his family deserves to know what happened, and I believe your story matters, because it shows something we all need to remember. We're human, doing the best we can in an impossible situation, and some memories can't be erased."

He agreed, though I could still hear the pain in his voice. "I dropped him," he said quietly. Those three words echoed with grief and guilt.

He went on to say how grateful he was that SSG Brister had been there that night. Brister had offered him quiet reassurance, telling him that SGT Deckard would've appreciated having him there in his final moments. He'd even thrown in a bit of lighthearted razzing, just enough to ease the weight of the moment and remind Derek that he wasn't alone in carrying it.

**24 July 2025 – Reflection of Derek Hauger, Bravo 4-64 Armor**

You and I were always trying to outdo one another, and SSG Brister often used this to poke at one of us. When you guys were down in the village going after the insurgents, he messed with me a little bit: "Harstad is down there shooting. How come he's always the one shooting, while you are up here with me?"

Well, I was busy doing other things, but I was glad he was keeping me distracted. He really helped me.

Later, I shared Hauger's story with Robert Roth, our battalion commander, who had a personal connection to SGT Deckard, having served at a different unit with him before.

**27 July 2025 – Reflection of Robert Roth, Commander of 4-64 Armor**

Thank you for sharing that story with me. It really humanizes what had to be done out there.

Before 4-64 Armor, I actually worked with SGT Deckard in the past. Knowing him, he probably would've laughed about it. Tell your buddy that Deckard would've just slapped him on the back, given him a big grin, and said it was okay. He was doing the best he could. Deckard would've understood. He always did.

I also spoke with Travis Nelson, the tank gunner aboard Red 2, the only Abrams in the original patrol that wasn't destroyed. Like Hauger, his tone shifted to a somber hush when he spoke of the events of September 16th. He, too, carried a quiet sense of regret.

**29 July 2025 – Reflection of Travis Nelson, Charlie 4-64 Armor**

In all that chaos, I never called up the suspicious jerry can to the other tanks on the radio. I just told my driver, not everyone else. If I had told them, maybe SGT Deckard would still be here. They might not have driven over that second IED.

This is where the story takes an unexpected and maybe even redemptive turn. For twenty years, Travis Nelson and Heath Hutchison remained the closest of friends. They talked often, stayed in each other's lives, and shared countless memories, but they didn't really speak of what happened on September 16th. That day remained locked away, too painful to touch, and neither one thought it was important to see it through the eyes of someone else. Hutchison had no idea that Nelson had been carrying a quiet, crushing guilt all these years, believing he'd failed to call out the jerry can marker.

Then, as I interviewed them for this book, I noticed something with their stories didn't quite add up. Their accounts of the event conflicted.

Hutchison told me, almost offhandedly, "Nelson called out the jerry can, and we tried to maneuver around it…"

I stopped him. "Wait. What? You *knew* about the jerry can? Nelson told me he never called it out. He's been blaming himself for that ever since."

"Yeah," Hutchison said, surprised. "He definitely called it out. We tried to steer clear, but the blast still got us."

I paused, stunned. "You need to tell him that. He's been living with this weight for two decades, thinking he did something wrong."

No conversation can undo what happened that day. It won't bring back the three men who were lost, nor will it erase the scars carried by those who survived. But if, in some small way, Travis Nelson can finally let go of a burden he never deserved to carry in the first place, then maybe that's something. Maybe there's healing in these quiet revelations.

For me, these conversations have been unexpectedly therapeutic, and my hope is that the men who lived through these moments might find the same comfort in speaking them aloud. It's one of the main reasons I've been writing this. After all this time, maybe we're finally ready to talk about what happened to us out there.

In the aftermath of the incident, the Army did what it could, given the circumstances, to support us. They brought in a chaplain and a mental health team, trying to offer us some form of comfort. But truthfully, it didn't go well. Looking back, I realize that many of us, myself included, lashed out at them, even though they were only trying to help.

I've carried some guilt about that over the years, not fully understanding at the time why I was so angry, so defensive, and so confrontational. I regret some things I said, especially to the chaplain, because they were way out of line. Despite his much higher rank, he didn't get angry or protest, though. Instead, he listened and allowed me to vent my frustration, even if I didn't share my feelings beyond overt hostility.

With the passage of time, I think I finally understand. The truth is, we didn't have the luxury of grief because we had to go right back into combat just a day or two later. There was no pause and no breathing room. We didn't have the chance to take a moment to reflect, or to make sense of what had happened. On the very day of the memorial, I machine gunned yet another car along Route Corvette. That was our reality. It was the reality of fighting in a war as a combat arms soldier.

To stop and grieve would have been a risk, one that could prove to be lethal. In that kind of environment, survival demanded numbness. So, we all buried our pain and pretended all was fine, not because we wanted to, but because we had to.

In their 2004 study, Pivar and Field explored unresolved grief in combat veterans with PTSD. Their findings suggest that when soldiers experience the loss of one or more comrades, the grief they feel is comparable to that of losing a spouse.[304] I certainly don't mean to diminish the pain of those who have actually lost a spouse, but rather wish to convey that each loss we endured felt just as profound. It was as if a part of us had been violently torn away, like a wife being murdered. The grief was overwhelming, yet there was no time to properly mourn, as we were immediately thrust back into the fight.

Through lived experience and later reflection, I came to understand a hard truth about death in combat: many in the combat arms community suffer from delayed or inhibited grief.[305] In the chaos of war, the natural response to loss is often postponed or suppressed entirely. The danger in this is that the grief doesn't disappear; it simply waits. And when it returns, it often does so with greater intensity, sometimes surfacing years later in unpredictable and deeply disruptive ways. I found this to be painfully true in my own life.

Because I didn't, or couldn't, properly process the trauma at the time, it began to seep through the cracks years later. I found myself confused by the intensity of my reactions to things that seemed unrelated. Certain smells, places, or sounds would trigger unexpected emotional storms. Something as ordinary as the sound of a helicopter approaching from the nearby hospital could send a chill through my entire body, raising goosebumps and flooding me with a sudden, vis-

---

[304] Pivar, Ilona, and Nigel Field. "Unresolved Grief in Combat Veterans With PTSD." *Journal of Anxiety Disorders* 18, no. 6 (2004): 745–55. https://www.sciencedirect.com/science/article/abs/pii/S0887618504000738?via%3Dihub.

[305] Rowe, Steven. "Delayed Grief: Causes, Symptoms, and How to Cope." Psych Central, January 31, 2022. https://psychcentral.com/health/delayed-grief#definition.

ceral alertness. It was as if I were about to literally explode, or be thrown back into the fight at any second.

But these weren't just memories. They were echoes of something unseen and unfinished. Somewhere deep beneath the surface of my mind, the enemy still lurked. He was waiting, watching, just biding his time for me to slip up, and I felt like I could never afford to let my guard down.

The problem was, my brain couldn't stay in a state of hyper-vigilance 24 hours a day; at least, not without consequences. Eventually, the pressure began to leak out in strange, involuntary ways. I started spinning my wedding ring between my fingers without thinking. I touched my face constantly. My fidgeting became noticeable to almost everyone, even if I wasn't fully aware of it myself. And every night, without fail, I would check the locks on my house over and over, slowly scanning the neighborhood for anything, or anyone, that didn't belong.

Some people turn to drugs or alcohol in search of relief. While these substances don't solve the problem, they can dull the pain, at least temporarily. Derek Hauger wrestled with this himself, though I'm proud to say my friend has since found sobriety. Travis Nelson, too, told me he had sought help in the past by turning to the bottle.

**29 July 2025 – Reflection of Travis Nelson, Charlie 4-64 Armor**

Before that day, I felt the tank was invincible. After everything that happened, I didn't want to go back out on patrol. It was terrifying, but I had to make my peace with it. If it was my time…

That second deployment to Iraq broke something in me. I stuffed it all down, and I began to have anger issues. Later, I started drinking and would get shit-faced drunk. I became a blubbering mess.

Though it may sound cliché, I discovered that confronting my subconscious fears helped ease the anxiety. The relief wasn't permanent, but it softened the sting. Rather than running from the fear, avoiding the numerous triggers, or numbing it with substances, psychologists often recommend voluntarily facing the "hidden dragon" that lurks in your dreams.[306]

In the opening scene of the 1979 movie *Apocalypse Now*, Martin Sheen stares blankly at the ceiling fan. The image of the fan blends with the haunting, dreamlike haze of a napalm attack and the relentless thumping of helicopter blades. Though I had seen the movie before joining the military, that scene took on a whole new meaning when I returned home. *That sound.* That sound carried weight, and it meant something.

A mental health professional once explained it to me this way: at some point during the deployment, my brain had learned to associate the sound of helicopter blades with death, destruction, and war. To break that connection, I needed to "rewire" my response by intentionally and repeatedly exposing myself to those sounds in a safe context.

---

[306] McWard, Noelle. "Facing Your Fears: The Path to Freedom From Anxiety Is Through It." *Psychology Today*, December 5, 2023. https://www.psychologytoday.com/us/blog/unpacking-anxiety/202312/facing-your-fears.

One night, while patrolling the streets of our city in my police uniform, I saw the hospital helicopter descending toward the downtown landing pad. I drove closer, parking near the fence that bordered the perimeter. I stepped out and gripped the cold metal tightly, my knuckles white as I watched it descend. The rotors thundered overhead, slicing through the air with that deep rhythm, and a chill surged down my spine. My heart pounded, and in an instant, I was no longer in Iowa. My brain was back in Iraq. I was back in those killing fields, even though I was physically standing thousands of miles away.

I did this whenever I had the chance. If the town wasn't falling apart and I spotted the helicopter approaching, I'd ease over and take my place by the fence. I would stand there, letting the rotor wash pour over me, feeling it ripple through my body like a distant memory returning home. And slowly, imperceptibly at first, the intensity began to fade. That old surge of fear and adrenaline gave way to something else. I no longer flinched at the distant growl of rotors cutting through the sky, and found that sound no longer owned me.

That method helped me tremendously. While I understand what others are going through, this is why I struggle to support the practice of veterans placing signs in their yards that say, "Please don't shoot fireworks. A veteran lives here." Similarly, I find myself shaking my head when I hear young college students requesting "trigger warnings." I do understand the instinct to avoid distressing experiences. I get it. But avoidance doesn't lead to healing and growth. In my experience, the only way to truly get better is to confront your fears and slay the dragon, not run from it.

There's more I want to say about PTSD, but I'll save that for the end of the book. For now, it is sufficient to say that we all struggled with forms of delayed grief, stress, survivor's guilt, and anger. No one is immune. As José Narosky once wrote, "In war, there are no unwounded soldiers."[307]

Sometimes I find myself reflecting on how many times we came terrifyingly close to death. A bullet snapping by just inches away. Nearly drowning in a canal filled with filth and waste. Hundreds of mortars exploding all around the FOB, shaking the earth and our sense of safety. Suicide bombers rushing the checkpoints, their chests strapped with explosives and ball bearings. And sometimes, it was something as silent and haunting as unknowingly driving over an IED that, for whatever reason, didn't detonate. But we all knew we had done just that. Apparently, it just wasn't our time.

**29 July 2025 – Reflection of Travis Nelson, Charlie 4-64 Armor**

The tank I was in drove over both of those IEDs. They never hit us. They hit the other tanks. That could have been us, but it wasn't.

---

[307] National Academies Press (US). "Gulf War and Health: Volume 8: Update of Health Effects of Serving in the Gulf War." National Library of Medicine - National Center for Biotechnology Information, 2010. https://www.ncbi.nlm.nih.gov/books/NBK220122/.

In our interview, Travis Nelson also reflected on a haunting detail: none of the three men who died that day were in the positions they were originally assigned. 1SG Gifford had volunteered to join the patrol, taking the place of another soldier who remained back at the FOB. SPC Ford, who usually gunned inside the turret, was driving instead. SGT Deckard, too, was normally in the turret, not the driver's position. Sometimes fate, or whatever force governs these things, can be unbearably cruel.

Everyone I spoke with shared heartfelt memories of all three men. They recalled the good times: the laughter, the camaraderie, the Halo tournaments, and the dinners together. They also expressed sincere hope that their families had found some measure of peace.

### 27 July 2025 – Reflection of Robert Roth, Commander of 4-64 Armor

When I met 1SG Gifford, I was impressed by his professionalism and dedication to his men. We talked a lot, and I just knew that this guy was good. He was a good man.

### 29 July 2025 – Reflection of Travis Nelson, Charlie 4-64 Armor

When I was a young private, I had a leave pass to go back home. I drove through this huge puddle in the parking lot and somehow broke a part on my Ford Ranger. I didn't have the money to fix it and wouldn't be able to drive home.

On the weekend, even though he could have just said I was a stupid private, SGT Deckard came in and fixed it. I helped him do it, but he never asked for any money. He just did it. He taught me a lot about being a tanker, and we had a lot of fun playing video games together.

Ford was so small and was the youngest of everyone. So, we teased him a lot, in a good way, but he was squared away, athletic, and very smart. He had a baby face and looked like a little kid.

We all walked away from that incident carrying the weight of irreplaceable losses. There was a void that couldn't be filled, but life demanded that we learn, grow, and somehow move forward. But it was a hard, unforgiving time.

I often think back to why I can't remember more of what happened in that village. It's strange because I can vividly recall entire "videos" of other moments, yet with that day, all I have are fragmented "photographs" in my mind. The image that stands out is the remnants of Red 3 in the distance. I also see the man flying backwards as we burst through the door. I see my arm tangled through his zip-tied limbs, my hand gripping his hair as I dragged him outside. I see the green illumination round drifting slowly toward the earth, with the tank sitting eerily still in the dark, and the flames from the rooftop licking the dark, humid night. But that's all. Everything else is gone.

Who was with me? How many of us were there? What was said? Where did the insurgents go, and where were they initially hiding? Did some escape along the Tigris River, or did we kill all of them? These questions remain unanswered, completely blank in my mind. I don't think I will ever find the answers to those questions.

I suspect part of the reason is that we were running on fumes, having had little sleep in the days leading up to that moment. On top of that, we were immersed in a chaotic, high-stress situation. It's a blur, one I wish I could reach into my brain and re-live, even though I know it would be painful. I just want to see again what I saw, but, unfortunately, that's not how it works.

In preparation for my story, I read David Anderson's short book, *Ride the Cyclone*, and many memories suddenly flooded back. It was a strange experience, like waking from amnesia, though I was still shrouded in a fog. My eyes teared up as I thought to myself, *I remember,* and it brought back fragments of details that had been lost to time. But even then, most of it remained elusive.

Some of the details came through in my interviews. While there are a few minor discrepancies between the individual stories, *Ride the Cyclone,* and my personal memories, I know the fog of war can easily explain these differences. It's strange how the mind works to protect itself. It's also strange how one person will see something completely different than someone standing a few feet away.

In the end, trying to find meaning in the chaos, CPT Anderson wrote something in his book that has stayed with me ever since: "True leaders create meaning out of difficult events or relationships, while others may be devastated or even defeated by them. Leaders come out of these experiences with something useful. Through a crucible type experience, leaders acquire new insights, new skills and new qualities of mind or character that make it possible to leap to a new, higher level."[308] I've done my best to live by those words.

---

[308] Anderson, *Ride the Cyclone.* Page 3.

# In Memoriam

1SG Alan Nye Gifford[309]
7 August 1966 (Fayetteville, North Carolina) – 16 September 2005 (Baghdad, Iraq)


SGT Matthew Lynn Deckard[310]
10 May 1976 (Elizabethtown, Kentucky) – 16 September 2005 (Baghdad, Iraq)


SPC David Harrison Ford IV[311]
10 September 1985 (Ironton, Ohio) – 16 September 2005 (Baghdad, Iraq)

---

[309] Military Times, "Army 1st Sgt. Alan N. Gifford," Honor the Fallen, https://thefallen.militarytimes.com/army-1st-sgt-alan-n-gifford/1116668.

[310] Military Times, "Army Sgt. Matthew L. Deckard," Honor the Fallen, https://thefallen.militarytimes.com/army-sgt-matthew-l-deckard/1114136.

[311] Military Times, "Army Spc. David H. Ford IV," Honor the Fallen, https://thefallen.militarytimes.com/army-spc-david-h-ford-iv/1116666.

# Chapter XVI

## Triangle of Death

In the aftermath of the three deaths on Route Corvette, the infantry began conducting far more foot patrols in the area. We discovered plenty of loose wires, unexploded rockets and mortars, and all manner of suspicious items, but the work was absolutely grueling. Every day in sector felt like a slow, punishing slog.

**A rare quiet moment on a rooftop, stealing a break whenever the war allowed one**

Under normal conditions, we operated from the Bradleys, riding in until it was time to dismount and conduct our mission. The downside to tracked vehicles, obviously, was their noise. Insurgents could clearly hear us coming from miles away *and* from a specific direction. To offset that, we began using Black Hawks, air assaulting into villages and specific neighborhoods with the hope of catching them off guard.

**Infantry inbound to the fight on Black Hawks**

**Photo courtesy of Jason Sturm**

"Kidnapping" insurgents in this way, as it turned out, was much easier. Grab onto some bad guys, smack them around a little bit if they resisted, and then throw them into the back of a Bradley that arrived a few minutes later. Even so, vehicles couldn't always reach where we needed to go. We sometimes had to go into thick vegetation or other areas where tracks and helicopters weren't well-suited.

So, we did foot patrols, but these were a different beast altogether. The 120-degree heat, relentless sun, heavy gear, and unforgiving terrain made each step a nightmare. By the time we returned to the FOB, our uniforms, underneath our stifling vests, would be stiff with thick, white chunks of salt. We had sweated so profusely that it couldn't evaporate, leaving actual piles of salt behind in our shirts. It was oddly fascinating.

**A long walk between objectives, searching for any signs of insurgent activity**

In the meantime, we were constantly and progressively inventing stupid, meathead games and pranks to outdo each other on base. They were childish, pointless, and, if I am being totally honest, absolutely hilarious. The entertainment value came from laughing our asses off at someone else's misfortune. At some point, a pair of nunchucks appeared out of nowhere, probably coming from the mischievous SGT Vance, and they were almost exclusively used for whacking unsuspecting grunts in the genitals. If you were at the FOB and not paying attention, your chances of getting ninja'd in the crotch were about 100%. Then there was a cheap acoustic guitar that someone tried to play until it inevitably got smashed over someone else's head like we were auditioning for the WWF.

Yes, it was all incredibly dumb. But dumb was fun, especially when we had to deal with the constant seriousness of combat, and hitting each other while causing general mayhem around the FOB was practically a pastime. It was also a great way to blow off some excess steam. Still, my personal favorite wasn't the guitar or the nunchucks; it was the legendary *salt trick* I invented in the chow hall. OK, it was probably only legendary in *my* eyes, but I still think it was great.

To really gross out the Fobbits in the chow hall, I'd break off a few of the bigger salt chunks from my shirt after a long day in the sun. Those slugs had no idea why I was filthy, caked in salt, and in *their* precious dining facility. Most clearly had never worn an Interceptor vest for more than a few minutes at a time, probably while they played a scary PlayStation game in their lair at night, so the salt thing from my uniform was a little confusing. Casually, I'd crumble the flaky white chunks from my dirty uniform onto my meal and start eating as if nothing was unusual. The infantrymen at my table would be doubled over with laughter, while the POGs gasped in horror and disgust, and scattered in search of more civilized company. Ah, the things we do to impress those around us.

### 21 September 2005 – Letter to the Harstad Family

We walked FOREVER today on a patrol. There was this chicken farm we had to search, and all these fields. It was probably four kilometers by four kilometers. The entire thing was flooded, and we kept getting stuck.

I took my team up onto that road where those three guys were killed (to look for more insurgents responsible), and we started searching cars. One car came right at us, and I lit his ass up. (This was after he ignored multiple stop signals and then drove at us like a maniac. He survived, but his car did not.)

I was so tired and dehydrated at the end.

### 5 August 2025 – Reflection of Derek Hauger, Bravo 4-64 Armor

I remember that. They wanted you to carry a SAW because we were out walking around, even though you normally carried a 240 or an M-16, and they put you in my squad temporarily. SSG Brister went over to check on you to make sure everyone was ok.  When he came back, he just looked at me, and I knew exactly what he was going to say. He goes, "Hauger, Harstad has had that SAW for like ten minutes, and he already shot a car with it. You've had that thing all deployment!" Always giving me a hard time.

Even if our temporary battalion still wanted us out kissing babies and walking around with big red targets on our backs, we weren't taking chances anymore, not with the thousands of rabid fighters who were in and around our area of operations. Ignore warnings and drive straight at us, flash a weapon in our direction, or move like you're about to set something off? That wasn't a misunderstanding, not out there, not then. We treated it like a threat and responded accordingly; quickly, decisively, and without a committee meeting in the middle of the road.

SPC Balicki, our FiSTer, always hunting for IEDs, command wires, and unexploded ordnance

Photo courtesy of Derek Hauger

### 3 January 2025 – Reflection of Cap Stanley, Command Sergeant Major of 4-64 Armor

In response to the killings, we had a UAV (Unmanned Aerial Vehicle, a drone) in the air, watching the route. We spotted a guy digging into the road, in preparation for another IED attack. We did a targeted strike on him.

And guess what? Just before the strike got to him, he looked up into the sky because he heard it coming. But it was too late, and he was abruptly turned into pink goo.

Charlie Company relied heavily on us as they worked to return to mission-ready status. A tanker platoon, after all, was only around sixteen men, so the incident on Corvette was particularly devastating. Three killed and three wounded was a significant blow to their fighting capabilities. Meaning, we understood exactly why we had to step it up. Until they were ready to return to the fight, the responsibility fell squarely on us.

### 25 September 2005 – Letter to the Harstad Family

We have been on quite a few missions, which include IED sweeps, meet-and-greet sessions, weapons disarmaments, and plenty of other activities. Our platoon has not had any losses so far, only minor injuries.

Charlie Company is still trying to get back on its feet since three members were killed a little over a week ago. It still feels strange, and I'm sure it will for the rest of our lives. To think that I was just talking to one of the individuals an hour before he was killed. It's hard on a lot of people. That's just the way things work.

In reality, and apparently much to the consternation of our National Guard task force, we became far more aggressive, shifting from simply waiting to be attacked to actively hunting the enemy. One of our most effective tactics was setting up flash traffic control points (TCPs). These were quick, surprise checkpoints designed to catch insurgents off guard. They proved far more productive in Dora than they ever had on the Karrada Peninsula, largely because insurgents were everywhere and they were freely transporting their goods. Here, it wasn't uncommon to actually catch them with weapons, explosives, and other contraband.

**The author conducting TCPs in Baghdad**

To pull this off, we would dismount from our vehicles and quietly move into a village or dark alleyway. Then, at the least expected moment, we'd surge out and storm random cars as they drove down the road. All of this was done at gunpoint, of course. It startled everyone, insurgent and civilian alike, but the results spoke for themselves. We quickly began seizing caches of smuggled weapons and, even better, started killing insurgents who thought they could actually fight.

### 4 October 2005 – Letter to the Harstad Family

We are headed out to do more TCPs today. Basically, we hide behind a wall and jump out at cars, randomly searching them. Yesterday, our platoon sergeant shot four guys that took off past us. They pointed their AK-47s at us just before he blasted them with the Bradley.

All of this was unfolding right as the Iraqis were preparing to sign the Constitution, and insurgent activity spiked to a fever pitch. We were running in every direction, barely catching our breath. Grab these guys: check. Secure that site: check. Escort the villagers: check. It felt like we were everywhere at once, chasing the next task before we'd even finished the last.

Working alongside us were many of the Iraqi Army and Iraqi Special Forces soldiers we had helped train. As I've mentioned before, some were competent, and others… not so much. Still, they were extra boots on the ground, and they proved useful when dealing with the locals. Their familiarity with the people and the terrain sometimes let them pick up on things we might have missed.

### 25 September 2005 – Letter from Robert Roth, Commander of 4-64 Armor

Yesterday, one of the Iraqi Army platoons went on a mission into a neighborhood to set up a TCP to search vehicles and people. As this patrol moved to its destination, they pulled off onto a small road, and several Iraqi Army soldiers leaped from the trucks and immediately stopped traffic coming from both directions.

Well, it happens that two of the Iraqi soldiers that went forward to stop traffic did so right in front of a car bomb.[312] Before they could get off a shot, the suicide bomber detonated his car and killed these two brave soldiers. Had it not been for their diligence, the car bomber could have driven closer to the patrol and decided to detonate next to the Iraqi and US Army trucks, causing many casualties. These two Iraqi heroes did their jobs, and did it well. They kept their Iraqi Army brothers and several Tusker soldiers safe.

More importantly, we believe this car bomb was headed to a heavily populated civilian area to kill as many men, women and children as possible. Thus, because of their actions, they protected the people of Iraq. There is no greater act of love than to lay down one's life for another.

---

[312] ABC News, "Two Killed as Suicide Bomber Targets Iraqi Army," September 24, 2005, https://www.abc.net.au/news/2005-09-24/two-killed-as-suicide-bomber-targets-iraqi-army/2110530.

**26 September 2005 – Letter to the Harstad Family**

In two weeks, they are going to sign the Constitution, so things are probably going to get worse. They are predicting that we are going to be mortared like crazy. So, we can only eat chow once a day (in the chow hall). This is going to last for three weeks.

Our more aggressive tactics, though effective, meant we were stuck in a constant tug-of-war with our adoptive battalion, causing perpetual heartburn. It was a never-ending circus of armchair quarterbacking and petty griping. Every move we made was second-guessed by people who wouldn't dare step into our boots, and the truth was, they were never going to be happy. Still, I think they slowly started coming around after seeing the results.

**30 September 2005 – Letter to the Harstad Family**

(Redacted) came to talk to us today, I guess to try and get us "motivated." We are perfectly motivated, we just don't want to do their retarded ideas.

(His speech) didn't work, but he informed us that we were (finally) changing our tactics. We are demolishing a lot of bridges, destroying a lot of their crops, and other stuff like that. It should really help us in sector. But still, (his plan) is basically nothing but a mission for us to try to stay alive. That's reassuring...

**4 August 2025 – Reflection of Heath Hutchison, Charlie 4-64 Armor**

I remember that "motivational speech" (redacted) did for us. He told us we needed to stop whining about our three friends who had just died and focus on the mission.

I walked away from that formation in disgust. Well, he accomplished one thing in his speech. He got pretty much everyone in the company to rally against him, even if we couldn't say it out loud.

Later, CPT Anderson got everyone in Charlie together and directly told us, "Listen, I'm going to back all of you. Do whatever you think needs to be done. Just make sure you have a legitimate reason for it, and I'll back you."

I was glad he was our leader. He wouldn't ask anyone to do something he wasn't willing to do himself. We were lucky to have him.

**21 July 2025 – Reflection of Juan Serrano, Platoon Sergeant, Bravo 4-64 Armor**

In that formation, (redacted) asked me what my mission was. I told him, "To bring all my men home." He didn't like that at all and chewed me out in front of everyone. It was constant fighting between us.

I had already been in a lot of trouble with them (National Guard battalion), and I had been formally written up twice. They told me, "Serrano, if you get written up one more time, we are taking the platoon away from you and are going to bust you back to staff sergeant."

I knew I had to make up my mind. What was more important? Doing what they told me to do, which I knew wasn't right and would end up with more people needlessly killed, or protecting all of you? I chose all of you.

But I never directly disobeyed an order. I played the Hispanic card. Since I spoke Spanish (as my first language), I would say, "Oh, I'm sorry, I didn't understand what you meant." I'd do what they wanted, but in my own way, that wouldn't get people hurt or killed. I had the courage to throw away my career, and I would do it all over again. I had to protect all of you from those idiots.

It was during this time that an unfortunate incident occurred. I can't remember exactly where it was, other than on a busy Baghdad street somewhere, perhaps in Dora. What I do remember is the atmosphere: a tense city that could turn on you in an instant. Because of that, it replays in my mind with unnerving clarity.

We were patrolling on foot through a hectic urban district, having been dropped off down the way, and the streets were alive with the constant hum of people and traffic. Because the street was partially blocked by logjammed vehicles, I had to temporarily cross into a wide center median, weaving in and out of traffic. Then he appeared, seemingly out of nowhere.

He was an Iraqi man, probably in his thirties, clearly mentally disabled, drooling and smiling, his gait awkward and unnatural. I still don't know where he came from. Maybe a doorway, a restaurant, or one of the narrow alleyways threading between buildings. But suddenly he was there with me, no more than fifteen feet away, right as a gap opened up.

Very quickly, I realized he was trying to get to me. At first, I tried to solve it without making a scene. I stepped laterally, zigzagging while staying with the staggered column, hoping he'd drift off or latch onto someone else. I made a quick attempt to beeline out of there, but everything was still so congested, with cars trying to squeeze through this nightmare knot of Baghdad. When I glanced back, he was literally right there next to me.

Now alarmed, I shouted for him to get back, but he didn't respond. He just had that glazed-over look, not understanding what I was saying. Nor did he comprehend that we were trained to see the hazard in what he was doing, getting that close to us.

Then, without warning, he lunged forward and tried to grab me. I don't know why he did it or what his plan was, but I immediately let go of my rifle and let the sling catch it against my body, and then shoved him away from me with everything I had. He was small in stature and flew backward, landing awkwardly. Over the noise and commotion, I heard the crowd gasp.

I felt terrible the instant I did it, but I turned and tried to get out of there. I glanced back one last time as a couple of Iraqi women helped him to his feet, and I saw his face. It wasn't fear, though. No, it was confusion and something like betrayal. The only way to describe it was like a child holding out a baseball card for their hero to sign, only for that hero to tear it in half right in front of them. In that split second, I was the villain on their street, the American just pushing over a helpless man.

So, why did I shove him and then sprint away? Because by then, we had learned what the enemy was willing to do. We heard the horror stories and watched their tactics evolve in real time. We knew they were turning civilians into delivery systems, and in their hands, a confused stranger could be a bomb with legs.

Slowly, these reports became something we talked about and took seriously: al-Qaeda using mentally disabled people as unwitting suicide bombers. As sick and twisted as it sounds, they would strap an explosives vest onto someone with a severe intellectual disability and send them toward an American patrol.[313] When that unsuspecting person got close enough, a hidden accomplice would detonate the explosives remotely.[314]

Dale Andradé described this tactical shift in *Surging South of Baghdad* as an "air of desperate improvisation," a grim adaptation as checkpoints and tighter security disrupted the insurgents' usual methods. As he recounted it, that improvisation led to increasingly hideous delivery tactics, including the exploitation of women and people with disabilities as bomb couriers.[315]

Thomas Ricks reported a similar evolution in *The Gamble*. He noted that when cars and young male bombers became harder to move through a landscape of checkpoints, attackers turned to other carriers, including bicycles, women, and even preteen boys. In the most grotesque accounts, the violence escalated into the coercion of girls with mental or physical disabilities, weaponizing vulnerability itself.[316]

Remember, this wasn't a war of clean, honorable fights. We were up against people who would use anyone, including women, children, the elderly, and even the disabled, if it meant killing us. So, when I shoved him away, it wasn't out of cruelty. It was instinct, a reflex born of self-preservation and the knowledge of what we were facing.

Still, none of that makes the memory any easier. No matter how many times I tell myself I did the right thing, at least in that split second, part of me keeps wondering if I could have handled it differently.

Maybe I could have just broken into a dead sprint, running in between cars and hoping not to get crushed. Maybe I should have stopped and given him a hug, even if it meant risking both our lives in a spray of nails and ball bearings. But the truth is, I'll never know what the right decision was, and I have to live with the one I made.

---

[313] Gaughen, Patrick. "Baghdad Neighborhood Project: Saydiyah." Understanding War, 2007. https://www.understandingwar.org/wp-content/uploads/2025/04/Backgrounder15.pdf.

[314] Howard, Michael. "Bombs Strapped to Down's Syndrome Women Kill Scores in Baghdad Markets." *The Guardian*, February 1, 2008. https://www.theguardian.com/world/2008/feb/02/iraq.international1.

[315] Andrade, Dale, Center of Military History, and United States Department of The Army. *Surging South of Baghdad: The 3d Infantry Division and Task Force Marne in Iraq, 2007-2008*. eBook edition. Military Bookshop, 2010, Page 307.

[316] Ricks, Thomas. *The Gamble: General Petraeus and the American Military Adventure in Iraq*. eBook edition. Penguin, 2010, Chapter 9.

Had it been a seemingly normal Iraqi, he probably would have been shot for trying to grab me. No, I take that back. He *would* have been shot, or, at the very least, stomped into the ground. But the moment I saw that something was wrong with him, I hesitated. I froze in that gray space between instinct and indecision, unsure of what to do, which is precisely why terrorists began using them against us in the first place. Afterward, though I felt awful, I realized it could have been much worse.

What bothered me most came when I replayed what was racing through my mind as it happened: *suicide bomber.* In a different heartbeat, I might have had to... I don't even want to think about that. So, if there was a silver lining, it was this: he walked away physically unharmed, maybe with nothing worse than a bruised tailbone.

And that is the kind of moment that crawls into your head and never leaves. There are soldiers who, in that suffocating heat of adrenaline and noise, squeeze the trigger and only learn afterward that they were wrong. The fear and the tension were real, but the threat was a mirage. Then they have to carry the weight of it forever, knowing that in one irreversible instant, they took an innocent life.

You find yourself caught in a relentless battle between compassion and survival, between the values you were raised with and the mission you've been ordered to carry out. Which one do you choose? Oh, and by the way, you have only 1.3 seconds to figure it out. *Go!*

It's in moments like this that war starts to bend your mind into shapes you didn't know it could take. And somewhere, far away from the blasts and the blood, there will always be those little basement-dwelling toads, hunched over their keyboards, dissecting every move like they're use-of-force experts. They'll never understand, though, because they weren't there. No matter how hard you try, you can't pour the smell, the heat, the fear, or the split-second choices into the head of someone whose only battle in life has been with a weak Wi-Fi signal.

When we weren't trudging through sketchy neighborhoods on foot patrol, we were hitching rides in Black Hawks to carry out "snatch and grabs." We'd teamed up with a SEAL platoon, and they were a blast to work with, though I'm pretty sure we fanboyed over them more than we'd like to admit. They often showed up with decked-out Humvees, bristling with miniguns and enough high-speed hardware to make an arms dealer wet himself with joy. It was fun being around them.

On October 9th, we were headed back into Arab Jabour, an extensive but spread-out neighborhood in southern Baghdad infamous for producing more than its fair share of so-called martyrs.[317] The general, macroscopic view of the area was like this: to the north lay dense urban sprawl, to the east the steady sweep of the Tigris River, and to the south and west a patchwork of farmland and small, poor villages. Together, this ground formed the edge of what we unofficially knew as the "Triangle of Death."[318] Meaning, it was hell on earth.

---

[317] DVIDS, "U.S. Aviation, Infantry Forces Hunt Down Terrorists," Defense Visual Information Distribution Service, October 11, 2005, https://www.dvidshub.net/news/3261/us-aviation-infantry-forces-hunt-down-terrorists.

[318] USIP Staff. "In Iraq's Former 'Triangle of Death,' a Decade of Stability." United States Institute of Peace, August 9, 2017. https://www.usip.org/publications/2017/08/iraqs-former-triangle-death-decade-stability.

**A Gambler assault team just before air assault, with one of our female interpreters and a dog team**

**Photo courtesy of Derek Hauger**

Moving to a closer view, Arab Jabour was home to about 100,000 people, but it wasn't a single, continuous urban district. Pockets of dense housing were broken up by farmland and groves. The satellite village we were headed for was one of the smaller clusters, maybe several dozen houses, loosely spaced, stretching almost perfectly east to west. A wide main alleyway cut through the center, hemmed in by stone walls and the occasional rusted metal gate.

Its overall layout mimicked the wider region, with a few minor exceptions. Less than half a kilometer to the east, the Tigris slid past, its banks tangled in thick vegetation, making it the perfect place to hide. To the north, a Y-shaped intersection fed into a busy road, Route Red Wings, that had seen more than its share of firefights and IEDs. Just beyond that, a broad grove of trees filled a massive open field.

To the south and west, there was little more than endless fields of crops. That's where we planned to insert, hitting the ground running toward the north. We aimed to catch the insurgents flat-footed before they had the chance to scatter. But, if they were going to flee anywhere, the most logical spot was to the Tigris.

**FOB Falcon below as we headed toward Arab Jabour and back into the Triangle of Death**

Six Black Hawks lifted off from FOB Falcon, blades whipping the air as we tore across the Baghdad skyline. With us, we had a dog team trained to detect explosives and, if need be, take down the enemy. Cruising between 150 and 200 mph, we hugged the deck at roughly 100 feet, skimming so low over the rooftops you could see laundry flapping on balconies and kids looking up in awe. Moving southeast, it wouldn't take long for us to arrive on target.

Meanwhile, a swarm of Gambler Bradleys and Cyclone Abrams had already rolled out, racing to link up at the objective moments after we hit the ground. Because the target village had lookouts, we needed a way to surprise and capture groups of men suspected of churning out wave after wave of VBIEDs and regular IEDs. The timing had to be just right.

I was right in the doorway as we came in fast and low, skimming just above a muddy field of crops. The helicopter never touched down; it just hovered about 5 feet off the ground. I jumped and instantly plunged into the muck, sinking up to my knees. That's when it hit me: I was stuck, wedged in like a human javelin. *Great.* Of all the places to be stuck. Well, at least I can add it to my résumé as a "special skill." *Survived being knee-deep in mud in the Triangle of Death. References available upon request.*

**Coming in low over the target, near our LZ that turned out to be a muddy field**

While I flailed, the pilots moved about twenty meters away to drop off the rest of the platoon. Someone eventually backtracked to help drag me out, finding me frantically cutting away thick mud from my legs with my hidden knife (yes, the same one I wasn't supposed to have). *But why didn't I bring my E-Tool?* With a little help, I pulled my legs free, a strange sucking noise coming from the ground as each leg was extricated. My boots were filled with stinking rot, but at least I was out!

Linking back up with the rest of the team, we cut through backyards, scaled fences, and pushed toward the target buildings. Bursting into one "backyard," which was really just a dirt field, we were suddenly charged by a furious ram who apparently considered trespassing a capital offense. This wool-covered lunatic, easily 200 pounds and nearly shoulder-high, came at us like he'd been saving up his anger for Americans for years. Several of us scrambled to escape while he tried to gore anything in reach. A few well-placed rounds to the head ended the threat, but then the radio crackled: *Who's shooting?* We didn't have time to explain that the ram had just returned from Osama bin Laden's training camp in Afghanistan, but the message got across.

Building after building, we cleared them out, rounding up every military-aged male we found. Their lookouts had failed to call ahead, leaving them wide-eyed and frozen as American grunts came crashing into their living rooms. The best fights are the ones where you don't have to fire a shot. You hit them so fast they can't react, or you scare them so badly they won't dare try, and that's exactly what happened here. No one put up a fight.

Well, almost no one. While the majority surrendered peacefully, a small group that slipped toward the Tigris started taking poorly aimed potshots at us. Their defiance didn't last long, though. The Bradleys and Abrams answered with thunderous bursts, cutting them down in seconds. Overhead, a pair of Apaches swooped in along the riverbank, raking the area with streams of 30mm chain gun fire that shredded everything in their path. That area went quiet.

While the other guys were having their fun shooting by the river, we were busy zip-tying and sorting out who we wanted to bring back for interrogation. We ended up with far more prisoners than we'd expected, nearly twenty, which slowed everything down. That's never good. In an operation like this, speed is everything. The longer you stand around, the more time the enemy has to regroup and hit back.

The call was made to have a Bradley or two pull up at the east end of the alley so we could load the captives. While that was getting set up, my fireteam worked roving security on the south side. We wandered into a backyard and found a middle-aged woman and several younger women, probably her daughters, stepping outside. They were polite, spoke broken English, and I tried to get information from them, but came up mostly empty. They didn't seem to be a threat, so I said "salam" and stepped back into the central alley.

Seconds later, a mortar landed right where I had just been standing. It wasn't a large round, but it was loud enough to rattle my teeth. I ducked and tried to look back through the gate to see if the women were alright, but they were gone, or at least out of my sight. *Where did they go?* If they were standing there when it hit, well… Another round slammed in about fifteen yards away, close enough for me to see the flash.

Now we had a serious problem on our hands: we had over a dozen detainees tied up in the alley, and mortars were starting to walk in on us. "Shove them in the back!" CPT Anderson yelled. We started piling prisoners into the Bradleys, stacking them on the floor like rows of Vienna sausages. It couldn't have been pleasant back there, but it beat getting shredded by whoever was lobbing those rounds at us. They obviously didn't care about their buddies. Once they were all loaded, we could get the hell out of there and call for the chopper extraction.

As all this was unfolding, SFC Serrano's Bradley rolled up onto the main road at the far east end of the alley, probably half a football field away from where I stood. His track was already packed with prisoners, so he told the grunts pouring out of the neighborhood to climb onto the top of the Brads. His plan was simple: quickly drive us to safety at the extraction point, where the Black Hawks would swing back in to pick us up. In the meantime, the Bradleys would whisk the prisoners away, and we would all rendezvous back at the base. Simple.

What happened next was baffling, to say the least. One individual from our National Guard battalion, who, as always, will remain nameless, pulled a Humvee directly in front of Serrano's Bradley, blocking it from moving. Then, *as mortars were coming in*, he got out and started screaming at Papi that what we were doing was unsafe, as if the exploding shells landing all around were filled with kittens and rainbows, not chunks of hot metal. According to him, everyone needed to *run* down the road, with no cover to be found, to the extraction zone rather than ride atop the Brads. I swear this actually happened.

### 21 July 2025 – Reflection of Juan Serrano, Platoon Sergeant, Bravo 4-64 Armor

As we were being attacked, (redacted) pulled his Humvee right in front of my Bradley, so I couldn't move. He made me get out of the Bradley, come down to his Humvee, and chewed me out right in the middle of the road as they were shelling us.

I told the gunner, "Shoot anybody that comes close to us," and climbed down. After he was done yelling at me, I told all of you, "Hurry the fuck up and run that way!" I was so pissed off.

As I neared the mouth of the alley, lungs burning, legs churning for my life, the ground sloped upward toward the main road, right where that ridiculous shouting match had just happened. Then another mortar slammed in, off to my right, close enough that I caught the blast in my peripheral vision. Little pieces of pea gravel and dirt flew across my path, with some of it lodging into my vest and going down the back of my shirt. Then, out of nowhere, something black and shapeless whipped down from the sky. I think it was a chunk of one of those heavy metal gates.

*Smack.* It struck the base of my left thumb, right where it curled around the handgrip of my M-16, with the kind of force you can feel in your bones. It didn't hurt, not yet, but the shock made my entire hand and wrist go instantly numb. The impact drove my weapon and numb hand hard into my pumping legs, buckling me mid-stride. I went down face-first, my nose crunching against the gritty, half-paved alley. A blinding white flash burst in my skull.

But there was no time to check for damage or to think. I was on my feet again before my brain caught up, running harder, because stopping meant dying.

We broke from cover and sprinted north down the road, the worst possible move tactically, as explosions chased us from behind. Looking back, I'm convinced someone in the village was acting as a forward observer; the rounds were too precise, too consistently landing just behind our retreat. Our only hope was to reach the cover of the trees clustered near the Y-intersection. Don't get tired now! Just run!

### 5 August 2025 – Reflection of Derek Hauger, Bravo 4-64 Armor

Those mortars came down, and Brister yelled, "Run!" I was like, "Fuck this, I'm outta here!"

I was right next to you on the other side of the alley when that thing hit you in the hand, whatever that black thing was. Then I realized you had blood all over the fucking place, and I asked, "What happened to you?"

I still don't know how it hit you, and nothing hit me. You know, sometimes it is really, really random. Even though we were deep in that village, I think you and I were the first ones to make it into the woodline.

Adrenaline was driving me forward, but my vision blurred as my eyes watered from the hit to my nose, making it hard to see where I was going. I felt what I thought was snot running down my face, so I wiped it away with the back of my left hand, the same one that had just been smashed. It wasn't mucus, but blood.

By the time I reached the tree line, crimson streaks covered the front of my vest and smeared across my knuckles. Hauger and the others assumed it was all from the earlier smash to my hand, but it wasn't. Not a drop had come from there. Every bit of it was from my nose.

Hunkered down in the woodline, we waited for the Black Hawks to swoop in and whisk us back to the FOB while the medic bandaged my hand. Naturally, our "beloved" battalion decided that instead of picking us up there, we'd need to abandon our nice, cozy cover and stroll two miles in the *opposite* direction. Changing the landing zone? Oh no, that would make too much sense. Better to have the grunts go on a scenic tour through potential mortar and sniper fire. So off we went, marching our happy asses across open terrain and down roads to the pickup site.

In the meantime, my nose had finally stopped bleeding, and I'd managed to wipe away most of the mess. After running miles down an open road, in which a couple of insurgents tried to ambush us, we finally made it to the landing zone. Fortunately, the flight back was uneventful, but I couldn't shake this strange electric tingle running through my wrist.

My first stop was the medical clinic. The medic examined the base of my thumb, where a small half-inch cut marked the impact site, and the skin around it was already turning a deep purple. Oddly, it never bled, even though the gash was deep enough to show the white of my knuckle, but the skin was really thin there. The doc figured the blunt force had compressed the tissue so hard it sealed itself, leaving only a small tear. X-rays thankfully showed nothing broken, though they suspected a touch of "nerve damage." That's why my hand felt weird.

The official narrative was that we captured 18 confirmed insurgents and killed one in Arab Jabour. Still, we were pretty confident there were bodies that never turned up along the overgrown banks of the Tigris. On our end, I was the only one with a minor injury. [319] All in all, a good day's work. Well, except for the part where we had to run for our lives while mortars tried to blow us to pieces.

My letter home the next day, sanitized for obvious reasons, still managed to let some sarcasm slip through.

---

[319] DVIDS, "4-3 AHB Air Assault," Defense Visual Information Distribution Service, October 9, 2005, https://www.dvidshub.net/image/10890/4-3-ahb-air-assault.

**10 October 2005 – Letter to the Harstad Family**

Yesterday, Oct. 9th, we went on a nice, pleasant mission in our sector. We have really been cracking down on these idiots and have been rounding up a lot of AK-47s, pipes, and wire. We were doing just that yesterday.

We air assaulted in, coming in six Black Hawk helicopters. As soon as we jumped off, we realized that we were in a giant field of crops, with water up to our knees. Not a good way to start. We had tanks on the ground, Bradley support, and Apaches in the sky ready to shoot the place up. My squad and I had a certain number of buildings to clear, which we did with great success.

Twice we had nerve-wracking moments, and both involved animals. A ram tried to kill us (probably because we ran into his yard), and we shot him in the head five times, finally killing the beast. We also had to shoot a dog that was trying to attack our dog team.

We rounded all the males up, searched and questioned them, and sent away the ones we deemed "innocent." At this time, we engaged in a firefight by the river. A few individuals were trying to use sniper fire on the dismounted infantry. This was suppressed by machine gun fire from tanks and Bradleys, as well as the 30mm cannons on the Apaches. They lit the entire river up, killing whoever was down there.

Just as we were finishing up the detainee in-processing, mortars started coming down. The first one landed approximately 30 yards from me, exploding on the other side of the wall that I was standing. I was in front of a gate, so I looked through, trying to see if it was a mortar or a rocket. Just then, another mortar hit, this time about 15 yards away. It was close enough that I saw the flash. As soon as that happened, we all ran.

Some people were able to cram into the back of the Brads. I was on foot with the rest of my squad as we tried to run for cover. They kept "walking" the mortars in to us, following the infantry as we tried to escape. Every time we moved, another mortar would land right behind us.

I made it up to the road and was struck in the hand by a piece of falling metal. It smashed my hand against my M-16, causing it to instantly go numb. My M-16 swung down and took out my legs.

My (nose) was bleeding, but I didn't realize it until we stopped about 500 yards down the road. We then walked for approximately two miles, trying to work our way over to the drop site. The tankers captured two individuals that were going to try and snipe us, so we proceeded without further incident. We were picked up and dropped off at FOB Falcon.

Paperwork had to be filled out, and I was forced to go to the hospital. They cleared me, saying that I may have slight nerve damage and a minor cut to the hand. I have to return in 10 days for a re-evaluation.

From what Dad told me, (they) were contacted by Washington and notified of my injury. But, to no worry, I'm fine.

When I got back to the barracks, the entire place erupted into a roast session over my "tragic" little injury. That was just our culture, though. If you came back missing a limb, they'd probably still tell you to walk it off. I'm pretty sure I could have staggered in with my spleen on a kebab skewer, and the reception wouldn't have changed. One guy even asked if I had sand in a certain female orifice. I'll let you figure that one out, but let's just say it wasn't a question for medical purposes.

### 5 August 2025 – Reflection of Derek Hauger, Bravo 4-64 Armor

When I was blown up in Little's track, the adrenaline kicked in, and I was able to chase that guy down. But afterward, once it wore off, my back really started to hurt.

I sat down next to Little as he was being treated, and the medic took a look at me. Everyone pulling security started making fun of me, randomly calling out, "Pussy!" That's just how it was. That's what we did to each other.

The doctors prescribed me some painkillers and sent me back to work, but everyone kept up with the hazing for about a week. Then they eventually found someone else to pick on.

I, too, was sent right back to work with the Army's universal cure-all, water and Motrin, but they also handed me this absurd-looking glass cast to wear at night. I guess it was supposed to immobilize my arm while I slept and help with the tingling. In reality, it just made me look like a discount cyborg and doubled the amount of crap I got from everyone. I only wore it at night for a short time before chucking it. No medical benefit could outweigh the sheer volume of shit I was catching.

### 23 October 2005 – Letter to the Harstad Family

I have had to wear this splint at night. My wrist has been acting up, and my hand goes numb and twitches a lot. I talked to the doc, and he says that my carpal nerve is inflamed, and that by putting it in the splint, the swelling should go down.

Can you believe that? I'm 21, and I'm already getting carpal tunnel syndrome. Though it's not from old age, of course.

As I briefly mentioned, my dad received a notification immediately after the incident from someone he believed was calling from Washington, D.C. It may have been a member of the Family

Readiness Group or Rear D, though, and he no longer recalls for certain. Understandably, the call caused significant alarm within my family.

A few days later, LTC Roth wrote home to everyone about the incident, keeping folks in the loop and casually slipping in details about IED clearance in the area. It was a wild time, to say the least. But hey, at least the helicopter rides were fun. Nothing like hurtling over rooftops at 150 miles per hour to make you forget, briefly, that people are trying to kill you.

**16 October 2005 – Letter from Robert Roth, Commander of 4-64 Armor**

CYCLONE has been very active. Leading up to the elections, they conducted a helicopter air insertion in coordination with ground forces to capture insurgents in some pretty rough areas. The mission went off well, and a dozen insurgents were detained and turned over to the authorities for further questioning.

Several insurgents attempted to ambush some of CYCLONE's tanks securing the outer perimeter of the raid site, but attack helicopters overhead were able to fend off the would-be attackers.

BEAST continues conducting route clearance operations along the most dangerous roads in the brigade's area of operations. They found several roadside bombs this past week with their Buffalo. The last mission, the bomb exploded, damaging the mechanical arm and flattening several tires. The crew inside was never at risk. However, I'm sure their blood level rose quickly.

We recovered the Buffalo and dragged it to FOB Liberty, where civilian contractors were waiting to repair it.

Later in November, several of us were awarded our Purple Hearts for wounds received in action, including Little, Hutchison, and myself. The ceremony was small but meaningful, with LTC Roth pinning the medals on us. What stood out most, though, was that both he and CSM Stanley took the time to speak with each of us individually. Their words were sincere and encouraging, and I genuinely appreciated the support. I think we all did.

Still, I couldn't help but feel a sense of imbalance. My Purple Heart came from what was, in the grand scheme of things, a relatively minor wound. Yet men like SSG Marek had paid with limbs, skin, and pain that would follow them long after the battlefield. Standing beside them in spirit, it felt almost absurd that our recognition was the same; that a medal pinned to my chest equated, on paper at least, to the cost they had borne. Sure, my hand tingled and occasionally sent a lightning bolt through my wrist, but I had nothing to complain about.

It reminded me of when I first received my CIB. I remember thinking there ought to be degrees to these things, some measure that reflected not only participation, but the depth of the sacrifice. But the Army doesn't really grade suffering. In the end, the award was the same, whether it marked a tiny scar or a life forever altered. But I guess that was just the nature of war's ledger.

**October 2005 – Letter from Benjamin Rider, Bravo 4-64 Armor**

I want to let you all know that you should be very proud of your soldiers. The men in this platoon have had a rough deployment. Your soldiers have endured numerous IEDs, mortar attacks, direct fire attacks, and anything else the insurgents could think of.

SGT Harstad and SPC Little have both been awarded their Purple Hearts for the wounds that they received during our operations. There are a few other soldiers who are pending theirs, as well.

Everyone in the platoon has been approved for the CIB.

LTC Roth delivers brief remarks at a later Purple Heart ceremony for the "walking wounded"

While cleaning my vest after the incident, I noticed several tiny metal shards that had pierced the outer fabric and lodged superficially in the Kevlar. I was lucky. The largest was a small, football-shaped fragment about an inch long, which I found near my right shoulder blade. It had worked its way through the covering and dropped into the pocket where my ceramic plate sat.

Looking back, I wish I had saved that piece as a memento, "To Jeff, with love – XOXO, Osama." But I just tossed it out with the rest.

**Covering a suspicious vehicle in the Triangle of Death**

I know it sounds hard to believe, and I've said it before, but this really wasn't a particularly crazy day. Aside from my hand hurting a little bit, there wasn't much that stood out. Unless you count cramming the dipshits into the back of the Bradleys like sardines, of course, which I thought was hilarious. But honestly, things like that day blended into the rhythm. It was just another ordinary entry in a long chain of days where insanity was the baseline, and it was only going to get crazier.

# Reflections

A lot of people have asked whether that day left me with emotional scars. Honestly, no, it didn't. The one exception is a recurring nightmare where I'm stuck in that mud, unable to move, while insurgents hammer the area with RPGs. Why RPGs? I have no idea. They didn't fire any at us that day, but that's what my brain keeps choosing. Either that, or I'm bogged down in the muck and can't find my rifle anywhere.

Looking back, I was more embarrassed that it had even happened than genuinely traumatized. Once the adrenaline faded, what lingered wasn't fear or distress, but shame. A quiet and depressing sense that I had somehow let everyone down.

From the beginning of training, we were taught that responsibility extended beyond ourselves. We carried each other. So, when that chunk of metal smashed my hand, my mind instantly filled with "what ifs." What if I had just run faster? What if I had gone to the other side of the alley? What if one small change in my choices could have prevented it altogether? I replayed every movement in my head, each time with a sharper edge of anger at myself. It was *my fault*.

Then the darker questions crept in: Why me? Why did it hit me and not Hauger, or one of my guys? Was I a weak link, a lesser soldier, singled out by fate because I wasn't good enough? Did I even belong out here? These doubts arrive uninvited, but once inside your mind, they settle in and make themselves at home.

And yet, I remembered the words of LTC Roth and CSM Stanley. More than once, they had said, *"The enemy always gets a vote."*[320] In other words, no matter how skilled, disciplined, or prepared we were, war always reserved the right to be cruel and unfair. Sometimes fate struck without reason, and sometimes survival or injury hinged on nothing more than pure chance. You could be a super soldier and still die out there. It wasn't my fault, but accepting that truth didn't stop me from feeling the weight of it all the same. So, in moments like these, it was absolutely crucial to have strong leaders who could bring you back into the fight with your head held high.

On top of everything else, I felt an almost immediate pull to return to duty, even though, in truth, I probably should have taken some time away from the line. That feeling wasn't unique to me, though. Heath Hutchison said nearly the same thing when I interviewed him for this book. After the explosion on Route Corvette wounded him, leaving him with internal bleeding and a concussion, he found himself temporarily stuck in the tactical operations center, listening over the radio as the rest of the company carried on. He told me, *"I couldn't take sitting by the radio any longer. I needed to get back out there!"*

I knew exactly what he meant. That urge wasn't about recklessness, or even duty in the formal sense, but about belonging to a group. Out there with your team, you felt like part of some-

---

[320] Pitts, Byron. "Live From Baghdad." US Marine Corps, 2003. https://www.marines.mil/Portals/1/Publications/U.S.%20Marines%20In%20Iraq%2C%202003_Anthology%20and%20Annotated%20Bibliography_8.pdf.

thing. Away from them, you felt incomplete. I think we all needed each other, almost in a stalker ex-girlfriend sort of way.

I suppose I shouldn't have been surprised when I was hit. We were operating in what many had already called the most dangerous place in Iraq, with some even saying the most dangerous place in the world.[321] Perhaps that was hyperbole, but it didn't feel far off. Other regions and countries, such as Anbar Province, Kandahar, Sudan, and even Somalia, could all have made their own claims to the title. Still, no one ever mistook the Triangle of Death for a weekend at the spa. It was, quite simply, an extraordinarily dangerous place.

In June 2006, less than a year after our own air assault mission, three soldiers of the famed 101st Airborne Division were manning a checkpoint near Yusufiyah, also within the so-called Triangle of Death, about twelve miles southwest of Arab Jabour. While the rest of their platoon operated nearby, insurgents launched an attack. SPC David Babineau was killed in the initial assault, while PFCs Kristian Menchaca and Thomas Tucker were wounded and taken captive.

When the sound of small-arms fire reached their comrades, the platoon rushed back roughly 20 minutes later, only to find the checkpoint overrun. A massive manhunt ensued, but it came too late. Menchaca and Tucker were brutally tortured and murdered in captivity. Reports described grotesque mutilation, including the removal of an eye and a tongue, followed by burning, beheading, and their bodies being dragged behind a vehicle.[322] To compound the horror, the insurgents filmed the atrocity. It is footage that, disturbingly, still circulates on the internet today.[323] Their bodies were eventually discovered by coalition forces near a village, tied together and left amid booby-trapped IEDs.[324]

As horrifying as it is, witnessing even a single video of a beheading or torture drives home the brutal reality of what we were, and still are, up against. That was our enemy. Savages. There's really no other way to put it.

This was also where Roy Hallums was rescued after being buried in a sweltering basement. Hallums, an American contractor, had been kidnapped in Baghdad on November 1, 2004, and remained in captivity for over ten months.[325] On September 7, 2005, just days before our three tanker

---

[321] Spinner, Jackie. "Marines Widen Their Net South of Baghdad." The Washington Post, November 27, 2004. https://www.washingtonpost.com/archive/politics/2004/11/28/marines-widen-their-net-south-of-baghdad/0bf27260-240f-4fd6-98d4-13d90aac02c5/.

[322] Investigative Project. "Virginia L. Foley, et al., V. Syrian Arab Republic, et al. (Civil Action No. 11-699 [CKK])," April 13, 2017. https://www.investigativeproject.org/case_docs/foley-et-al-v-syrian-arab-republic-et-al/3311/memorandum-opinion.pdf.

[323] DVIDS, "U.S. Officials Condemn Video of Mutilated Soldiers," Defense Visual Information Distribution Service, July 10, 2006, https://www.dvidshub.net/news/541966/us-officials-condemn-video-mutilated-soldiers.

[324] Associated Press. "Al-Qaida Says New Leader Killed Kidnapped GIs." NBC News, June 21, 2006. https://web.archive.org/web/20140211230204/http://www.nbcnews.com/id/13432770/#.UvqsCWTP1qY.

[325] Balda, Dan. "Soldiers Honored for Hallums Hostage Rescue," Defense Visual Information Distribution Service, October 4, 2005, https://www.dvidshub.net/news/3178/soldiers-honored-hallums-hostage-rescue.

brothers lost their lives on Corvette, a Delta Force air assault mission finally freed him from the Triangle of Death.[326] Although Charlie Company wasn't tasked with that mission, we knew the terrain intimately, having spent most of our time operating in and around the area.

Writing about that time, Dale Andradé captured the feeling perfectly when he wrote: "Horror stories abounded, from the power-drill torture of opponents to cutting off children's heads and delivering them in coolers to their parents' doorstep…" [327] That violence was meant to be witnessed, repeated, and feared, because terror traveled faster than any patrol. That was the point. As if all that were not evil enough, he reported that these were the same people who "opened dams along the major canals, periodically flooding the low-lying Shi'ite neighborhoods." [328]

Andrew Lubin described the same strategy of terror from another angle, stressing public brutality designed to frighten locals into compliance. In one incident, he wrote that AQI aimed to "deliver a lesson" by slaughtering ten siblings and forcing their surviving brother to witness it. In the same blog, he also referenced a separate episode in which four residents were beheaded. [329]

Still, looking back on it two decades later, I don't really carry any personal baggage from my day in Arab Jabour. What lingers most isn't the memory of being wounded, but the thought of my parents. The worst part was knowing they would hear about it and imagining their fear when the phone rang. I knew their hearts would drop before they even knew the details. For me, it was just another moment in a long war. For them, it must have felt like the world had come apart.

**19 August 2025 – Reflection of my father, Bill Harstad, Jr.**

I think I blocked it all out. It's like I'm trying to wake up from amnesia. The whole thing rings a bell, but it's not really there.

I vaguely remember a very brief, to-the-point phone call from someone in Washington telling me, "…has been hurt in a mission, not serious… will be okay… CLICK." Huh? And then trying to communicate to your mom what I started to think was a dream. You know, you get that surreal "glow," like it's not real.

That was really, really tough on Mom. That look on her face, a look I'll never forget.

---

[326] Hallums, Roy. *Buried Alive: The True Story of Kidnapping, Captivity, and a Dramatic Rescue.* Thomas Nelson, 2012.

[327] Andrade, History, and Army, *Surging South of Baghdad: The 3d Infantry Division and Task Force Marne in Iraq, 2007-2008*, Page 212.

[328] Andrade, History, and Army, *Surging South of Baghdad: The 3d Infantry Division and Task Force Marne in Iraq, 2007-2008*, Page 299.

[329] Lubin, Andrew. "With 1-30 Infantry Reg. South of Baghdad." The Military Observer, January 27, 2008. https://themilitaryobserver.blogspot.com/2008/01/with-1-30-infantry-reg-south-of-baghdad.html.

The only thing that has always bothered me far more than that incident, believe it or not, was the way I pushed down that mentally handicapped man days before our mission. I still feel terrible about it. Yet, it became a lesson that stayed with me. It was a reminder to remain vigilant about anyone who entered into close physical proximity and to address the threat immediately, preferably before they could get that close.

Years later, when I became a police officer, I learned something I had known back then, but didn't know how to articulate. In 1983, Salt Lake City police lieutenant Dennis Tueller conducted a study on edged weapon defense. His findings, published in an article, demonstrated just how dangerous close proximity can be when facing a weapon, such as a knife. The drill that bears his name, the Tueller Drill, became a cornerstone of law enforcement training.

What Tueller discovered was simple, but sobering. An average, trained police officer could draw a service pistol and fire two rounds in about 1.5 seconds. In that same span of time, an average adult male could sprint 21 feet with a knife and strike before the officer's weapon was on target.[330] The implications were crystal clear. If someone with hostile intent gets within that twenty-one-foot envelope, you're already in grave danger, whether they have a weapon or not.

For me, it reframed that earlier moment entirely. Back then, I had only sensed the danger of letting people get too close; later, I understood the science and mathematics behind it. The Tueller Drill confirmed what my instincts already suspected: allowing potential threats into your personal space is not just risky; it can be fatal. Though extremely helpful and groundbreaking, these trials lacked the rigor of scientific studies and were closer to real-world field testing.

Later, true scientific studies were conducted, and their findings were even more concerning. Many concluded that the 21-foot rule should actually extend farther, perhaps beyond 30 feet.[331] And remember, this benchmark applies to a *properly trained* police officer. Now imagine the distance required for someone without adequate training, which unfortunately includes many police officers, military personnel, and civilian concealed-carry holders.

However, all of this applies to edged weapons, not to someone wearing a bomb strapped to their chest. The question then becomes: how close is too close? The answer depends on the size of the explosive and other factors, but according to a DHS/FBI bomb threat standoff card, a 20-pound suicide vest requires a minimum evacuation distance of 110 feet, with a preferred distance of 1,700 feet or more.[332]

Which means that by the time I even saw the guy lumbering toward me, I could already have been dead. I was just lucky he wasn't rigged with explosives. But perhaps it was a dry run, a

---

[330] Jake. "The Science Behind the Tueller Drill: Why Knives May Be More Dangerous Than You Think." Tier Three Tactical, September 12, 2020. https://www.tierthreetactical.com/the-science-behind-the-tueller-drill-why-knives-may-be-more-dangerous-than-you-think.

[331] Sandel, William L., M. Hunter Martaindale, and J. Pete Blair. "A Scientific Examination of the 21-foot Rule." *Police Practice and Research* 22, no. 3 (May 31, 2020): 1314–29. https://doi.org/10.1080/15614263.2020.1772785.

[332] Director of National Intelligence. "Bomb Threat Standoff Distances." Accessed December 21, 2025. https://www.dni.gov/nctc/jcat/references.html.

test to see how a live soldier would react. It's those kinds of details that stick with you. But still, *why did I have to push him so hard?*

The final thing I've reflected on from this time is just how grateful I am for CPT MacKinnon and 1SG Barnes' foresight in pushing us through air assault training at Fort Stewart and FLETC. No, it wasn't the officially sanctioned Army school with shiny wings, but it was exactly the kind of training that mattered when it counted. That preparation helped us succeed on missions where the difference between doing it right and doing it wrong wasn't a certificate. It was life or death.

Years later, I was talking with a fellow police officer who proudly told me he'd attended Air Assault School at Fort Campbell during his time in the National Guard. I told him I would have liked the chance to attend more official schools myself, but I was happy with the training I got and the leadership that pushed it upon us. Still, I could tell he was wearing those wings a little like an invisible crown, the kind you can't see but everyone is expected to respect.

So, I asked him, as casually as I could, "That's awesome. How many times did you conduct an air assault mission in combat?" Not in a gotcha way, but just the most basic follow-on question in the real world.

The look on his face said plenty. While he had the badge, I had the memory of doing it for real. And it wasn't that I was "better" than him, or that his school didn't count, it was just that his implied hierarchy didn't survive contact with the obvious. A chest full of flair is nice, but it doesn't magically outrank the dirt under your boots.

The military can be goofy like that. We really love our badges and our tabs. We love our little symbols so much that they sometimes start to matter more than the thing they're supposed to represent. And when it comes to promotions, it's often those ribbons and badges that tip the scale, not what you've actually done. Meaning, a soldier could have carried out dozens of real-world combat missions and not earned a single promotion point for his troubles. But spend ten days in a school at Fort Campbell and suddenly you're "more qualified." See how backward that can get?

As I mentioned in a previous chapter about college degrees, it can all sound impressive on paper. But until you've been in the fight, it doesn't mean shit.

# Chapter XVII

## Bandit 6

I was minding my own business when I heard a familiar voice call out, "Hey, Harstad!" I turned and saw CPT MacKinnon, that big goofy grin on his face, and a fat dip tucked in his bottom lip. His eyes immediately went to the sergeant's chevrons on my collar, and he practically lit up with pride.

"Sir," I said, returning the smile, "it's great to see you again! We really miss having you around."

It was October 26th, the first time I'd seen him since we'd left FOB Prosperity and he'd taken command of Alpha Company, 1-184. We stood just outside the line, catching up on the past few months. True to form, he inquired about my mom and dad, checked in on how the platoon was doing, and even remembered to ask about Stormin' Norman, my dog. I was happy to tell him all was well.

"You know, making sergeant is a big step," he said. "I'm proud you went to the board. Just remember, your guys always come first."

I nodded in agreement. I had learned that lesson well from the many remarkable leaders I'd been fortunate to serve under. Standing before me was the greatest of them all, the one I admired most.

"How's the new command treating you? Hate it yet?" I asked with a half-joking grin. Deep down, I wanted him to say he wished he had us all back.

Without missing a beat, he replied, "No, I actually like these guys a lot. They had some issues before, but we're working through them. The thing is, all these guys have civilian jobs, so they bring different skills. One of them is an E-4, but he has a doctorate! I just have to figure out how to put all their talents to use."

He was just that way, always focused on the positives. I might have felt a bit jealous knowing that Alpha Company was now under his command, but mostly I just missed having him around and our conversations in the armory.

From our chats, I knew he was born in Helena, Montana, in 1975. His mother passed away shortly after his birth, and he was raised by his father, John, a former Army infantry officer who had himself attained the rank of captain. Alongside his three sisters, Mike grew up in a household that valued resilience, duty, and service, and from the beginning, he embodied those values.

In high school, Mike was the kind of young man who seemed to stand out at everything he tried. He excelled in football, swimming, and band, all while maintaining the highest academic standards. He graduated as valedictorian, proof that he pushed himself to be the best in every arena.

That drive carried him to the U.S. Military Academy at West Point, where he earned a Bachelor of Science in 1997. He went on to conquer some of the Army's most grueling courses, including Airborne, Air Assault, and Ranger Schools, before earning his first assignment at Fort Bragg, North Carolina, with the 82nd Airborne Division. As a young platoon leader, he led soldiers on deployments to both Haiti and Egypt, always leading from the front.

Eventually, his career trajectory brought him to Fort Stewart, Georgia, and into the fight during the invasion of Iraq. That was where our paths initially crossed. I still remember the first time I saw him in action and the way he carried himself. He balanced authority and humor effortlessly, and he was someone everyone wanted to follow.

When he took command of Bravo Company, everything seemed to click. His presence made the unit stronger, and he had this knack for making the hard days lighter, often by turning the simplest things into competitions. Who could do the most pushups? Who could jump the farthest? Who could knock a specific pinecone out of the tree with a rock? He had a way of turning that sweat and strain into camaraderie, and it bonded us together like family.

That family wasn't just soldiers, either. His wife, Bethany, and their two young children, Madison and Noah, sometimes came by to visit. His kids would play in the back while we worked, and Noah, wide-eyed and curious, was especially fascinated by the weapons, hanging around outside the armory in hopes of catching a glimpse of a machine gun or night-vision goggles.

That was the kind of leader CPT Mike MacKinnon was for us. He wasn't distant or unapproachable. He was present, engaged, and fully committed to his people, both his family at home and his family in uniform. Watching his actions, I realized I didn't just respect him. I wanted to be exactly like him, and to me, he was the very definition of what an Army officer should be: tough, funny, brilliant, and deeply human.

So, when he left Bravo Company for another command, it hit hard. We had built something special under his leadership, and the truth was, I wanted him back, because leaders like him don't come around too often. To me, and to many of us, he was more than a commander. He was our actual hero.

Our conversation in the chow hall then turned to what we'd been through in the weeks since Arab Jabour. He asked how we were holding up with Charlie Company and what the tempo had been like. The truth was, those weeks had been some of the hardest yet, especially since the three tankers had been killed. As the voting process approached, our pace only accelerated. There was real fear that terrorists would do anything possible to derail the democratic process by murdering voters, bombing polling sites, whatever it took. That meant our job was to secure as many polling stations as possible.

A couple of weeks before my conversation with CPT MacKinnon, the Gamblers had set up shop in a run-down, abandoned school that was more of a shell than a building, using it as a makeshift defensive position. At the polling sites, we were basically reduced to playing cops: frisking people at the entrance, searching bags, and scanning for anything suspicious. It was tedious, miserable work, and people weren't too thrilled about the assignment. When night fell, we'd trudge back to the school, where rotating shifts of grunts pulled guard duty while the rest of us tried to grab some shuteye. Apparently, though, the terrorists didn't want us to get our beauty sleep.

**15 October 2005 – Letter to the Harstad Family**

The first night went without incident. The second night, we were hit.

I was asleep inside when the first three mortars were launched. The first two landed several hundred yards away from the school, while the third landed just outside the walls.

After the mortars stopped, I had to take a shit. I grabbed my M-16 and a battle buddy, headed down towards the river, and started doing my business. Well, I guess some fuckheads thought it would be funny if they started mortaring us again.

I had to scramble up the bank to get away while PFC Chism, my battle buddy, ran beside me. We jumped into the back of the Bradley until they stopped.

Thanks for ruining my shit, guys.

It's easy to laugh about now, but at the time, I was anything but amused. Imagine trying to claw your way up a muddy riverbank in the pitch dark, with one hand grabbing for a hold, the other fighting to pull your pants back up before you face-plant, all the while explosions are going off in every direction. The real punchline came the following night, when Chism went through the same ordeal. The only difference was that this time, he was the one caught in the middle of doing his business. We scrambled, yet again, back into the safety of the Bradleys, laughing hysterically. "Can you believe that!? Two nights in a row!"

**15 October 2005 – Letter to the Harstad Family**

After the mortars, a firefight broke out down the road.

We scanned our sector, trying to pick out any hostiles. Unfortunately, I couldn't really see anything from where I was at. Then, I ended up chasing an "invisible car" that probably never existed through the woods.

Yeah, it was one of those nights.

Someone reportedly spotted a car driving without headlights in a nearby, heavily wooded area. My squad and I were dispatched to investigate. Throughout our search, we received radio reports saying it was "over here" or "over there," but every time we tried to track it down, we found nothing. It felt like we were chasing Bigfoot.

**Chism and I safely inside a Bradley - he and I dodged incoming mortar rounds by the river like pros**

After several hours of aimless wandering, we eventually returned to the school, frustrated and covered in dust, with dozens of mosquito bites. None of us was particularly happy about the outcome of our search. Nor were we happy to have been sent out there in the first place, as none of us had seen any evidence of this mystical car at all.

**15 October 2005 – Letter to the Harstad Family**

Later on, when I was sleeping again, we were mortared. Most of us didn't even bother getting up. I just tried to sleep through it.

On October 10th, tragedy struck again. SSG Jerry Bonifacio of the 1-184 had been part of the National Guard company we relieved at FOB Falcon. In the swap, his company was temporarily placed under LTC Roth and 4-64 Armor, tasked with manning those same dangerous checkpoints we had handed over months prior.

As an American convoy reentered the Green Zone, an Iraqi civilian vehicle attempted to follow. SSG Bonifacio intervened on foot, halting its progress. Moments later, the car erupted in a massive VBIED detonation, taking his life and that of an interpreter.[333] Though tragic, his decisive actions prevented the attacker from entering the Green Zone, meaning he safeguarded countless soldiers and civilians who otherwise would have been exposed in unfortified positions. Additionally, our command revised checkpoint procedures to help ensure something like this never happened again.

### 16 October 2005 – Letter from Cap Stanley, Command Sergeant Major of 4-64 Armor

We had a tragic event happen at the checkpoint. SSG Bonifacio was taken from us in a failed attempt by the bad guys to breach the International Zone. That will NEVER happen as long as the Tuskers and those that support us are on post!

Your soldiers, my soldiers... OUR soldiers were on top of things as soon as the incident took place. I wonder what the terrorists really think they accomplished that day. One killed himself, killed a fellow countryman, and ended the life of one heroic American soldier.

As I traveled around the area just a few short days later, the referendum vote was beginning. I could not help but notice the smiles on the faces of the Iraqi people in the polling centers, and then later that evening as the people of Baghdad filled the streets for a night of festivities to celebrate their country's historic vote.

More smiles per mile than I could count... I just wondered if the terrorists saw those smiles. I know SSG Bonifacio did!

After the voting activities concluded, we resumed patrols and raids. By this stage in our rotation, everyone was exhausted and running on fumes. We had already spent ten months in country, dealing with nearly every homicidal idiot the region could throw at us.

### 19 October 2005 – Letter to the Harstad Family

Today is Wednesday the 19th, but it feels like Sunday the 42nd.

---

[333] Tempest, Rone. "Army National Guard Staff Sgt. Jerry L. Bonifacio Jr., 28, of Vacaville; Killed in Suicide Car Bomb Attack." *Los Angeles Times*, October 16, 2005. https://www.latimes.com/archives/la-xpm-2005-oct-16-me-bonifacio16-story.html.

Monday, we had to respond to a call that there were some people on the streets (in our area of operations) at 2 AM with AK-47s. This is a problem for several reasons.

First, Iraqis have a curfew from 11 PM to 6 AM. Anyone caught outside can be arrested. Second, AKs are for home protection only. Third, a lot of terrorists have been spotted in that neighborhood.

We started at the end of the street and worked our way in.

Dismounting from our Bradleys, we slipped into the narrow alleys of Dora, moving quickly from one shadowed corner to the next. The night was thick, and it was difficult to see, so we relied on our monocular night vision to guide us. Whenever we hit brighter patches, which rendered night vision inoperable, we'd switch to the other naked eye. In practice, like every infantryman on night patrol, we were really just navigating the battlefield with one eye at a time.

Up ahead, I caught sight of what looked like a teenage boy carrying an AK-47. He hadn't seen us. Without hesitation, he slipped through a nearby doorway and vanished inside. Seconds later, we stacked on that same door, ready to breach. What happened next unfolded in only a few heartbeats.

The door was kicked in, and a dozen of us surged inside. The entryway was tight, and we quickly realized it wasn't a single house at all but a partitioned unit inside a larger apartment complex. Straight ahead was an empty dining area and kitchen. To the immediate left and right were closed doors.

By the angle of our entry, I ended up as the point man at the left door. Like clockwork, one team flooded the room to the right as we smashed through the door to the left.

Storming inside, I was hit with a sight I'll never forget. A man was sprawled on his bed with a young woman, both half-naked, and a pornographic film played in the background. The expressions on their faces froze in my mind, with their eyes bulging wide and mouths hanging open in sheer horror. I can only imagine what it looked like from their side: a team of futuristic soldiers bursting through the door, goggles glowing, weapons leveled, ammo pouches hanging off us, looking like something straight out of a nightmare. It must have been a hell of a climax!

We cuffed everyone and pulled bags over their heads, a safety procedure done until we could secure the area and figure out what was going on, while others tore the place apart looking for the rifle. We eventually found one stashed under the bed, buried behind a pile of junk, but it didn't look like it had been outside recently. Just as we began pressing them about why they'd been seen with a weapon, the radio crackled that our suspect was now spotted on the roof.

That's when we discovered the apartment had a narrow, hidden stairwell leading upward. Apparently, it was being used as a pedestrian thoroughfare. We climbed it, weapons ready, and came out face-to-face with not one, but multiple suspects on the rooftop.

**19 October 2005 – Letter to the Harstad Family**

Toward the middle of the block, we entered a building and cleared it. SPC Mata, one of the drivers, spotted three individuals on the roof. We ran up the stairs to find all three of them lying on top of the AKs, trying to hide them. As soon as we confronted them, they got really jumpy.

Our squad split apart instantly to handle the threats. Two of the men were steamrolled under a wave of soldiers, cuffed before they could even process what was happening. No shots were fired. The third, crouched and hidden in the dark right beside me, didn't reach for his rifle. Nope. Instead, he jumped up and inexplicably punched SGT Tchad Dvorak square in the face.

That was a serious mistake. Dvorak, who outweighed him by at least a couple of Thanksgiving turkey dinners and was built like a brick wall, scooped him up and slammed him into the floor like a rag doll. I tried to land a punch of my own to the scrawny guy's face, but completely whiffed as Dvorak went postal. He pummeled the guy so fast the poor bastard frantically rolled over onto his stomach, which only left one thing exposed. And, well, you'll see...

**19 October 2005 – Letter to the Harstad Family**

The first two were quickly taken down by an overwhelming amount of soldiers. The third, crouching right next to myself, jumped and punched SGT Dvorak in the face. Dvorak responded by smashing him and punching him in the grill. I had no place to hit, except for his waist and ass, so I hit where I could. I must have punched him in the ass 50 times. My hand smelled so bad afterward!

We fought like mad to get this guy under control and cuffed before he could grab his rifle. Then that smell hit us. It turns out the guy literally shit his pants while Dvorak and I were beating the crap out of him (pun absolutely intended). My precious Mechanix gloves were collateral damage, and no amount of scrubbing could get that stench out.

In all honesty, Dvorak probably saved the guy's life. He took an ass-whupping, sure, but that beat the alternative. If he'd gotten to that rifle, he likely would've been shot dead on the spot.

But why he punched Dvorak in the first place, I have no idea. Once he did, though, we had to move fast, neutralize the threat, and get him under control. Until the cuffs went on, he fought like a rabid wolverine. The moment they clicked, he quit.

Dvorak deserves the credit for saving the day. I just happened to be there. I did what I could, but with nothing but his ass in front of me, my punches didn't accomplish much. Afterward, we brought the three men to jail, where we discovered they were on our blacklist. So, it turned out to be a worthwhile raid that ended with no shots fired.

A couple of days later, we ran across the SEAL platoon again, and when I told that story to the Humvee gunner, one of them just laughed and handed me a brand-new pair of Mechanix gloves. Same color and everything!

The day after the ass-punching incident, we resumed our patrol in Dora. It was there that East experienced something that deeply unsettled him, but rather than letting him dwell on it, we relied on our usual defense mechanism: humor. He laughed with us, though it was evident that he was still troubled by it.

**19 October 2005 – Letter to the Harstad Family**

We were eating lunch inside this bombed-out palace. We heard a lot of machine gun fire right outside the gates, so we scrambled to the site.

When we got there, we immediately realized that it was a carjacking. One individual was shot in the back of the head, and several others were slightly wounded.

We quickly secured the area, rerouting the traffic and questioning people, and put the dead guy in a body bag. Unfortunately, we were not able to catch those responsible.

My team was pulling security around the body sprawled on the street, and the entire back of his head was gone. Apparently, he had been carjacked and then executed. I shouted over the noise to East, "Grab him and put him in the body bag!"

A second later, East called back, "Hey, SGT Harstad, come here. Quick."

I broke from my position and trotted over. The man was on his back, blood pooling beneath his ruptured skull. That's when I saw what had East frozen. The man's eyes were still moving, slowly drifting left, pausing, then sliding back to the right. But they weren't *seeing* anything. They were just stray nerve impulses firing off in the last moments before shutting down for good. I had seen it before.

"He's dead," I told him flatly. "It's just nerves. Grab his shoulders and let's bag him."

East bent down, slid his hands under the body, and started to lift. One hand slipped. His finger sank into the ruined back of the man's skull and came out through the corner of his eye socket. I saw the color drain from East's face instantly.

"Go pull security!" I ordered, shoving him away from the mess. SSG Thaler and I finished the job while East tried to process what had just happened. Later, we gave him hell for it, because that's just what we did. The jokes were dark and highly inappropriate, but they made the horror survivable, and after a while, East was laughing with us again.

Earlier in the tour, East had another memorable and briefly terrifying moment while we were explaining what RPGs looked like in flight. Young and green on his first deployment, he was trying hard to absorb everything. RPGs, we told him, streaked through the air like sparklers on the Fourth of July. As one of the platoon's M-240B gunners, he lay prone in a dirt field during a raid, scanning his sector for just such a threat. Apparently, what we told him left a lasting impression.

Suddenly, as if by cosmic design, sparks exploded across his vision, and a loud electrical crack tore through the air. Convinced we were under attack, East nearly laid waste to everything in front of him. From behind, the rest of us immediately saw the truth and stopped him before he violated the Geneva Convention. He had been lying directly beneath a stretch of overhead power lines, and one of them had short-circuited, sending a shower of sparks raining right over his head.

Once the danger passed and East had rolled away, we couldn't stop laughing. Poor East tried to collect himself, admitting that for a split second, he thought the enemy had us dead to rights with RPGs. His confession only made the laughter worse. Sometimes, there's nothing else to do but laugh about it.

The day after the carjacking, we got hit by another IED, though nothing compared to the blast that tore up Little's track. The attacks kept coming, one after another, but each time our Bradleys managed to limp back to the FOB, where we'd spend hours patching them together again.

The author covering an entry team on a raid with his M-240B machine gun

**19 October 2005 – Letter to the Harstad Family**

Today, we were out on patrol when our Bradley was hit by a roadside bomb. No one was hurt, just slightly jarred around. It messed up the front right, but did no serious damage. We were lucky on that one.

**22 August 2025 – Reflection of Matt Turcotte, Platoon Leader, Bravo 4-64 Armor**

The worst IED attack on our platoon was out on Mercedes, but we had all been hit a number of times. I think SSG Rider's track alone was hit eight or nine times during the deployment. We were all very lucky.

In the back of the Bradleys, with nothing to see, I usually just tried to sleep. When an IED went off, the pressure slammed our ears, the loud bang instantly woke everyone up, and the whole vehicle violently shook. Somehow, though, the shrapnel never penetrated the walls to reach us in the troop bay. That had only ever happened to Little.

In the meantime, we were continuously searching various fields for wires and other suspicious items that could be used against us. While it may sound like our days were filled with nonstop action, the reality is that much of our time was spent enduring the boredom of scouring large fields with little to show for it. There would be brief moments of excitement, or sometimes terror, and then followed by more extended periods of monotony. This pattern repeated itself frequently, but occasionally, something entirely unexpected would occur…

The author in a Dora field sweep – 0 IEDs found, 1 infantryman marinated from the waist down

**19 October 2005 – Letter to the Harstad Family**

I did something really stupid today.

We were walking around a giant field, and I found this manhole. I decided to try and look inside of it. I got it a little ways off, but it was really heavy. So, what did I do? I was going to try and jump on it (to knock it the rest of the way off), like a dumbass.

Well, I managed to fall through and landed right in the sewer. I had shit all over my legs, and I felt like a total jackass. Sometimes I wonder how I made it past the third grade.

While we were still operating out of Falcon, the rest of the 4-64 unit was responsible for maintaining safety in the Green Zone during the trial of former Iraqi President Saddam Hussein. In all honesty, our platoon had no involvement in that effort. All the credit for their successes belongs to them.

**24 October 2005 – Letter from Robert Roth, Commander of 4-64 Armor**

On the 19th of October, 4-64 AR played a small part in the Saddam Hussein trial.[334] Nothing much was different from what we do each day here in Baghdad. We secure the IZ (International Zone), and we do it very, very well! I'm sure you watched the news media and probably know more about what happened than we do. We don't watch much TV around here, only news once in a while. It was a one-day success, and we are proud to play our part.

I was talking to a journalist today. She recently watched a movie about the Nurnberg Trials following World War II. She made a comment that the actors who portrayed the accused Nazi war criminals sounded like Saddam Hussein. I thought that was quite interesting, and the more I thought about it, the more it made sense. Mr. Hussein and the leaders of the Nazi Party have many things in common. One being that they come from the same mold of evil. Just as we ridded the world of the Nazis in 1945, we removed the Ba'ath Party regime in 2003, and the mission continues today. Perhaps that is why I feel a special pride in the accomplishments of the Soldiers and Marines here in Iraq.

As LTC Roth recounted, Saddam Hussein and seven co-defendants appeared before the tribunal to enter their pleas. As expected, they pleaded not guilty to charges of torture and murder connected to the 1982 crackdown in Dujail, a town in Iraq where nearly 150 people were executed.

The trials continued over the following months, drawing heavy publicity with each session. Ultimately, on November 5, 2006, Saddam was found guilty and sentenced to death. Several co-defendants were acquitted, while others received convictions ranging from lengthy prison terms to

---

[334] Garrels, Anne. "Saddam Trial Begins in Baghdad." *North County Public Radio*, October 19, 2005. https://www.northcountrypublicradio.org/news/npr/4965005/saddam-trial-begins-in-baghdad.

additional death sentences. Two were condemned to hang alongside Saddam. On December 30, 2006, Saddam Hussein was executed by hanging.[335]

Fast-forward back to the present day in the story, where a tray clinked behind me. CPT MacKinnon had been listening intently as I explained everything, the regular bustle of troops getting chow all around us. After wrapping things up, we exchanged farewells and went our separate ways. Every conversation with him felt like walking out of a motivational seminar. No problem ever seemed too big to tackle.

As our tour dragged on, monumental history unfolded all around us, yet we remained largely out of the loop of the bigger picture. Such is the reality of life as an enlisted soldier. Each day became a matter of survival, balanced with whatever distractions we could find to keep ourselves entertained. The following is one such lighthearted example.

**26 October 2005 – Letter to the Harstad Family**

I was out playing catch (baseball) with Hauger a few hours ago. All of a sudden, we heard a bunch of loud explosions. We thought we were being attacked, so we jumped inside, taking cover.

Turns out it was our Paladins, giant artillery pieces, that were firing at the enemy. Scared the fuck out of us. It's funny now, but when it happens, you kind of go into a "caveman survival mode."

On October 27, I sent a brief email home, stating only that we were standing by at the base, waiting to load onto the Bradleys for a mission that was deliberately left vague. It was a simple note, but even as I typed it, something about the day felt unsettled. What would the rest of it have in store for us?

Nothing from the morning lingers in my memory, as if the hours themselves dissolved into nothing. The afternoon, though, couldn't have been more different.

Driving down Route Buick near Hawr Rajab, Alpha 1-184, led by CPT MacKinnon, was headed toward a village inside our area of operations. In their usual lightly-armored Humvees, he sat in the front passenger seat, carefully directing the convoy and keeping a watchful eye out for any enemy activity. Also riding in the convoy was the new battalion surgeon. Although this was completely outside his usual daily activities, in which he typically worked in the safe confines of a medical station or hospital, he had courageously agreed to accompany the unit on a special mission.

They navigated the dusty roads through poverty-stricken neighborhoods and villages. Large groves of trees clumped together near sources of water, offering a slight reprieve to the depressing sights of squalor. But they were on an important task and were willing to brave their way to the ultimate destination. It was just up ahead.

---

Suddenly, a hidden IED detonated beside CPT MacKinnon's Humvee. The blast struck the front passenger side, the very spot where our former commander sat, shearing through armor and reinforced glass as if the vehicle were nothing more than tin beneath a can opener. Shrapnel sprayed out in every direction as the Humvee came to a smoking rest, permanently knocked out of the fight. Dust had been thrown into the sky, temporarily blocking the sun and making it impossible to see. It was utter chaos.

As Alpha Company scrambled to react to the hostilities, a few other patrols were already in the area when the calls for help came in. CPT Anderson and several tanks from Charlie Company were near a road called Chicken Run.[336] They temporarily halted their activities to listen to the radio traffic from the ambushed convoy, then quickly redirected their focus to Route Buick. Help was on the way.

Also monitoring the radio was LTC Wood, the new commander of 1-184, who was likewise out in the sector. He, too, immediately ordered his vehicles to turn toward Buick. If no other ambushes took place along the way, both patrols would likely meet at roughly the same time at the site.

Back at the FOB, in a scene eerily reminiscent of September 16, elements of Charlie Company and the entire Gambler Platoon would suddenly be rushed forward to provide support. As the battalion's de facto QRF, it always seemed to fall on us to drag others back from the brink whenever things went bad, but that was our area of expertise.

### 3 September 2025 – Reflection of Matt Turcotte, Platoon Leader, Bravo 4-64 Armor

You guys in Gambler Platoon were unreal at getting spun up and rolling out on a moment's notice. More than once, we beat the actual QRF to the fight. If anyone called for help, we were already moving, trying to be the first ones there.

"Get your gear!" someone yelled. "Captain MacKinnon's been hit!" Like dozens of times before, we scrambled to put on our equipment and load into the Bradleys, sprinting the entire way. But wait. *What does this mean? Was his unit hit? Or was he hit?*

As we raced out of the FOB, mere minutes after being notified, those questions reverberated in everyone's mind. Inside the deafening roar of the troop compartment, words were useless; we sat in uncomfortable silence, slammed from side to side as our thoughts spiraled through every grim possibility. The dread sat heavy in our guts, a cold weight that only grew with each dusty mile.

At the explosion site, Alpha Company was engaged in frantic activity to secure the scene, search for the enemy, and render first aid. A MEDEVAC was already inbound to the site. In the meantime, they needed to protect the disabled Humvee and hunker down until help could arrive. In the fog of war, anything could still happen.

---

336 Anderson, *Ride the Cyclone.* Page 25.

The first to quickly respond was LTC Wood's patrol, which halted several hundred meters short of the shattered Humvee. They scanned the area, weapons ready, searching for the enemy and a way forward. Then LTC Wood stepped out onto Route Buick, striding quickly down the road toward CPT MacKinnon's vehicle. He was determined to take control of the chaotic scene and assist his men.

Wood had covered only part of the distance when a second hidden IED erupted beside him, the explosion tearing through the air. On foot, he was exposed, stripped of the precious armor of a vehicle. Only his combat vest and helmet offered protection, but against an IED, they were essentially useless. In a split second, he completely disappeared from Route Buick.

CPT Anderson saw the second IED erupt in the distance just as his tanks approached the area, a plume of brown and black debris flung violently into the air. He, too, brought the column to an abrupt halt a few hundred meters short of this second blast, then ordered a soldier to grab his M4 and follow. Together, they dismounted and sprinted down the road toward the unfolding horror.

**1 August 2025 – Reflection of David Anderson, Commander of Charlie 4-64 Armor**

We were running down the road when I suddenly froze. I thought, *What am I doing? How could I be so stupid? We are running on top of a road that has already had two IEDs go off!* We quickly made our way off Buick and started going through people's yards, trying to get around homes, walls, and fences.

Then the radio crackled with chilling news: LTC Wood was reported Missing in Action (MIA). The last sighting placed him on Route Buick moments before the second IED tore through the road. Now, in the smoke and confusion, he was simply gone. No one could find him. He had vanished.

Moments later, the Gamblers and I spilled out of our Bradleys near the outer perimeter of the blast site. Word spread like wildfire that LTC Wood was missing, and we had to find him *immediately*. This was head-chopping country, and the thought of one of our own in enemy hands was unthinkable. To make matters worse, he was a high-ranking officer, and the sensitive knowledge he carried could endanger the entire battalion if it fell into insurgent hands.

Doors were ripped from their hinges, walls were collapsed, people were dragged out of their homes, and we literally tore everything apart in search of LTC Wood. There was no time to stop and think. It was only a matter of finding him. But even then, in the back of my mind, I kept thinking: *I hope CPT MacKinnon is ok*.

Our furious activity erupted into pure pandemonium, like a hornet's nest kicked apart, bodies surging in every direction at once. It was Route Corvette all over again, only magnified a hundred times. Shouts collided with the crackle of radios, the grind of engines, and the metallic slam of hatches. The air vibrated with tension, every sound and movement threatening to tip us into complete mayhem. Yet through the madness, we somehow, thankfully, kept our composure, forcing discipline to hold against the pull of barbarity. But how easy it could have been to succumb to the allure of brutality.

While we moved door to door in our determined search, CPT Anderson, also looking for our colonel, was drawn by a sudden commotion. He hurried to a nearby canal, its waist-deep water now stained red, and saw a grim sight: LTC Wood floating face down. Anderson, SSG Santiago, 1LT Turcotte, and several others slid down the bank to reach him. But when they turned him over, it was clear he was already gone. The blast had killed him instantly, severing a leg and leaving his face unrecognizable. In that moment, the mission shifted from rescue to recovery.

After a few moments of taking in the dreadful sight, CPT Anderson then climbed back up the steep bank to retrieve a body bag. They needed to figure out how to get him out, as pulling him back up from the canal would be nearly impossible.

### 1 August 2025 – Reflection of David Anderson, Commander of Charlie 4-64 Armor

I got down into the water where LTC Wood was. He was absolutely obliterated. We tried putting him into a body bag, but it filled with water, which made it so heavy.

While we were doing this, there was a soldier down there with us, and he had his head down in despair. I thought it was someone from the colonel's security detachment, maybe a lieutenant. He kept staring at the boots sticking out of the body bag, as one of them was pointed up and the other down.

I slapped him on the back of his helmet and told him, "Come on, you need to focus on the mission and the soldiers alive around us." He finally lifted his head, looked at me, and grimaced. I realized it was actually the new battalion surgeon, and he outranked me as a major. He had been one of the first guys into the water, trying to save LTC Wood. But he couldn't hide his emotion. I later found out he initially rushed to Mike MacKinnon's aid right after his Humvee had been hit, too. He was having a hard time processing everything that had just happened.

### 22 August 2025 – Reflection of Matt Turcotte, Platoon Leader, Bravo 4-64 Armor

The explosion had launched him quite a distance into the canal. It was basically a giant dirt ditch, partially filled with water. We put him into the body bag, but it kept filling up with water, and we couldn't move him out. The banks had turned to mud, and everything was slippery. In the end, we had to wrap the bag around him, place him on a rescue hoist, and lift him out with a Humvee.

After extricating the colonel's lifeless body from the canal, the men were left with the harrowing duty of gathering what remained of him, and then placing those remains on a Black Hawk. In his book, David Anderson recalled the gruesome task, one of countless scenes repeated by soldiers across Iraq's thousands of blasts. It was a horror no one could ever erase, the sort that lingers in the dark corners of sleep, destined to echo in nightmares.

While most of us were in the nearby homes and fields, it took some time for us to learn that the colonel had been found and that CPT MacKinnon had been urgently flown to the hospital. When I reflect on that time, I still feel a weighty sense of manic desperation from our efforts to find Wood. We took greater risks than we normally would have; we were sloppier, infinitely more aggressive, and willing to cut corners, all to find him. By sheer luck, no one else was injured or killed during our frantic search.

Just like on Route Corvette, every Iraqi in the area played stupid, pretending not to know anything. This act was growing old very quickly. The rage spread through all of us, and each lie cut what little compassion I had left. I started to feel hatred.

We spent endless hours raiding houses, tearing through their belongings for signs of bomb-making materials and tools. Others swept the fields, dragging up wires and probing for hidden charges. Still, it all felt empty. No one would talk, and the constant cycle of answering our friends' injuries and deaths was wearing thin on every one of us.

Later that evening, once the site was secured and the wreckage hauled off, we returned to the FOB. We dropped our gear in our rooms and impatiently waited. No one seemed to know anything, and an uneasy silence hung over us as we aimlessly wandered and waited for news.

A short time later, word spread through the barracks for everyone to gather in a large room with a ping-pong table pushed to the side. Papi, our platoon sergeant, stood in the middle of the circle we formed around him. His face was grim, and the room went silent as we waited.

"Guys," he began, his voice low, "I've got bad news. I think most of you know this already, but LTC Wood was killed by an IED while walking down the road. He was headed to help CPT MacKinnon, who'd also been hit. I just got word that he passed away a short while ago."

The room collapsed into an uneasy stillness, and it felt like a punch to the gut. Everyone was left winded and absolutely stunned. No, this couldn't be happening. There had to be a mistake. *How could he die? He was better than all of us!*

After a few short follow-up questions, most met with a hollow, "I don't know," we were left with a suffocating silence, and the sad truth hit us: there would be no real answers, not now, maybe not ever. That was just how it went in this place. Numb and broken, we drifted away like aimless ghosts, each of us retreating to our rooms to wrestle with the news alone.

SGT Howerton and I ended up spending the next couple of hours with a young soldier who looked like he'd just had the ground ripped out from under him. He barely spoke, but he didn't have to. The emptiness in his eyes told the whole story. It was the same look mirrored in all of us, a raw ache that words couldn't touch.

Slowly, over the days, weeks, months, and even years that followed, fragments of the truth drifted our way. Every time we thought we had finally laid it to rest, another detail surfaced, ripping open the wound all over again. Each revelation dragged the sorrow back, heavy and unrelenting. Yet, in all that grief, there was one small, determined light: the sheer number of people who mourned him. They cared deeply for Mike MacKinnon because he had always cared just as deeply for us.

**1 August 2025 – Reflection of David Anderson, Commander of Charlie 4-64 Armor**

Mike knew of this little kid in a village who had been injured in an insurgent mortar attack. The battalion had just received a new surgeon, the major I slapped on the helmet, so Mike asked him if he would go out to visit the kid. They were going to see if there was anything that could be done to help relieve his pain.

That's where they were going, out to this village to help a little kid.

**3 January 2025 – Reflection of Cap Stanley, Command Sergeant Major of 4-64 Armor**

LTC Roth and I went to see him as soon as we heard the news. A small piece of shrapnel had hit him under his helmet, right by his eyebrow. If you didn't know any better, it just looked like he was sleeping.

**22 July 2025 – Reflection of Mark Barnes, 1st Sergeant of Bravo 4-64 Armor**

When I got to the hospital, the general and brigade commander were already there. They were standing around the gurney they had him on. The general looked at me and then looked down, shaking his head, as if this was much worse than what we were told.

They were originally trying to get him to Germany so Bethany (his wife) could be there. He was solely alive because they had him on the machine, but he was showing no brain activity. We all gathered around, touched him, and prayed. It was soul crushing. I'll never forget the sound of the respirator breathing for him, and then he passed away. We went outside to the vehicles and broke down.

**24 June 2025 – Reflection of Donald Cyr, Alpha 3-7 Infantry**

It didn't surprise me that CPT MacKinnon turned out to be a good commander. While I wasn't in Bravo Company, I remembered him from our first days in Iraq in 2003. He was a soldier I highly respected. Someone told me he had been killed right after it happened, and I remember saying, "No. I just talked to him yesterday in the gym." I didn't want to believe it. He always said hi to me.

**1 August 2025 – Reflection of Noor "Nora" Aljassim, Interpreter for the U.S. Military**

When CPT MacKinnon died, I just lost it. I couldn't take the war any longer. He was a good man and always protected us (interpreters). I still wear his name on my bracelet. I never take it off.

Days later, we gathered for a ceremony to honor our fallen comrades. The loss struck our adoptive battalion with devastating force for several reasons. The 1-184 had just lost its active-duty commander and would need its third leader in only a few months. LTC Wood, posthumously promoted to full-bird colonel, was, at that time, the highest-ranking U.S. service member killed in the war.[337] Alpha Company had also just lost its commander, Mike MacKinnon, a soldier who, to most of us, stood as the finest we had ever known. And beyond rank or title, we all lost friends who could never be replaced.

**LTC Roth and CSM Stanley pay their respects at four battlefield crosses for our men – two others were included in the memorial that were KIA not long after the incident on Route Buick**

**Photo courtesy of Robert Roth**

That faint metallic chime of dog tags in the breeze has returned to me in dreams more times than I can count. They aren't quite nightmares, but they're far from dreams of peace. They linger in a place somewhere in between, echoing like a distant bell swallowed by dense fog on un-

[337] Bynum, Russ. "Army Bids Farewell to Colonel Killed in Iraq." *Lawrence Journal-World*, November 5, 2005. https://www2.ljworld.com/news/2005/nov/05/army_bids_farewell_colonel_killed_iraq/.

charted seas. Out there, just beyond reach, our fallen seem to be waiting, almost calling to us, whispering that it is all right.

The tags, suspended from a vertical rifle, would sway gently, one twisting in the Iraqi wind until it gave that sharp *clink*. That sound. We all knew its meaning.

### 6 January 2026 – Reflection of Derek Hauger, Bravo 4-64 Armor

After that, with so many casualties, CSM Stanley started coming over regularly to check on us. He knew how rough we had it out there. Over time, he got really close with the platoon, and whenever he stopped by, he'd end up playing a bunch of us in ping-pong.

In our own ways, we each tried to honor them by writing about their lives. Few of us, myself included, were accomplished writers or poets, yet that didn't matter. What mattered was letting our families know how much we cared, and how deeply their loss had wounded us.

### 30 October 2005 – Letter from Robert Roth, Commander of 4-64 Armor

Most of you have already been told of the death of CPT Mike MacKinnon, who commanded B/4-64 AR. Mike was in line to take command of HHC/4-64 AR in September, but things changed. Our sister battalion to the south, 1-184 Infantry, was having a tough time dealing with several issues. The battalion commander and company commander of A Company, 1-184 Infantry, were relieved of their commands. CPT Mike MacKinnon was selected to go and take command of A Company for the remainder of the deployment.

On Thursday last week, as he was on patrol, a powerful roadside bomb exploded next to his armored vehicle. Mike was MEDEVAC'd to the Army Hospital in the IZ, where I linked up with him, as well as many other Tusker soldiers and leaders. Soon after our arrival, Mike was pronounced deceased. Mike's wife, Bethany, is grieving now, and I would ask that all of you give her time to deal with the loss of her husband. Mike was a warrior, a leader, a husband, a friend, and a father. You simply cannot replace the loss of this great American. He meant so much to so many people.

If that wasn't bad enough, the battalion commander who assumed command of 1-184 Infantry in August was a friend of mine. LTC Bill Wood assumed command of the National Guard Battalion from California. When Bill heard of the explosion and the critical condition of Mike, he rushed to the scene. As he surveyed the site, another powerful bomb exploded near LTC Wood. LTC Wood was killed instantly. CPT Dave Anderson personally carried LTC Wood to the MEDEVAC helicopter. Everyone was in shock for the first 24 hours, and many are only now beginning to deal with the losses of both men.

**10 September 2025 – Reflection of Derek Hauger, Bravo 4-64 Armor**

I'd run cross-country track pretty much my whole life, so I was cocky when we had to do company runs. Nobody could really touch me, and I thought I was better than sliced bread.

One morning at Fort Stewart, someone told me, "Hauger, report to CPT MacKinnon's office for PT this morning." I remember thinking, *Oh man, what the fuck did I do?* So, I reported to him, and he ended up taking me out on a long run. We ran for ten miles or something, I can't remember, but it was really far, and he ran me into the ground. The whole time, too, he was talking with me like it was no big deal. We were moving so fast I could barely keep up. I could barely breathe.

But it was a really special moment, because I had one-on-one time with the company commander. We talked about everything. Well, I tried talking, but I was out of breath for most of it.

Later, when we were in Kuwait, I saw him and asked if he would go on a run with me again. He said he would, but it never happened, just because everyone was so busy. After he died, I kept thinking about that day we ran together, and how much it meant to me. I wear his name on my bracelet, and I never take it off. I won't even take it off to go through a metal detector.

**27 October 2005 – Letter from Mark Barnes, 1st Sergeant of Bravo 4-64 Armor**

Many of us were with him in his final moments of life, and I believe he knew we were there, and that we loved him. And there were dozens that simply weren't able to be there because of missions. But not one man was thinking of anything other than him during that time. I have never seen so many people come to the hospital to see a wounded warrior. He was as good as they get.

To him, I say... when I got to the company, you were already there. When I got to JRTC, you were there first. When we went to Kuwait, you were there before me. You beat me to Iraq. And you are now the first of us to Glory. Keep leading the way... just like you always have.

**31 October 2005 – Letter to the Harstad Family**

This letter is difficult to write, but it is certainly the least I can do to honor a great man. CPT Michael MacKinnon was one of the most high-speed soldiers I ever met. He was a Ranger, Airborne, Air Assault, Master Parachutist, and had his Combat Infantryman Badge from OIF I. He was just one of those guys where you liked to stand by him... simply because someone else might see you with him. It's hard to describe, but he carried himself in a way that demanded respect.

He wasn't a normal soldier by any standards. He was a little goofy, always had a dip in his mouth, and quite often had a wisecrack to say. He looked more like a college nerd than a typical special operations soldier, to be honest. But he could PT like crazy, though! He was fast as hell and always challenged anyone and everyone to pushup matches, fights... anything to make us better.

CPT MacKinnon loved his company. He attended every football game and every softball game. Anything the company was doing, he was there. He wasn't the best at softball, but he TRIED his hardest every single time he got up to bat, and helped win the post championship for the Bandits. That's how he was: always giving more to the team.

A fond memory I have was during his first field problem as company commander. We were initiating all our new soldiers by throwing them into a creek by the rubber shoot house. He was standing there, watching, when we all decided he was next. He fought the entire way to the creek, but in the end, he was tossed in. He tried to chase us all down, but there were far too many to catch. CPT MacKinnon did it all with a big smile on his face. We accepted him as a Bandit, because it was then that he became THE Bandit.

I worked in the arms room for my platoon. He would regularly come in and chat with us, asking me about my parents, about where I lived, and how my bulldog was. He knew one of the things I treasured was Stormin' Norman, my bulldog. He knew more about me than a lot of the soldiers from my platoon did.

When I was promoted (to sergeant), I saw him in the chow hall. He had this giant grin on his face, I'll never forget it. He talked with me for quite a while, congratulating me on the big step. He gave me words of encouragement and motivated me. This was a day before he died.

I also remember one day, I was sick as hell but still at work (at Fort Stewart). I was running with the platoon, throwing my guts up, lagging behind, and he ran up beside me. I figured he was going to yell at me. Instead, he kept pushing me to run harder. I ran faster and faster until I caught back up with the platoon. I almost passed out, but I finished, all because of him.

The day he passed was incredibly hard. We tried our best to help him, but it was too late. At first, I felt absolute anger. Then, I felt total helplessness. Right now, I'm starting to deal with it, but I don't think it will ever go away. We have a memorial service for him tomorrow, where I will pay my respects.

This letter doesn't do him justice, but I want all of you know that he meant a lot to quite a few people… to me. His wife, his two young children, his family, and his soldiers will miss him dearly. He was, and always will be, Bandit 6, company commander. Goodbye, Captain MacKinnon. Thank you for your guidance and the memories.

Years later, after he was nominated Secretary of Defense, I read Pete Hegseth's book *The War on Warriors*. He had also served in the Army and, as I previously mentioned, had been stationed at FOB Falcon with us in 2005. In it, he described with painful accuracy the "human meat grinder" that was our area of operations, and he briefly mentioned the deaths of LTC Wood and CPT MacKinnon.[338] After reading this, I wrote him a letter, part of which is shared here. I wished him success in repairing our Army, and I told him about CPT MacKinnon, enclosing a photograph so

---

[338] Hegseth, *The War on Warriors: Behind the Betrayal of the Men Who Keep Us Free*, Chapter 10.

that his mission would always be remembered. Whatever one's politics, it is a powerful book, one that captured much of what was truly happening in the military.

**11 March 2025 – Excerpt of Letter to Secretary of Defense Pete Hegseth**

In "Chapter 10: More Lethality, Less Lawyers," I must confess that I cried upon reading it. You see, it was a mixture of sadness over the unexpected, brief story on the loss of my hero, and relief to finally see someone put into words what many of us should have said long ago. I was there with you at FOB Falcon in 2005, though we didn't know each other, and I was on my second combat deployment as an infantryman to Iraq with the 3rd Infantry Division. It was my unit and friends that were being devastated in the Baghdad "meat grinder," as you so accurately stated.

My hero was CPT Michael MacKinnon, who was tragically killed on the 27th of October, 2005. You wrote of him in your book, though not by name. Rushing to his aid was LTC William Wood, who, like CPT MacKinnon, was killed by an improvised explosive device. I was part of QRF on that horrible mission, and we frantically kicked in the doors of each and every nearby house, shed, and barn. The reason was simple: we believed LTC Wood had been taken captive, as we could not find his body. We scrambled to rescue him from the clutches of the bloodthirsty insurgents, who had been torturing and defiling bodies left everywhere around the vicinity of FOB Falcon. It was only later that we discovered the remnants of his body, as he had been thrown a great distance away from the site because of the tremendous blast that he was practically standing over. That day still haunts me as we near the twenty-year mark.

Before I close this letter, I would like to briefly tell you about my hero, CPT MacKinnon. I met him in Iraq before he was our company commander, and was immediately impressed by his competitiveness, devotion to his troops, intelligence, and kindness. He was always demanding excellence from those around him and was an inspiration on and off of the battlefield.

CPT MacKinnon resembled a college student and didn't look like the proverbial soldier's soldier, but that was merely a façade, as he was the true embodiment of the spirit. He was an infantryman. When he took command of the Bravo Company Bandits, he became the Bandit. We all looked up to him because we knew there was simply no one better. We became faster, smarter, and deadlier, all because of his guidance and leadership. We wanted to make our leader proud.

Early in my enlisted career, I became Bravo's unit armorer, in addition to my regular duties as a grunt. CPT MacKinnon would sometimes randomly show up in the armory and always made time to sit and talk. Somehow, I was closer to him than many of those in my own platoon. As a young PFC and Specialist, it boggled my mind that the company commander would take time out of his busy day to talk about my bulldog at home, ask about my parents, and truly make me feel appreciated.

While I would like to believe that I was the only special soldier he did this for, I know that he developed close relationships with many of his subordinates. He knew how to connect to his troops and make them feel as though they were valued. Because we were. We truly became a family... a band of brothers.

When he was killed in action, I went through the different phases of emotional grief over the loss of our commander. No, not just our commander, my *friend*. It was shock and denial, which eventually mutated into blinding anger. I then became depressed about his loss, and have only started to accept it the past few years.

After we returned home, his wife and children had cards printed up for all of us to keep. I display one of them in my house, to serve alongside the bracelet that bears his name on my wrist. It is the reminder of the love that we had for our commander, and how our lives are a bit emptier with him no longer leading us. It also gives me a sense of motivation when times are tough. After all, he would have wanted us to complete the objective, even if he wasn't there to see it.

I humbly ask of you to accept this card. Perhaps you will occasionally look at his face, and the faces of his family, and remember that he was a true warrior and a good man. We all lost a lot that day, though his wife and children lost everything. If nothing else, my hope is that his photograph will inspire you to do the very best for our beloved Army and the rest of the military.

Days later, our tactics underwent a radical improvement. We soon found ourselves becoming a deadly thorn in the side of the insurgents running rampant south of Baghdad. Then this thorn turned into a sledgehammer.

# Reflections

When warriors fall in combat, sometimes moments of unexpected beauty rise from the tragedy. Small gestures, often from complete strangers, can ease the weight of grief, if only a little. One such gesture came after the death of SSG Jerry Bonifacio, the soldier killed in a VBIED attack, likely saving many lives in the process. Jerry had always been passionate about comic books, video games, and heavy metal music. At his burial, his family received something they would never forget.

Gene Simmons, bassist, songwriter, and one of the vocalists of KISS, long known for his support of American troops, personally donated a custom-made KISS casket to Jerry's grieving family.[339] Jerry was laid to rest in that casket, a reflection of his lifelong love of rock music. In death, a lasting bond was forged with one of rock and roll's most iconic figures, the Demon himself.

CPT MacKinnon, too, was honored posthumously through a project called *Bridges for the Fallen*. Across the Missouri River, which eventually winds its way past my hometown, stands a bridge nestled among rugged hills and scattered trees near Craig, Montana. In 2019, it was dedicated as the Michael John MacKinnon Memorial Bridge.[340]

While searching online, I came across a KTVQ News video of the ceremony. Friends, family, and ordinary people came to pay their respects at the event. His daughter, Madison, and his son, Noah, spoke to reporters afterward.[341] Watching them, I couldn't help but smile. Their strength and resilience were unmistakable, and I knew, as surely as I know my own heart, that their father would have been proud.

It's strange that twenty years have passed, yet I still find myself thinking of him, and of so many of our fallen brothers, almost daily. Whenever I face a problem, I catch myself wondering what he would have done. As I reached out to others for interviews for this book, I discovered they do the same. Again and again, I heard the same refrains: *"You know, he'd be a general by now,"* or *"These idiots at my job drive me crazy. Captain M would never have stood for that."* Over and over, the message was clear. We are all still thinking about the man. He left an impression, a lasting mark of greatness.

There's a popular saying often attributed to Ernest Hemingway: "Every man has two deaths, when he is buried in the ground, and the last time someone says his name." It's widely cited in blogs and quote collections, but no primary source confirms that Hemingway ever wrote it, at least that I could find.

---

339 Blabbermouth. "GENE SIMMONS Donates Custom-Made KISS Casket to Family of Soldier Killed in Iraq," October 22, 2005. https://blabbermouth.net/news/gene-simmons-donates-custom-made-kiss-casket-to-family-of-soldier-killed-in-iraq.

340 The Historical Marker Database. "Michael John MacKinnon: An American Hero - Historical Marker," October 29, 2019. https://www.hmdb.org/m.asp?m=141858.

341 KTVQ News. "Michael John Mackinnon Memorial Bridge," June 22, 2019 https://www.youtube.com/watch?v=8olcPWyiRsk.

Others attribute it to neuroscientist David Eagleman, as seen in his book *Sum: Forty Tales from the Afterlives*, although it is slightly altered. He wrote, "There are three deaths. The first is when the body ceases to function. The second is when the body is consigned to the grave. The third is that moment, sometime in the future, when your name is spoken for the last time."[342]

Perhaps both of these are simply reflecting an older, communal meditation on mortality, which reflects shared wisdom rather than definitive statements of single authorship.

Regardless, Mike MacKinnon remains alive and well in the only way that truly matters in the grand scheme of things, through the lessons he taught and the guidance he gave, which still live within us. His children carry his legacy forward, and so do we, the men who served under him. I remind myself of that often.

After his death, I naturally lost all contact with his family. They moved away from Fort Stewart, and to be honest, I didn't know what to say. We were trained to be grunts, never to express our emotions or share how we truly felt. Still, I felt tremendous guilt for not doing more, for not taking action. Instead, I kept to myself and remained silent, locking my feelings away.

A few years later, after I had returned to Sioux City, I looked up the address for his father, John. I wrote him a short letter, telling him how much I admired his only son. Knowing that John was a former infantryman himself, I figured he would understand how I felt. He sent me a letter in response, thanking me for my words. That was all. I just couldn't bring myself to do any more, though I should have. Perhaps this book is my attempt to make amends for the silence I kept all those years. A way to write everything I should have personally said to his family.

In a conversation for this book with Matt Turcotte, the former Gambler Platoon Leader whom I deeply respect, I admitted that I had carried anger in my heart for years. I needed someone to blame, yet the people who had actually caused the loss were nameless and faceless. So instead, I turned my anger toward the 1-184. I convinced myself that if they hadn't been in trouble, LTC Wood and CPT MacKinnon would never have been there, and they would still be alive. The same went for Gifford, Deckard, and Ford, too.

With time, my views have softened. Age and reflection have shown me that what I really needed back then was simply someone to hold accountable. Matt offered me an insight that struck me deeply. He said, "You can't do that to yourself. You can't think of it that way. Instead, ask yourself: how many people did he save in 1-184 by being there? How many people did he touch and inspire, just like he did with us? That's how you have to think of it." He was right. We will never know all the small things Mike did to help those around him.

After his death, an elderly Iraqi man from a nearby village asked us, "Where is Captain Mike?" When he was told that Mike had been killed, the man began to cry. His death had left ripples far beyond our platoon, company, and battalion. It deeply hurt many, including this poor Iraqi farmer, and he wasn't the only one. Many other Iraqis expressed the same grief.[343]

---

342 Eagleman, David. *Sum: Forty Tales From the Afterlives*. Vintage, 2010 Chapter 8.

343 Dan Balda, "Soldiers Eulogize 4 Fallen Comrades at Ceremony," Defense Visual Information Distribution Service, November 9, 2005, https://www.dvidshub.net/news/3686/soldiers-eulogize-4-fallen-comrades-ceremony.

In recent years, I have also reflected on the randomness of war. On that fateful day, in his rush to help one of his soldiers, LTC Wood made a fatal tactical mistake. Rather than moving through yards and around houses, he ran down the roadway and was struck by an IED, likely remotely detonated by a distant insurgent. This isn't a criticism, but an example of the kinds of errors that can and do occur in the heat of combat. We all made mistakes. Ninety-nine times out of a hundred, though, we walked away from them, recognized what we had done, and swore never to repeat them. CPT Anderson admitted that he had done the exact same thing, but he had caught himself in the moment. For LTC Wood, the odds were simply not in his favor that day.

And yet, he died a hero. As a lieutenant colonel commanding an entire battalion, he could have easily ordered someone else to take that risk. Instead, it was *he* who took the initiative and charged into the chaos.

After he was flown home, CPT MacKinnon was buried with full military honors at West Point. A photograph from that day shows his wife and young daughter outside the chapel as the casket is carried to the hearse.[344] The look on his daughter's face is heartbreaking. It is of pure anguish, the kind no child should ever have to endure, and there wasn't anything any of us could do to relieve the pain.

For years, I've told myself that I would visit him at West Point soon, yet somehow, I always find an excuse not to go. Now, in Montana, his bridge waits for me as well. Perhaps it's cowardice. Or perhaps it's the weight of not knowing what to say. How do you thank someone like him? *Thank you for everything, then and now.* One day, I will stand at West Point. One day, I will embark on a journey into the Montana wilderness. And one day, I will tell him.

In the meantime, a family photograph of Mike, Bethany, Madison, and Noah still hangs in a frame on the wall above my treadmill. It's the same one I sent to Pete Hegseth. Beside it rests a blue infantry cord. During the long and brutal Iowa winters, when running outside is nearly impossible, I find myself staring at his face as I run indoors. On the last half-mile, I break into a full sprint, pushing as hard as I can until I feel like I might collapse. I remember that day he ran beside me, urging me to go faster. Even now, after I push through that final stretch, gasping for breath, I can't help but smile through the pain.

---

[344] Bruno, Greg. "Family, Friends Mourn Loss of Army Captain." *Times Herald-Record*, November 8, 2005. https://www.recordonline.com/story/news/2005/11/08/family-friends-mourn-loss-army/51113875007.

Bandit 6: Gone ahead of us to Glory – still leading the way

Photo courtesy of Robert Roth

# In Memoriam

SSG Jerry Lee Bonifacio, Jr.[345]
13 May 1977 (Vacaville, California) – 10 October 2005 (Baghdad, Iraq)


COL William Wesley Wood[346]
12 October 1961 (Lubbock, Texas) – 27 October 2005 (Baghdad, Iraq)


CPT Michael John MacKinnon[347]
26 May 1975 (Helena, Montana) – 27 October 2005 (Baghdad, Iraq)

---

[345] Military Times, "Army Staff Sgt. Jerry L. Bonifacio Jr.," Honor the Fallen, https://thefallen.militarytimes.com/army-staff-sgt-jerry-l-bonifacio-jr/1168702.

[346] Military Times, "Army Col. William W. Wood," Honor the Fallen, https://thefallen.militarytimes.com/army-col-william-w-wood/1220922.

[347] Military Times, "Army Capt. Michael J. Mackinnon," Honor the Fallen, https://thefallen.militarytimes.com/army-capt-michael-j-mackinnon/1213018.

# Chapter XVIII

## Thorn

Starting on October 28th, immediately after the deaths of COL Wood and CPT MacKinnon, the battalion launched a series of missions deep into the heart of Dora called Operation Clean Sweep.[348] On the 29th, Gambler Platoon pushed toward the objective at the very front of a massive convoy, the tip of the spear, which was designed to clear out insurgent strongholds in a series of major combat assaults. We would be used as the battalion's sledgehammer.

Far back in the same formation, a Humvee rolled along the identical route, 27th in line out of more than 60 vehicles.[349] Unbeknownst to us at the time, that vehicle carried an insurgent bounty on it, as did SSG Ellis "Jerry" Majetich, a Psychological Operations Tactical Team Leader with the 7th Psychological Operations Group.[350] They accompanied us on this vital mission.

His team's assignment wasn't about firepower, but about persuasion and compassion. They worked with Iraqi civilians to bridge cultural divides, alleviate daily hardships, and guide vulnerable communities away from extremist influences. The concept was simple, yet powerful and effective. Gaining allies through understanding and problem-solving was far more desirable than attempting to eliminate enemies through force alone. It was our way of using "carrot and stick" diplomacy.

Eight insurgents lay in wait, eyes fixed on the endless line of vehicles, hidden from sight, biding their time for the perfect target. When the Humvee with the loudspeakers came into view, a vehicle that stood out from everything else in the convoy, they triggered a massive IED. The explosion hurled CPL Charles DuPree nearly fifty meters through the air, out of the gunner's hatch. He

---

[348] DVIDS, "U.S., Iraqi Forces Sweep up 49 Terror Suspects," Defense Visual Information Distribution Service, October 31, 2005, https://www.dvidshub.net/news/3522/us-iraqi-forces-sweep-up-49-terror-suspects.

[349] Tennessee Titans. "Love and Theft Unveil Song Written With Veteran Jerry Majetich | Salute to Service," November 17, 2021. https://www.youtube.com/watch?v=U3A9b7MEHV0.

[350] CFA Society Tampa Bay. "Ellis J. Majetich, (Retired) US Army." Accessed December 26, 2025. https://cfatampabay.org/net/frmPeoBio.aspx?id=2585.

struck the ground with a shattered right elbow, burns covering almost ten percent of his body, and an open fracture tearing through his lower right leg.[351] Incredibly, against all odds, he was still alive.

### 10 August 2025 – Reflection of Juan Serrano, Platoon Sergeant, Bravo 4-64 Armor

My Bradley was at the very front of the huge convoy. We had to cross through a small Iraqi checkpoint, which meant we kept stopping to let everyone catch up. I sat there for a while, waiting, before finally moving forward again.

A few minutes later, the IED went off behind us, and we raced back to the scene. It was at the exact spot where our Bradley had been sitting. We had been there for probably ten minutes, parked right on top of it, never knowing it was there. The insurgents didn't hit us. They waited for the Humvee instead.

SSG Jerry Majetich was still trapped inside the wreckage, the propane tank used in the explosion having fused the remaining Humvee doors shut, while the rear of the vehicle lay torn apart. Flames engulfed the ruins, searing his body as he endured catastrophic burns. Amid the chaos, fellow soldiers fought their way to him. One of them was CPT David Anderson, who rushed forward through the fire to pull him from the inferno as we worked to secure the scene and find those responsible.

### 1 August 2025 – Reflection of David Anderson, Commander of Charlie 4-64 Armor

I grabbed Jerry and helped pull him out. He had been terribly burned, and my hand stuck to the back of his head as I dragged him from the fire. He was still conscious and asked me, "How bad is it? Is my face OK?"

I couldn't tell him the truth. I tried to tell him he was going to be fine, but I could see he was in rough shape.

Unbelievably, Majetich survived the blast. Yet the toll on his body was devastating. He sustained a traumatic brain injury, burns covering more than 37% of his body, including complete burns of the skin on his face and scalp, three fractured vertebrae, the loss of both ears and his nose, severe damage to one hand, a fractured foot, and ruptured internal organs. As if that were not enough, he also endured bullet wounds to his right shoulder and leg.[352]

---

[351] U.S. Central Command. "Wounded Warriors Return to Battlefield," July 1, 2010. https://www.centcom.mil/MEDIA/NEWS-ARTICLES/News-Article-View/Article/884096/wounded-warriors-return-to-battlefield.

[352] Alexander, Rachel. "Jerry Majetich: The Story of a Severely Injured Iraq War Veteran." Politanalysis, June 21, 2008. https://politanalysis.com/2008/06/21/jerry-majetich-the-story-of-a-severely-injured-iraq-war-veteran-5008256.

Two of the vehicle's other occupants were not so lucky, if one can call it that. CPT Raymond Hill, serving as the battalion's Public Affairs Officer, and SGT Shakere Guy, attached to the Psychological Operations detachment, were both terribly injured in the blast and would later succumb to their wounds. Their Humvee had been enroute to deliver care packages to Iraqi children, a small gesture of goodwill amid the broader battalion operations, when their mission was cut short, the gifts scattered and burning across the roadway.[353]

Though the scene was horrifying, another entry in a seemingly endless chain of nightmare-inducing traumas, we had to push forward and complete the mission. Although the deaths of our men were devastating, they were not in vain. After several days of fighting and searching, dozens of insurgents were captured, along with large caches of weapons and money. Hundreds of houses were systematically searched, and, if insurgent activity was discovered, some were ultimately reduced to rubble.

This reflected a monumental change in the way we did business. In the aftermath of CPT MacKinnon and the battalion commander's deaths, our rules of engagement shifted dramatically. The restraints that had held us back at every turn were suddenly gone. Responsibility now fell heavily on individual companies, platoons, and even squads, each of which was forced to make the hard decisions of killing in the field. But this was far preferable to being handcuffed in a literal fight to the death.

### 8 November 2005 – Letter to the Harstad Family

Since our colonel was killed, we have stepped up operations in our sector. They have changed our rules of engagement, and we have total use of all our weapons now.

Today, we destroyed several cars and a house with the 25mm main cannon. They were being used by insurgents. We were then hit by a roadside bomb on the way out. It slightly damaged the Bradley we were in, but there were no injuries.

### 18 July 2025 – Reflection of Juan Serrano, Platoon Sergeant, Bravo 4-64 Armor

After all the deaths, we became really aggressive. We burned overgrown vegetation, blew up bridges, ran over houses, destroyed cars, and made the insurgents pay.

How did we know who was responsible? Most of you guys were in the back of the Bradleys, so you couldn't see what I saw from the turret. I paid attention to everything. Wasn't this road paved yesterday? Today, it has dirt on it, so we aren't going to drive down that way. Didn't that balcony on the house over there have a dirty window yesterday? Today it's clean, so I watch it carefully to see if they're spying on us through it.

---

[353] State of Hawai'i Department of Defense. "Honor Roll: 29th Infantry Brigade, Operation Iraqi Freedom," September 11, 2025. https://dod.hawaii.gov/blog/honor-roll-29th-infantry-brigade-operation-iraqi-freedom/.

I had the gunners look at all the roadways with our thermal sights, scanning for heat signatures and looking for recently buried IEDs. I watched and learned from the clues. You had to pay very close attention to everything. Otherwise, we would die.

**The side of our Bradley after the IED strike - every reactive armor block had to be swapped out before we rolled**
**Photo courtesy of Derek Hauger**

Elsewhere in central Baghdad, insurgents staged a coordinated assault on the Palestine Hotel, long known as a hub for foreign journalists. A car bomb and a cement truck packed with explosives tried to tear through the blast walls guarding the compound, while gunmen opened up with small arms to cover the suicide drivers.[354] The defenses held, though barely, keeping the attackers from fully breaching the perimeter. Had they made it inside, it likely would have been catastrophic.

---

[354] The Guardian, "Bombers Target Journalists' Hotel in Baghdad," October 24, 2005. https://www.theguardian.com/media/2005/oct/24/iraq.internationalnews.

Moments later, a third VBIED detonated beside a mosque on the far side of the complex. The blasts killed at least 17 people, though the toll could easily have been far higher.[355] Even so, the triple explosions ripped through the surrounding neighborhoods, leaving another scar on an impoverished area already battered by war.

**21 November 2005 – Letter from Robert Roth, Commander of 4-64 Armor**

Several days ago, two car bombs detonated outside a hotel filled with Western reporters. The car bombs failed to penetrate the perimeter security walls of the hotel, but when they detonated their bombs, they were next to an apartment building and a housing complex filled with sleeping Iraqi men, women, and children.

The force of the explosion was tremendous, and it killed several Iraqi citizens as they slept. One housing complex crumbled under the force of the explosion. Others had broken glass in every window, which caused many injuries. It was a horrendous act against innocent victims in one of the poorest neighborhoods in our area.

It was also around this time that our platoon responded to a drive-by shooting in yet another violent neighborhood skirmish. That day I made another choice I've regretted ever since. It has kept me awake at night more times than I care to count.

Masked gunmen had rolled up to a home in Dora, opening fire on a father as he washed his family's minivan. Rifle rounds tore through the glass, leaving him crumpled on the ground, bleeding out on the driveway. Why he had been targeted, no one seemed to know. He was just another casualty in a city drowning in madness.

We arrived minutes later, the attackers already gone, though we didn't know this immediately. Shards of glass glittered across the pavement, and blood pooled near the open garden hose that still spilled water onto the ground. The man lay motionless, one sandal flung aside and resting in the gore. Around him wailed women in black burkas, children clinging to them and to each other, their fear saturating the air. The screaming only magnified the chaos, making an already volatile scene harder to control. At the time, I felt their cries as more of a distraction than grief. It was irritating, and I just wanted them to shut up. It sounds brutally harsh to admit, especially now, but in that moment, our focus was narrowed to securing the area, searching for the gunmen, and staying alive.

One boy, perhaps no older than ten, stepped toward me as I rolled the dead man over, a wet *chug* of blood escaping his belly like the last gasp from a milk carton. I looked up and saw his eyes wide with fear, his arms raised, fingers spread as though reaching for assistance. Tears streaked down his face as he pleaded in Arabic for me to help his father. But there wasn't anything left to salvage. The man was gone.

---

[355] RFE/RL. "Suicide Bombers Attack Baghdad Hotel, Killing at Least 17." *RadioFreeEurope / RadioLiberty*, October 24, 2005. https://www.rferl.org/a/1062374.html.

By then, after so much death among my own friends, my heart had hardened to their plight. I felt no sympathy, no urge to comfort, and no desire to carry their grief further. Instead of kneeling to that boy, instead of doing the human thing by offering the simplest gesture of kindness and compassion, I turned around and walked away. I climbed back into the Bradley, and the ramp lifted, sealing him and his dead father out forever.

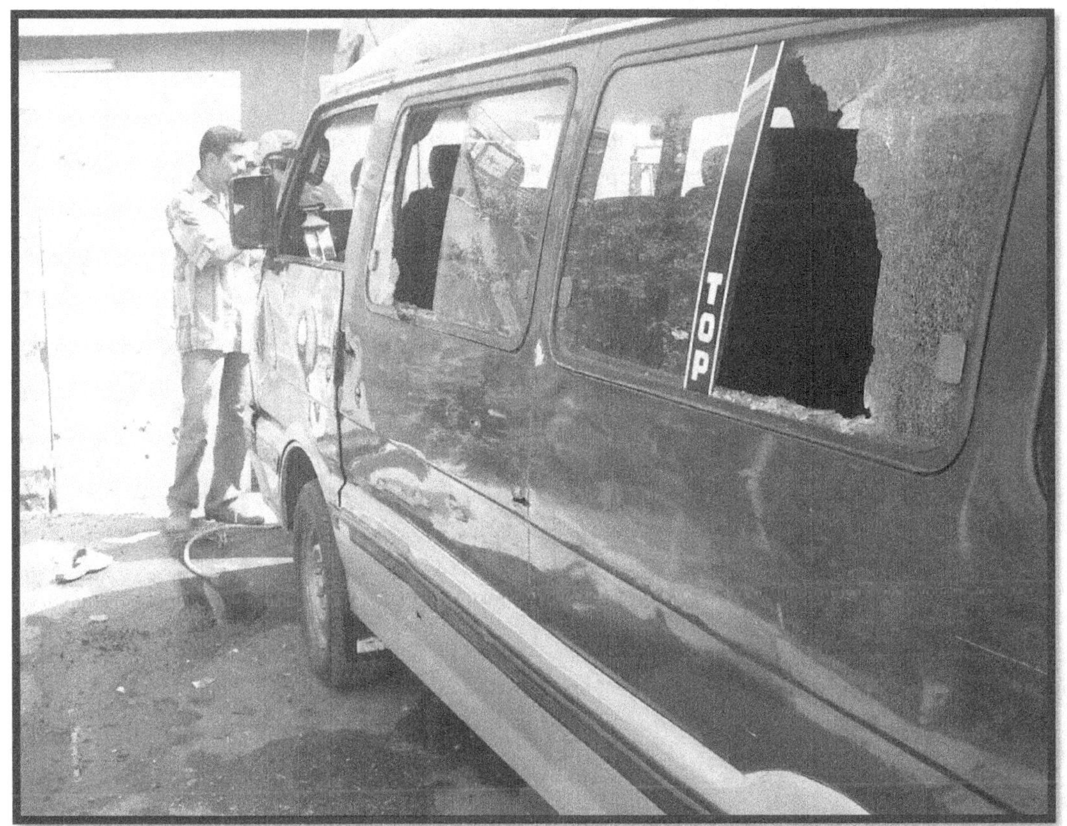

**At the scene of yet another shooting, where blood and gore covered the ground**

I left him standing alone at the end of that driveway. He was the Iraqi Police's problem now, I figured. Although I never saw him again, I can still see his face in my mind's eye. I'm certain he remembers me, not as a protector, but as the one who turned away when he needed help the most. I can't rewind that moment, and the shame of it still lingers.

But what should I have done? I can't say for sure, but I know, deep down, that what I did, or failed to do, wasn't the right thing. To me, it was just another day in the blur of war, but for that boy, it was a life-altering moment that would shape him forever. Only years later, with distance, decompression, and reflection, did I truly understand the weight of what I had done. But in the meantime, we still had a war to fight, and we couldn't allow our emotions to creep inside.

It was like that, day after day, lurching from one crisis and firefight to the next. Jason Sturm, who served alongside me in the Gambler Platoon, remembered that period just as viscerally.

**15 September 2025 – Reflection of Jason Sturm, Bravo 4-64 Armor**

There was this small girl, no more than six or eight years old, who would come out to meet us every time we patrolled. We gave her candy, food, and water, and she was always smiling.

One day, we were assigned QRF for the area when another unit was hit by an IED right by where she lived. We responded to the scene. She had been killed in the blast.

I've often blamed myself, thinking that if I hadn't given her candy or been so friendly with her, she wouldn't have gone near American troops. That was one of the worst days of my life.

To take the fight to the enemy, it was decided to establish an observation post (OP) called Thorn. We joked that the bigwigs at FOB Falcon had finally gotten tired of being mortared day in and day out. Apparently, some of the Fobbits on base were so rattled they wouldn't even walk to the chow hall. That's when someone floated an idea that, surprisingly, turned out to be brilliant: instead of chasing insurgents on their terms, why not make them come to us?

As you might recall, the insurgents operated deep inside the neighborhoods, which made it nearly impossible for us to engage them without restrictions. In practice, that meant we couldn't return fire without risking countless civilian lives. The only real solution was to draw them out.

We found it in the abandoned shell of a partially finished mansion on the southern edge of Baghdad, said to belong to "Chemical Ali" Hassan al-Majid, perched along the Tigris and surrounded by a massive wall. Even better, it was relatively isolated, right in the heart of Arab Jabour, the insurgent stronghold where it felt like every other dipshit was trying to kill us. It was literally in the backyard where I had been hit in the mortar attack, less than half a kilometer away. From the front gates of Thorn, you could see the grove of trees we had used for cover before running miles down the roadway.

All we had to do was take control of the place and make sure everyone and their brother knew it, but that wouldn't be too hard. After we set up the OP, it'd be like Black Friday shoppers storming a Walmart. Only, this Walmart would be laced with booby traps and guarded by a bunch of sleep-deprived infantrymen and tankers who thought "customer service" meant burst fire.

Arriving at the front gate one morning, we used a Bradley to break open the towering metal door. Dozens of us poured out of the tracked vehicles, rapidly spilling into a vast courtyard lined with hundreds of orange trees. The fruit blazed in the sunlight, brilliant bursts of color set against the dust-stained earth.

We swept through quickly, finding no one squatting inside. Before the bulldozers arrived, curiosity got the better of me, and I plucked one of the oranges. It looked perfect, the kind of produce you'd expect to see in a supermarket bin, but when I peeled it open, the illusion collapsed.

The inside was dry, hollow, nothing but a withered outline where juice and pulp should have been. It almost seemed a reflection of Iraq itself, or, maybe even me toward the end of the deployment.

As the engineers tore out the rows of orange trees and shoved them to one side of the vast rectangular courtyard, we went to work filling sandbags and stacked them into the gaping holes where windows should have been. The building itself was quite large, but little more than a concrete skeleton. It had no railings, no plumbing, no glass, nothing that hinted at real habitation. Just stark slabs of poured gray and dusty bricks, which we quickly transformed into a fortress wrapped in razor wire.

**Courtyard of OP Thorn: uprooted trees stacked at left, a MEDEVAC for a wounded soldier, and the same nearby grove we used for cover when I was injured in Arab Jabour**

On the roof, we mounted several MK-19 automatic grenade launchers, their lines of fire sweeping across the grounds below. Behind the complex lay an empty swimming pool, set beside a broad boat ramp that sloped into the Tigris. Out in mid-current, an island choked with trees loomed in the near distance. Appropriately, we nicknamed it Devil's Island.

For the next two months, infantrymen and tankers lived in that cozy OP, our new home away from home. Our platoon and company were split into two groups: half remained at FOB Falcon, running missions and performing QRF, while the other half was at OP Thorn, serving as bait. Every three or four days, we rotated; the cycle never ended.

**OP Thorn was turned into a reinforced kill zone**

It didn't take long for the mortars to start pelting us, though the teams that shot at us kept their distance for obvious reasons. From the rooftop, I could put 7.62mm rounds out well past a kilometer, so I spent nights picking off vehicles and thermal heat signatures through the scope. Anyone who pushed too close was fair game for the machine gunners, and we constantly lobbed rounds at anything that moved.[356]

Before we ever moved in, though, we made sure the surrounding neighborhoods got the message: this place was an armed bear trap. If you weren't an insurgent looking for trouble, you

---

[356] "M249 Saw Night Vision." Uploaded by Derek Hauger, January 6, 2011. https://www.youtube.com/watch?v=eIuNIOo8XZc.

needed to stay the hell away. But if you knew of any dipshits that wanted to fight, send them our way.

**25 November 2005 – Letter to the Harstad Family**

Tomorrow, I head out for a three-day operation. We have to maintain security until their elections pass. My job will be to sit on a roof with my machine gun for three days, and shoot anyone that tries to get close. From what I hear, we are going to shoot a lot of stuff out there.

At night, the morons, brave as they were, would sometimes try to sneak out to Devil's Island in boats to snipe or lob mortars around the trees. That tactic never paid off. We parked Bradleys and tanks on the ramp and used main-cannon fire to cut down scrambling heat signatures; trees erupted in geysers of wood, and debris rained across the river. When it escalated, an Air Force JTAC on the net would call in 500-pound strikes from the jets above, and the place would change in an instant: craters, shattered limbs of trees, and a beautiful silence that meant nobody had survived. At least one Iraqi vehicle was destroyed by a main round from a tank.

Gamblers at OP Thorn: firing at the endless supply of attacking insurgents from sandbagged windows
Photo courtesy of Derek Hauger

At the rear boat ramp, usually quiet during the day, we had a couple of close calls. One morning, SPC Franzen was nearly taken out by a sniper as he moved from a parked Bradley back toward the main compound. Dust kicked up around him as rounds smacked into the dirt, and he zigzagged his way to safety while dozens of us laid down blind suppressive fire. We couldn't see the shooter, but the goal was simple. We had to buy him the seconds he needed to make it back alive.

**Behind OP Thorn: the empty pool, the boat ramp into the Tigris, a Bradley parked on overwatch, and the thick Devil's Island across the current**

Another incident at the same spot turned out to be darkly hilarious. SGT Heath Hutchison, one of the tankers wounded in the September 16th attack, had returned to duty and was now back in the fight alongside us at OP Thorn. Having a wonderful sense of humor, he thought it would be a good time to make all of us laugh. And, well, he did… but it nearly cost him dearly.

**4 August 2025 – Reflection of Heath Hutchison, Charlie 4-64 Armor**

One calm morning, I grabbed one of those Iraqi newspapers, which I couldn't read, and casually walked down to the boat ramp. Right in front of everyone, I dropped my pants by the river and started taking a shit. I could hear everyone behind me laughing.

Then I heard this weird noise… *brrrrrrrrr*. It almost sounded like a go-kart engine. I looked south and saw a little boat coming up the river. It wasn't even really a boat, more like a tiny canoe. Three guys were in it, slowly creeping upstream toward me.

Before they even made it to the compound, one of them whipped out an RPG launcher and fired it right at me. I scrambled away as the grenade sailed high over my head and off to the north. All of you infantry guys on the roof heard it, but you couldn't see the boat yet because it was still behind that huge wall.

A second or two later, it popped out from behind the wall and ran straight into a stream of machine-gun fire and that MK-19 grenade launcher. You blew all three of them apart so fast, and with so much ammo, the boat sank, and they floated back down the river. That was the last time I pulled that kind of stunt. You really did save my ass!

At the front gate, we'd built a bunker out of sandbags and coils of concertina wire, overlooking the heavy metal doors. It sat directly above an old chicken coop still packed with dried feces, and the area was infested with enormous rats. They scurried through the bunker at all hours, utterly fearless, stealing any scrap of food and leaving piles of shit-pellets everywhere until our fighting position felt more like a disease-ridden nest than a fortification. At night, while scanning the horizon through night vision and thermals, my attention would often break as I swung an E-Tool at the rodents darting across my boots.

One morning, just after finishing a long night shift, I started back toward the main compound, dead-tired and ready for a few hours of sleep. That's when I heard the unmistakable *whump* of mortars in the distance. I was stuck right in the open, caught almost exactly halfway between the bunker and the compound, with nothing but bare ground around me. I had two choices: drop flat and hope for the best or run like hell.

I chose forward and to run like hell. Legs pumping under the drag of my machine gun and full kit, I sprinted toward safety, lungs burning as I pushed harder. Two rounds landed in the courtyard to my right, yellow-black flashes flaring in my peripheral vision. They were close enough to shake the earth but far enough that not a single fragment reached me, not even a pebble. I dove inside the compound just as the rest of my platoon opened fire from the windows, pouring rounds into the most likely avenues of attack.

**5 August 2025 – Reflection of Derek Hauger, Bravo 4-64 Armor**

The mortar rounds would sometimes land right next to us, and you could see the impact of them. We're just out running around, trying to get back into the tracks or inside. It could've landed four feet this way, or four feet that way, and gotten us.

So, you feel a bit of survivor's guilt when you reflect on it later, when nothing happens to you. Those other people died, but not me. Nothing happened to me.

With Chism at the front gate, where we machine-gunned anything that moved

To deal with what we nicknamed the "mad mortarmen," those insurgents endlessly lobbing shells our way, LTC Roth ordered in a team of indirect fire infantrymen, 11C grunts. They threw up sandbagged pits in the center of the courtyard and set their tubes and gear in place, ready to launch a counterfire barrage the moment the next attack came.

**20 July 2025 – Reflection of Robert Roth, Commander of 4-64 Armor**

After Bill Wood was killed, I started involving myself more with what was going on at FOB Falcon. I'm not taking credit for it, as there were a lot of people involved, but we assisted with the establishment of OP Thorn.

I then had the mortar platoon go in to assist in countering the "mad mortarmen," trying to give the platoons on the ground more room to respond to the threats.

These newly arriving grunts were trained mortarmen, our go-to for lobbing portable high-explosives back at whoever was shooting at us, and they worked with gear they said could pinpoint incoming fire. According to what I was told at the time, their high-tech systems could narrow the target area from a single round. If three rounds came in quick succession, even better, they could zero in fast enough to lay down a counter-barrage with deadly and devastating precision.

OP Thorn: mortar pits in the courtyard as the unfinished mansion looms in the background

Once the 11Cs joined us, the response drill became routine. We'd take mortar fire, they'd fire back almost instantly, and then a Bradley or two would roar out of the gate toward the grid square where the rounds had originated. If anyone was still on site, our orders were simple. We were to capture them if possible, kill them if necessary.

The first few missions I rode on ended in nothing more than scorched dirt. We found only fresh craters, including proof that the insurgents had launched and then sped off in a waiting vehicle before we could close in. But eventually, luck ran out for one of the mad mortarmen teams.

On one response, I saw the aftermath of a counter-barrage that had scored a direct hit. A fighter's remains were scattered across the road, with his upper body mangled and his head obliterated. What stood out, though, was his face. It lay on the pavement beside him, amazingly intact, like a Halloween mask set down in the roadway. The best I could figure was that a round had landed directly on his head, shredding his body while flinging the skin of his face outward, where it landed flat against the concrete like a grotesque pancake. It was awesome to see and wildly motivating, giving us proof that our tactics were effective.

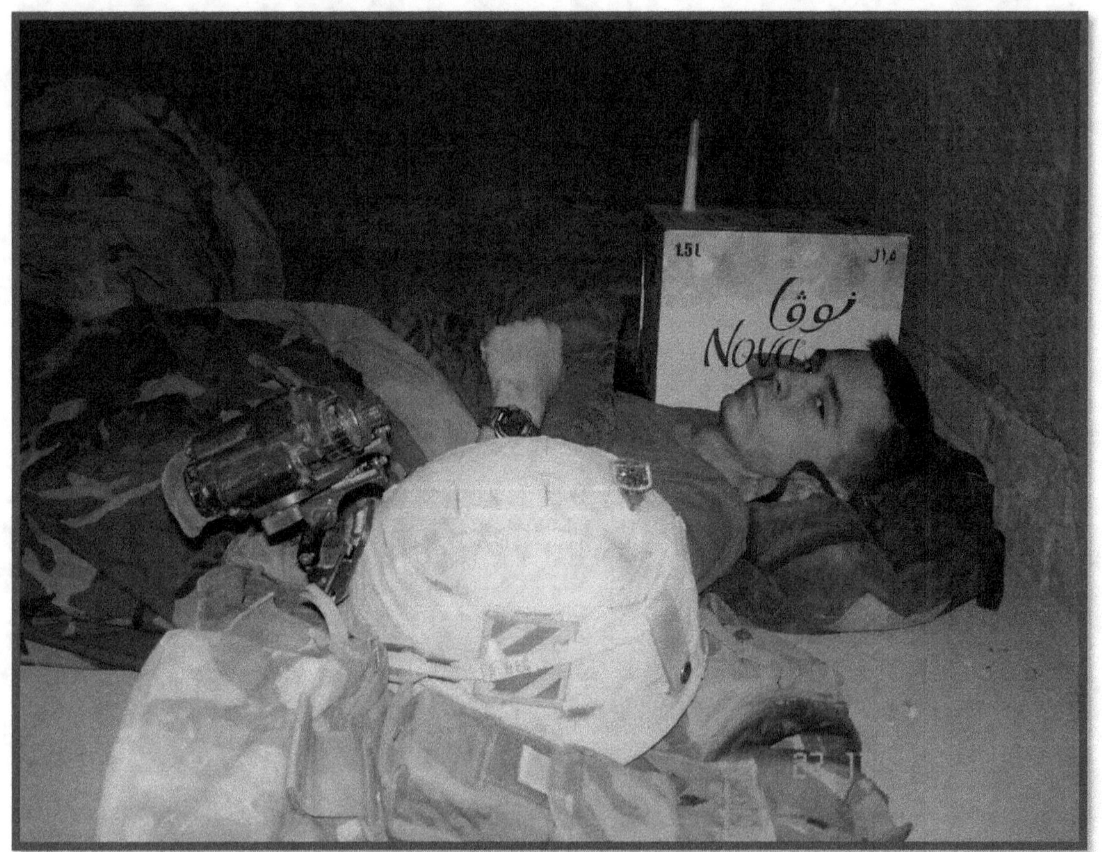

**Sleep, if you can call it that - constant mortar attacks and outgoing machine-gun fire**

And that was the rhythm of it, every single day. In one unpublished video I have, I'm filming from a north-side window while overwatching a field with my machine gun. Insurgents hammered us with a mortar barrage and opened up from the southwest. Your first instinct is to sprint to the angle of attack, but you can't. You must maintain 360° discipline and cover every approach. You trust your brothers to do the same. If one of them needs help, they'll call for it.

As nothing was happening in my sector of fire, I took a few moments to capture a bit of the action. On the video, you can hear the last of the incoming rounds and my voice: "We are gettin' fuckin' attacked right now." Radio chatter cuts in, then, about five seconds later, a mortar detonates just off to my left and knocks the camera hard to the right. "Ah, there goes a fuckin' mortar! That one was close," I say, and we answer with a barrage of rifles and machine guns, pouring fire into the surrounding areas. Then the video ends, as I'm needed elsewhere.

The same applied at night, or whenever it was time for you to sleep. In the beginning, the first crack of a rifle or the heavy *whump* of a mortar would have me on my feet before my brain had even shaken off sleep, but you can't sustain that forever. At some point, exhaustion wins. You have to force yourself to trust the guys pulling watch, because when it's your turn to rest, you *must* rest. Again, if they need you, they'll call.

All this goes against every instinct. You're trying to shut down your body while a fight to the death rages all around you, and explosions and gunfire echo through the walls. Training yourself to sleep in that chaos feels completely unnatural, but I guess that was life as a grunt.

Fighting this way forges an intimacy with your buddies that words can barely touch. From the front gate, peering through night vision toward the main compound more than a football field away, I could instantly recognize someone in a window. It was by the tilt of his shoulders, the rhythm of his stride, or some other barely perceptible gesture. When my relief came for guard duty, I often knew who it was before I even turned around. The cadence of his footsteps on the gravel, or even his scent in the night air, gave it away. It sounds crazy, and I don't expect anyone outside that world to understand it, but in that environment, we became so attuned to one another and to surviving that recognition went far beyond sight.

**12 January 2026 – Reflection of David Anderson, Commander of Charlie 4-64 Armor**

One afternoon, I got back to OP Thorn and walked inside. Since we'd been hit by mortars so often, I yelled as a joke, "You guys ready for some mortars? I think it's about time for mortars!" And like I had scheduled it on the calendar, here they came. The guys stared at me with huge eyes like, "Did you plan that?" Nope. That's just how it was out there.

In November, we celebrated a couple of holidays, though we did so in our own way. The first was Veterans Day, which passed without much fanfare. However, my dad, always eager to offer motivation and support, sent a series of encouraging emails to many of us.

**11 November 2005 – Letter from my father, Bill Harstad, Jr.**

I salute each and every one of you guys today on Veterans Day. You guys are my heroes, and you make me proud!

Thanksgiving at FOB Falcon was quite the affair. From what I read in an email, the dining facility rolled out platters of prime rib, shrimp, roasted pig, pecan and pumpkin pie, and all the traditional fixings, making it sound like a holiday back home. Out at OP Thorn, though, we had no such luxury, just the usual stack of MREs.

**Dealing with insurgents and local politics one block at a time**

Still, to this day, even in the chaos, it remains the best Thanksgiving I've ever had. Sitting in that sand-blown outpost, sharing MREs with those men, I felt more thankful than I ever had before. They weren't just my platoon anymore; they were my family. That day has stayed with me ever since.

**24 November 2005 – Letter to the Harstad Family from Donald Cyr, Alpha 3-7 Infantry**

Happy Thanksgiving to everyone. I went to see Jeff to give him a hug for his mom. Worry not, your baby still looks the same, and I'm doing just fine.

At the end of November, I found myself on yet another reconnaissance patrol, one of the countless foot movements we conducted to keep insurgents from creeping too close to Thorn. These sweeps were also designed to deny them the opportunity to emplace IEDs or prepare ambush sites, but they were nerve-wracking by nature. Every step beyond the wire carried the sense that danger could be waiting just a few feet ahead, hidden behind a wall, in the shadows of a doorway, or beneath the soil under our boots.

SSG Santiago led the patrol while I took point. We moved in wide, staggered intervals along both sides of the narrow roadway, the only viable path through terrain still choked with overgrowth and wild vegetation. The silence was heavy, broken only by the crunch of boots and the faint rattle of gear. As we exited the front gate, turned north, and advanced a few hundred meters, the stillness shattered.

An explosion erupted to our left in a grove of trees, a sharp, concussive crack that ripped through the air. The blast was small, no more powerful than an M203 grenade, but it was close enough to make our hearts pound and instinct take over. We dove for cover, scanning the tree line, rifles shouldered, adrenaline coursing.

Cautiously, after nothing happened, we regrouped and probed forward, but the source of the detonation remained elusive. No wires, no trigger man, and no crater that could confirm what we'd faced was found. It left us with only speculation. Later, when we spoke about it back inside Thorn, two theories emerged. Perhaps someone had tossed a grenade or small explosive from behind the foliage, and it had struck a branch or some unseen obstacle, falling short of its target. Or, and this seemed more likely, it had been a crudely built IED that failed to function as intended, detonating prematurely or incompletely.

Either way, the conclusion was the same: luck had walked beside us that day. Whether by faulty wiring, bad aim, or divine providence, we'd escaped with nothing more than frayed nerves. On foot, with no armor to shield us, surviving an explosion, even one small and inconsequential, felt like cheating death.

**5 August 2025 – Reflection of Derek Hauger, Bravo 4-64 Armor**

They made me a driver toward the end of the deployment. I remember driving past old IED ambush sites. The holes in the road were still there, and you had to try to drive around or over them as quickly as possible. I'd sort of close my eyes as we went through and blurt out, "Oh, God!"

There was one IED that hit the track right behind me. The insurgent had hidden it on the other side of a guardrail, so it didn't really do much damage when it exploded. He ran into a group of people, and we never caught him. There were so many explosions like that happening to us every day.


In December, as the Christmas season approached, we'd been shelled so often that the aftermath became routine. The stabilizing fins from the exploded rounds started piling up, scattered around the outpost like ugly little souvenirs. Some were twisted, some intact, and many came in different colors.

At some point, the absurdity of it flipped a switch, and someone, I'm not sure who, came up with a brilliant idea. If the world insisted on giving us nothing but shrapnel and scrap, then we'd make something out of it. We gathered those fins from far and wide and built a makeshift Christmas tree out of them, right there at OP Thorn. Then, we hit it with spray paint to make it look less like a war trophy and more like a holiday joke we could live with. We certainly didn't have tinsel or ornaments, but we had plenty of what combat had left us: endless piles of machine gun brass, links, and spent casings. We turned those into decorations and hung them wherever we could, stringing them from coils of concertina wire like it was garland.


### 11 December 2005 – Letter from Robert Roth, Commander of 4-64 Armor

I deployed the Mortar Platoon to assist CYCLONES with their fight. The platoon responded magnificently!

CYCLONE has established an outpost in a tough part of the outskirts of Baghdad. They have their own Christmas tree decorated in this new outpost. The local Iraqi people in the area have responded with thanks and praised their efforts for providing better security.


It certainly wasn't pretty in any normal sense. It was crude, improvised, and a little bit insane, the kind of "Christmas cheer" only a place like that could produce, but it was ours. For a little while, it made the outpost feel less like a target and more like a home, or at least the closest thing to one we could manage. After all, it was the most festive time of the year, and if we were going to be stuck there, we were going to decorate with whatever made us happy.


### 4 December 2005 – Letter from Robert Roth, Commander of 4-64 Armor

Over the past month, terrorist activity has been almost non-existent. With the exception of one car bomb several weeks ago, insurgent activity has virtually fallen off the scale.


What LTC Roth said was true. We had driven terrorist activity down to the point where it was nearly non-existent. They all seemed to be too busy fighting *us*. The sense of accomplishment was undeniable, not just because the attacks on civilians and the surrounding FOBs had diminished, but because we were finally being trusted to carry out the mission in the way we knew best, and it had worked. Even when random attacks occurred, they were often weak and ineffective.

Many locals, who understood the risks of approaching Thorn, expressed their gratitude, albeit cautiously and from a distance, to our teams. Most were frustrated when foreign fighters and rival insurgent groups transformed their neighborhoods into battlegrounds. Meaning, drawing them out and into a kill zone we controlled had been a wise move by our leaders, something even the local residents could get behind.

**A postcard-perfect home next to OP Thorn, which was abandoned by its owners and then taken over by an Iraqi Police unit – it suffered a direct mortar hit during our stay there, though somehow no one was killed**

On the night of December 10, half of my platoon was rotated back to FOB Falcon for showers and a much-needed night in a real bed. Stepping out of the shower trailers, towel slung over my shoulder, I trudged sleepily toward the barracks, already thinking about the relief of uninterrupted sleep.

Then the night suddenly erupted. In an instant, the air came alive with a deafening storm of gunfire, louder and more chaotic than anything I had ever heard. Tracer rounds streaked across the sky in every direction, crisscrossing above the compound in a kaleidoscope of red and green light. For a split second, I was certain the FOB was under siege by an incomprehensibly huge military force. Either that, or aliens were invading.

Heart pounding, I sprinted back to my room and shouted for everyone to get ready. Adrenaline surged as I braced for the worst, but within moments, the truth filtered in. We weren't under attack at all…

**11 December 2005 – Letter to the Harstad Family**

Last night was crazy! I guess the Iraqis won some soccer game or something, so everyone in Baghdad just started shooting up into the air at about 10 PM. I was outside in one of the trailers taking a shower, and bullets started coming down. I ran like crazy and got back inside.

Too bad I didn't get it on tape, as it looked just like the Fourth of July. Sometimes these people are stupid! Stupid, stupid.

As it turned out, Iraq had just defeated its rival, Syria, to claim the gold medal in the West Asian Games, but none of us had the faintest clue this was happening. No one watched Iraqi television, none of us cared much about soccer, and we were far too busy fighting a war to track local sports scores. But the moment the match ended, Baghdad erupted. Thousands of Iraqis poured into the streets, firing wildly into the air with anything they could get their hands on. Machine guns, pistols, rocket launchers… heck, even grandpa's old musket was let loose.

According to the Los Angeles Times, 46 people were accidentally wounded by celebratory gunfire raining from the skies that night. The article included a line that made me laugh later: "Some residents of the Iraqi capital, unaware of the soccer victory, thought they were in the middle of a gunfight."[357] Yeah. Exactly. That was me!

Christmas at OP Thorn didn't arrive with lights, music, or gifts under the tree. It arrived as everything else there did: dust, fighting, and the usual routine of playing tag with mortars. We were close to the end, and we all knew it. That made everything feel more intense, but you also start counting the days and realize you are going to miss the people you have been surviving with.

Then someone, and I truly cannot remember who, called a "meeting," but this one was different. It wasn't a standard briefing or a lecture. It was a quiet decision, made by men who'd been shot at together long enough to stop asking for approval. And somehow, the idea found overwhelming support. The plan was simple: we were going to steal a goat and have a feast.

Now, I'm *pretty sure* the statute of limitations has run out on this particular farmyard felony, so I'm probably safe admitting I played a part in the scheme. And if it hasn't, then just like my pinning ceremony, I'll stick to the story that I acted alone. If anyone asks, I was the only goat-sniper

---

[357] Times Staff Writer. "Celebratory Gunfire Injures 46." *Los Angeles Times*, December 12, 2005. https://www.latimes.com/archives/la-xpm-2005-dec-12-fg-bullets12-story.html.

on the dusty knoll. Besides, the goat came from a villager with "friends" in the wrong crowd. You know, the kind who tried to kill us on a daily basis. And what can I say? We were hungry.

**OPERATION YULETIDE GOAT**

Mission Type: Special operations "Snatch & Grab"

Task & Purpose: Restore Christmas spirit through unauthorized livestock acquisition

Risk Level: Medium, including possible encounters with head-choppers, rabid goats, and higher headquarters

Intelligence: Target located loitering in open field, displaying high levels of obliviousness

Outcome: Target acquired and repurposed for morale operations; decisive victory

Operation Yuletide Goat: SSG Thaler and SGT Foutch providing dinner

Photo courtesy of Derek Hauger

We built the plan with the kind of seriousness usually reserved for raids and human targets. Roles were assigned, including a lookout, a shooter, and a grab team. We then moved out like the professionals we were, conducting the most absurd operation in Iraq on Christmas Eve. When we found the biggest goat, we acted just like we had located a high-value target. It was quick, clean, and there was no debate. We dropped it, hauled it, and hustled back to the wire as if we had just captured Zarqawi.

**Papi, upgrading field rations with vegetables, spices, and his makeshift Puerto Rican cooking**

**Photo courtesy of Derek Hauger**

Back inside the wire, everything became a team effort in the way only soldiers understand. A few guys took over the butchering while somebody else got a fire going. Others were in charge of "seasoning," which meant acting like we had options beyond what was available from our MRE packets and items rat-fucked from the chow hall. Before long, the smell of roasting goat drifted through the outpost and, for the first time in a long time, it didn't smell like burning trash or gunpowder. It smelled like food.

We ate right there in the sand, shoulder to shoulder, swapping stories and laughing at funny jokes over dinner. Every incident told was exaggerated to maximum effect. In between pulling guard duty shifts, we roasted marshmallows, made a big fire in a burn-barrel, and thought about spending the holidays with our families back at home.

So, it wasn't a normal Christmas. It wasn't even a sensible Christmas, but it was ours, and it was made by the rough men who had carried me through the worst days. Sitting at Thorn with roast goat and those guys, I felt something I hadn't felt in a while. The war, if only temporarily, went silent.

By the end of the month, Gambler Platoon rotated back under the Tuskers, this time as the battalion's permanent QRF. We were worn down and battered, but we were still standing. Every one of us had made it through that hellhole.

Aside from the pre-Christmas "gift" of a broken nose I earned at Arab Jabour, when a thrown rock caught me square in the face, the Gamblers escaped without any other serious injuries. In a place that seemed designed to punish everything and reward nothing, that felt almost impossible. It was a rare stroke of luck in a wildly dangerous war where fortune rarely smiled.

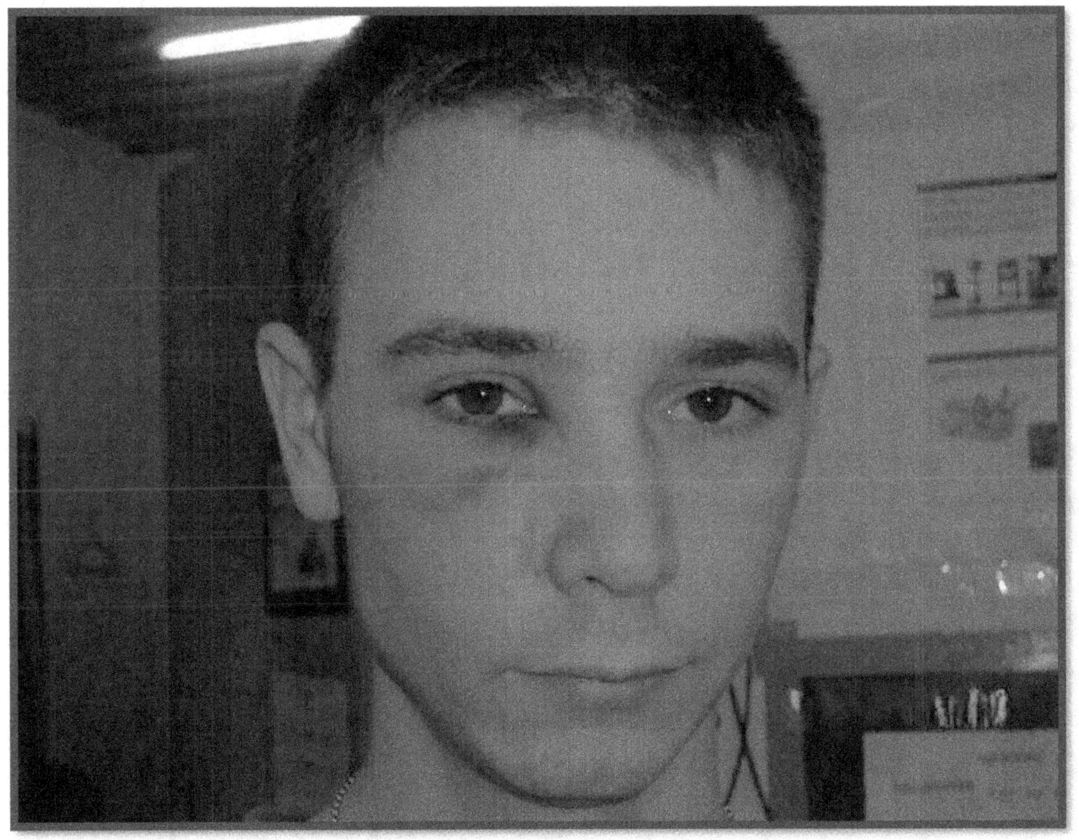

**I survived Iraq, but my nose didn't - when they said "incoming," this isn't what I pictured**

**Photo courtesy of Derek Hauger**

**22 August 2025 – Reflection of Matt Turcotte, Platoon Leader, Bravo 4-64 Armor**

Do you know how lucky we were, Harstad? People were dying all around us, and somehow our entire platoon made it back. It doesn't even make sense when I say it out loud. After everything that happened, how did we all survive?

While I was busy getting a "nose job" so I could finally break into Hollywood, Geraldo Rivera from Fox News dropped by FOB Falcon. OK, not a *real* nose job, but that's what I'm calling it anyway. In reality, it was just a medic ever-so-gently pushing on my crooked nose to restore some semblance of symmetry.

The Geraldo part, however, was completely real. That was extra funny, considering he got yanked out of Iraq in 2003 after drawing American troop movements in the sand on live TV.[358] This time, though, he was on his best behavior. He rubbed elbows with the Gambler Platoon and even did a photo op. So, while my nose was busy squeaking every time I tried to inhale, they were over there living it up with the rich and famous. Then, for my troubles, a few years later, an ENT went in and carved out a significant portion of my sinus cavity and shaved away damaged bone to fix the rest of it. In the end, they got a photo with Geraldo Rivera, while I got an interior designer for my face. But such is life.

From what I was told, Geraldo's visit was far more pleasant than a previous VIP's stop had been. During that earlier incident, a bunch of us walked down to the chow hall with the apparently radical intention of eating. I know, right? The nerve. We made it all the way to the front door before a herd of security guys in civilian clothes blocked the entrance like we'd stumbled into a nightclub with the wrong shoes on.

One of them hit us with, "You can't bring any weapons or ammo in here."

Evidently, the VIP inside was an American politician, or so we were told. We were given two choices: walk all the way across the FOB, drop off our weapons and ammo, then walk unarmed back across the FOB again to get lunch, or just not eat there at all. Because nothing says "common sense in a combat zone" like taking away the one thing that keeps you alive, just so you can eat chicken cordon bleu.

We responded with a respectful, professional, military-grade, "Fuck off, we'll eat MREs." So, a large group of us turned around and left, hungry and annoyed, which is basically the Army's natural state anyway. Maybe the politician was worried about getting fragged. I don't know. All I know is we weren't going to disarm ourselves for a tray of rubber chicken.

Geraldo, on the other hand, was totally fine with our guys being armed. He didn't seem to care at all. Hell, he looked like he'd joined the Gambler Platoon for a minute, posing around the Bradley with heavily armed infantrymen and tankers like it was a recruiting poster. Then he asked to go see OP Thorn, much to the surprise of every sane human being there. After all, it sat in about

---

[358] Collins, Dan. "Geraldo Hit by Unfriendly Fire." CBS News, April 1, 2003. https://www.cbsnews.com/news/geraldo-hit-by-unfriendly-fire/.

the worst place imaginable. Not exactly a wise career move either, because being blown up really kills the ratings. Still, I'll give him this: the man had guts.

**Geraldo Rivera with the Gambler Platoon – Photo Op:** *Yes,* **Triangle of Death Tour:** *Maybe*

**Photo courtesy of Derek Hauger**

**6 January 2026 – Reflection of Derek Hauger, Bravo 4-64 Armor**

It was a last-minute thing. SSG Brister asked if we wanted to be on TV, and I said, "Yes!"

Fox was putting together a big story about us losing the battalion commander, and about the injuries and deaths we'd taken. They wanted footage of guys who actually went out and ran patrols in our dangerous area (Triangle of Death).  So, we went over, did the interview, and I think it ended up on *Fox & Friends* or something like that.

They even let us call our families beforehand on a satellite phone. We told them, "Turn on Fox News! Thirty minutes!" Then, boom. They hit record.

**12 January 2026 – Reflection of David Anderson, Commander of Charlie 4-64 Armor**

Geraldo Rivera from Fox News showed up and said he wanted to see OP Thorn. We were a little hesitant about taking him out there, but he was determined, so we brought him anyway. He was recording "From the Front" for *At Large with Geraldo Rivera*.

Once we got there, he started walking around like he owned the place, checking everything out and stopping wherever he felt like it. At one point, I said, "Uh, Geraldo, you might not want to stand right there." I motioned to the chunks of brick blown out of the wall. "Those are bullet holes from where they've been taking sniper shots at us." He got down really fast after that.

While we were still making the rounds, SGT Wilson came up behind him, gave him a hard slap right on the ass, and yelled, "Good game!" Geraldo froze and looked completely stunned, like, *Who does that to a stranger*? He clearly didn't get our humor. I had to jump in and explain, "Sir, it's just a stupid thing we do in the military. We do it to each other all the time."

While our platoon had come through mostly unscathed, the 1-184 wasn't as fortunate. On New Year's Eve, the Gamblers were called out to a frightful accident in which an Abrams tank had launched off a bridge and crashed onto its side. Meanwhile, the rest of the Cyclones, along with a detachment of grunts from the 1-184, remained back at Thorn, holding the line. We complained about being stuck at the Abrams site, pulling security in the middle of a cold, black night, but it could have been so much worse.

**1 January 2006 – Letter to the Harstad Family**

As of January 1st, 2006, Gambler Platoon is currently assigned to the quick reaction force for the battalion. We have been responding to anything of significance in the area.

A few things have been exciting. Last night, we were called out to the scene where an Abrams tank threw its tread and then fell off a bridge. They fell about 20 feet to the ground, and it landed on its side. Somehow, no one was injured.

We had to stand outside and pull security for 6 hours while they tried to dig it out, thus making me freeze my ass off on New Year's.

At the same time, at the front gate of OP Thorn, where the rats scurried in thick numbers, several infantrymen from the 1-184 were manning the same bunker we had spent so much time in ourselves. Insurgents, usually so wildly inaccurate and prone to lobbing rounds blindly, somehow managed to drop a mortar directly into their position. Sometimes, bad luck needs no explanation.

SPC Marcelino Corniel, just 23 years old, was struck by the exploding mortar and killed.[359] In only a matter of days, he and the rest of his unit were scheduled to rotate home, and Corniel had already set a wedding date for January 23. Instead of celebrating his marriage, his family was left to prepare for his funeral.[360] He would be posthumously promoted, though that was little relief for his devastated family.

Several other soldiers were wounded in the attack as well. CPT Anderson, who had arrived on scene to help manage the tank accident at the bridge, now found himself racing back to the hospital. He described seeing Corniel and admitted to the heavy weight of guilt, a burden familiar to every battle-hardened commander. Leading men in combat is more difficult than most can ever imagine, and its toll often lingers long after the shooting stops.

**1 August 2025 – Reflection of David Anderson, Commander of Charlie 4-64 Armor**

I made it to the hospital, but Corniel was already gone.

A couple of other infantrymen were in the emergency room with severe wounds to their lower bodies. It was horrific to see. At least two of them had their scrotums flayed open. Even with painkillers already in their systems, the usual dark infantry humor still surfaced. One soldier called across the room to his buddy, "Hey, it's no big deal. At least I already have kids! It's a free vasectomy."

I think about things like that. Or the woman who was shot in the chin, and her face looked like the Predator. Or the aftermath of that car bomb, when a little kid was running around with part of the back of her head blown off. You can't forget those things. They all happened under my command.

Toward the end of January, the 3rd ID finally began its slow march back to the United States. The 1-184 returned to California, and the various attached units scattered in their own directions. Some of us limped more than others, but all of us left behind too many brothers on the battlefield.

Before we departed Iraq, CPT Anderson, who had personally led us through more than our fair share of hell in southern Baghdad, stood before us and gave one last heartfelt speech. He thanked us for fighting beside him, and we loved him for it. He had proven himself to be a solid commander, the kind of leader you'd actually follow into the fire. But he also seemed to have an invisible target painted on his back. He wasn't reckless, though, nor was he unlucky by choice. Sometimes, the universe just picks a person to dump all the chaos on, and Anderson drew that

---

[359] Dan Balda, "Soldiers Honor Memory of Fallen Troop," Defense Visual Information Distribution Service, January 10, 2006, https://www.dvidshub.net/news/5036/soldiers-honor-memory-fallen-troop.

[360] Finnegan, Michael, and Jean Merl. "A Funeral Instead of a Wedding." *Los Angeles Times*, January 15, 2006. https://www.latimes.com/la-me-marcelino-corniel-story.html.

short straw. Around the company, we joked he was a "shit magnet," not because he did anything wrong, but because bad things seemed to be drawn to him, like moths to a flame.

**1 August 2025 – Reflection of David Anderson, Commander of Charlie 4-64 Armor**

We were at the airport, getting ready to fly back home. Because so many terrible things had happened under my command, someone stood up and joked, "Can CPT Anderson take a different flight home?" Everyone exploded with laughter.

I guess we had to laugh about it. What else could we do?

The final lineup of Heavy Weapons Squad before heading home: (L-R) SSG Wiseman, SSG Gasman, the author, PFC East, SGT Dvorak, SGT Franklin, SPC Chism

My last clear memory of Iraq is standing alone in a desolate field at night. The wind was still, and for a moment, so was I. I remember realizing, with a kind of quiet shock, that I didn't want to go home. Life here, for all its danger and misery, felt stripped down to its bare essentials. There was no endless noise. No petty complaints about things that didn't matter. Here, my purpose was clear. My brothers were my family. My only job was to survive one more day. How could I ever want to leave that behind?

**With SSG Brister, my mentor, in front of our Bradley, celebrating the end of our tour with Presidential cigars… even though I wasn't ready to go home**

I stared out across the dark expanse, the green glow of my night vision pouring ghostly shapes into my eyes. It felt like we were abandoning this place, abandoning the friends who would never make it home. Any sane person, I suppose, would see the bodies, the blood, and the loss, and want nothing more than to get as far away as possible. But for me, it was the opposite. The thought of leaving left a hollow ache, like walking away from a grave you're not ready to leave.

In a family travel book I published years later, I wrote something that still feels like the truest expression of that moment. "Saying goodbye to a place is sometimes difficult. Though, in today's world of cheap and expedient travel options, it is entirely possible to return to many places that I found particularly enjoyable. One day, my true hope is to return to Iraq, and settle a few things. Say goodbye to some ghosts and welcome new memories. I vividly remember the night before I left, while pausing just outside of Baghdad. I looked through my night-vision goggles at oil refinery flames in the distance. It wasn't lost upon me that I was standing at the doorstep of Babylon, gazing across the centuries of momentary time. Everything was slowly but steadily being reclaimed by the desert sea, giving us but a fleeting chance to peer into the past."

**Gambler Platoon: a chaotic collection of personalities that turned into one hell of a team**

# Reflections

Our bloody battles in southern Baghdad managed, for a time, to rein in the raging insurgency. But the units that followed us were stretched thin with too few men and too few resources, paired with far too much ground to hold. Bit by bit, the belts surrounding Baghdad, those lifelines that funneled fighters and weapons into the city, slipped back under insurgent control. It wasn't until years later, with the arrival of the troop surge, that those areas were finally beaten into a fragile calm. Even then, "pacified" was only ever a relative term. Iraq continues to bear the scars of that war, and the shadow of terrorism still lingers.

There's a powerful documentary by Australian journalist Michael Ware titled *"Only the Dead"* that captures the atmosphere of Iraq during the very years I was there.[361] It features American combat troops, including soldiers from my own 3rd Infantry Division, and portrays the war with an honesty that is both raw and unsettling. I recommend it without hesitation, but offer a fair warning: for those unprepared, it will be deeply disturbing and terrifying. Still, if you want to see what it was really like, beyond the limits of my words on these pages and the imagined movie in your head, this film is about as close to reality as you will ever come.

That being said, I have to admit that my time in Iraq, especially at OP Thorn, was, in a strange way, the most fun I'd ever had in my life. I realize how little sense that makes, but it's the truth. For once, I wasn't boxed in by the rules and restraints of polite society. Instead, I found myself in a place that felt like a lethal playground, where every move carried the weight of life or death. One mistake could prove fatal, yet the rush, the intensity, the sheer adrenaline of it all was unlike anything I had ever experienced before, or since.

Still, I needed time away to decompress. I realize that now. I had begun to grow jaded toward the very people I was supposed to be helping, and a year of nonstop fighting had worn me thin in ways I didn't fully understand at the time. Some units stayed even longer, eighteen months or more, which only deepened the strain. I'm not sure enough research has been conducted on the effects of that on a person, but I can say with certainty that the consequences are real. We've already seen the long-term effects playing out in the lives of veterans all around us.

Sleepless nights, constant shifts in schedule, too few calories, dehydration, the unrelenting need to stay on guard, all of it compounded by the daily stress of combat, are not things that can be quantified neatly on a chart. The toll is more like taking repeated blows to the head in a fight. Damage is being done, but it doesn't always show on the surface like a punch to the face.

There's a fitting paradox for this, I think, known as the Ship of Theseus. First described by Plutarch, it asks us to imagine a great wooden vessel weathering the passage of time. As the years wear on, its planks begin to rot and are replaced, one by one, until eventually not a single original board remains. The question then arises: after all boards have been replaced, is it still the same ship?

---

[361] *Only the Dead*. Screen Australia, 2016. Directed by Michael Ware and Bill Guttentag.

Later, Thomas Hobbes added another twist. Suppose all those discarded, rotting planks were gathered unceremoniously into a pile. Once the original ship had been fully restored with new timber, someone took those old, weathered boards and reassembled them into a second ship.[362] Which, then, is the true Ship of Theseus?

It may sound like a clever word game between philosophers, but for me it feels painfully real. I went into the Army and fought in Iraq as Jeff Harstad, an infantry sergeant, but the person who came out the other side wasn't the same. The skin and bones remained, but the man within had been replaced, piece by piece, until I was someone else entirely. Unlike in the Hobbesian variation, that original person can never return.

After coming home, most of these stories were locked away in a vault deep in my mind. They were buried beneath layers of other memories, gathering dust and cobwebs, until I forced myself to open that vault and sift through them for this book. Inevitably, some stories are gone for good, lost in the haze of exhaustion, sleep deprivation, or are too deeply buried to retrieve.

What I never expected, once I returned to the U.S., was the question that strangers seemed so eager to nonchalantly ask: "How many people did you kill?" The first time, I was stunned by the bluntness of it, the sheer callousness from someone who didn't even know me. Over time, I realized it wasn't meant as cruelty; it was morbid curiosity, the same impulse that makes people slow down to stare at a fatal car crash.

But how do you even answer a question like that? The truth is, I don't know. How many insurgents and terrorists did the Gamblers, the Cyclones, the Bandits, the Tuskers, and the Night Stalkers put into the ground? I can't give you a specific number. What I can say is that it was many. We were effective in our jobs… once we were allowed to do them, of course. But that isn't really what some people are after.

I've found that people often want more of the inside story, as if we were lifelong combat buddies swapping tales over a beer. They lean in for the gore, the shock value, and the grisly details, but they rarely have the patience to sit through the whole truth. And without the full story, the fragments only distort the tale. I'm not willing to hand those memories out in casual conversation; they're too heavy, too layered, and too easily misunderstood. That's why writing this book has been therapeutic. It gives me the space to tell the story the way it needs to be told, with the weight and context it deserves.

The other question I often get is, "If you were firing a machine gun from a distance, how do you know you killed them?" My answer is always the same, and it's the only one that feels both true and sufficient: "They were shooting at us before, we returned fire, and then they stopped shooting."

That is the truth, as far as most firefights are concerned. We weren't out there with clipboards and measuring tape, and we weren't running CSI: Baghdad. In the moment, you're not

---

[362] Britannica Editors. "Encyclopaedia Britannica." Ship of Theseus, November 14, 2025. https://www.britannica.com/topic/ship-of-Theseus-philosophy.

counting bodies. You're trying to stay alive, keep your friends alive, and end whatever is trying to end you.

And no, we didn't always go out and search for bodies afterward. People love to imagine that part, like there was some neat, official procedure where we calmly walked over, confirmed everything, and then marked notches on our rifles. That's not how it usually worked. Sometimes you could see exactly what happened. Sometimes you couldn't. Distance, darkness, smoke, walls, alleyways, and other terrain can hide a lot. And plenty of times, going to check would have been stupid, suicidal, or both. If the shooting stopped, we treated the threat as ended, and we went back to doing whatever it was that we were doing beforehand.

What happened after that was often handled by someone else, or by nobody at all. Sometimes, family members hauled their dead away. Sometimes, the Iraqi Police got called in. Sometimes, everything stayed quiet, the heat did what heat does, and then nature took its course.

So, if someone wants more gruesome details than what I've shared here, too bad. This book is all they're ever going to get.

All of us returned from that deployment carrying wounds, some visible and others hidden away. Once the welcome-home parties faded and the flags were packed away, most of us simply slipped into the background, trying to piece our lives together in silence. Yet every so often, a veteran emerges who leaves a lasting impression. Not because their injuries define them, but because their resilience shines all the brighter through the scars.

A powerful example is Jerry Majetich, who miraculously survived the near-fatal attack in Iraq. He has endured more than 80 surgeries, including the amputation of his severely damaged hand. Numerous articles and videos document his journey, and his story resonates deeply with me, partly because I was there with him when it happened and partly because it is simply that powerful. In his soft-spoken way, Jerry recounts his history, struggles, obstacles, and hard-won triumphs. His mannerisms lend his testimony an authenticity that makes it even more compelling.

### 1 August 2025 – Reflection of David Anderson, Commander of Charlie 4-64 Armor

Jerry came in and told his story to one of my classes a few years ago, and it was great to catch up with him. We talked about that day and how traumatic it was. He said, "It's probably worse for you than it is for me, since you remember it all."

That put a lot into perspective. Even though he was the one who got hurt, he was still thinking about my well-being.

Because of the unrelenting pain in his injured hand, Jerry elected to undergo an experimental amputation procedure. In a groundbreaking collaboration, surgeons, scientists, and engineers created an artificial limb that responds directly to his brain signals, restoring a sense of movement and position known as proprioception.[363] By paving the way for similar advancements, he has become a beacon of hope and innovation for fellow veterans.

Jerry has also shared his experiences through motivational speeches delivered to audiences across the country, many of which are available on YouTube. These talks are profoundly moving, including one where he recounts the horrific execution of Iraqi friends he had worked alongside, and I strongly recommend watching them.[364] Unfortunately, these types of incidents were all too common in our area of operations.

Perhaps part of Jerry's resilience stems from his participation in Operation Proper Exit, a program that brought wounded service members back to the battlefields where they were injured.[365] For him, that meant returning to FOB Falcon. One can only imagine the emotions he and others felt as they retraced steps to the very places where their lives were forever changed. It may echo what Vietnam veterans experience when revisiting the sites of their own suffering and loss.

I often wonder about this. What would it feel like to stand again in Arab Jabour, along Route Corvette, or at OP Thorn, gazing across the Tigris? I don't yet know. However, I do know that something continues to draw me back. One day, if the stars align, perhaps I will be able to return and properly say goodbye to those ghosts still lingering there.

---

[363] Gil, Gideon. "'I Can Still Feel It': Groundbreaking Arm Amputation Surgery Makes a 'Phantom' Hand Seem Real." *STAT News*, September 14, 2021. https://www.statnews.com/2021/09/14/i-can-still-feel-it-new-arm-amputation-surgery-makes-a-phantom-hand-seem-real.

[364] Ponte Vedra Beach Toastmasters Club. "Jerry Majetich," October 21, 2024. https://www.youtube.com/watch?v=iehIb61Mkas.

[365] DVIDS, "Operation Proper Exit VII Brings Wounded Warriors Back to Iraq for Closure," Defense Visual Information Distribution Service, June 16, 2010, https://www.dvidshub.net/image/292403/operation-proper-exit-vii-brings-wounded-warriors-back-iraq-closure.

# In Memoriam

CPT Raymond Dwayne Hill II[366]
14 April 1966 (Pittsburgh, Pennsylvania) – 29 October 2005 (Baghdad, Iraq)


SGT Shakere Taffari Guy[367]
18 July 1982 (Kingston, Jamaica) – 29 October 2005 (Baghdad, Iraq)


SGT Marcelino Ronald Corniel[368]
21 October 1982 (Los Angeles, California) – 31 December 2005 (Baghdad, Iraq)

---

[366] Military Times, "Army Capt. Raymond D. Hill II," Honor the Fallen, https://thefallen.militarytimes.com/army-capt-raymond-d-hill-ii/1220974.

[367] Military Times, "Army Sgt. Shaker T. Guy," Honor the Fallen, https://thefallen.militarytimes.com/army-sgt-shaker-t-guy/1220978.

[368] Military Times, "Army Spc. Marcelino R. Corniel," Honor the Fallen, https://thefallen.militarytimes.com/army-spc-marcelino-r-corniel/1449540.

# Chapter XIX

## What Am I Gonna Do Now?

Redeploying back to Georgia in early 2006 felt like a blur. I took some R&R at home in Iowa, then returned to face the predictable and massive effort of cleaning weapons, updating gear, and reorganizing everything in preparation for a new training cycle.

It wasn't easy. Many soldiers were mentally checked out, knowing a huge turnover was coming as veterans left the service or transferred to other bases, much like in 2003. The atmosphere was chaotic, and even some of the best soldiers resisted cleaning or training.

Their reasoning, while problematic, was understandable. They had already endured combat, and many of the drills we were running had proven less effective in real fighting. To them, it felt pointless. The training procedures needed to be revised, and, well, they were leaving anyway.

That left the NCOs and officers with little choice but to enforce discipline. No one enjoyed cracking down, but order had to be maintained, especially with a wave of new, impressionable soldiers arriving. The unit couldn't afford to slip into bad habits, not with the next deployment already on the horizon.

At the start of our deployment, prior to the unfortunate events involving the 1-184 and the death of CPT MacKinnon, I had decided to re-enlist in the Army. My intention was to participate in the Green to Gold program, which would allow me to attend college full-time and ultimately earn a commission as an officer. However, those incidents left me with a bitter taste, and I quickly changed my mind. I, too, would be leaving for good.

It was incredibly difficult for many of us, myself included, to readjust to life back in the United States. At the time, I didn't recognize it, but I was already showing clear signs of PTSD. My temper could ignite from zero to explosive without warning, and I was constantly on edge, hyper-aware of everything around me. Rationally, I knew the garbage cans lined up at the end of suburban driveways were nothing more than trash, yet some part of my mind whispered: *Be careful. Don't drive that way.*

**3 October 2025 – Reflection of my wife, Jasmin Harstad**

When we first started dating, there were moments when I honestly felt helpless. I didn't know you much before the Army, but I could tell something was wrong. You never talked about it, so I didn't know what to do or how to help.

It showed up most clearly when you snapped at people. I'm not going to lie, it could be scary sometimes. This look would come over your face, and then you'd go hyper-focused and intense, like you were suddenly someone else. You're normally such a patient, kind person, so when that switch flipped, it really alarmed me.

My buddies and I pooled our money to rent a newly built house just off post, but that's when additional cracks started to show. Sleep became a major problem. The slightest noise would jolt me awake, such as a car door shutting outside, its hollow *whump* too much like the distant thud of a mortar tube. Before I even registered what was happening, I'd be on my feet beside the bed, my heart racing. I could hear *everything*, even when it wasn't a threat. My brain simply wouldn't turn off for a few minutes. In fact, it *never* turned off.

Having access to alcohol for the first time, now that I was of legal age, I quickly overindulged. On my 22nd birthday, I drank so much that I blacked out in the shower after vomiting up my spaghetti dinner all over the floor. My friend and roommate, Brian Torres, eventually found me passed out with the water still running cold over me. We laughed about it at the time, but looking back, I realize it was less about celebration and more about compensation. It was an attempt to numb things I didn't want to face, including my worsening sleep problems.

Eventually, I took to sleeping in my walk-in closet, where the walls dulled the noise. I never told anyone because I was embarrassed, but most nights I unrolled a sleeping bag on the floor in that closet instead of lying in my comfortable bed. Even there, though, the subtle noises of everyday life still found me.

During the day, while driving my Jeep around post, I would sometimes almost hallucinate that it, and I, were about to detonate, flipping end over end in the aftermath of a massive explosion. Deep down, I knew it wasn't really going to happen, but that small, insistent voice kept warning me: *Something's wrong.*

**3 October 2025 – Reflection of my wife, Jasmin Harstad**

Right after you got out of the Army, I remember we came up to a red light. You looked around, like something bad was about to happen, and you sped right through it. When I asked you why you did that, you didn't even realize you had. You just did it without thinking.

I remember thinking, *OK, something isn't right here*. And it wasn't just that one moment. There were a lot of other little things like that, too.

Everywhere I looked, there were too many variables. There were too many people walking around, too many moving pieces, and too many blind spots where an ambush could come from. I was like a kid with sensory processing disorder being shoved into a bright and loud casino on the Las Vegas strip. Or, as Karl Marlantes wrote in *What It Is Like to Go to War*, "The Marine Corps taught me how to kill, but it didn't teach me how to deal with killing."[369] The very same could be said for just dealing with normal life.

I never spoke about it to anyone at the time. I carried it quietly, thinking it was just me and that I was exaggerating my perceived issues. Only years later did I realize that nearly everyone else had been fighting the same battle in their own silent way. Some were worse than others, but everyone had at least a touch of it. But we never talked about it. *Never.*

**5 August 2025 – Reflection of Derek Hauger, Bravo 4-64 Armor**

It took me forever to calm down once we got back home. I was constantly scanning rooftops. And the fireworks. The fireworks were brutal.

I kept telling my mom, "I just feel naked. Something's not right. Something's off. I don't have my SAW with me." I'm sure you struggle with that too. I still scan rooftops everywhere I go. It never really goes away.

The problem was the sheer number of new stimuli that could trigger radical shifts in behavior. It wasn't just one, easily identifiable source, though. It was a litany of things, each capable of provoking strange, unwanted reactions. And while you'd think they would be obvious, they often weren't.

Helicopter blades, for example, were a significant trigger. Every time I heard them, goosebumps rose on my skin and sweat prickled across my back, convincing me something bad was about to happen. The problem is that on a military base, helicopters are a constant background noise, so it took me a long time to realize that particular sound was fueling my hyper-awareness. Only later, when I could isolate it far away from the military, did I understand.

Other triggers piled on, such as the sound of fireworks, the sight of unattended packages, the smell of certain deodorants, the acrid stench of burning trash, or the odors of a diesel engine. Any one of them could set me off.

But it wasn't like the movies. There was no dramatic breakdown where the veteran suddenly believed he was back in a jungle prison. Instead, it was a relentless flood of awareness. I could hear everything, see everything, all at once, at blistering speed. The problem was that my mind got stuck in the "on" position and wouldn't switch off, sometimes for days or even weeks. Naturally, this became a problem in everyday life, whether shopping for groceries, eating at a restaurant, or

---

[369] Marlantes, Karl. *What It Is Like to Go to War.* eBook edition. Atlantic Books Ltd, 2011, Chapter 1.

simply trying to go to a crowded place. On top of that, sleep, when it came, was naturally fractured and shallow, never restful.

### 3 October 2025 – Reflection of my wife, Jasmin Harstad

It was hard to figure all of this out at first. I always knew you didn't like crowded places, and that includes restaurants. Sometimes it feels like you don't even realize I am sitting there across from you. You're constantly "sniping" things out, always watching, always on guard.

Then, when I can tell you're starting to get uncomfortable, you go really quiet. You eat fast, and then you want to leave. So, we don't eat in public very often. It's been a challenge, but I know you're just doing your thing.

At Fort Stewart, I vaguely recall that psychiatric and counseling sessions were technically available, but I don't remember anyone actually using them. Regardless, none of us would have been caught dead walking into something like that, though, even if it might have helped. At the time, it was simply unthinkable.

In addition to the stigma, I didn't even actually realize at the time that anything was wrong, though I know that may sound crazy. I had grown so accustomed to life in a combat zone that it never occurred to me my behavior was problematic. I was simply blind to it. Only later, when civilians cautiously pointed it out, did I finally start to notice.

### 15 September 2025 – Reflection of Jason Sturm, Bravo 4-64 Armor

I go to a VA group, and hearing from the old timers, they stress that they wish they had talked about stuff sooner. A lot of their issues are the same as ours. It doesn't matter the conflict. They all have the same stuff going on. It's mind-blowing.

Then one day, the entire Gambler Platoon and our old Cyclone brothers were ordered to a mandatory meeting. We filed into a cramped room, unsure why we'd been called there. Eventually, it became clear. The widow of a fallen soldier was still in the depths of her grief, quite understandably, and she wanted to ask us questions about how her husband had died.

We sat in a semi-circle around this grieving widow as she asked about details, such as his last meal, who had been with him, where it happened, had he felt any pain, and a variety of other questions. I sat toward the back, fighting an overwhelming urge to vomit. Every word brought back memories I'd buried, things I didn't want to recall. I stayed quiet, sad, and deeply uncomfortable. I think most of us did. The limited conversation, though heartfelt, was almost certainly unhelpful to her. It wasn't out of cruelty, but because almost none of us were ready, or willing, to open up.

I hated being in that room, and I hated even thinking about talking. I never actually said a word, but in my head I kept rehearsing what I might say if I did. The strangest part was the urge it stirred in me, in that I just wanted to walk out and go back to Iraq. Again, as brutal as it had been, life there had felt simpler. It was predictable.

After things had finally settled down within the unit, the Tuskers gathered in Savannah for a formal celebration, the Task Force 4-64 AR Ball. We put on our best uniforms, raised toasts to one another, drank grog, and honored our fallen. The night carried a bittersweet weight, with far too many empty chairs in that room.

**With SSG Brian Torres before the Task Force 4-64 AR Ball**

Bravo Company hadn't forgotten CPT MacKinnon's instructions before deployment: bring everyone home. In a cruel twist of irony, he pulled off his end of that order. We all made it back, every last one of us… except for *him*. Sadly, we couldn't reciprocate, so there were a lot of toasts in his honor that night, and to the others that fell, but none of them felt like enough.

One bright moment, however, came when Donald Urbany, who had lost an eye in that early IED blast, made a surprise appearance. They'd let him slip away from the hospital ward just long enough to join us. His presence lifted the room, reminding us all of resilience amid loss.

486

That emotional celebration marked the end of something sacred. There was a sense of finality, as though we had passed through a rite of passage reserved for only a rare few. Soon after, week by week, one by one, people began to disappear. As I watched each person slowly leave the unit, their time in the military complete, I felt the terrible pain of separation. Even with those I hadn't been especially close to, their departure hit hard. It was a strange, hollow ache, almost like losing a family member.

Worse still, I began pulling away from the people who mattered most. Cyr had wanted us to meet in Florida one last time, to celebrate the simple fact that we'd made it home alive. I'm ashamed to admit I never returned his calls or made the trip. I just couldn't bring myself to face another goodbye.

When my turn to leave finally came, Papi formed the platoon up and offered a few kind words about my service. I knew I wasn't some legendary warfighter destined to live on in 3rd ID history, but what he said struck a chord. Papi's words echoed something my dad had told me years earlier as I entered basic training, something I had always worked to achieve. "You may not have been the fastest or the strongest in the platoon, but you had something more important. You showed up, you were dependable, and you kept your word. No one ever outworked you, and you never quit. We are going to miss you."

He also pointed out that, despite being the smallest guy in the platoon, I was always volunteering, whether it was hauling the heaviest machine gun, heading out on patrol, shoveling shit, or working long hours in the armory. As the Infantryman's Creed put it, *I was always there.* Hearing it said aloud in front of my brothers nearly broke me, and I almost cried.


**26 October 2012 – Reflection of (Redacted)**

You know, you were the smallest gunner I knew who carried that thing (the M-240B) like a champ. You could have passed it off to someone else, but you never did.


And then I did the one thing I regret most. Instead of personally saying goodbye to Mata, to Hauger, and to all the men who had become closer than brothers, I just quietly walked away from the formation. There were no handshakes, heartfelt words, or grand farewells. I don't really know why I did it. I only know that I left.

Years later, I came across a line in former Ranger Mat Best's book *Thank You For My Service* that rang true: "When you've lost track of the person you were proud to be for all those years, who cares what happens to whatever is left in his place?"[370] It struck me because it captured exactly how I had felt. I was officially lost without my brothers, my job as an infantryman, my unit… and I knew it.

After driving a U-Haul halfway across the country, I finally arrived back in Sioux City. Readjusting to civilian life in the Midwest proved just as tricky as it had been at Fort Stewart. Before

---

370 Best, Mat, Ross Patterson, and Nils Parker. *Thank You for My Service.* eBook edition. Bantam, 2019, Chapter 13.

separating from the Army, they had sent me to a few transition classes, with emphasis on résumé writing, interview prep, things like that. While somewhat helpful, those lessons didn't really prepare me for the brick walls I hit in the job market.

**23 April 2025 – Reflection of my father, Bill Harstad, Jr.**

You came over to the house as soon as your time in the service was up. And when you finally pulled away for your new apartment, Brad (my youngest brother) just crumpled to the ground and started crying. "My brother is home," was all he could manage. Those four years took a lot out of all of us.

I applied to several places and went through interviews that seemed to go well, but the outcomes were almost always the same. The responses tended to fall into two categories: "Infantry, huh? I'm not sure what we can do with you here." Or, "You're overqualified for this role, so I don't think we can offer you anything. Best of luck." It didn't seem to matter that I had survived a buzzsaw, kept my men alive, and earned a stack of awards on the way out. None of that translated into the civilian world, or at least it didn't according to them. Employers didn't know what to make of me, or what to do with me… and most didn't seem to care.

Worse still were the ones who seemed unsettled by my background. I could see it in their eyes the moment they realized I was a combat veteran. Maybe they were intimidated, afraid I was there to take their job. I never knew exactly what the reasoning was, but I could always tell, usually within seconds, that I had already been written off simply by using the word *veteran*.

**30 March 2025 – Reflection of Donald Cyr**

I did defense contracting work after the Army, and that made the transition pretty smooth at first. But when that ended, I struggled to land a job anywhere. Most employers focused on what I didn't have: civilian experience, a degree, and a résumé that translated my service into something they understood. Every application started to feel like another closed door. It just killed my morale.

Eventually I did find work, but it didn't come from the usual channels. It came from fellow veterans, or from people who had veterans in their families. They were the ones willing to give me a shot when nobody else would. I'll never forget that. And like always, I worked my ass off for them. I really do owe them for opening that door.

Being rejected over and over was demoralizing, especially after the fanfare I'd grown used to while in uniform. Back then, people were quick to shake my hand, shout "Thank you for your service! USA! If you ever need anything…" But once the uniform came off, the silence was deafening. It made me wonder how much of that support was real. Unfortunately, I came to realize much of it had been little more than virtue signaling and slacktivism.

**30 March 2025 – Reflection of Donald Cyr**

After years of working construction and maintenance, which I genuinely enjoyed because it gave me meaning and purpose, I still thought about the job I'd wanted since I was a kid. Eventually, I became a Park Ranger, and I still enjoy it every day.

I'm grateful it came along when it did, because that stretch before it was frustrating. Getting turned down over and over wears you down. And unless you go into military contracting, law enforcement, or start your own business, it can feel like not many companies are eager to hire someone from combat arms.

Missing the camaraderie, I let an acquaintance talk me into visiting a nearby Guard unit to watch some of their training. He hoped I'd join. I lasted about ten minutes before politely telling him, "Ah, I don't think this is for me. Thanks, though." What I was really thinking was, *There's no fucking way I'm doing this! Are you kidding me?*

After everything I'd been through, the air assaults, the combat, the high-speed training, what I saw that day only reinforced my decision to stay away from part-time roles. I couldn't imagine going backwards like that. Some of it, I'll admit, was probably ego, but part of it was brutal reality. I knew what combat really looked like. I didn't want to deploy with those troops, many of whom were overweight and looked like shit, using those tactics, with that equipment. No thanks.

Fortunately, my home life was better than my professional life. My girlfriend, Jasmin, who would later become my wife, was incredibly supportive. She helped me transition from that crazy lifestyle, kept me motivated, and gave me direction when I needed it most. I was lucky. Many others weren't. For a lot of guys, leaving the military meant going from a hundred miles per hour every day to the slow crawl of civilian life, and that shift could be unbearable. Even worse, not having someone in your corner could be debilitating. As I've often said, "The military was easy. Coming back to civilian life was soul-crushing."

As luck would have it, the local police department was hiring new recruits, and Jasmin encouraged me to give it a shot. She hoped, as I did, that I might finally find the sense of purpose and belonging I had been searching for. If not, well… at least they paid pretty well for our area.

I breezed through the tests and was quickly hired, soon reporting to the police academy. It wasn't anything at all like Infantry School; instead, it was more of a gentleman's course mixed with a college classroom atmosphere. The pace was easy, the environment relaxed, and it bore little resemblance to the military. Still, I excelled, kept to myself, and graduated without issue.

The real training began after the academy. I was paired with veteran officers for weeks at a time, learning through hands-on instruction that I found both challenging and enjoyable. After more than six months in that training cycle, I was finally cleared to patrol the streets of Sioux City on my own.

In those early years, I earned the nickname "Shit Magnet," much like CPT Anderson had back in Iraq. Apparently, it transferred from him to me. I never went looking for trouble, but it always seemed to find me. I had a deranged man try to drive an icepick through my skull, traded blows in a knock-down fight after taking a punch to the face, crawled into a hostage situation with my assault rifle to shield a woman and her child from a man with two knives duct-taped to his hands, and found myself in high-speed pursuits, brawls, Tasings, and a constant churn of chaos that seemed to erupt every other night. And yet, all of that paled in comparison to 2005. All of that was easy.

But, while I mostly enjoyed my time on the force, I never felt at ease in my own skin there. The best way I can describe it is this: being a cop was like being a fisherman. You cast your line, hoping for a big catch, reeling in what you can, and tossing the smaller ones back into the water with a warning. But as a grunt, I wasn't fishing. I was hunting. I stalked prey that was every bit as lethal as I was, prey that was also stalking me. Both roles dealt with pursuit, capture, and survival, but the difference was absolute. One profession skimmed the surface, the other lived and died in the kill zone.

Most of the people I worked with at the police department were solid. They were loyal, hardworking, and genuinely committed to protecting their community. A few would have even made excellent grunts. And while I'd say I was good at the job, maybe even great for a stretch, I never truly fit in. I could handle the work itself, but the culture of being a cop never felt like mine, and I felt like an outsider.

Part of it was philosophical. There were laws I enforced that I had real qualms about; not because I wanted criminals to get away with things, but because some of them felt arbitrary. It sometimes felt more like control dressed up as "protection." Sure, I wanted to protect my community, but I didn't want to be the tool used to enforce whatever made a politician look tough in a campaign ad or generate revenue. That isn't normally the fault of the average cop on the street, as most cops don't write laws or set priorities, so don't take it that way. Instead, they're handed a set of rules and a radio and told to make it work. The blame for questionable laws belongs where they originated.

The other part was my mindset. Infantry trains you to solve problems with speed and force, and the mission always comes first. If it takes gouging out eyes and ripping out guts, so be it. You sometimes bend the rules and fight like your life depends on it, because sometimes it does. You live hard, talk hard, play hard, and you get conditioned to a world where the stakes are clear and the consequences are immediate.

Policing was different, though. The mission wasn't to "win at all costs," but to "manage." I was taught to be patient, be polite, take disrespect with a smile, and keep your hands to yourself unless absolutely necessary. Oh, and try not to say the wrong thing, try not to be interpreted the wrong way, and try not to end up on the news because a five-second clip makes you look like a villain. And that was a terribly hard thing to do, I'm not going to lie. Being a cop today is often a thankless profession.

So yeah, I could do the job and even excel at it, but I just never felt like that's who I was. I was a soldier temporarily doing cop work, nothing more.

I've said this before, and I mean it with every fiber of who I am. I could earn ten doctorates, become President of the United States, fly into space, even cure cancer. And still, if you asked me what I was, whether it was a cop, scientist, or politician, I wouldn't hesitate. I was a grunt. That is who I was then, and that is who I remain. Though I may never again find anything as deeply satisfying as being an infantryman, I will always keep searching, and I remain endlessly optimistic about the future.[371]

But I would be remiss if I didn't say that the greatest highlight of my story, woven with a strange twist of fate, was how my relationship with my wife brought everything full circle. She was born in Rasht, Iran, the "City of Rain," during the Iran–Iraq War. The hospital had no electricity, and she was delivered by candlelight. She took her first breath in the middle of that conflict, with shelling nearby, while her parents and the medical staff held the line between fear and hope.

Her father was Iranian, her mother German, and they moved to Europe when she was still a little kid. She grew up in Heppenheim, a small city in western Germany, with two cultures under one roof. Even in Europe, the Middle Eastern spirit stayed strong: hospitality, tradition, and family first. Later, she and her family immigrated to the United States.

Then, somehow, by sheer chance or maybe even destiny, she and I found each other. She entered the world in the middle of a war, and I came of age in one basically right next door. And yet the arc of it bent toward something better. It feels like the universe took two people shaped by the shadow of chaos and handed us a chance to build peace anyway.

Years later, we were blessed with two beautiful daughters, Laylah and Alexandria. I will never forget the awe I felt when I first held them in my arms, looking into their tiny faces. My heart stopped when I realized what I was seeing. Yes, they were of me, but they also carried distinctly Middle Eastern features. In that moment, the past came rushing back. The memories of handing out candy and clothes at checkpoints, of trying to care for children in the dust and heat of Iraq, of small gestures of compassion and humanity in the middle of war. Somehow, improbably, those experiences had found their way back into my own life. A part of the Middle East now permanently lived within my own family, in the very faces of my wife and daughters. And as I held them close, I understood: what had once been distant and war-torn was now forever bound to me by the greatest gift of all.

Love.

---

[371] McNett, Jared. "Sioux City Man Has Served in Iraq and the Police Force, Now He's Back in School Looking to Change," November 11, 2021. https://siouxcityjournal.com/news/local/military/article_d6fd5dec-d3d7-53ec-a79a-5695c62821a5.html.

**The Harstad Family in 2025**

Photo courtesy of Talat Heinen

# Final Reflections

When I left the Army, I made the worst mistake I could have. I abandoned many of the bonds forged in battle, the very ties of communication that had once held us together. I still don't fully know why I did it. Maybe it was a defense mechanism, or perhaps it was just me trying to cope with the upheaval of my entire life. Likely it was both.

Fortunately, a few people refused to let me disappear. Donald Cyr and Toby Glass, in particular, kept the thread alive, and for that I will always be grateful. Still, they were only two among dozens of strong connections I had, ties that eventually withered as everyone scattered to new lives in different places.

So, when I finally sat down to seriously begin this book, I was struck by how incomplete many of my memories were. They were often hazy at best and completely missing at worst. I knew I needed the perspectives of others who had been there. But by then, I had lost contact with almost everyone, and the thought of reaching out again terrified me. Would they forgive me for walking away? And after so much time had passed, how was I supposed to find the words to reestablish those connections?

After tracking down a few people on social media and getting their numbers, I hesitantly began dialing. More than once, I couldn't bring myself to finish, hanging up before the call went through. But when I finally did, almost every time I was greeted with pure excitement, sometimes even by an old nickname that made me laugh.

"Hardstud! Where the fuck have you been, man? It's so great to hear from you!"

What amazed me was how quickly the years seemed to vanish. The conversations picked up as if we had seen each other yesterday. Those bonds I had feared were gone forever hadn't disappeared at all; they'd simply been waiting. We laughed, we cried, and we relived the good times. It was one of the best feelings I've ever known.

So, if you're a veteran facing the same hesitation, trust me on this. Find your buddy's number and make the call. Time cannot break those bonds. It's not too late. And maybe this doesn't apply only to veterans. If you're estranged from someone you love, pick up the phone. Do it today. I should have done it a long time ago.

In many of those conversations, I cautiously brought up the subject of PTSD, still uneasy speaking about it with men who had once been killers. What I discovered shocked me. Nearly everyone had faced, or was still facing, the same symptoms I thought were mine alone.

That realization opened the door to deeper talks with my wife, specifically about the struggles we've endured, and the ways my unseen battles have sometimes made life harder for us both. In its own way, those conversations have brought us closer. They've given us a better understanding of each other, and a way to steer together through those dark waters.

**3 October 2025 – Reflection of my wife, Jasmin Harstad**

When you were writing some of the tougher chapters, I noticed your behavior would change. You ran on the treadmill a lot. You'd get hyper-focused, and sometimes you'd be fidgety and distant. To someone who didn't know you, it probably looked like, "Wow, that guy has a serious case of ADHD." But I've learned that communication is everything. When you were ready, we'd go on long walks and talk it out. If I had to give one piece of advice to the spouse of a veteran, it would be this: be willing to communicate, don't judge, and *truly* listen.

Over time, I also learned to take the lead when you were going through one of your "things." It was a learning process for all of us. I'll tell the kids, "Hey, dad's dealing with something, so let's keep it down." They've figured it out, too, and they usually do a great job helping out. It seems that if we can dial back the stress, noise, and chaos of daily life, you get back to yourself a lot faster. It's tough sometimes, but we stick together, and we've made it work.

After years of living with it, I finally came to understand that my so-called "quirks," the fidgeting, the irritability, the constant scanning of my surroundings, weren't normal at all. I had always shrugged it off as just part of who I was, but eventually, I let myself get talked into meeting with someone at the VA. The organization has its flaws, sure, but when it came down to it, they became a solid resource. Still, those first sessions were awkward.

I remember one in particular. I was sitting across from a VA mental health provider, tugging nervously at the end of my shoelace, wishing I could be anywhere else.

"Okay, Jeff," she said gently, notebook balanced on her knee. "When you think back to your traumatic experiences, what stands out the most about *the incident*?"

I blinked. "The incident?" I repeated, confused.

"Yes," she replied, steady but expectant. "The one that was disturbing to you."

I shook my head. "Which *one*? There were dozens. IEDs, explosions, mortar strikes, firefights, dead bodies. Which one are you talking about?"

Her eyebrows lifted, but she didn't interrupt.

"Dozens," I said, then paused, feeling my chest tighten. "Actually, that's probably an understatement. Probably hundreds, if I try to count."

She leaned back slightly, her voice still calm. "Hundreds?"

And there it was, the look on her face, like she couldn't quite wrap her head around it. For her, trauma was an isolated event, something sharp and definable. For me, though, it was a blur of time where every day carried the possibility of ending violently. Saying it out loud made me realize how different those worlds were, and how much I had convinced myself that living this way was somehow normal.

But I stuck with the sessions, and in the process, I learned a great deal. It was actually my counselor who first suggested that I go to the hospital's helicopter landing pad as a way of confronting my fears. I took her advice.

Later, while working with the police department, I volunteered with Project Lifesaver, a program dedicated to locating missing children and elderly patients with conditions such as autism and Alzheimer's disease. The role eventually required me to go aboard the hospital helicopter, which was something I never thought I'd be able to do again. I received training on specialized tracking equipment, learning how to operate it from the air, where the signal was clearer and less obstructed by buildings, trees, or terrain. It was great! I was so glad I faced my fear and overcame it.

Next, one of the most important lessons I took away from all of this was learning to temper my arrogance. Like many infantrymen, I carried a fierce pride in the role. After all, we were the ones on the ground, closing with and destroying the enemy. But that pride can harden into a chip on your shoulder if you're not careful.

I was fortunate to serve alongside some of the most badass non-infantry professionals in the Army. I saw tankers who could bring overwhelming firepower at just the right moment, medics who ran into danger without hesitation to save a brother's life, helicopter pilots who flew into hot zones to pull us out, scouts who crept into the unknown so the rest of us had a fighting chance, and countless others across every MOS. These men carried their own kind of courage and skill that demanded respect.

Serving with them stripped away the illusion that any one branch, specialty, or individual carried the fight alone. The battlefield doesn't work that way. Every success we had was because of the team's collective effort, each role fitting into the others like gears in a machine. That realization humbled me in the best possible way.

To this day, I remain proud of my identity as an infantryman, but I'm just as proud to have stood shoulder to shoulder with tankers, medics, aviators, scouts, and so many others. We were one team, bound together by shared struggle and sacrifice. They are my brothers, and I will honor them always.

And finally, there is one peculiar detail I feel compelled to mention, though it may seem small in the grand scope of war. When I was living at FOB Prosperity, Howerton and I came across a little-known film being sold by an Iraqi vendor in the form of a pirated DVD. Sorry, Hollywood. It was a science-fiction psychological thriller called *The Jacket*, and from the very first viewing, it did something to me that I still struggle to understand fully.

The film follows a Gulf War veteran, grievously wounded, who finds himself locked away in an insane asylum and subjected to experimental treatments. Through these ordeals, he discovers a strange ability to slip through time. The story isn't my story, not exactly, of course, and yet I couldn't shake the eerie resonance. There were parallels, fragments of my own life reflected in distorted ways. His trauma, his dislocation, and his struggle to make sense of the broken pieces all seem to be reflections. And then there was the figure of a woman reaching for him, trying to pull him back from the abyss, just as Jasmin so often has done for me.

What stayed with me most was the ending. As he bleeds out, a hauntingly beautiful song playing in the background, his fate uncertain, the woman asks him a simple, but profound question: "How much time do we have?" The screen fades as Iggy Pop's voice carries over the credits, singing *We Have All the Time in the World*.[372] It was a beautiful paradox. The character may not have had time at all, yet the message was one of love enduring beyond fear, beyond wounds, and even beyond death.

I don't entirely understand it, and I don't expect you to, either. For me, though, that ending became less about the film and more about life itself. It reminded me that, despite everything, despite Iraq, despite trauma, despite the scars that still echo through my nights, Jasmin and I *do* have all the time in the world. Not because time is infinite, but because when it is filled with love, even the briefest moment can stretch into eternity.

---

[372] *The Jacket*. Warner Independent Pictures, 2005. Directed by John Maybury.

# Chapter XX

## If You Are Reading This

For those of us whose business was war, many wrote letters home under the instruction: "Only open if…" Known as an "If You Are Reading This Letter," their purpose was simple. It was to say goodbye. Most were destroyed, still sealed upon return, but sometimes the inevitable came to pass, and families were forced to open them. In those last moments, the words carried across the divide, a final message spoken from beyond. And just perhaps, you should read one. Because once you do, it will never leave you.

**Goodbye letter to his family from Michael MacKinnon[373]**

Dear Bethany, Madison, and Noah,

If you are reading this, then I failed to be fast enough, smart enough, or lucky enough. Writing this is very difficult for me, so bear with me if I ramble along. I needed to write this letter because there are some things that I needed to say and that I wanted you to hear.

Bethany, you are the love of my life. I'll never forget the first time I saw you. You were so beautiful. I had so much fun falling in love with you. Our long walks at West Point, our trips to New York City, and Niagara Falls. The times we spent in Montana, even if we almost froze on the river that day. I loved every second that I spent with you. I know that through the years we have had some rough times, but our love and enjoyment of each other was so much stronger.

A funny little thought I always had. You know that all the formals at West Point were nothing but a competition of who had the prettiest date? I'm not going to lie, but I always checked the competition. Every time, you were the most strikingly beautiful woman in the room. I know I told you that every time, and you brushed it off, but I truly believed it. I think ring weekend was the most beautiful I had ever seen you, second only to watching you hold and sing to babies Madison and Noah.

---

373 Grateful Nation. "Honoring Montana's Fallen." Accessed January 12, 2026. https://archive.legmt.gov/bills/2013/Minutes/Senate/Exhibits/eds48a01.pdf.

I want you to know that my biggest regret during my relationship with you is that I never gave you the wedding ring you/we both deserved. I have never doubted that someday we would make it happen and that we would have our day in the sun. I hope someday you will find someone who loves you as much as I do. I want you to be with someone who sees you as beautiful as I did. Don't ever feel guilty if you fall in love again. You don't deserve to be alone and unhappy the rest of your life. I understand.

I believe there is a heaven, and I think I have been a good enough man to be there. I'll watch you, and I'll watch the kids grow up with a tear in my eye, wishing I was there to see with my own eyes. Some day we will all be together again. I don't think words could ever describe how much I love you. I want you to know that I loved my life. You made me very happy. I became a better person after meeting you – more complete.

When you think of me in years to come, I want you to celebrate my life and not mourn my death. Don't think about all that you, Madison, Noah, and I missed, but think of the good times we had together.

Madison, I'm sorry I broke the promise I made you when I said I was coming back. You were the jewel of my life. You made me so happy every time I saw you. You are the sweetest little girl. Don't ever change. I wanted nothing more in my life than to be with you as you grew up. I want you to know that I loved you more than I can ever imagine.

If you can do anything for me, you can take school seriously and do well. With a good education, you can become anything you want. Stay away from drugs. They will ruin your life. Finally, stay away from bad men. You deserve too much. I don't think that anyone would ever be good enough for you.

I have to tell you that you really broke my heart that day when I left. You said, "Don't go, I need my daddy, I don't want my daddy to go." I nearly cried in front of all the Army guys. Stay beautiful, stay sweet. You will always be Daddy's little girl. I loved every minute of my life that I spent with you. From the first time I held you, to pushing you on the swings, playing steamroller, and walking you to school. I always felt that that was our little thing. I loved holding your hand and walking you to class. You are a good girl. I'm so proud of you. You made me so happy. I will always be with you.

Noah, you are my miracle son. I loved you very much. I regret that we never got the bond together that we so much wanted. I know I was hard on you, but only because I saw so much potential. You made me very happy and proud. I admired how tough you were when you had your surgery. That was the first time that I realized you would always be a fighter and never give up on anything.

I have the same dying wish for you as Maddie. Take school seriously and do well. Stay away from drugs. Try to live the rest of your life as the best man you can be. I remember holding you after you were born. I was so happy. You loved it when I stroked your head. You are now the man of the house. Take care of your mom and your sister. You now have the same burden as your father had, to carry on the MacKinnon legacy. Try hard to keep it alive.

Noah, I loved you very much and am proud to have you as a son. You brightened up my life every time we spent together. I will always be with you.

I want you all to know that I loved my family very much. Every time I was away, all I could think about is all of you. I know I wasn't the best husband or father in the world, but I always want to be remembered as a family man. I don't care to be remembered as a smart guy, an athletic guy, or a good soldier. I want people to say, "That Mike really did love his family."

I want you to know that my last thoughts in my life were of you. All of you need to get on with your life and do the things that make you happy. It's OK to be sad that I'm gone, but be happy for me that I got to spend time with a loving wife, a beautiful daughter, and an exceptional son. I got to see both your births, I got to see your first steps. I got to watch you grow into perfect children. I'm very honored to have the three of you as my family. My greatest regret in life is that I can't watch you grow up. I will be with you all forever.

I consider myself the luckiest man ever to have lived. I couldn't have asked for a better soul mate or better children. I don't think that in any letter I could ever say all that needs to be said, so I will end it here. I'm proud, I'm happy, I have been filled with so much love. I will always be there in your hearts.


Love always and forever,

Michael – Daddy


    I still read that letter from time to time, staring at the MacKinnon family photograph. They are bittersweet memories, reminders of my friend and the brief time we shared in uniform. I'm grateful we were there with him at the end.

    Those moments always carry me back. I remember the early morning PT sessions at Fort Stewart, the strange odors of the swamp drifting through the heavy dawn air. I hear Reveille in the distance and feel the swell of pride as the flag rises over the post. I remember the bone-deep chill of hypothermia, trying to sleep in a flooded hole carved into Georgia mud. And I smell the grit and grime of Iraq, the kind that burrows into your skin, seeps into your very soul, and never lets go.

    But most of all, I think of my brothers, and how deeply I still love them. Some of their names and faces have slipped into the haze of memory, like a dream fading on a warm summer's eve. Yet no matter how much time erodes the details, we will always share those days in the desert. I can still see us there. We are standing as one, together, willing to kill, bleed, and die for one another, looking out across the endless dunes of Babylonia.

    Deep in the distance, I still can't help but reflect with wonder at the mirage in the sand.

# Suggested Further Reading

*12 Rules for Life: An Antidote to Chaos*
Jordan Peterson

*Act of War: Lyndon Johnson, North Korea, and the Capture of the Spy Ship Pueblo*
Jack Cheevers

*All Quiet on the Western Front*
Erich Maria Remarque

*All Secure: One Delta Force Operator's Fight From the Battlefield to the Homefront*
Tom Satterly & Steve Jackson

*Band of Brothers: E Company, 506th Regiment, 101st Airborne from Normandy to Hitler's Eagle's Nest*
Stephen Ambrose

*The Bloody Battle for Suribachi: The Amazing Story of Iwo Jima that Inspired Flags of Our Fathers*
Richard Wheeler

*Bravo Two Zero*
Andy McNab

*Call Sign Chaos: Learning to Lead*
Jim Mattis and Bing West

*D-Day, June 6, 1944: The Battle for the Normandy Beaches*
Stephen Ambrose

*Eye of the Tiger: Memoir of a United States Marine, Third Force Recon Company, Vietnam*
John Delezen

*Generation Kill: Devil Dogs, Iceman, Captain America, and the New Face of American War*
Evan Wright

*It Doesn't Take a Hero: The Autobiography of General H. Norman Schwarzkopf*
Norman Schwarzkopf and Peter Petre

*On Killing: The Psychological Cost of Learning to Kill in War and Society*
Dave Grossman

*Ordinary Men: Reserve Police Battalion 101 and the Final Solution in Poland*
Christopher Browning

*Ride the Cyclone*
David Anderson

*The Rifle: Combat Stories from America's Last WWII Veterans, Told through an M1 Garand*
Andrew Biggio

*Surprise, Kill, Vanish: The Secret History of CIA Paramilitaries, Operators, and Assassins*
Annie Jacobsen

*Takedown: The 3rd Infantry Division's Twenty-One Day Assault on Baghdad*
Jim Lacey

*Thank You for My Service*
> Mat Best, Ross Patterson, and Nils Parker

*The Things Our Fathers Saw – The Untold Stories of the World War II Generation From Hometown, USA – Volume I: Voices of the Pacific Theater*
> Matthew Rozell

*War Is a Racket*
> Smedley Butler

*The War on Warriors: Behind the Betrayal of the Men Who Keep Us Free*
> Pete Hegseth

*We Few: U.S. Special Forces in Vietnam*
> Nick Brokhausen

*We Were Soldiers Once… and Young: Ia Drang – The Battle that Changed the War in Vietnam*
> Harold Moore and Joseph Galloway

*What It is Like to Go to War*
> Karl Marlantes

*With the Old Breed: At Peleliu and Okinawa*
> E.B. Sledge

# Acknowledgments

Although my name is on the cover, this book belongs to the community around it, especially the people listed here. I hope seeing your name in these pages means as much to you as it does to me, and I hope I've done your contributions justice. This book was shaped by conversations, shared memories, honest feedback, and the quiet support that showed up when it counted. I'm grateful to everyone who helped carry this project, whether you provided a key detail or simply helped keep me grounded while I wrote. The following people deserve recognition, and I'm proud to acknowledge them here.

**The original Harstad family, who were with me from the beginning and made my military dream possible:** Joyce Harstad, Bill Harstad Jr., Dan Harstad, Mike Harstad, Brad Harstad, Bill Harstad Sr., Toni Harstad, and Mark Harstad.

**The "new" Harstad family, who offered constant support and motivation, in addition to the endless supply of patience:** Jasmin Harstad, Laylah Harstad, and Alexandria Harstad.

**The extended Harstad, Grage, and Youssefpour families.**

**Childhood friends:** David Lohry, John Steele, and Drew Campbell.

**Influential teachers a cut above the rest:** Mrs. Elizabeth Engle, Mr. Bret Reed, and Dr. Nathan Bates.

**Infantry School leaders who made a real difference:** CPT McCormick, DS Ross, DS Nash, DS Parkins, DS Clements, DS Whiteley, and DS Schneider.

**Infantry School comrades:** Brad "Opie" Greenlee, Scott Harper, and Logan Hastings.

**Military mentors and leaders who had a huge impact on my life:** Michael MacKinnon, Marcus Brister, Juan Serrano, Harold Hill, Mark Barnes, Robert Roth, Cap Stanley, Matt Turcotte, and David Anderson.

**My Army brothers:** Donald Cyr, Toby Glass, Brian Torres, Zack Howerton, Derek Hauger, Noel Mata, Izonia Chism, Casey Enos, and Jihad Franklin.

**Units and brotherhood:** The 3rd Infantry Division, 3-7 Infantry "Cottonbalers," 4-64 Armor "Tuskers," and all the beautiful bastards from Gambler Platoon.

**Readers, editors, and advisors:** Jasmin Harstad, Bill Harstad, Dan Harstad, Mike Harstad, Brad Harstad, Melissa Harstad, Jeff Deignan, Donald Cyr, Phil and Cyndi Ellis, Paiton Robertson, Jane and Barry Snyder, Todd Petersen, Mary Christine Begovich, and Sahand Yousefpour.

**Photos:** Derek Hauger, David Anderson, Jason Sturm, Talat Heinen, Chicago Honey Bear Dancers, Visual Touch Photography, Cap Stanley, Bill Harstad, Phillip Cornell, Donald Urbany, Travis Nelson, Noor Aljassim, and Robert Roth.

**Special note to Chicago Honey Bear Dancers:** DS Whiteley may have kept his drill-sergeant façade during the Honey Bear Dancers' routine, but the rest of us were grateful for their patriotism, energy, and support. Thank you for your support of America's troops.

*Since 1986, the "Chicago Honey Bear Dancers" have brought theme song and dance shows to over 180 Military Bases worldwide, dedicated to boosting morale for our Heroes. Their mission extends to the youngest on base with anti-bullying dance workshops that encourage kindness, teamwork, and the positive message, "Be a Buddy, not a bully." Produced & Directed by Greg Schwartz – ChicagoHoneyBearDancers.com*

**Interviews and excerpts:** The Harstad Family, Bill Harstad, Joyce Harstad, Jasmin Harstad, Mark Harstad, Toni Harstad, Logan Hastings, Toby Glass, Donald Cyr, Noor "Nora" Aljassim, Kim Linafelter, Casey Riewaldt, Jane Snyder, Derek Hauger, Juan Serrano, Robert Roth, Cap Stanley, Donald Urbany, Mark Barnes, Michael MacKinnon, David Anderson, James Barnett, Phillip Cornell, Harold Hill, Matt Turcotte, Brian Dugan, Travis Nelson, Heath Hutchison, Benjamin Rider, Simon Tumlinson, and Jason Sturm. Students at Bryant Elementary School (Sioux City, Iowa), Students at Sunny Slope School (Omaha, Nebraska), and Students at Kluckhohn Elementary School (Le Mars, Iowa).

**Special thanks:** Tanja Franke, George and Lou Ann Lindblade at Sioux City Gifts, Zach Lewis, Jason Fleckenstein, Marie Divis, Jeremy Vis, and attorney John Pickerill.

**Organizations:** The American Legion and the Wounded Warrior Project.

**Cover, editing, and research tools:** Grammarly, ChatGPT, and Clipchamp. The words here are mine, but these tools and programs helped me work faster, check clarity, and streamline research while shaping the manuscript into a cohesive whole.

**A quiet thank you:** Trinity Heights (Sioux City, Iowa), where I spent a lot of time thinking, reflecting, and trying to find peace while writing the harder parts of this book.

**Thank you to author and podcaster Michael Malice:** You helped push me over the edge into actually writing a book. I once heard you joke that a quick lap through a big bookstore can make any aspiring author feel better about their odds. Consider this my humble contribution to that tradition. I, too, now have a terrible book!

**Finally, a heartfelt thank you to the MacKinnon family**: For trusting me with stories and some of Mike's words. Your husband and father was a good man, and I've tried to honor him in these pages.

**Captain Michael John MacKinnon**
**Photo courtesy of Noor Aljassim**

# In Memoriam

---

Fallen Soldiers of the 3rd Infantry Division[374]
Remembered at the 3rd ID Warriors Walk


Fallen Soldiers of Task Force Night Stalker[375]
Remembered at the National Guard Honor Roll

---

[374] U.S. Army. "Warriors Walk," https://home.army.mil/stewart/my-fort/community/warriors-walk.

[375] State of Hawai'i Department of Defense. "Honor Roll: 29th Infantry Brigade, Operation Iraqi Freedom," September 11, 2025. https://dod.hawaii.gov/blog/honor-roll-29th-infantry-brigade-operation-iraqi-freedom/.